Computational Biology

AF789276

The *Computational Biology* series publishes the very latest, high-quality research devoted to specific issues in computer-assisted analysis of biological data. The main emphasis is on current scientific developments and innovative techniques in computational biology (bioinformatics), bringing to light methods from mathematics, statistics and computer science that directly address biological problems currently under investigation.

The series offers publications that present the state-of-the-art regarding the problems in question; show computational biology/bioinformatics methods at work; and finally discuss anticipated demands regarding developments in future methodology. Titles can range from focused monographs, to undergraduate and graduate textbooks, and professional text/reference works.

Author guidelines: springer.com > Authors > Author Guidelines

For other titles published in this series, go to http://www.springer.com/series/5769

Jianfeng Feng • Wenjiang Fu • Fengzhu Sun
Editors

Frontiers in Computational and Systems Biology

Springer

Editors
Dr. Jianfeng Feng
Centre for Scientific Computing
Dept. Computer Science & Mathematics
Warwick University
CV4 7AL Coventry
UK
Jianfeng.Feng@warwick.ac.uk

Prof. Fengzhu Sun
Dept. Biological Sciences
University of Southern California
90089 Los Angeles, CA
USA
fsun@usc.edu

Wenjiang Fu
Dept. Epidemiology
Michigan State University
48824 East Lansing, MI
USA
fuw@msu.edu

ISBN 978-1-4471-2570-9 ISBN 978-1-84996-196-7(eBook)
DOI 10.1007/978-1-84996-196-7
Springer London Dordrecht Heidelberg New York

British Library Cataloguing in Publication Data
A catalogue record for this book is available from the British Library

Cover design: VTEX, Vilnius

Printed on acid-free paper

Springer is part of Springer Science+Business Media (www.springer.com)

Preface

Biological and biomedical studies have entered a new era over the past two decades thanks to the wide use of mathematical models and computational approaches. A booming of computational biology, which sheerly was a theoretician's fantasy twenty years ago, has become a reality. Obsession with computational biology and theoretical approaches is evidenced in articles hailing the arrival of what are variously called quantitative biology, bioinformatics, theoretical biology, and systems biology. New technologies and data resources in genetics, such as the International HapMap project, enable large-scale studies, such as genome-wide association studies, which could potentially identify most common genetic variants as well as rare variants of the human DNA that may alter individual's susceptibility to disease and the response to medical treatment. Meanwhile the multi-electrode recording from behaving animals makes it feasible to control the animal mental activity, which could potentially lead to the development of useful brain–machine interfaces. Embracing the sheer volume of genetic, genomic, and other type of data, an essential approach is, first of all, to avoid drowning the true signal in the data. It has been witnessed that theoretical approach to biology has emerged as a powerful and stimulating research paradigm in biological studies, which in turn leads to a new research paradigm in mathematics, physics, and computer science and moves forward with the interplays among experimental studies and outcomes, simulation studies, and theoretical investigations. In the current collection of papers, which are mini-reviews written by leading experts in their own areas of computational and systems biology, we attempt to summarize and share with the readers some of the most recent thriving developments.

The conference from which this book results was to celebrate the 70th birthday of Qian MinPing, a Professor in Mathematics and Theoretical Biology at Peking University. She is one of the people who foresaw the forthcoming tide of the computational and systems biology more than 20 years ago. "She is an amazing woman" said Mike Waterman, one of the conference attendees. Most contributing authors of the book are her students and are proud of being members in the "Qian-School." Below is a brief biography of Prof. Qian, written by herself.

The editors of the volume asked me to write something about myself instead of contributing a research article. Reflecting on this request, it finally daunted on me that I am now at the age of telling grandma's story which might be indeed more interesting to the younger generation. The life of Chinese scientists of my generation could be quite colorful and their stories multitudinous. We experienced two wars (The World War II and the civil war between the communists and the nationalists), numerous political movements including "The Cultural Revolution" of more than ten years, and finally the past thirty years of reform era with rapid changes in every aspect of Chinese life.

In 1979, after the end of the nightmare like "The Cultural Revolution," absolutely out of my expectation even in dream, I was fortunate enough to be a selected member of the first group of people sent to the US as visiting scholars. When I arrived in the US, I was impressed by the affluent life in America, with great contrast to that in China at that time. However, my strongest feeling was about the expectation and enthusiasm for the future of China and the lately started academic career of myself. That seems to be so long ago and might not be easily understandable by today's youngsters. But it concerns almost everything with the contemporary Chinese history in general and my own family background in particular.

My father was a Professor in polymer science educated in England. Growing up under the family influence, I decided in high school that I would devote myself to science. I entered the Peking University in 1956 and studied mathematics. Even though officially I was a student and later a faculty member in a good university, I could hardly fulfill what I wished. We often could only do research in our spare time after "work"; Even for that we were criticized as holding "illegal" seminars on mathematics in evenings and on Sundays. Thus, any opportunity to me for learning was really like food in starvation and water for thirstiness. Even though I was already 40 years old, what I was thinking the most was to take the advantage of the opportunity to study as much as one could, so that I would be able to catch-up scientific development of the world. It would be silly to expect myself making first class contributions to mathematics or science starting that late. The realistic goal to us was to play a connecting-role between the preceding generation, such as Professor Xu, Bao-Lu (Pao Lu Hsu), and the coming generation that entering undergraduate studies at the time. Therefore, the job I assigned myself was to introduce the most important new development in the field of stochastic processes to my students and to guide them to the frontier of scientific research. Still, if one wants to teach the students about modern research, he or she self definitely needs to have experience with such. Hence, even though we had not had much opportunity in doing research before, I, and most others like me in China, tried very hard to get a flavor of original research. Now looking back to my last 30 years, I could say I feel happy and satisfied, since I had made a good decision on my job and I tried my best for it. Many friends have asked me if I have ever thought about staying in the US, and the answer was always "No," since my motto is that "To marry a person who loves you instead of one you admire": China indeed needed me the most at that moment of the history.

To me, working with students and enjoying their achievements were really the greatest pleasure. In fact, I have been growing with my students in my academic life: Often before I teach them, I do not really know much more than they do; the only advantage I have is my experience. When I teach a course, no matter whether it is familiar or new to me, I always try to bring something new into the class, and to improve my previous teaching and to adjust it according to the changing situation. I feel very sad if no one among the students really wants to understand what I teach. However, if the students truly enjoy my teaching, especially those I made a special effort, the pleasure to me is as great as I obtain a nice result in research.

If I ever had some good influence on my students, I think it is mostly through my enthusiasm and persistence in work and the attitude of pursuing the scientific truth and originality with self-confidence, irrespective of being in a familiar field or in an early stage of other new research. I myself obtained such influence mostly from my farther, my brother, and my teachers, such as P.L. Hsu and Z.P. Jiang, not through their preaching but from their deeds. In fact, my father rarely talked to me on serious topics but teased me as his little girl. His influences on me were from his own hard working days and nights, and life-long enthusiasm toward scientific research. My brother Min Qian, twelve years older, has been my real scientific mentor since my university days. As a faculty member in the same department, he has never taught me any formal courses. But he often talked to me about mathematics, and even physics, in the way like oral exams between teachers and students. I have learnt what it means to thoroughly understand something and how to grasp the essence of a problem this way. Learning from Professor P.L. Hsu's lectures notes on general topology (not on probability and statistics) in a complete new presentation and writing reports to him several times had given me life-time benefit, since that made me understand the importance to always pursue one's own originality in study, no matter in teaching or in research. I have learnt to be a devoted scientist without seeking fame and wealth from Professor Z.P. Jiang.

Since 1998, I have become increasingly attracted to the fascinating development of genetics and computational molecular biology. Starting at the age of 59 from knowing almost nothing, I, together with my students, have learnt a great deal of biology. I teach them what I know, but more importantly, how to obtain and understand the new biological knowledge fast by taking advantage of our background of mathematical training. At the meantime, I have learnt from them a great deal through presentations and discussions in seminars. Indeed, working together with students keeps me updated in biology, which, unlike mathematics, is such an extremely rapidly changing field. Together we continuously build connections between the wide range of new theoretical and experimental results.

Here I would like to talk a little bit about my feeling for leaning and studying biology as a mathematician. We all see that right now it is a time of tremendous new discoveries in biology. Significant experimental facts and novel data are obtained in biomedical sciences almost every month. Mathematics can play some important roles in explaining and understanding them. Real implications of observations and data are often far from what they seem to be. For instance, huge amount of microarray expression data have been obtained in recent years, and many have public access. Some conclusions have already been drawn, such as which genes or genomic parts are involved for what complex diseases or phenotypes. Still, because the above conclusions are drawn from statistical analysis, rarely a mechanistic understanding is offered on why and how the genes are involved. The questions why so many statistically significant SNPs appear in the genomic desert and how SNPs in regulatory regions affect the phenotypes remain to be elucidated. What we know and understand is still very limited. The situation reminds us of the early 20th century when the great progress in theoretical physics followed a large amount of new observations and data, such as those in X-rays radiations, photoelectrical effect, and electron diffraction. Revolution or paradigm shift in physics eventually led to great applications in technology that have affected the life of mankind ever since. I often wonder whether a deep and thorough theoretical biology will come in the 21st century based on all these recent observations and data. I feel strongly that whatever the final theoretical edifice will be, the statistical genetics and bioinformatics will be integral parts of it.

To reach a thorough understanding of something from what one observes, theoretical induction and integration with imagination are not only important but actually also are necessary. I believe that mathematics can play precisely this indispensable role in this respect. However, this may not be just simple use of existing models and methods from existing mathematics. Real-world biological systems are extremely complex, and one needs first to grasp their most essential elements before representing them in terms of mathematical models. Furthermore, one also needs new mathematical concepts, tools, and methods for modeling and integrating simpler components, and to characterize more and more complex systems. Thus, a mathematician coming into contact with biological problems should position him- or herself as a scientist instead of considering him- or herself as merely a mathematical tool and model provider. He or she should learn and study biology together with biologists. The ultimate real truth would be obtained as a whole, rather segmented into biology, chemistry, physics, and mathematics as isolated disciplines.

This volume consists of 19 chapters and covers a wide spectrum of topics. In Chap. 1, S. Zhong and his colleagues used thermodynamic models to analyze gene regulatory mechanisms. In Chap. 2, Y. Ding reviewed major algorithms for RNA secondary structure prediction, with a focus on ensemble-based approaches that have proved to be advantageous in many applications since they provide complete statistical characterizations of the Boltzmann ensemble of RNA secondary structures. He described applications of an RNA structure sampling algorithm to the rational design of short interfering RNAs for gene silencing by RNA interference and to target identification for microRNAs that play important roles in posttranscriptional gene regulation. In addition to sequence features, incorporation of target mRNA secondary structure is an important consideration in these applications. The microarray technology has developed rapidly and has advanced our knowledge of the genomes of various species and the understanding of complex diseases. Particularly, the oligonucleotide microarrays have received increasing attention in biological and biomedical research. However, many aspects of the oligo arrays have not been thoroughly studied or fully understood, which lead to issues related to the array data quality control. In Chap. 3, W.J. Fu and his colleagues demonstrated that new developments in these areas of the oligo arrays lead to better understanding of the array mechanism and improvement in the microarray data analysis. In Chap. 4, H. Ge tried to apply the models of stochastic processes into two very active fields now, nonequilibrium thermodynamics and biological signal transduction. Many essential concepts and relations related to classical thermodynamic laws have been put forward and discussed in details. Besides it, stochastic approach is also used to model biological signal transduction pathways and modules. Here, he focused on the phosphorylation and dephosphorylation module and mainly investigate its sensitivity against external signals. It was found that at least to some extent stochastic models could explain the mechanism producing ultrasensitivity better than the corresponding deterministic one. In Chap. 5, J.F. Feng and his colleagues reviewed some of recent progresses in applying Granger causality to recover network structures: gene networks, protein networks, and neuronal networks. Some successful applications are included to demonstrate the power of the approach. Phylogenetic footprinting is one of the most effective approaches for transcription factor binding site identification. In the past decade, many phylogenetic footprinting methods have been developed and have demonstrated their power in predicting binding sites. In Chap. 6, X.M. Li and his colleagues differed from other reviews on phylogenetic footprinting and presented a few representative methods based on whether these methods depend on alignments. They also pointed out a few challenging problems for future directions. In Chap. 7, P. Wang and her colleagues introduced penalized regression-based methods, `space` and `LogitNet`, for constructing genetic interaction or regulatory networks from high-dimensional continuous and binary array data. They also introduced `remMap` for constructing networks using two different types of high-dimensional array data. These methods are illustrated through both simulated and real data examples.

Protein domains are parts of the protein that can function independently of other parts. Thus, domains form the basic units of proteins, and domain–domain interactions are the fundamental causes of protein interactions. Although large amounts of protein interaction data sets from many different organisms are available, our knowledge of domain interactions is limited. Several computational methods have been developed to predict domain–domain interactions from protein interactions and other information including gene coexpression, gene annotation, and domain fusion. In Chap. 8, F.Z. Sun and his colleagues reviewed several computational approaches to achieve this objective, including a maximum likelihood estimation method, a likelihood ratio based method, maximum parsimony, and methods integrating interaction data from multiple organisms, gene annotation, co-evolution, etc.

Now with the developed mathematical theory of irreversible stochastic processes carried out by Min-Ping Qian, Min Qian, and their colleagues at Peking University, it is clear that the irreversible stochastic processes are applicable to many of the interesting open-system phenomena in chemistry and biochemistry. In Chap. 9, H. Qian and his colleagues started with the simple Michaelis–Menten enzyme kinetics from a purely stochastic perspective and then turned to an irreversible Markov process called coupled diffusion, which could be used to model motor protein, fluctuating enzymes in a living cells, and self-regulating genes. They also found that a bifurcation, saddle-node or pitchfork, occurs in certain coupled diffusion systems while decreasing the rates of jump processes. In Chap. 10, D.F. Wu and her colleagues briefly reviewed the current status of the probability model and the statistical methods in cancer screening and their limitations.

The smallest confidence interval for a given class of intervals was defined to be the intersection of all intervals in the class. If this intersection belongs to the given class, we say the smallest interval exists in the class, and this interval is simply the best in that class. In Chap. 11, W.Z. Wang introduced a general method to construct the smallest one-sided $1-\alpha$ confidence interval when there exist nuisance parameters. In Chap. 12, J. Xie and her colleagues introduced the idea of group variable selections in a regression model and applied the method to genomic data. In Chap. 13, inspired by the Granger causality idea in time series, W.Q. Yang and his colleagues extended the notation to static data and applied it to protein data. In Chap. 14, N.R. Zhang reviewed the computational and statistical problems that arise in DNA copy number data and surveyed recent advances in their treatments. In Chap. 15, T.L. Zhang outlined a number of cluster detection approaches and disease mapping approaches.

Treating mRNA transcript abundances as quantitative traits and mapping gene expression quantitative trait loci for these traits has been studied in many species from yeast to human. There has been significant success in finding associations between gene expression and genetic markers. These eQTL studies have been used to identify candidate causal regulators, to construct gene regulation networks, to identify hot spot regions, and to better understand clinical phenotypes. Because of the large number of genes and genetic markers in such analyses, it is extremely challenging to discover how a small number of eQTLs interact with each other to affect mRNA expression levels for a set of (most likely co-regulated) genes. In Chap. 16, J.S. Liu and his colleagues reviewed a few methods for studying eQTL data and outlined a new Bayesian method they recently developed for eQTL mapping. In Chap. 17, H.Y. Zhao and his colleagues first constructed a weighted gene co-expression network and then extracted gene modules from the constructed network based on some topological measure. To interpret the biological meaning of the extracted modules, they used information from Gene Ontology, Kyoto Encyclopedia of Genes and Genomes, and genome-wide location data to study whether each module is enriched for certain categories. Furthermore, they compared the utility between topological overlap and Pearson correlation similarity measures to define modules. Additionally, to study the relationships between modules derived from different expression data sets for the same species, they compared the consistency of gene modules inferred using different expression data sets. Lastly, they performed expression Quantitative Trait Loci (eQTL) analysis to gain a better understanding of the genetic basis of gene modules. In Chap. 18, X.J. Zhang developed a rigorous approach to decode spike trains in a single neuron and an ensemble of neurons with or without interactions. Finally in Chap. 19, Y. Zhang proposed a new method that can accurately approximate the statistical significance of peaks adjusting for multiple testings.

Finally we would like to thank Zhang QianYi, the secretary of the Computational Systems Biology Centre in Fudan University, for her hard work to go through all chapters several times to unify all references and paper style.

Fudan, Warwick/Michigan/California *Jianfeng Feng, Wenjiang Fu, Fengzhu Sun*

Acknowledgements

This is a collection of papers when we celebrated Prof. Qian MinPing's 70 Birthday. We learned so much from her and often failed to give adequate attribution.

Contents

Contributors

Chieh-Chun Chen Department of Bioengineering, University of Illinois, Urbana-Champaign, USA

Ting Chen Molecular and Computational Biology Program, Department of Biological Sciences, University of Southern California, Los Angeles, CA, USA

Ye Ding Wadsworth Center, New York State Department of Health, New York, USA; Center for Medical Science, 150 New Scotland Avenue, Albany, NY 12208, USA, yding@wadsworth.org

Jianfeng Feng Mathematics and Computer Science College, Hunan Normal University, Changsha 410081, P.R. China; Department of Computer Science and Mathematics, Warwick University, Coventry CV4 7AL, UK

Wenjiang J. Fu The Computational Genomics Lab, Department of Epidemiology, Michigan State University, East Lansing, MI 48824, USA

Hao Ge School of Mathematical Sciences, and Center for Computational Systems Biology, Fudan University, Shanghai 200433, People's Republic of China

Shuixia Guo Mathematics and Computer Science College, Hunan Normal University, Changsha 410081, P.R. China

Li Hsu Division of Public Health Sciences, Fred Hutchinson Cancer Research Center, Seattle, WA, USA

Haiyan Hu School of Electrical Engineering and Computer Science, University of Central Florida, 4000 Central Florida Blvd, Orlando, FL 32816, USA, haihu@cs.ucf.edu; Harris Corporation and Engineering Center, University of Central Florida, 4000 Central Florida Blvd, Orlando, FL 32816, USA

Christophe Ladroue Department of Computer Science and Mathematics, Warwick University, Coventry CV4 7AL, UK

Hyunju Lee Department of Information and Communications, Gwangju Institute of Science and Technology (GIST), Gwangju, Republic of Korea

Ming Li The Computational Genomics Lab, Department of Epidemiology, Michigan State University, East Lansing, MI 48824, USA

Xiaoman Li Biomolecular Science Center, University of Central Florida, 4000 Central Florida Blvd, Orlando, FL 32816, USA, xiaoman@mail.ucf.edu; Harris Corporation and Engineering Center, University of Central Florida, 4000 Central Florida Blvd, Orlando, FL 32816, USA

Jun S. Liu Department of Statistics, Harvard University, Cambridge, MA, USA

Qiang Luo Department of Mathematics, National University of Defense Technology, Changsha, Hunan 410073, China

Jie Peng Department of Statistics, University of California, Davis, CA, USA

Likit Preeyanon The Computational Genomics Lab, Department of Epidemiology, Michigan State University, East Lansing, MI 48824, USA

Hong Qian Department of Applied Mathematics, University of Washington, Seattle, WA 98195-2420, USA

Gary L. Rosner Department of Biostatistics, MD Anderson Cancer Center, University of Texas, Houston, TX 77030, USA

Pei-Zhe Shi Department of Applied Mathematics, University of Washington, Seattle, WA 98195-2420, USA

Fengzhu Sun Molecular and Computational Biology Program, Department of Biological Sciences, University of Southern California, Los Angeles, CA, USA

Pei Wang Division of Public Health Sciences, Fred Hutchinson Cancer Research Center, Seattle, WA, USA

Weizhen Wang Department of Mathematics and Statistics, Wright State University, Dayton, OH 45435, USA, weizhen.wang@wright.edu

Yalu Wen The Computational Genomics Lab, Department of Epidemiology, Michigan State University, East Lansing, MI 48824, USA

Dongfeng Wu Department of Bioinformatics and Biostatistics, University of Louisville, Louisville, KY 40292, USA

Jun Xie Department of Statistics, Purdue University, 250 N. University Street, West Lafayette, IN 47907, USA

Wenqiang Yang Department of Mathematics, Hunan Normal University, Changsha, Hunan 410875, China; Department of Mathematics, National University of Defense Technology, Changsha, Hunan 410073, China

Lingmin Zeng MedImmune, 1 MedImmune Way, Gaithersburg, MD 20878, USA

Nancy R. Zhang Department of Statistics, Stanford University, 390 Serra Mall, Stanford, CA 94305-4065, USA

Tonglin Zhang Department of Statistics, Purdue University, 250 North University Street, West Lafayette, IN 47907-2066, USA, tlzhang@purdue.edu

Wei Zhang Department of Statistics, Harvard University, Cambridge, MA, USA

Xianghua Zhang Biomedical Engineering Institute, Department of Electronic Science and Technology, University of Science and Technology of China, Hefei, Anhui 230027, P.R. China; Department of Epidemiology and Public Health, Yale University, New Haven, CT 06520, USA

Xuejuan Zhang Mathematical Department, Zhejiang Normal University, Jinhua, 321004, Zhejian Province, P.R. China

Yu Zhang Department of Statistics, The Pennsylvania State University, 325 Thomas Bldg, State College, PA 16803, USA

Hongyu Zhao Department of Epidemiology and Public Health, Yale University, New Haven, CT 06520, USA; Department of Genetics, Yale University, New Haven, CT 06520, USA

Sheng Zhong Department of Bioengineering, University of Illinois, Urbana-Champaign, USA; Institute for Genomic Biology, University of Illinois, Urbana-Champaign, USA

Chapter 1
Analysis of Combinatorial Gene Regulation with Thermodynamic Models

Chieh-Chun Chen and Sheng Zhong

1.1 Introduction

Transcriptional control is a key regulatory mechanism for cells to direct their destinies. A large number of transcription factors (TFs) could simultaneously bind to a regulatory sequence. With the constellation of TFs bound, the expression level of a target gene is usually determined by the combinatorial control of a number of TFs.

Thermodynamics was first introduced in physics to study the conversion of energy into work or heat of a system from a macroscopic point of view. Statistic mechanics incorporating statistical tools with thermodynamic principles provides a powerful framework to model and further to predict the collective motion of molecules at the microscopic level on the basis of known characteristics and interactions of a system. The statistic thermodynamic concept [12] was first adopted on the study of molecular mechanism for gene regulation in Bacteriophage Lambda. Later it was further utilized on modeling TF–DNA and TF–RNA polymerase (RNAP) interactions in bacteria [3–5]. These models brought the stochastic interactions of TFs, regulatory sequences, and RNAP together and enabled a quantitative model for the transcription rate in prokaryotes.

Recently, several attempts to employ thermodynamic models in the study of eukaryotic gene regulation were made. With thermodynamic analysis of synthetic promoters, certain phenomena in gene regulation, such as cooperativity and the effects of weak binding sites, were uncovered [9]. Applying a thermodynamic model under a fixed time point in Drosophila development successfully predicted the spatial expression patterns of segmentation genes in Drosophila [11]. In differentiating embryonic stem cells (ESCs), the interaction types of the TFs could be predicted

C.-C. Chen · S. Zhong (✉)
Department of Bioengineering, University of Illinois, Urbana-Champaign, USA
e-mail: szhong@ad.uiuc.edu

S. Zhong
Institute for Genomic Biology, University of Illinois, Urbana-Champaign, USA

J. Feng et al. (eds.), *Frontiers in Computational and Systems Biology*,
Computational Biology 15,
DOI 10.1007/978-1-84996-196-7_1,

Table 1.1 The Boltzmann distribution for the two states of a TFBS

State	TF	Weight
Free	0	1
Attached	1	q_{TF}

from the temporal response of the target gene, and further a transcription network composed of 34 TF–TF interactions and 185 TF-target relationships were identified [6, 7]. These successes made thermodynamic models an applicable route to analyze gene regulatory mechanisms.

1.2 Thermodynamic Models for TF–DNA Binding

Here we introduce the fundamental model to integrate combinatorial signals at the level of *cis*-regulatory transcription control in bacteria through the thermodynamics of TF–DNA and TF–RNAP–DNA interactions [5]. These interactions can be quantified by several parameters that are tuneable by the selection and placement of various protein-binding DNA sequences.

1.2.1 TF–DNA Interactions

At a given time in a cell, there are only two states for a transcription factor binding site (TFBS): attached with or free of a TF. Let q_{TF} denote the ratio of the probability of a TFBS in the attached state to that in the free state (Table 1.1).

The probability that the TFBS of a target gene is bound with a TF could be denoted as

$$P(\mathrm{TF}_{\mathrm{binding}}) = \frac{q_{\mathrm{TF}}}{1+q_{\mathrm{TF}}}.$$

On the other hand, let [TF − DNA] represent the cellular concentration of the promoter bound by the TF. The binding process can be denoted as

$$[\mathrm{TF}] + [\mathrm{DNA}] \rightarrow [\mathrm{TF} - \mathrm{DNA}].$$

Then the probability that the TFBS of a target gene is bound with a TF can be formulated as

$$P(\mathrm{TF}_{\mathrm{binding}}) = \frac{[\mathrm{TF} - \mathrm{DNA}]}{[\mathrm{DNA}] + [\mathrm{TF} - \mathrm{DNA}]}.$$

At equilibrium state, the concentrations of the substrates can be described as

$$P(\mathrm{TF}_{\mathrm{binding}}) = \frac{[\mathrm{TF}]}{[\mathrm{TF}] + [K_{\mathrm{TF}}]} = \frac{\frac{[\mathrm{TF}]}{K_{\mathrm{TF}}}}{\frac{[\mathrm{TF}]}{K_{\mathrm{TF}}} + 1},$$

Table 1.2 The Boltzmann distribution of a promoter with one TF and one RNAP

State	TF	RNAP	Weight
1	0	0	1
2	0	1	q_p
3	1	0	q_{TF}
4	1	1	$\omega_{\mathrm{TFp}} q_p q_{\mathrm{TF}}$

where [TF] is the cellular concentration of the activated TF targeted by this site, and K_{TF} is the effective dissociation constant (relative to the genomic background) representing the concentration required for half of the TF binding to the promoter. Thus, we can obtain

$$q_{\mathrm{TF}} = \frac{[\mathrm{TF}]}{K_{\mathrm{TF}}}.$$

RNAP-promoter binding (without any TF present) can be described by the same form,

$$P(\mathrm{RNAP_{binding}}) = \frac{q_p}{1+q_p},$$

where the ratio of the probability of an RNAP in the attached state vs. in the free state is denoted as $q_p = [\mathrm{RNAP}]/K_p$.

1.2.2 *TF–RNAP–DNA Interactions*

1.2.2.1 One TF

If we consider the case of a TF interacting with an RNAP, there are four possible states for a promotor: (1) bound by both the TF and the RNAP; (2) bound by the RNAP only; (3) bound by the TF only; (4) free from either the TF or the RNAP (Table 1.2).

The probability of the promoter of the target gene bound with an RNAP can be represented as

$$P(\mathrm{RNAP_{binding}}) = \frac{q_p + \omega_{\mathrm{TFp}} q_{\mathrm{TF}} q_p}{1 + q_p + q_{\mathrm{TF}} + \omega_{\mathrm{TFp}}},$$

where

$$\omega_{\mathrm{TFp}} = \begin{cases} 1, & \text{no interaction,} \\ 10\text{–}100, & \text{activation,} \\ 0, & \text{repression.} \end{cases}$$

Different settings of ω reflect different roles a TF could play (Table 1.2). If ω is set to 1, it represents that there is no interaction between the RNAP and the TF. They bind independently to the promoter. If ω is set to 10–100, it represents that the

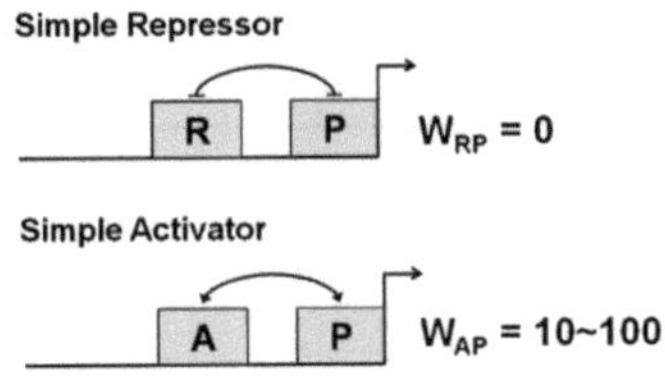

Fig. 1.1 Forms of TF–RNAP interactions and their corresponding parameters for modeling the probability of RNAP binding [6]. A is a transcription factor acting as an activator of genes. R is a transcription factor acting as a repressor of genes. P represents RNAP. The *curve* with a *dot* at the end represents an repression effect; the *curve* with an *arrow* in the end indicates either cooperation between transcription factors or activation of gene by transcription factors

Table 1.3 The Boltzmann distribution of a promoter with its RNAP and two TFs

(TF_1, TF_2)	(0, 0)	(1, 0)	(0, 1)	(1, 1)
RNAP				
0	1	q_{TF_1}	q_{TF_2}	$\omega_{TF_1 TF_2} q_{TF_1} q_{TF_2}$
1	q_p	$\omega_{TF_1 p} q_p q_{TF_1}$	$\omega_{TF_2 p} q_p q_{TF_2}$	$(\omega_{TF_1 p} + \omega_{TF_2 p}) \omega_{TF_1 TF_2} q_{TF_1} q_{TF_2} q_p$

TF helps recruit the RNAP binding to the promoter. The larger ω is, the larger the synergism is. If ω is set to 0 or close to 0, it represents that the TF blocks the RNAP binding to the promoter, and thus the TF serves as a repressor (Fig. 1.1).

1.2.2.2 Two TFs

The case of two TFs capable of binding to a promoter together with an RNAP could be represented in the same fashion (Table 1.3).

The probability of RNAP binding to the promoter can be denoted as

$$P(\text{RNAP}_{\text{binding}}) = \frac{\sum_j \sum_k P(1, j, k)}{\sum_{i,j,k \in \{0,1\}} P(i, j, k)},$$

where $P(i, j, k) = P(\text{RNAP} = i,\ \text{TF}_1 = j,\ \text{TF}_2 = k)$. The parameters ω could be set differently to reflect the nature of these interactions between two TFs or the interactions between one TF and one RNAP (Fig. 1.2).

1.2.2.3 Multiple TFs

A general form for multiple regulatory TFs able to bind to a promoter with an RNAP could be represented as the following. Let Z_{ON} be the partition sum of the Boltzmann weights W over all states of TF binding to the promoter bound by RNAP, and Z_{OFF} not bound by the RNAP on the contrary.

$$P(\text{RNAP}_{\text{binding}}) = \frac{Z_{\text{ON}}}{Z_{\text{OFF}} + Z_{\text{ON}}}.$$

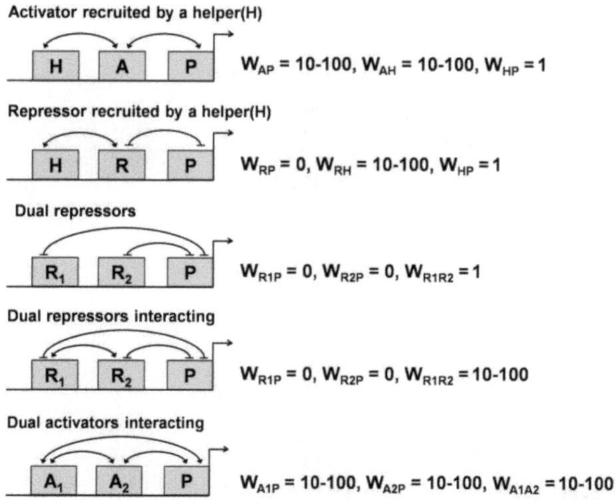

Fig. 1.2 Forms of interactions between two TFs and one RNAP, and their corresponding parameters for modeling the probability of RNAP binding [6]. A_1 and A_2 are activators. R_1 and R_2 are repressors. P represents RNAP. The *line* with a *dot* at the end represents an repression effect; the *line* with an *arrow* at the end indicates either cooperation between two TFs or activation of a gene by a TF

With multiple TFBSs, different configurations of site occupation can be formed. The Boltzmann weight W for a configuration could be simply represented by the product of q_i and w_{ij}, where q_i reflects the TF–DNA interaction (i.e., the binding affinity of a TFBS to the TF), and w_{ij} reflects the interaction between two TFs on sites i and j. Let $\sigma_i = 1$ if site i is occupied and $\sigma_i = 0$ otherwise. Then

$$W[\sigma_1, \ldots, \sigma_L] = \prod_{i=1}^{L} q_i^{\sigma_i} \prod_{i<j} \omega_{ij}^{\sigma_i \sigma_j}.$$

Thus, Z_{OFF} can be obtained by summing over all configurations without the RNAP binding on the promoter:

$$Z_{\mathrm{OFF}} = \sum_{\sigma_1=0,1} \cdots \sum_{\sigma_L=0,1} W[\sigma_1, \ldots, \sigma_L].$$

Z_{ON} can be further represented as

$$Z_{\mathrm{ON}} = \sum_{\sigma_1=0,1} \cdots \sum_{\sigma_L=0,1} Q[\sigma_1, \ldots, \sigma_L] \cdot W[\sigma_1, \ldots, \sigma_L],$$

where $Q[\sigma_1, \ldots, \sigma_L]$ reflects the interaction between the RNAP and each bound TF,

$$Q = q_p \prod_{i=1}^{L} [1 - \sigma_i \delta(\omega_{0i}, 0)] \cdot \left[1 + \omega \sum_{j=1}^{L} \sigma_j \delta(\omega_{pj}, \omega) \right].$$

The first part makes sure that the RNAP could bind to DNA without any repressor present (i.e., $\omega_{pi} = 0$). The second part represents the additional weights from the cooperative interactions between the RNAP and each bound TF, respectively.

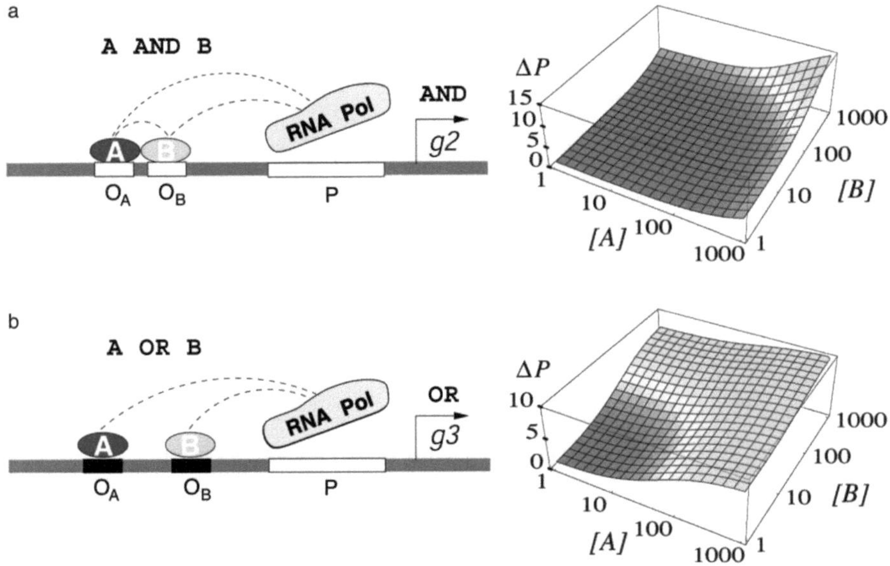

Fig. 1.3 *Cis*-regulatory constructs and response characteristics of the AND(a) and OR(b) gates [5]. *Filled* and *open boxes* denote strong and weak binding sites, respectively. *Dashed lines* indicate cooperative interaction with $\omega_{ij} = 20$. Plotted to the right of each construct is the fold change in RNAP-binding probability for typical cellular TF concentrations [A] and [B] (in nM). Qualitative features of these plots are insensitive to the precise values of the parameters used

Given the binding strengths K_i and the cooperativity factors ω_{ij} for all the DNA sites, the binding probability of the RNAP to the promoter can be computed straightforwardly. Different regulatory functions could be implemented by arranging TFBSs in *cis*-regulatory regions with appropriate settings of the parameters, such as interaction parameters ω and TF dissociation constants K, under various TF concentrations. Take two TFs binding to a promoter with an RNAP for example (Fig. 1.3). By tuning the parameter $w_{\mathrm{TF_1TF_2}}$, different logic functions can be implemented. Figure 1.3(a) shows the implementation of the logic function AND, where weak binding sites for TFs A and B are designed to be next to each other so that only with the presences of both TFs the additional cooperative interaction enables them to bind onto the promoter. Similarly, the OR gate could be constructed, where no interaction between TFs A and B is needed for binding.

1.3 Models for Gene Expression

With the above thermodynamic model of TF–DNA interactions, we now know how to quantify the equilibrium binding probability of the RNAP to the promoter, given the cellular concentrations of all the TFs. However, the bridge connecting from the binding probability of RNAP to the gene expression levels is still missing.

In general, it is often assumed in thermodynamic models that the degree of gene transcription is proportional to the binding probability of the RNAP to the promoter. In the following section, we discuss two different routes to further model gene expression. A kinetic model was proposed to analyze the dynamics of gene expression over times [7]. A logistic regression model was also used to associate RNAP binding with gene expression levels. In particular, it could handle the situation that beyond a certain number of activators or repressors, maximal or minimal transcription levels could be made.

1.3.1 Kinetic Model

Assume that the changes of TF concentrations can be inferred from the changes of mRNA levels of TFs and that the mRNA degradation rates are linearly dependent on the mRNA concentration. Thus, based on the principle of thermodynamic models that the transcription rate is proportional to the binding probability of RNAP, in [7], the following ordinary differential equation was proposed to mimic the dynamics of gene expressions:

$$\frac{dG}{dt} = K_g\big(P(\text{RNAP}_{\text{binding}})\big) - K_d G,$$

where G denotes the transcript concentration, K_g represents the maximal synthesized rate of transcripts, and K_d is the degradation rate of transcripts.

Although gene expressions should be continuous signals throughout the time, an assumption should be made that gene expressions are measured when the transcriptional system is in its equilibrium state at each time point, which is satisfied by all time course microarray data. Under this circumstance, the expression can be represented by

$$G = \alpha P(\text{RNAP}_{\text{binding}}),$$

where

$$\alpha = \frac{K_g}{K_d}.$$

1.3.2 Logistic Model

A statistical framework was developed [11] to predict the expression of a target DNA sequence. The main idea is to sum over expression levels predicted by a logistic model under all possible configurations of TFs on a given sequence.

Let each possible configuration of TFs be c, and the RNAP binding probability be $P(E)$. Then the whole probability of RNAP binding is the weighted sum of

RNAP binding probability for every configuration, where the weight of each configuration is the probability of the configuration,

$$P(E) = \sum_{c \in C} P(c)P(E|c),$$

where $P(c)$ denotes the probability of a configuration on a DNA sequence, and $P(E|c)$ denotes the probability of RNAP binding under a configuration c. Note that although $P(E)$ is the probability of RNAP binding, $P(E)$ is proportional to the expression level under the thermodynamic principle.

1.3.2.1 Sequence Component $P(c)$

To compute the probability of a certain configuration c on a given sequence, all possible configurations of TFBSs should be considered. Note that no TFBSs can overlap each other in a configuration. Just as the Boltzmann distribution mentioned in Sect. 1.2, the probability of each configuration $P(c)$ is given by

$$P(c) = \frac{W(c)}{\sum_{c' \in C} W(c')},$$

where $W(c)$ represents the statistical weight associated with configuration c.

Intuitively, assuming TFs bind independently to the sequences, the statistical weight of a configuration c should be the product of the contribution of each TF binding on c. Moreover, two factors can influence a TF binding to its binding site, the concentration of the TF, denoted by τ, and the binding affinity of the TFBS.

A standard position specific scoring matrix (PSSM) is utilized to represent the binding affinity of a TFBS. It defines a separate probability distribution over the four nucleotides at each position of the binding site recognized by the TF. The simple assumption of a PSSM is that all positions within a binding site are independent. Thus, for a TF i that binds to a site of length $L(i)$, the binding affinity can be represented as

$$\mathrm{BA}_i(S_1, \ldots, S_{L(i)}) = \prod_{j=1}^{L(i)} \frac{P_i^j(S_j)}{P_B(S_j)},$$

where P_B is a background distribution, such as a uniform Markov order zero background (i.e., $P_B(A) = P_B(C) = P_B(G) = P_B(T) = 0.25$). The background model serves as a scale against a binding site being measured. Compared to the probability from the background model, the larger the probability from the PSSM, the higher the binding affinity.

Thus, the statistical weight of the entire configuration c could be derived by simply multiplying together the contributions of all TFBSs in c. Let a configuration c has k TFs $f(1), \ldots, f(k)$ binding on their TFBSs at positions $p(1), \ldots, p(k)$, respectively, the statistical weight $W(c)$ of the configuration can be represented as

$$W(c) = \prod_{i=1}^{k} \tau_{f(i)} \mathrm{BA}_i(S_{p(i)}, \ldots, S_{p(i)+L(i)}).$$

To further incorporate the cooperativity interactions between TFs, the effect of the binding site arrangement needs to be considered. Different arrangements of binding sites, such as distances and orientations, might affect the nature of TF interactions. Moreover, the effects of hetero-cooperativity and homo-cooperativity are regarded as essential factors to help TF binding onto DNA.

To simplify the model, here only adjacent TF pairs are assumed with such cooperative effects. Second, no orientation difference affects the cooperativity. Third, only homo-cooperativity effect is applied. Based on these general assumptions, the statistical weight of c can be represented in an extended form with binding cooperativity as follows:

$$W(c) = \left[\prod_{i=1}^{k} \tau_{f(i)} \mathrm{BA}_i(S_{p(i)}, \ldots, S_{p(i)+L(i)})\right] \times \left[\prod_{i=1}^{k-1} \gamma\big(f(i), f(i+1), p(i+1) - p(i)\big)\right],$$

where function γ defines the binding cooperativity between adjacent TFs $f(i)$ and $f(i+1)$ separated by distance d. Intuitively, the closer the TF pair is, the stronger the cooperativity interaction is. For any pair of adjacent different TFs, $\gamma(f(i), f(i+1), d) = 0$, where $f(i) \neq f(i+1)$. Nevertheless, γ can be easily generalized by other functions incorporated with reasonable features of cooperativity interactions.

1.3.2.2 Expression Component $P(E|c)$

For a configuration c, the logistic function is used to infer its ability of recruiting RNAP. The contribution of each TF on c is assumed to be independent of the expression outcome, where activators contribute positively, and repressors contribute negatively. With the unique saturation property of the logistic model, maximal or minimal transcription is achieved beyond a certain number of bound activators and repressors, respectively.

Thus, given a configuration c with k TFs $f(1), \ldots, f(k)$ binding to their TFBSs at positions $p(1), \ldots, p(k)$, respectively, the probability of RNAP binding, $P(E|c)$, can represented as

$$P(E|c) = \mathrm{logit}\left[\omega_0 + \sum_{i=1}^{k} \omega_{f(i)}\right] = 1/\left[1 + \exp\left[-\left[\omega_0 + \sum_{i=1}^{k} \omega_{f(i)}\right]\right]\right],$$

where w_0 represents the basal level of the expression, and w_i represents the contribution of TF i on the expression levels. A positive value for w_i reflects TF i to be an activator, while a negative one represents TF i to be a repressor.

1.4 Reconstruction of Regulatory Networks

The interactions among regulatory proteins and their regulatory sequences collectively form a regulatory network, which controls the fate of cells. A major challenge in the study of gene regulation is to identify the interaction relationships within a regulatory network. In the following section, we introduce a computational framework based on thermodynamic modeling to reconstruct regulatory networks.

Based on Interaction-Identifier [7] to select for the thermodynamic model that best describes the TF–TF and TF–RNAP interaction for each target gene, the Network-Identifier method [6] was further developed for inferring regulatory networks from time course gene expression data.

1.4.1 Interaction-Identifier

The Interaction-Identifier method models how different TF interaction forms (Figs. 1.1 and 1.2) affect the expression levels of a target gene at steady states. First, a thermodynamic model is used to translate a TF interaction form with the TF concentrations into the probability of RNAP binding onto the promoter of the target gene (see Sect. 1.2). Next, a kinetic model derives the gene expression profile across times for each TF interaction form (see Sect. 1.3.1). By searching the space of TF interaction forms, Interaction-Identifier identifies the underlining TF interaction form of each target gene, which minimizes the difference between the model-derived expression profile and the observed expression data (Fig. 1.4).

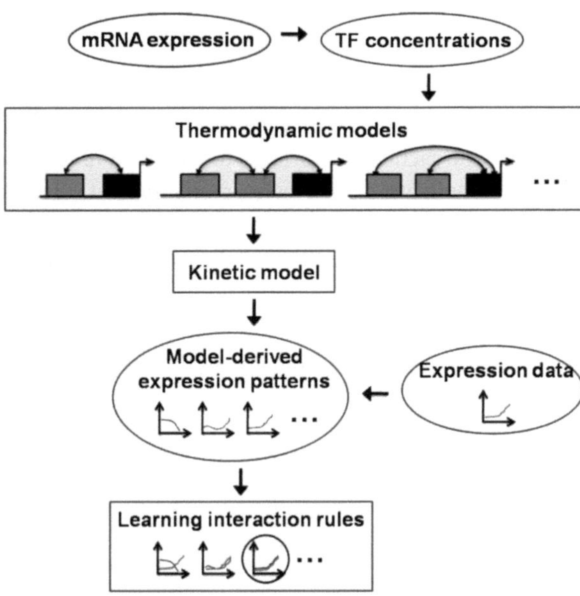

Fig. 1.4 Flowchart of the Interaction-Identifier algorithm [7]

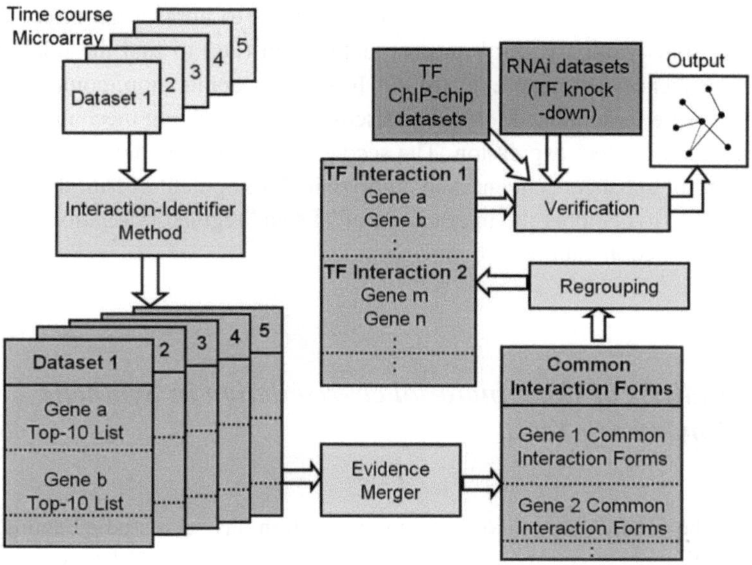

Fig. 1.5 Flowchart of the Network-Identifier algorithm [6]

1.4.2 Network-Identifier

Network-Identifier utilizes Interaction-Identifier to find common TF interaction forms of target genes across multiple time course microarray datasets and then incorporates those predicted regulatory relationships supported by independent datasets into a regulatory network. The method has three components: (1) Interaction-Identifier [14], (2) Evidence merger, and (3) Verification component (Fig. 1.5).

For each time course dataset, Interaction-Identifier first evaluates the fitness of each interaction form on each target gene and returns its Top-10 most-likely TF interaction form. Next, Evidence merger identifies the most frequently appeared interaction form across multiple datasets for each target gene. The verification component groups target genes based on their most frequently appeared interaction forms. Chi-square tests are used to examine whether the identified TF-target relationships are enriched with regulatory relationships identified from independent experimental data, such as ChIP-chip and RNAi data. Finally, Network-Identifier will report the regulatory relationships confirmed by both two independent data sources.

1.5 Applications

Thermodynamic models of gene regulation have shown promising results in eukaryotic systems. We introduce three applications in yeast (*Saccharomyces cerevisiae*), Drosophila, and Mouse embryonic stem cells, respectively, in the following sections. The common goal for these applications is to unravel the effects of *cis*-regulatory transcription control on gene expression (i.e., to find the relationships between sequence and gene expression). In the first application, compositions of sequences were manipulated to further discover the underlying mechanism of how sequences could affect expression. The second application predicts spatial expression pattern of enhancer sequences in embryonic development of fruit fly, while the third one tried to decipher the interactions of TFs and regulatory relationships from temporal expression data.

1.5.1 Analysis of Combinatorial cis-regulation in Synthetic Promoter in Yeast

Although the fundamental theory of gene regulation has been studied and defined, the connections between regulatory information (*cis*-motifs and transcription factors) and gene expression profiles is still unclear [13]. Several studies developed in silico promoter models [8, 14] demonstrated the associations between promoter modules and gene expressions. A ground-breaking study in Yeast [2] achieves the relatively high accuracy of prediction from conserved *cis*-motif logics to expression. This made it tempting to design synthetic promoters that allow refined and targeted modifications of promoter architecture. Through synthetic promoter engineering [1], *cis*-motif logic, including orientation, binding energy, and position could be clearly elucidated and served as control variables to study gene expression and gain insights of regulatory complexity.

In order to learn how *cis*-regulatory mechanisms affecting gene expression in yeast, a strategy of combinatorial engineering was utilized to construct the synthetic promoter libraries [9]. All random combinations of three or four TFBSs as building blocks were placed upstream of a core-promoter attached with yellow fluorescent protein. Then those synthetic promoters were integrated into the yeast genome. By quantifying florescent intensities, the level of gene expression can be observed.

In the L1 promoter library, 429 promoters were analyzed with five fitting parameters. The thermodynamic method enabled to explain 49% of the variance in expression (Table 1.4), which is more than double the amount of variance explained by the best models of genome-wide expression data. Another independent data, the L1-test library, composed of novel combination of the L1 building blocks, was used as testing data to assess the predictive power of the model for L1 library. With the same parameter settings as the L1 library's, the model still captured 44% of the variance in expression, which suggested that the model was not overfitted. The results of the thermodynamic approach suggest that modeling the biophysical principle of TF–DNA and TF–TF interactions can generally depict the expression driven by different combinations of TFBSs.

Table 1.4 Summary of synthetic promoter libraries [9]

Library	# of promoters	# of fitted parameters	Fraction of variance explained (R_2)
L1	429	5	0.49
L1-test	83	0	0.44

1.5.2 Predicting Spatial Expression Patterns from Sequence in Drosophila Segmentation

Drosophila melanogaster is a model organism for genetics research because of its short life cycle, the relatively small genome, and easily manipulation in laboratory. Moreover, since its embryos grow outside the body, it provides an excellent means of studying embryonic development in eukaryotes. Studying its notable segmentation network has helped accumulate most of our knowledge about the mechanisms of segmentation in arthropods [10].

The well-characterized segmentation gene network involves a cascade of gene regulation (Fig. 1.6(a)). It consists of a four-tiered hierarchy of maternal and zygotic factors that define the antero-posterior (A-P) axis in a stepwise refinement of expression patterns. First, maternal transcripts of the segmentation genes in an oocyte are specifically targeted to the anterior (bicoid(BCD), hunchback(HB)) and posterior (Nanos(NOS), caudal(CAD)). Those maternal proteins together activate certain zygotic gap genes, such as kruppel (Kr), giant (Gt), knirps (Kni), and tailless (Tll), at specific positions along the A-P axis. All transcripts of gap genes, together with maternal proteins, activate periodic patterns of seven pair-rule genes (even-skipped (eve), fushi tarazu (ftz)) that finally activated fourteen segmental polarity genes, resulting in establishment of segment boundaries.

With the spatial expression patterns for eight key transcription factors, including BCD, HB, CAD, Kr, Gt, Kni, Tll, Torso-response element (TorRE), and their binding-site preferences as inputs, a computation framework [11] (see Sect. 1.3.2) was applied to model the process of transcriptional regulation and further to predict the spatial expression of 44 gap and pair-rule gene modules with known patterns collected from literatures.

The results show that expression patterns predicted by the model trained on these data exhibit good or fair agreement with the measured patterns for most modules (Fig. 1.6(b)). The expression of gap gene modules is generally predicted well, suggesting that the model has adequately captured the input and interaction rules. Prediction of pair-rule gene modules seems more mixed suggesting that some input factors may be missing, and some higher-level interaction rules are not captured.

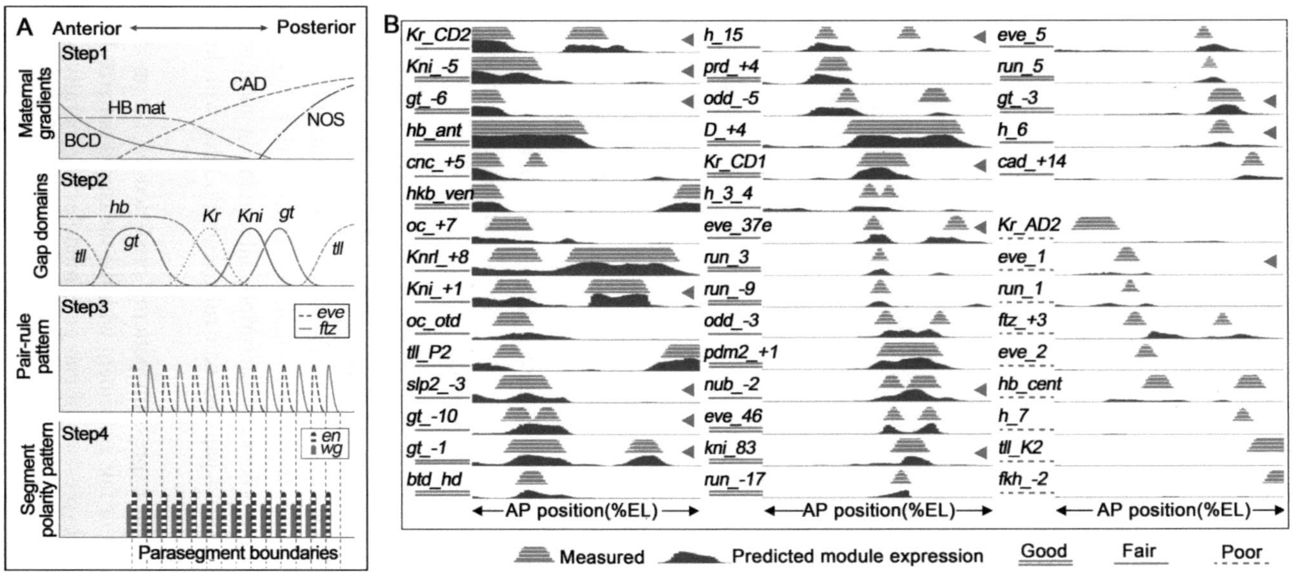

Fig. 1.6 (Color online) (**a**) Segmentation steps in Drosophila melanogaster [10]. (**b**) Observed module expression patterns (*red*) v.s. Predicted expression patterns (*blue*) [11]

1.5.3 Inferring Gene Regulatory Networks in Mouse Embryonic Stem Cells

Embryonic Stem Cells (ESCs) are derived from early mammalian embryos. ESCs possess two important characteristics that define their importance in scientific and medical fields. First, they are capable of self-renewal through apparently unlimited, undifferentiated proliferation in cultured cell lines; second, they have remarkable pluripotency potentials to give rise to many different cell types in the body that may contribute to the study of body development and regenerative medicine. A few transcription factors have shown to be key transcriptional regulators in ESCs. These include Oct4, Sox2, Nanog, Klf4, Esrrb, and Tcl1. Large-scale genomic data have been generated for these regulators. Microarray data are allowed to measure changes in expression levels across different time or different experimental conditions. Chromatin immunoprecipitation (ChIP)-chip data enable to determine the binding loci to identify the targets of TFs, while RNA inference (RNAi) is used for shutting down a TF to help distinguish its target genes. By systematically analyzing the high-throughput genomic data, the mystery of regulatory circuit in ESCs is gradually unraveled.

Network-Identifier was developed [6] to analyze the combinatorial control of the key transcription factors and to further infer the regulatory network in mouse ESCs. Five time series microarray datasets of mouse ESCs were used, including a dataset for retinoid acid-induced differentiation and four datasets for spontaneous differentiation of four ESC lines. Six known keys TFs, Oct4, Sox2, Nanog, Klf4, Esrrb, and Tcl1 served as regulators of this system. 747 genes annotated by Gene Ontology term "Transcription Regulator Activity" are used as target genes.

Network-Identifier reported an ESC transcription network with 87 regulators and target genes (Fig. 1.7). Several interesting regulatory relationships are revealed. In particular, the mutual regulation of Klf2 and Klf4 were recently shown to be an important module for maintaining the undifferentiated state of ESCs. Mtf2 has only recently been implied to inhibit differentiation by recruiting the polycomb group of transcription repressors. The results further indicate that Klf4 and Sox2 can synergistically activate Mtf2 in ESCs. The regulatory relationships for a number of genes involved in lineage specific differentiation were also identified. These include Gata6, Gata3, Sox17, and FoxA2. Inhibiting these lineage specific differentiation genes in ESCs is critical to maintain an undifferentiated state. Among the predicted network, there were a number of transcription repressors, including Ctpb2 and Rest. Ctpb2 was predicted to be activated by Oct4. Rest was predicted to be jointly regulated by Oct4 and Sox2. These results suggest that Oct4 and Sox2 can indirectly inhibit differentiation genes by activating transcription repressors such as Ctpb2 and Rest.

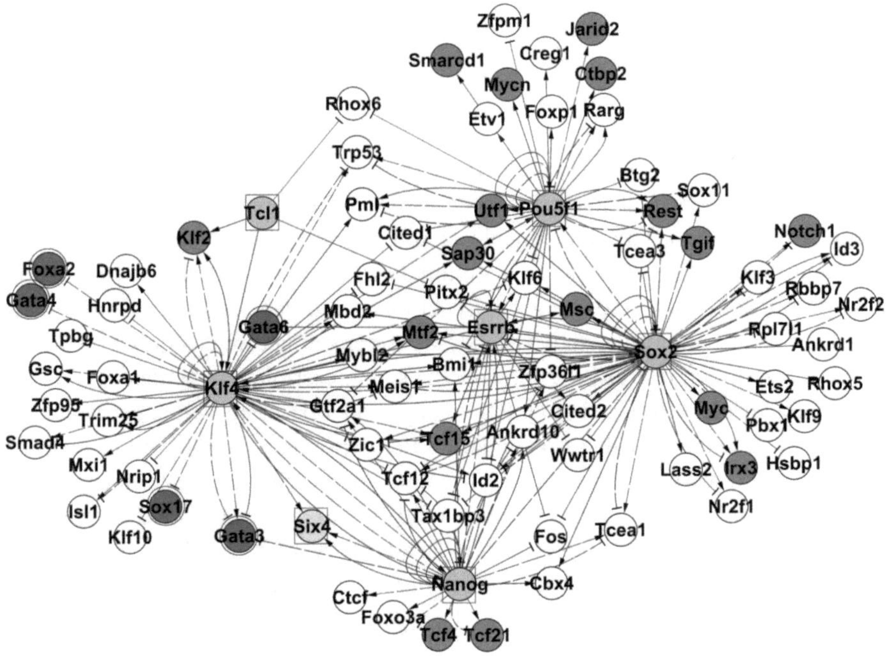

Fig. 1.7 (Color online) The gene regulatory network identified by Network-Identifier [6]. Nodes with *squares* (*yellow*) represent regulators. Nodes with *double circles* (*red*) represent genes used for differentiation. The rest *filled* nodes (*green*) represent genes promoting self-renewal and pluripotency. *Sharp* and *blunt arrows* represent activation and repression effects, respectively. The *solid lines* (*red* and *green lines* showing activation and repression activities, respectively) are with RNAi evidence. The *dash lines* (*blue* and *black lines* showing activation and repression activities, respectively) denote regulatory relationships with ChIP-chip evidence

1.6 Concluding Remarks

Thermodynamic models based on depicting the interactions between TF–TF and TF–DNA to predict RNAP binding probability have shown their applicability to capture the underlying relationship between regulatory sequence and gene expression in either prokaryotic or eukaryotic systems.

However, molecular events are much more complicated in reality. There are still a number of simplification made in modeling the biophysical properties of gene regulation. Many mechanisms, such as the cooperativity interactions for more than two TFs, long-range interaction of enhancer binding TFs and RNAP, DNA methylation, and chromatin structure, are not included in current methods. Future work that takes these molecular features and events into account will potentially provide us with a thorough understanding of combinatorial gene regulation.

References

1. E. Andrianantoandro, S. Basu, D.K. Karig, and R. Weiss. Synthetic biology: new engineering rules for an emerging discipline. *Mol Syst Biol*, **2**: 2006–2006, 2006.
2. M.A. Beer and S. Tavazoie. Predicting gene expression from sequence. *Cell*, **117**(2): 185–198, 2004.
3. L. Bintu, N.E. Buchler, H.G. Garcia, U. Gerland, T. Hwa, J. Kondev, T. Kuhlman, and R. Phillips. Transcriptional regulation by the numbers: applications. *Curr Opin Genet Dev*, **15**(2): 125–135, 2005. Chromosomes and expression mechanisms.
4. L. Bintu, N.E. Buchler, H.G. Garcia, U. Gerland, T. Hwa, J. Kondev, and R. Phillips. Transcriptional regulation by the numbers: models. *Curr Opin Gen Dev*, **15**(2): 116–124, 2005.
5. N.E. Buchler, U. Gerland, and T. Hwa. On schemes of combinatorial transcription logic. *Proc Natl Acad Sci USA*, **100**(9): 5136–5141, 2003.
6. C.C. Chen and S. Zhong. Inferring gene regulatory networks by thermodynamic modeling. *BMC Genomics*, **9**(2) (2008).
7. C.C. Chen, X.G. Zhu, and S. Zhong. Selection of thermodynamic models for combinatorial control of multiple transcription factors in early differentiation of embryonic stem cells. *BMC Genomics*, **9**(1) (2008).
8. S. Fessele, H. Maier, C. Zischek, P.J. Nelson, and T. Werner. Regulatory context is a crucial part of gene function. *Trends Genet*, **18**(2): 60–63, 2002.
9. J. Gertz, E.D. Siggia, and B.A. Cohen. Analysis of combinatorial cis-regulation in synthetic and genomic promoters. *Nature*, **457**(7226): 215–218, 2009.
10. A.D. Peel, A.D. Chipman, and M. Akam. Arthropod segmentation: beyond the drosophila paradigm. *Nat Rev Genet*, **6**(12): 905–916, 2005.
11. E. Segal, T. Raveh-Sadka, M. Schroeder, U. Unnerstall, and U. Gaul. Predicting expression patterns from regulatory sequence in drosophila segmentation. *Nature*, **451**(7178): 535–540, 2008.
12. M.A. Shea and G.K. Ackers. The OR control system of bacteriophage lambda. A physical-chemical model for gene regulation. *J Mol Biol*, **181**(2): 211–230, 1985.
13. M. Venter. Synthetic promoters: genetic control through cis engineering. *Trends Plant Sci*, **12**(3): 118–124, 2007.
14. T. Werner, S. Fessele, H. Maier, and P.J. Nelson. Computer modeling of promoter organization as a tool to study transcriptional coregulation. *FASEB J*, **17**(10): 1228–1237, 2003.

Chapter 2
RNA Secondary Structure Prediction and Gene Regulation by Small RNAs

Ye Ding

2.1 Introduction

RNA molecules are involved in some of the cell's most fundamental processes that include catalysis, pre-mRNA splicing and RNA editing, and regulation of transcription and translation. To a large degree, the function of a regulatory RNA molecule is determined by its structure. Computational methods for modeling RNA secondary structure provide useful initial models for solving the tertiary structure by crystallography or nuclear magnetic resonance (NMR). The problem of computational prediction of secondary structure for a single RNA sequence dates back to the early 1970s [99]. Free energy minimization has been an important method for such prediction. The partition function approach by McCaskill enables rigorous computation of base-pair probabilities and heat capacity [70]. In recent years, there has been increasing interest in ensemble-base approaches that extend the pioneering work of McCaskill. In this chapter, we briefly review these developments. Gene silencing by RNA interference and posttranscriptional gene regulation by microRNAs are fundamental discoveries in molecular biology. Rational design of short interfering RNAs for improving potency of gene silencing and regulatory target prediction for microRNAs are two important computational problems. We here review work from our group and others to show that target mRNA secondary structure is important for both efficient gene silencing and microRNA target recognition.

Y. Ding (✉)
Wadsworth Center, New York State Department of Health, New York, USA
e-mail: yding@wadsworth.org

Y. Ding
Center for Medical Science, 150 New Scotland Avenue, Albany, NY 12208, USA

J. Feng et al. (eds.), *Frontiers in Computational and Systems Biology*,
Computational Biology 15,
DOI 10.1007/978-1-84996-196-7_2,

2.2 RNA Secondary Structure Prediction

RNA plays a variety of important functional roles that include catalysis, RNA splicing, and regulation of transcription and translation. These roles are typically carried out at specific RNA structural sites, often through molecular interactions or conformational change. Hence, the function of an RNA molecule is primarily determined by its secondary and tertiary structures. RNA tertiary interactions involve secondary structure elements and are substantially weaker than secondary interactions. Thus, to a large extent, the free energies in secondary structure represent the thermodynamics of RNA folding. The tendency for RNA folding to be primarily driven by secondary structure features is a tremendous advantage for structural and functional studies on RNAs. Furthermore, computational RNA tertiary structure prediction without experimental information is an intractable problem, and the thermodynamics of tertiary interactions have not been well characterized. In addition, RNA secondary structure is well conserved in evolution. For these reasons, computational algorithms have focused on RNA secondary structure prediction in the last several decades. Given an RNA sequence, a secondary structure is simply defined by a list of base pairs, typically Watson–Crick (G•C or A–U) and Wobble G–U. As shown by Fig. 2.1 for a predicted minimum free-energy structure for *Xlo* 5S rRNA, helices and loops of various types represent basic structural elements of RNA secondary structure.

2.2.1 Free Energy Minimization

In structural computational biology, free-energy minimization for prediction of macromolecular folding is a long-established paradigm. It assumes that, at equilibrium, the solution to the underlying molecular folding problem is unique and that the molecule folds into the lowest-energy state. Also, it is implicitly assumed that the free energies of individual structural motifs are additive. This paradigm had been the foundation for prediction of RNA secondary structure for several decades [67, 68, 74, 99, 116]. For RNA secondary structure prediction, free-energy parameters for basic structural motifs are estimated or extrapolated from chemical melting experiments [67, 68, 110]. The discrete optimization problem is ill-conditioned, in that the prediction is sensitive to small changes in the energy parameters [53, 115]. Furthermore, there is substantial uncertainty in the energy parameters, particularly for loops. For these reasons, efficient algorithms have been developed for not only computing the minimum free energy (MFE) structure, but also for generating a heuristic set of suboptimal structures [67, 68, 116]. An alternative approach computes all suboptimal foldings within an energy increment above the MFE [109]. The exponential growth in the number of these foldings motivated the development of the RNAshapes method for the efficient representation of the near-optimal foldings [36]. The complete suboptimal approach addresses the low-energy end of the

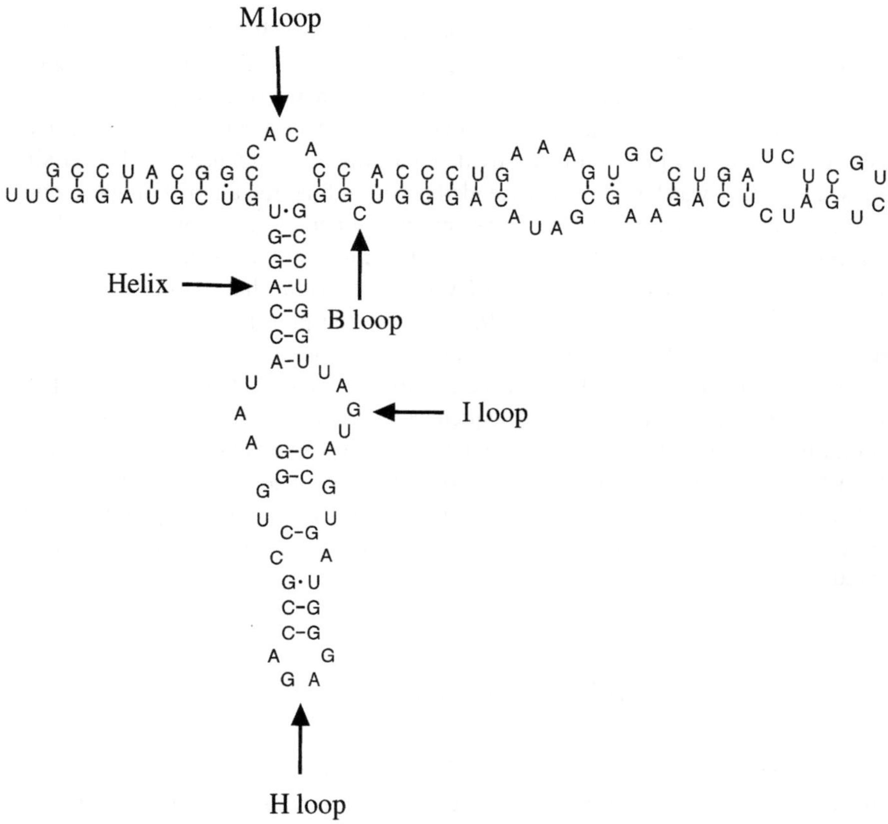

Fig. 2.1 The minimum free-energy structure for *Xlo* 5S rRNA and all types of secondary structural elements: helix (formed by stacked base pairs), bulge loop (B loop), interior loop (I loop), hairpin loop (H loop), and multibranched loop (M loop)

unweighted energy landscape. Neither approach guarantees an unbiased representation of the Boltzmann-weighted ensemble. The free-energy minimization algorithm [116] and the algorithm for computing suboptimal structures [109] have been extended for two or more interacting RNAs [3].

2.2.2 Partition Function Approach

In a drastic departure from free-energy minimization, the partition function approach pioneered by McCaskill (1990) [70] laid the foundation for statistical characterizations of the equilibrium ensemble of RNA secondary structures. In particular, base-pair probabilities can be calculated. Similar to its MFE counterpart, the algorithm for computing partition function and base-pair probabilities is cubic and requires quadratic storage. The significance of base-pair probabilities has been further demonstrated in two studies. For base pairs in the MFE structure, those with higher probabilities have higher predictive accuracy measured by positive predictive value [64]. The positive predictive value is the percentage of base pairs in the predicted structure that are in the structure determined by comparative sequence analysis. Thus, base-pair probabilities provide measures of confidence for MFE predictions. That study was based on an extended partition function algorithm that accommodated coaxial stacking and more recent energy parameters. Furthermore, base-pair probabilities are found to be less affected by uncertainties in energy parameters than is the MFE structure [53]. The McCaskill algorithm has also been extended to include a class of pseudoknots [29, 30]. Like the partition function, the mean and variance (and any moments in general) of the Boltzmann-weighted free-energy distribution can be calculated, and these ensemble characteristics are reported to be useful for distinguishing biological sequences from random sequences [71]. A partition function algorithm for k-point mutants of an RNA sequence has recently been described [17]. For modeling the hybridization of two nucleic acid molecules, the Zuker group was the first to compute partition function and base-pair probabilities [21]. These developments are indicative of a paradigm shift towards ensemble-based approaches.

2.2.3 Statistical Sampling Approach

In the traceback step of an RNA folding algorithm, base pairs are generated one at a time according a chosen principle (e.g., energy minimization or probabilistic sampling as discussed below) to form a secondary structure. The long-standing problem of a statistical representation of probable foldings can be addressed by a sampling extension of the partition function approach [24]. In the traceback step, the conditional probabilities computed with partition functions are used to sample a new base-pair or unpaired base(s), given partially formed structure. Thus, the essence of the sampling algorithm is stochastic traceback. The Boltzmann distribution in statistical mechanics gives the probability of a secondary structure I at equilibrium as $\exp[-E(I)/RT]/U$, where $E(I)/$ is the free energy of the structure, R is the gas constant, T is the absolute temperature, and U is the partition function for all admissible secondary structures of the RNA sequence, i.e., $U = \sum_I \exp[-E(I)/RT]$. The sampling algorithm generates a sample of secondary structures in proportion to their Boltzmann probabilities, guaranteeing a statistical representation of the Boltzmann-weighted ensemble.

A statistical sample of the ensemble allows sampling estimates of the probabilities of any structural motifs, from the simplest elements of base pair and unpaired base, to loops of various types, to more complex structures consisting of stems and loops that may be of special interest in a given application. In particular, probability profiling of single-stranded regions in RNA secondary structure is directly applicable to the rational design of mRNA-targeting nucleic acids [22–26]. The Boltzmann-weighted density of states (BWDOS) [24] characterizes the weighted energy landscape, whereas a density-of-states algorithm [19], applicable only to short sequences, describes the *unweighted* landscape. A structure sample can also be used for computation of other characteristics of the Boltzmann ensemble. For example, the mean and the variance of the free-energy distribution can be estimated by a sample, whereas exact calculations require laborious algorithm development [71]. In principle, a sampling extension can also be developed for a partition function algorithm including pseudoknots. In this case, base-pair probabilities can be estimated by a sample, and the estimates should closely approximate those computed by a high-order algorithm [30].

A sample of moderate size drawn from the ensemble of an enormous number of possible structures is sufficient to guarantee statistical reproducibility in the estimates of typical sampling statistics. The reproducibility is best demonstrated when two independent samples do not have a single structure in common [24, 28]. These seemingly surprising observations are fully expected for an exact sampling algorithm. The sampling algorithm is the basis of the Sfold RNA software package [26] and has been implemented into other RNA folding software including UNAfold [63], Vienna RNA package [42], and RNAstructure [65]. Sampling was adapted for probabilistic representation of structure shapes for RNA sequences of moderate length or longer [107]. A method has been presented to speed up the sampling step [80].

2.2.4 Cluster and Centroid Representation of Boltzmann Ensemble

In the sampled ensemble, distinct structural clusters were observed [24]. This observation suggested that the Boltzmann ensemble could be efficiently represented by clusters. Automated clustering procedure and tools have been developed for this purpose [13, 27, 28]. The procedure returns three to four clusters on average. Another advantage of clustering is that the centroid structure, as the single best representative of the cluster, can be easily identified with little computational cost. The centroid of any set of structures is defined as the structure in the whole ensemble that has the shortest total distance to structures in the set. For the base-pair distance between two structures, the centroid is simply the structure formed by all base pairs having a frequency > 0.5 in the structure set [27]. The clusters, together with their probabilities (estimated by frequencies in the sample) and their centroids, present a complete and efficient statistical characterization of the Boltzmann ensemble. Similar to the reproducibility of ensemble-level sampling statistics [24], the clusters and centroids are also statistically reproducible from one sample to another, even when the two independent samples do not share a single structure [28]. The centroid of the sampled ensemble and the best cluster centroid provide alternative structural predictions. It was a surprising finding that these predictions are substantially improved over the minimum free energy predictions [27], a result that further validates ensemble-based approaches. The idea of centroid has generated considerable interest. Generalized centroid estimators for bioinformatics problems in particular RNA secondary structure prediction have been proposed [12, 38].

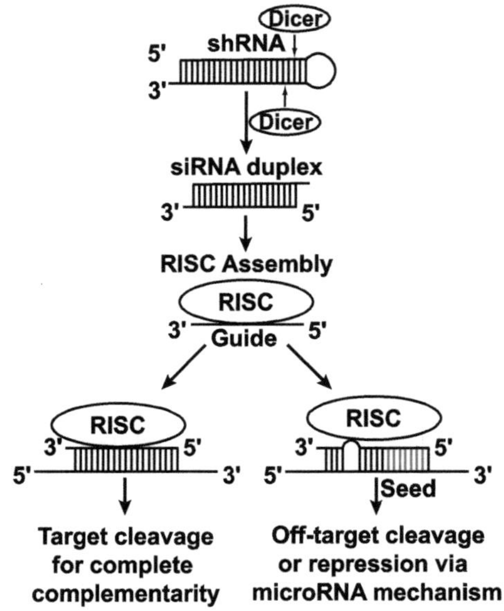

Fig. 2.2 Post-transcriptional regulation by shRNAs or siRNAs. An shRNA (with a typically 19–29 bp stem) can be processed by Dicer into an siRNA. The guide strand in the assembled RISC guides target recognition by complementary base-pairing. Target cleavage by RNAi machinery is triggered by perfect complementarity. Partial complementarity can induce off-target mRNA cleavage or repression of gene expression via microRNA pathway, for which the seed base-pair match (in red) involving nt 2 to nt 7 or 8 of the 5′ end of the guide strand is reported to be important

2.3 Gene Silencing by Small Interfering RNAs

RNA interference (RNAi) is a sequence-specific gene silencing mechanism that is induced by double-stranded RNA (dsRNA) homologous to the target gene [35]. RNAi can be mediated either by small interfering RNAs (siRNAs) of about 21 nt with two-nucleotide 3′ overhang [33], or by stably expressed short hairpin RNAs (shRNAs) that are processed by Dicer into siRNAs [9, 76]. During activation of the RNA-induced silencing complex (RISC), the guide (antisense) strand of the siRNA duplex is preferentially assembled into the RISC when the stem formed by the 5′ end and its complement is less stable than the one formed by the 3′ end and its complement [49, 90]; the "passenger" (sense) strand is cleaved by Argonaute2 (Ago2), the catalytic component of RISC [69, 82]. The antisense strand guides Ago2 to cleave mRNA by perfect base-pairing with the complementary site in the target (Fig. 2.2). In comparison with antisense oligos or trans-cleaving ribozymes for gene knockdown, RNAi generally offers greater potency and target specificity. As the method of choice for loss-of-function studies in mammalian systems and drug target validation, RNAi has revolutionized basic biology study and drug discovery research. In addition, novel RNAi-based therapeutic agents for treating a variety of human diseases have been under development, most notably by Alnylam Pharmaceuticals.

2.3.1 Design Rules for Improving Potency

Large variation in the efficiency of siRNAs for different sites on the same target is commonly observed [43]. Usually, only a small proportion of randomly selected siRNAs are potent. Thus, there has been a great interest in determining rules for the improvement of RNAi design. A number of empirical rules on siRNA duplex features have been reported. These include the asymmetry rule for siRNA duplex ends, which requires that the 5′ end of the antisense strand forms a stem with its complement that is less stable than the stem formed by the 5′ end of the sense strand [49, 90]. The asymmetry rule is strongly related to the requirements of high A/U content at the 5′ end of the antisense strand and high G/C at the 5′ end of the sense strand [84, 101]. A number of position-specific nucleotide preferences and other siRNA sequence features have been proposed [78, 84]. In addition, the importance of target secondary structure and accessibility has been suggested by several studies based on computational modeling of target structure and accessibility [41, 50, 51, 61, 62, 89, 94, 97] and was supported by compelling evidence based on experimentally assessed accessibility [2, 7, 54, 75, 106, 108].

2.3.2 Structure Based Assessment of Target Accessibility

A number of approaches have been published for quantifying target site accessibility for rational design of RNA-targeting nucleic acids. Based on target structures predicted by RNA folding algorithms, these methods are either probabilistic or energetic. Probabilistic methods assess the probability that a base or a block of bases is single stranded [23, 70, 73], whereas energetic methods model the energy exchanges of the hybridization process [59–61, 66, 93–95], arguably providing more refined measures of accessibility. For example, consider two target sites with (nearly) equal probability of being single stranded. If one site has high AU and the other has high GC, then the energetic costs for disrupting the target structure and the stabilities of the hybrid could be quite different for the two sites. In data analysis for some of our studies, energy measures were observed to give improved correlations than probabilistic measures. Thus, our efforts have focused on energetic models. Below, we briefly discuss several major methods.

The Sfold structure sample [24, 26] allows computation of both probabilistic measures [23] and energetic measures of target accessibility [59, 60, 93–95]. It is well established that a single-stranded block of 4–5 nts can facilitate the nucleation step of the hybridization [40, 112]. Thus, a moderate structure sample is sufficient for revealing potential effective sites by using block size of 4 nts for accessibility profiling [23]. The major advantage of using the structure sampling algorithm is that the time-consuming partition-function calculation for the whole target sequence only needs to be computed once. Folding constraints such as maximum nucleotide distance L for two bases to form a pair can be imposed for "local" folding. Such local folding was found to be significant for prokaryotic applications [93]. For prokaryotes, transcription and translation are tightly coupled events so that the target mRNA is unlikely to be able to fold globally. In contrast, eukaryotic mRNAs are first transcribed in nucleus and then transported to cytoplasm where they can conceivably fold globally before they engage in interactions with other molecules

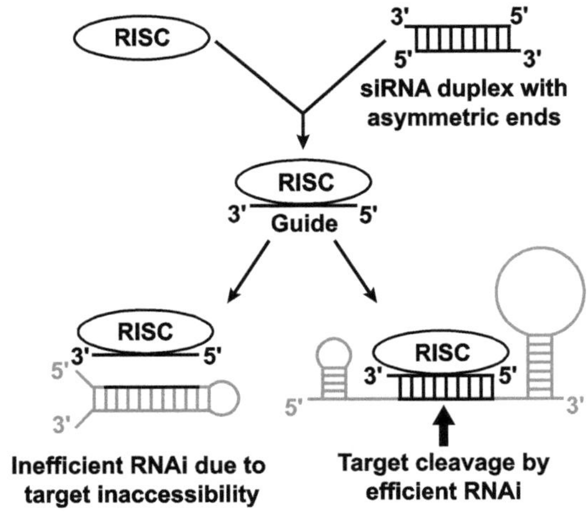

Fig. 2.3 A proposed simple model for efficient RNAi. RISC assembly is facilitated by asymmetric ends of siRNA duplex; target recognition via intermolecular base-pairing is aided by structural accessibility at the target site. The combination of the upstream effect of duplex asymmetry and the downstream effect of target accessibility is generally essential for potent gene silencing

in the cytoplasm for regulation of gene translation. Global folding using Sfold sampling algorithm can reveal highly unstructured sites that are well "conserved" in the likely mRNA structure population. These well-predicted sites can be valuable for the selection of effective target sites.

Target site disruption energy, $\Delta G_{disruption}$, is the energy cost of local disruption of the mRNA structure so that the binding site becomes completely single stranded [94]. A largely single-stranded (i.e., structurally accessible) site does not require substantial structure alteration for the guide siRNA strand to bind to the target. $\Delta G_{disruption}$ is a quantitative measure of the structural accessibility at the target site and is calculated based on target secondary structures predicted by Sfold [26] to address the likely population of mRNA structures. We found in data analysis, as illustrated by Fig. 2.3, that target accessibility is an important determinant of RNAi activity and the asymmetry of siRNA duplex asymmetry is important for facilitating RISC assembly [94]. We also found that the commonly observed negative effect of high siRNA GC-content on RNAi potency is due to generally poor target accessibility for a high GC target site which is likely to have stable secondary structure [14], rather than the likelihood that the high GC siRNA guide strand may form stable intramolecular secondary structure as previously suggested [78].

An alternative to the local disruption assumption is the global disruption model. For this model, as a result of siRNA:mRNA hybridization, the base pairs outside the target site can be rearranged so that the mRNA adopts a new globally altered structure. In this case, the free energy of the target secondary structure after siRNA binding must be recalculated by refolding the mRNA with the binding site constrained to be unpaired. This constraint option has been implemented in Sfold and available through the Sfold web server [26]. However, refolding will cost a hefty computational price. This global model is essentially equivalent to an approach based on exact calculation of ensemble free energies from initial folding and refolding [61]. This approach makes the assumption that the target will reestablish structure equilibrium after siRNA binding. The analysis of siRNA datasets in our study suggests that target cleavage by RNAi machinery appear to be rather rapid so that the target may not have time to refold before cleavage [94]. While this issue warrants further investigation, data analysis using ensemble energies also confirmed the importance of target secondary structure in RNAi activity [61].

An extension of the McCaskill algorithm [70] can compute the probability that a block of nucleotides is single stranded [73]. However, for each block, this extension requires recomputation of the partition functions for the entire RNA and is too time consuming to be efficient for scanning through all possible blocks of a long RNA in the search of best target sites. To handle this problem for RNAi application, a short local RNA folding window of size W was used, along with L and block length u [97]. These treatments introduce substantial uncertainty in computational analysis. Indeed, for u, the empirically selected optimal values are quite different for two training datasets [97], raising the concern of the general applicability of optimal parameter values learned from one source of data. For a specific mRNA, because it is not possible to have accurate information on its independent folding domains which may be better predicted individually, the overall prediction accuracy would be compromised by a prespecified local folding window length that does not suit this specific mRNA. The major findings from this study are the same as we previously reported [94], i.e., target accessibility as a down stream factor in the RNAi pathway and duplex asymmetry for facilitating RISC assembly [49, 90] are two most important factors for RNAi efficiency.

2.3.3 Specificity and Off-targeting

Gene silencing by RNAi can be highly gene-specific [16, 92]. Single base-pair mismatches could drastically alter RNAi efficacy [33, 43], and siRNAs can be designed to discriminate the wild-type and mutant alleles of many genes that differ by just s single nucleotide [91]. However, off-target gene regulation by RNAi has been observed [45, 88]. Each strand of an siRNA duplex can possibly be assembled into the RISC to guide recognition of both fully and partially complementary mRNAs [79]. Off-target activity results from partial complementarity for nontargeted genes. Off-targeting can induce measurable phenotypes [58] and thus represents a major impediment to large-scale phonotypic screening applications of RNAi. While chemical modification of siRNA duplexes may reduce off-target effects [15, 46], it is essential to take into account the issue of target specificity in the design of siRNAs or shRNAs. Microarray studies suggest that off-targeting is mainly associated with perfect $3'$ UTR matches for nucleotide positions 2–7 or 8 (hexamer or heptamer "seed" [57]; see Fig. 2.2) of the $5'$ end of the siRNA guide strand [5, 47]. The seed region is an important determinant for target recognition by microRNAs [57]. However, in these microarray studies, the number of mRNAs with seed matches is far greater than the number of actual off-targets. In addition, two recent studies reported either a lack of enrichment for either $3'$ UTR or seed matches [100] or a substantial number of off-targets that do not have a seed match [103]. These observations strongly indicate that additional factors responsible for off-target effects remain to be identified. In general, it is advisable that an siRNA or shRNA contains at least three mismatches to any other genes in the genome of the species under study, that known single-nucleotide polymorphism should be avoided, and that common exons of alternatively spliced mRNAs should be avoided as well [77]. For a complete suite of RNAi design tools, it is essential to address both the issue of gene silencing potency and the issue of targeting specificity. This is particularly important for large-scale loss-of-function screens by using siRNA libraries [32] or shRNA libraries [10, 87, 96].

2.4 Posttranscriptional Gene Regulation by MicroRNAs

MicroRNAs are endogenous noncoding RNAs (ncRNAs) of ~22 nt and are among the most abundant regulatory molecules in multicellular organisms. microRNAs typically negatively regulate specific mRNA targets through essentially two mechanisms: (1) when a microRNA is perfectly or nearly perfectly complementary to mRNA target sites, as is the case for most plant microRNAs, it causes mRNA target cleavage [85]; and (2) a microRNA with incomplete complementarity to sequences in the 3′ untranslated region (3′ UTR) of its target (as is the case for most animal microRNAs) can cause translational repression or mRNA destabilization [34]. microRNAs regulate diverse developmental and physiological processes in animals and plants [1, 6, 11, 31, 102]. Besides animals and plants, microRNAs have also been discovered in viruses [18].

2.4.1 Target Identification Using Sequence Features

Identification and experimental validation of microRNA targets are essential for understanding the regulatory functions of this important class of ncRNAs. The targets and functions of plant microRNAs are relatively easy to identify due to the near-perfect complementarity [85]. By contrast, the incomplete target complementarity typical of animal microRNAs implies a huge regulatory potential but also presents a challenge for target identification. A number of algorithms have been developed for predicting animal microRNA targets. A common approach relies on a "seed" model based on a critical observation by Lai [52], wherein the target site is assumed to form strictly Watson–Crick (WC) pairs with bases at positions 2 through 7 or 8 of the 5′ end of the microRNA (see Fig. 2.2). In the stricter, "conserved seed" formulation of the model, perfect conservation of the 5′ seed match in the target is required across multiple species [56, 57]. One well-known exception to the seed model is the interaction between *let-7* on *lin-41* in *C. elegans*, as shown by Fig. 2.4, for which G–U pair and unpaired base(s) are present in the seed regions of two binding sites with experimental support [104]. While the seed model is supported as a basis for identifying many well-conserved microRNA targets [81], two studies suggest that G–U or mismatches in the seed region can be well tolerated and that conserved seed

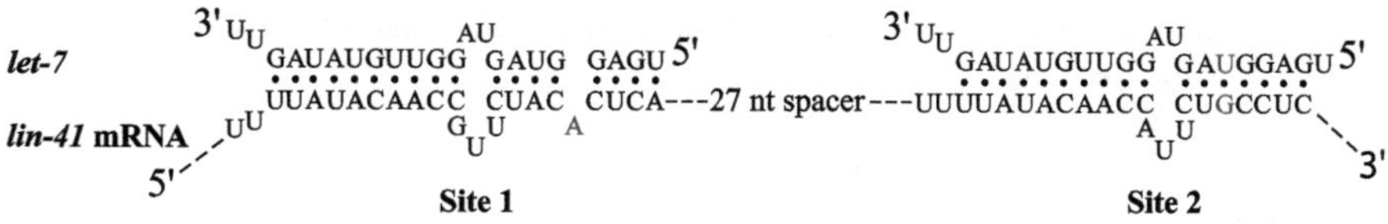

Fig. 2.4 *let-7* regulates *lin-41* by complementary base-pairing at two sites in the 3′ UTR of the *lin-41* mRNA [104, 105]. Neither the bulged A in the seed region for site 1 (in *red*, at position 5 from the 5′ end of the 27 nt spacer) nor the wobble G–U pair in the seed region for site 2 (in *red*, with U at position 6 of the 5′ end of *let-7*) meets the requirements of the seed model [56, 57] that bases 2 to 7 or 8 of the miRNA 5′ end must form Watson–Crick pairs with its target (for the color version, see Color Plates on p. 389)

match does not guarantee repression [20, 72]. These suggest that the seed model may represent only a subset of functional target sites and that additional factors are involved in further defining target specificity at least for some cases with conserved seed matches. A comprehensive study led to the proposal of three classes of target sites: "canonical", "seed", and "3′ compensatory" [8]. A canonical site pairs well with a microRNA on both the 5′ end and the 3′ end; a seed site has strong pairing to the 5′ end of the microRNA, with little or no pairing required on the 3″ end to stabilize the hybrid; a 3′ compensatory site requires strong pairing to the 3′ end of the microRNA to compensate for weak pairing on the 5′ end. Most genetically validated target sites appear to be of the canonical configuration, including the sites for *let-7: lin-41* (see Fig. 2.4). In addition to seed match, a number of features of site context have been proposed for enhancing targeting specificity [37]. More recently, functional target sites within the protein coding region of mouse mRNAs have been reported, and four of five validated mouse targets do not contain sites with seed match [98]. Also interestingly, a new class of human microRNA targets was reported to contain interaction sites in both the 5′ UTR and the 3′ UTR, and the 3′ end of the microRNAs are primarily involved in target binding for 5′ UTR sites [55].

2.4.2 A Target Structure-Based Model for MicroRNA: Target Hybridization

To attempt to understand the exceptions to the seed model and to develop target prediction methodology that does not rely on but can incorporate sequences features such as seed match, we considered the secondary structure of target mRNA that has been found to be important for other types of mRNA-targeting nucleic acids including siRNAs. We developed a model for modeling the interaction between a microRNA and a target as a two-step hybridization reaction (see Fig. 2.5): nucleation at an accessible target site, followed by hybrid elongation to disrupt local target secondary structure and form the complete microRNA-target duplex [59]. Nucleation potential and hybridization energy are two key energetic characteristics of the model. In this model, the role of target secondary structure on the efficacy

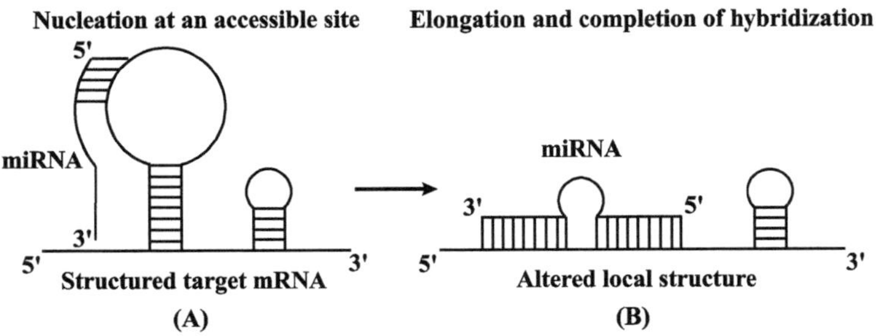

Fig. 2.5 Two-step model of hybridization between a small (partially) complementary nucleic acid molecule and a structured mRNA: (1) nucleation at an accessible site of at least 4 or 5 unpaired bases (**A**); (2) elongation through "unzipping" of the nearby helix, resulting in altered local target structure (**B**)

of repression by microRNAs is taken into account, by employing the Sfold program to address the likelihood of a population of structures that coexist in dynamic equilibrium for a specific mRNA molecule. This model can accurately account for the sensitivity to repression by *let-7* of both published and rationally designed mutant forms of the *Caenorhabditis elegans lin-41* 3′ UTR, and for the behavior of many other experimentally tested microRNA-target interactions in *C. elegans* and *Drosophila melanogaster*. The model is particularly effective in accounting for certain false positive predictions obtained by other methods. The model also performed well in a study of mammalian and viral microRNA targets [60].

In a more recent study [39], we analyzed a set of 3404 transcripts in *C.* elegans that were suggested by immunoprecipitation (IP) to be the targets for worm microRNAs [111]. Enrichment analyses by comparing targets and nontargets (i.e., transcripts absent in the IP dataset) revealed several important parameters. These include 5′ seed and modifications, structural accessibility of both the target site and the 25 nt-region upstream of the target site as assessed by target structures predicted by Sfold, the nucleation potential and the total energy change of the hybridization described in our previous work [59]. We developed a method to incorporate these significant parameters into worm microRNA target predictions. This method was found to make much better predictions than several well-known algorithms. Surprisingly, for this large target dataset, there was a lack of correlation for the contextual features based on analysis of microarray data for a small number of microRNAs [37]. In an independent study, three prediction parameters were analyzed for 6,387 candidate microRNA-target interactions between 114 human microRNAs and 890 mRNA transcripts, with patterns of expression across 88 human tissue samples [44]. These three parameters are the total energy change of the hybridization [59], the context score based on contextual features [37], and a core for measuring site conservation. It was found that only the total energy change of the hybridization is predictive of paired microRNA and mRNA expression data. Thus, the results from analyses of these large datasets not only further support our structure-based hybridization model but also cast doubts on the general applicability of the contextual features proposed from analysis of a relatively small microarray dataset. Results from microarray data may not be highly reliable, due to inherent limitations and difficulty in the interpretation of the microarray data. For example, it has been shown that the secondary structure of the target is important for microarray probe design and data interpretation [83]. However, this issue has been largely overlooked in the analysis of microarray data. The importance of target structural accessibility is also supported by several other studies [48, 86, 113, 114].

2.5 Concluding Remarks

The paradigm-shifting work by McCaskill has inspired the developments of extended partition function algorithms for modeling single molecular folding and hybridization of two nucleic acid molecules, sampling extension and clustering representation of sampled ensemble. These methods enable characterizations of the equilibrium structure ensemble that are not possible with the use of free energy minimization.

For improving the potency of RNAi, target structure is clearly an important factor in the design of siRNAs. Several existing methods use different assumptions and treatments in parameter calculations for RNAi design. It is not clear whether one approach is superior to the other. Clearly, analyses of large datasets would be needed to compare these methods and to further investigate relevant issues such as the validity of global or local target folding. Off-targeting by RNAi is a major impediment for large-scale RNAi screening. $3'$ UTR seed match can explain some but not all of observed off-target effects. It remains a challenge to identify additional factors responsible for off-target effects for improving the specificity of gene silencing.

It has been established that microRNAs can also target protein coding regions and $5'$ UTR, in addition to $3'$ UTR, and target binding can primarily involve the $5'$ end or the $3'$ end, or both ends of the microRNA. Seed match may represent a major class of target sites; however, it remains a challenge to estimate how large this class and other classes of targets are, which will require large-scale carefully designed experiments and analysis. Because seed pairing and contextual features are learned from small number of highly expressed microRNAs [4], the ratios of different classes of targets may well depend on the abundance of the microRNAs. It is conceivable that strong pairing for both $5'$ end and the $3'$ end can be essential for a microRNA of low abundance. The strength of microRNA-target hybridization would depend on the expression levels of the microRNA and its target. Incorporation of concentrations of microRNAs and target mRNAs will be a logical step for extension of hybridization modeling, as some data for expression levels have become available. Since target binding by microRNA can lead to two regulatory outcomes, translational repression or target mRNA degradation, it is an open question whether it is possible to predict the two outcomes.

Acknowledgements The Computational Molecular Biology and Statistics Core at the Wadsworth Center is acknowledged for providing computing resources. This work was supported in part by National Science Foundation grant DBI-0650991 and National Institutes of Health grant GM068726 to Y.D.

References

1. V. Ambros. The functions of animal microRNAs. *Nature*, **431**(7006):350–355, 2004.
2. S.L. Ameres, J. Martinez, and R. Schroeder. Molecular basis for target RNA recognition and cleavage by human RISC. *Cell*, **130**(1):101–112, 2007.
3. M. Andronescu, Z.C. Zhang, and A. Condon. Secondary structure prediction of interacting RNA molecules. *J Mol Biol*, **345**(5):987–1001, 2005.
4. D.P. Bartel. MicroRNAs: target recognition and regulatory functions. *Cell*, **136**(2):215–233, 2009.
5. A. Birmingham, E.M. Anderson, A. Reynolds, D. Ilsley-Tyree, D. Leake, Y. Fedorov, S. Baskerville, E. Maksimova, K. Robinson, J. Karpilow, W.S. Marshall, and A. Khvorova. 3′ UTR seed matches, but not overall identity, are associated with RNAi off-targets. *Nat Methods*, **3**(3):199–204, 2006.
6. M. Boehm and F. Slack. A developmental timing microRNA and its target regulate life span in C. elegans. *Science*, **310**(5756):1954–1957, 2005.
7. E.A. Bohula, A.J. Salisbury, M. Sohail, M.P. Playford, J. Riedemann, E.M. Southern, and V.M. Macaulay. The efficacy of small interfering RNAs targeted to the type 1 insulin-like growth factor receptor (IGF1R) is influenced by secondary structure in the IGF1R transcript. *J Biol Chem*, **278**(18):15991–15997, 2003.
8. J. Brennecke, A. Stark, R.B. Russell, and S.M. Cohen. Principles of MicroRNA-target recognition. *PLoS Biol*, **3**(3):e85, 2005.
9. T.R. Brummelkamp, R. Bernards, and R. Agami. A system for stable expression of short interfering RNAs in mammalian cells. *Science*, **296**(5567):550–553, 2002.
10. F. Buchholz, R. Kittler, M. Slabicki, and M. Theis. Enzymatically prepared RNAi libraries. *Nat Methods*, **3**(9):696–700, 2006.
11. G.A. Calin, M. Ferracin, A. Cimmino, G. Di Leva, M. Shimizu, S.E. Wojcik, M.V. Iorio, R. Visone, N.I. Sever, M. Fabbri, R. Iuliano, T. Palumbo, F. Pichiorri, C. Roldo, R. Garzon, C. Sevignani, L. Rassenti, H. Alder, S. Volinia, C.G. Liu, T.J. Kipps, M. Negrini, and C.M. Croce. A MicroRNA signature associated with prognosis and progression in chronic lymphocytic leukemia. *N Engl J Med*, **353**(17):1793–1801, 2005.
12. L.E. Carvalho and C.E. Lawrence. Centroid estimation in discrete high-dimensional spaces with applications in biology. *Proc Natl Acad Sci USA*, **105**(9):3209–3214, 2008.
13. C.Y. Chan, C.E. Lawrence, and Y. Ding. Structure clustering features on the Sfold Web server. *Bioinformatics*, **21**(20):3926–3928, 2005.
14. C.Y. Chan, C.S. Carmack, D.D. Long, A. Maliyekkel, Y. Shao, I.B. Roninson, and Y. Ding. A structural interpretation of the effect of GC-content on efficiency of RNA interference. *BMC Bioinform*, **10**(1):S33, 2009.
15. P.Y. Chen, L. Weinmann, D. Gaidatzis, Y. Pei, M. Zavolan, T. Tuschl, and G. Meister. Strand-specific 5′-O-methylation of siRNA duplexes controls guide strand selection and targeting specificity. *RNA*, **14**(2):263–274, 2008.
16. J.T. Chi, H.Y. Chang, N.N. Wang, D.S. Chang, N. Dunphy, and P.O. Brown. Genomewide view of gene silencing by small interfering RNAs. *Proc Natl Acad Sci USA*, **100**(11):6343–6346, 2003.
17. P. Clote, J. Waldispuhl, B. Behzadi, and J.M. Steyaert. Energy landscape of k-point mutants of an RNA molecule. *Bioinformatics*, **21**(22):4140–4147, 2005.
18. B.R. Cullen. Viruses and microRNAs. *Nat Genet*, **38**:S25–30, 2006.
19. J. Cupal, C. Flamm, A. Renner, and P.F. Stadler. Density of states, metastable states, and saddle points exploring the energy landscape of an RNA molecule. *Proc Int Conf Intell Syst Mol Biol*, **5**:88–91, 1997.
20. D. Didiano and O. Hobert. Perfect seed pairing is not a generally reliable predictor for miRNA-target interactions. *Nat Struct Mol Biol*, **13**(9):849–851, 2006.
21. R.A. Dimitrov and M. Zuker. Prediction of hybridization and melting for double-stranded nucleic acids. *Biophys J*, **87**(1):215–226, 2004.

22. Y. Ding. Rational statistical design of antisense oligonucleotides for high throughput functional genomics and drug target validation. *Stat Sin*, **12**:273–296, 2002.
23. Y. Ding and C.E. Lawrence. Statistical prediction of single-stranded regions in RNA secondary structure and application to predicting effective antisense target sites and beyond. *Nucleic Acids Res*, **29**(5):1034–1046, 2001.
24. Y. Ding and C.E. Lawrence. A statistical sampling algorithm for RNA secondary structure prediction. *Nucleic Acids Res*, **31**(24):7280–7301, 2003.
25. Y. Ding and C.E. Lawrence. Rational design of siRNAs with the Sfold software. In K. Appasani, editor, *RNA Interference: from Basic Science to Drug Development*, pages 129–138. Cambridge University Press, Cambridge, 2005.
26. Y. Ding, C.Y. Chan, and C.E. Lawrence. Sfold web server for statistical folding and rational design of nucleic acids. *Nucleic Acids Res*, **32**:W135–141, 2004. (Web Server issue)
27. Y. Ding, C.Y. Chan, and C.E. Lawrence. RNA secondary structure prediction by centroids in a Boltzmann weighted ensemble. *RNA*, **11**(8):1157–1166, 2005.
28. Y. Ding, C.Y. Chan, and C.E. Lawrence. Clustering of RNA secondary structures with application to messenger RNAs. *J Mol Biol*, **359**(3):554–571, 2006.
29. R.M. Dirks and N.A. Pierce. A partition function algorithm for nucleic acid secondary structure including pseudoknots. *J Comput Chem*, **24**(13):1664–1677, 2003.
30. R.M. Dirks and N.A. Pierce. An algorithm for computing nucleic acid base-pairing probabilities including pseudoknots. *J Comput Chem*, **25**(10):1295–1304, 2004.
31. D.V. Dugas and B. Bartel. MicroRNA regulation of gene expression in plants. *Curr Opin Plant Biol*, **7**(5):512–520, 2004.
32. C.J. Echeverri and N. Perrimon. High-throughput RNAi screening in cultured cells: a user's guide. *Nat Rev Genet*, **7**(5):373–384, 2006.
33. S.M. Elbashir, J. Harborth, W. Lendeckel, A. Yalcin, K. Weber, and T. Tuschl. Duplexes of 21-nucleotide RNAs mediate RNA interference in cultured mammalian cells. *Nature*, **411**(6836):494–498, 2001.
34. W. Filipowicz, S.N. Bhattacharyya, and N. Sonenberg. Mechanisms of post-transcriptional regulation by microRNAs: are the answers in sight? *Nat Rev Genet*, **9**(2):102–114, 2008.
35. A. Fire, S. Xu, M.K. Montgomery, S.A. Kostas, S.E. Driver, and C.C. Mello. Potent and specific genetic interference by double-stranded RNA in Caenorhabditis elegans. *Nature*, **391**(6669):806–811, 1998.
36. R. Giegerich, B. Voss, and M. Rehmsmeier. Abstract shapes of RNA. *Nucleic Acids Res*, **32**(16):4843–4851, 2004.
37. A. Grimson, K.K. Farh, W.K. Johnston, P. Garrett-Engele, L.P. Lim, and D.P. Bartel. MicroRNA targeting specificity in mammals: Determinants beyond seed pairing. *Mol Cell*, **27**(1):91–105, 2007.
38. M. Hamada, H. Kiryu, K. Sato, T. Mituyama, and K. Asai. Prediction of RNA secondary structure using generalized centroid estimators. *Bioinformatics*, **25**(4):465–473, 2009.
39. M. Hammell, D. Long, L. Zhang, A. Lee, C.S. Carmack, M. Han, Y. Ding, and V. Ambros. mirWIP: microRNA target prediction based on microRNA-containing ribonucleoprotein-enriched transcripts. *Nat Methods*, **5**:813–819, 2008.
40. M.R. Hargittai, R.J. Gorelick, I. Rouzina, and K. Musier-Forsyth. Mechanistic insights into the kinetics of HIV-1 nucleocapsid protein-facilitated tRNA annealing to the primer binding site. *J Mol Biol*, **337**(4):951–968, 2004.
41. B.S. Heale, H.S. Soifer, C. Bowers, and J.J. Rossi. siRNA target site secondary structure predictions using local stable substructures. *Nucleic Acids Res*, **33**(3):e30, 2005.
42. I.L. Hofacker. Vienna RNA secondary structure server. *Nucleic Acids Res*, **31**(13):3429–3431, 2003.
43. T. Holen, M. Amarzguioui, M.T. Wiiger, E. Babaie, and H. Prydz. Positional effects of short interfering RNAs targeting the human coagulation trigger Tissue Factor. *Nucleic Acids Res*, **30**(8):1757–1766, 2002.
44. J.C. Huang, B.J. Frey, and Q.D. Morris. Comparing sequence and expression for predicting microRNA targets using GenMiR3. In *Pacific Symposium on Biocomputing, volume 13*, pages 52–63, 2008.

45. A.L. Jackson, S.R. Bartz, J. Schelter, S.V. Kobayashi, J. Burchard, M. Mao, B. Li, G. Cavet, and P.S. Linsley. Expression profiling reveals off-target gene regulation by RNAi. *Nat Biotechnol*, **21**(6):635–637, 2003.
46. A.L. Jackson, J. Burchard, D. Leake, A. Reynolds, J. Schelter, J. Guo, J.M. Johnson, L. Lim, J. Karpilow, K. Nichols, W. Marshall, A. Khvorova, and P.S. Linsley. Position-specific chemical modification of siRNAs reduces "off-target" transcript silencing. *RNA*, **12**(7):1197–1205, 2006.
47. A.L. Jackson, J. Burchard, J. Schelter, B.N. Chau, M. Cleary, L. Lim, and P.S. Linsley. Widespread siRNA "off-target" transcript silencing mediated by seed region sequence complementarity. *RNA*, **12**(7):1179–1187, 2006.
48. M. Kertesz, N. Iovino, U. Unnerstall, U. Gaul, and E. Segal. The role of site accessibility in microRNA target recognition. *Nat Genet*, **39**(10):1278–1284, 2007.
49. A. Khvorova, A. Reynolds, and S.D. Jayasena. Functional siRNAs and miRNAs exhibit strand bias. *Cell*, **115**(2):209–216, 2003.
50. R. Kretschmer-Kazemi Far and G. Sczakiel. The activity of siRNA in mammalian cells is related to structural target accessibility: a comparison with antisense oligonucleotides. *Nucleic Acids Res*, **31**(15):4417–4424, 2003.
51. I. Ladunga. More complete gene silencing by fewer siRNAs: transparent optimized design and biophysical signature. *Nucleic Acids Res*, **35**(2):433–440, 2007.
52. E.C. Lai. Micro RNAs are complementary to 3′ UTR sequence motifs that mediate negative post-transcriptional regulation. *Nat Genet*, **30**(4):363–364, 2002.
53. D.M. Layton and R. Bundschuh. A statistical analysis of RNA folding algorithms through thermodynamic parameter perturbation. *Nucleic Acids Res*, **33**(2):519–524, 2005.
54. N.S. Lee, T. Dohjima, G. Bauer, H. Li, M.J. Li, A. Ehsani, P. Salvaterra, and J. Rossi. Expression of small interfering RNAs targeted against HIV-1 rev transcripts in human cells. *Nat Biotechnol*, **20**(5):500–505, 2002.
55. I. Lee, S.S. Ajay, J.I. Yook, H.S. Kim, S.H. Hong, N.H. Kim, S.M. Dhanasekaran, A. Chinnaiyan, and B.D. Athey. New class of microRNA targets containing simultaneous 5′-UTR and 3′-UTR interaction sites. *Genome Res*, 2009.
56. B.P. Lewis, I.H. Shih, M.W. Jones-Rhoades, D.P. Bartel, and C.B. Burge. Prediction of mammalian microRNA targets. *Cell*, **115**(7):787–798, 2003.
57. B.P. Lewis, C.B. Burge, and D.P. Bartel. Conserved seed pairing, often flanked by adenosines, indicates that thousands of human genes are microRNA targets. *Cell*, **120**(1):15–20, 2005.
58. X. Lin, X. Ruan, M.G. Anderson, J.A. McDowell, P.E. Kroeger, S.W. Fesik, and Y. Shen. siRNA-mediated off-target gene silencing triggered by a 7 nt complementation. *Nucleic Acids Res*, **33**(14):4527–4535, 2005.
59. D. Long, R. Lee, P. Williams, C.Y. Chan, V. Ambros, and Y. Ding. Potent effect of target structure on microRNA function. *Nat Struct Mol Biol*, **14**:287–294, 2007.
60. D. Long, C.Y. Chan, and Y. Ding. Analysis of microRNA-target interactions by a target structure based hybridization model. In *Pacific Symposium on Biocomputing*, volume *13*, pages 64–74, 2008.
61. Z.J. Lu and D.H. Mathews. Efficient siRNA selection using hybridization thermodynamics. *Nucleic Acids Res*, **36**(2):640–647, 2008.
62. K.Q. Luo and D.C. Chang. The gene-silencing efficiency of siRNA is strongly dependent on the local structure of mRNA at the targeted region. *Biochem Biophys Res Commun*, **318**(1):303–310, 2004.
63. N.R. Markham and M. Zuker. UNAFold: software for nucleic acid folding and hybridization. *Methods Mol Biol*, **453**:3–31, 2008.
64. D.H. Mathews. Using an RNA secondary structure partition function to determine confidence in base pairs predicted by free energy minimization. *RNA*, **10**(8):1178–1190, 2004.
65. D.H. Mathews. RNA secondary structure analysis using RNAstructure. In *Curr Protoc Bioinformatics*, Chapter 12: Unit 12.16, 2006.
66. D.H. Mathews, M.E. Burkard, S.M. Freier, J.R. Wyatt, and D.H. Turner. Predicting oligonucleotide affinity to nucleic acid targets. *RNA*, **5**(11):1458–1469, 1999.

67. D.H. Mathews, J. Sabina, M. Zuker, and D.H. Turner. Expanded sequence dependence of thermodynamic parameters improves prediction of RNA secondary structure. *J Mol Biol*, **288**(5):911–940, 1999b.
68. D.H. Mathews, M.D. Disney, J.L. Childs, S.J. Schroeder, M. Zuker, and D.H. Turner. Incorporating chemical modification constraints into a dynamic programming algorithm for prediction of RNA secondary structure. *Proc Natl Acad Sci USA*, **101**(19):7287–7292, 2004.
69. C. Matranga, Y. Tomari, C. Shin, D.P. Bartel, and P.D. Zamore. Passenger-strand cleavage facilitates assembly of siRNA into Ago2-containing RNAi enzyme complexes. *Cell*, **123**(4):607–620, 2005.
70. J.S. McCaskill. The equilibrium partition function and base pair binding probabilities for RNA secondary structure. *Biopolymers*, **29**(6–7):1105–1119, 1990.
71. I. Miklos, I.M. Meyer, and B. Nagy. Moments of the Boltzmann distribution for RNA secondary structures. *Bull Math Biol*, **67**(5):1031–1047, 2005.
72. K.C. Miranda, T. Huynh, Y. Tay, Y.S. Ang, W.L. Tam, A.M. Thomson, B. Lim, and I. Rigoutsos. A pattern-based method for the identification of microRNA binding sites and their corresponding heteroduplexes. *Cell*, **126**(6):1203–1217, 2006.
73. U. Muckstein, H. Tafer, J. Hackermuller, S.H. Bernhart, P.F. Stadler, and I.L. Hofacker. Thermodynamics of RNA–RNA binding. *Bioinformatics*, **22**(10):1177–1182, 2006.
74. R. Nussinov and A.B. Jacobson. Fast algorithm for predicting the secondary structure of single-stranded RNA. *Proc Natl Acad Sci USA*, **77**(11):6309–6313, 1980.
75. M. Overhoff, M. Alken, R.K. Far, M. Lemaitre, B. Lebleu, G. Sczakiel, and I. Robbins. Local RNA target structure influences siRNA efficacy: a systematic global analysis. *J Mol Biol*, **348**(4):871–881, 2005.
76. P.J. Paddison, A.A. Caudy, E. Bernstein, G.J. Hannon, and D.S. Conklin. Short hairpin RNAs (shRNAs) induce sequence-specific silencing in mammalian cells. *Genes Dev*, **16**(8):948–958, 2002.
77. P.J. Paddison, M. Cleary, J.M. Silva, K. Chang, N. Sheth, R. Sachidanandam, and G.J. Hannon. Cloning of short hairpin RNAs for gene knockdown in mammalian cells. *Nat Methods*, **1**(2):163–167, 2004.
78. V. Patzel, S. Rutz, I. Dietrich, C. Koberle, A. Scheffold, and S.H. Kaufmann. Design of siRNAs producing unstructured guide-RNAs results in improved RNA interference efficiency. *Nat Biotechnol*, **23**(11):1440–1444, 2005.
79. Y. Pei and T. Tuschl. On the art of identifying effective and specific siRNAs. *Nat Methods*, **3**(9):670–676, 2006.
80. Y. Ponty. Efficient sampling of RNA secondary structures from the Boltzmann ensemble of low-energy: the boustrophedon method. *J Math Biol*, **56**(1–2):107–127, 2008.
81. N. Rajewsky. microRNA target predictions in animals. *Nat Genet*, **38**:S8–13, 2006.
82. T.A. Rand, S. Petersen, F. Du, and X. Wang. Argonaute2 cleaves the anti-guide strand of siRNA during RISC activation. *Cell*, **123**(4):621–629, 2005.
83. V.G. Ratushna, J.W. Weller, and C.J. Gibas. Secondary structure in the target as a confounding factor in synthetic oligomer microarray design. *BMC Genomics*, **6**(1):31, 2005.
84. A. Reynolds, D. Leake, Q. Boese, S. Scaringe, W.S. Marshall, and A. Khvorova. Rational siRNA design for RNA interference. *Nat Biotechnol*, **22**(3):326–330, 2004.
85. M.W. Rhoades, B.J. Reinhart, L.P. Lim, C.B. Burge, B. Bartel, and D.P. Bartel. Prediction of plant microRNA targets. *Cell*, **110**(4):513–520, 2002.
86. H. Robins, Y. Li, and R.W. Padgett. Incorporating structure to predict microRNA targets. *Proc Natl Acad Sci USA*, **102**(11):4006–4009, 2005.
87. D.E. Root, N. Hacohen, W.C. Hahn, E.S. Lander, and D.M. Sabatini. Genome-scale loss-of-function screening with a lentiviral RNAi library. *Nat Methods*, **3**(9):715–719, 2006.
88. P.C. Scacheri, O. Rozenblatt-Rosen, N.J. Caplen, T.G. Wolfsberg, L. Umayam, J.C. Lee, C.M. Hughes, K.S. Shanmugam, A. Bhattacharjee, M. Meyerson, and F.S. Collins. Short interfering RNAs can induce unexpected and divergent changes in the levels of untargeted proteins in mammalian cells. *Proc Natl Acad Sci USA*, **101**(7):1892–1897, 2004.

89. S. Schubert, A. Grunweller, V.A. Erdmann, and J. Kurreck. Local RNA target structure influences siRNA efficacy: systematic analysis of intentionally designed binding regions. *J Mol Biol*, **348**(4):883–893, 2005.
90. D.S. Schwarz, G. Hutvagner, T. Du, Z. Xu, N. Aronin, and P.D. Zamore. Asymmetry in the assembly of the RNAi enzyme complex. *Cell*, **115**(2):199–208, 2003.
91. D.S. Schwarz, H. Ding, L. Kennington, J.T. Moore, J. Schelter, J. Burchard, P.S. Linsley, N. Aronin, Z. Xu, and P.D. Zamore. Designing siRNA that distinguish between genes that differ by a single nucleotide. *PLoS Genet*, **2**(9):e140, 2006.
92. D. Semizarov, L. Frost, A. Sarthy, P. Kroeger, D.N. Halbert, and S.W. Fesik. Specificity of short interfering RNA determined through gene expression signatures. *Proc Natl Acad Sci USA*, **100**(11):6347–6352, 2003.
93. Y. Shao, Y. Wu, C.Y. Chan, K. McDonough, and Y. Ding. Rational design and rapid screening of antisense oligonucleotides for prokaryotic gene modulation. *Nucleic Acids Res*, **34**(19):5660–5669, 2006.
94. Y. Shao, C.Y. Chan, A. Maliyekkel, C.E. Lawrence, C.E. Roninsonx, and Y. Ding. Effect of target secondary structure on RNAi efficiency. *RNA*, **13**(10):1631–1640, 2007.
95. Y. Shao, S. Wu, C.Y. Chan, J.R. Klapper, E. Schneider, and Y. Ding. A structural analysis of in vitro catalytic activities of hammerhead ribozymes. *BMC Bioinf*, **8**(1):469, 2007.
96. D. Shirane, K. Sugao, S. Namiki, M. Tanabe, M. Iino, and K. Hirose. Enzymatic production of RNAi libraries from cDNAs. *Nat Genet*, **36**(2):190–196, 2004.
97. H. Tafer, S.L. Ameres, G. Obernosterer, C.A. Gebeshuber, R. Schroeder, J. Martinez, and I.L. Hofacker. The impact of target site accessibility on the design of effective siRNAs. *Nat Biotechnol*, **26**(5):578–583, 2008.
98. Y. Tay, J. Zhang, A.M. Thomson, B. Lim, and I. Rigoutsos. MicroRNAs to Nanog, Oct4 and Sox2 coding regions modulate embryonic stem cell differentiation. *Nature*, **455**(7216):1124–1128, 2008.
99. I. Tinoco, Jr., O.C. Uhlenbeck, and M.D. Levine. Estimation of secondary structure in ribonucleic acids. *Nature*, **230**(5293):362–367, 1971.
100. C. Tschuch, A. Schulz, A. Pscherer, W. Werft, A. Benner, A. Hotz-Wagenblatt, L.S. Barrionuevo, P. Lichter, and D. Mertens. Off-target effects of siRNA specific for GFP. *BMC Mol Biol*, **9**:60, 2008.
101. K. Ui-Tei, Y. Naito, F. Takahashi, T. Haraguchi, H. Ohki-Hamazaki, A. Juni, R. Ueda, and K. Saigo. Guidelines for the selection of highly effective siRNA sequences for mammalian and chick RNA interference. *Nucleic Acids Res*, **32**(3):936–948, 2004.
102. E. van Rooij, L.B. Sutherland, X. Qi, J.A. Richardson, J. Hill, and E.N. Olson. Control of stress-dependent cardiac growth and gene expression by a microRNA. *Science*, **316**(5824):575–579, 2007.
103. S. Vankoningsloo, F. de Longueville, S. Evrard, P. Rahier, A. Houbion, A. Fattaccioli, M. Gastellier, J. Remacle, M. Raes, P. Renard, and T. Arnould. Gene expression silencing with 'specific' small interfering RNA goes beyond specificity—a study of key parameters to take into account in the onset of small interfering RNA off-target effects. *FEBS J*, **275**(11):2738–2753, 2008.
104. M.C. Vella, E.Y. Choi, S.Y. Lin, K. Reinert, and F.J. Slack. The C. elegans microRNA let-7 binds to imperfect let-7 complementary sites from the lin-41 3′UTR. *Genes Dev*, **18**(2):132–137, 2004.
105. M.C. Vella, K. Reinert, and F.J. Slack. Architecture of a validated microRNA:target interaction. *Chem Biol*, **11**(12):1619–1623, 2004.
106. T.A. Vickers, S. Koo, C.F. Bennett, S.T. Crooke, N.M. Dean, and B.F. Baker. Efficient reduction of target RNAs by small interfering RNA and RNase H-dependent antisense agents. A comparative analysis. *J Biol Chem*, **278**(9):7108–7118, 2003.
107. B. Voss, R. Giegerich, and M. Rehmsmeier. Complete probabilistic analysis of RNA shapes. *BMC Biol*, **4**:5, 2006.
108. E.M. Westerhout, M. Ooms, M. Vink, A.T. Das, and B. Berkhout. HIV-1 can escape from RNA interference by evolving an alternative structure in its RNA genome. *Nucleic Acids Res*, **33**(2):796–804, 2005.

109. S. Wuchty, W. Fontana, I.L. Hofacker, and P. Schuster. Complete suboptimal folding of RNA and the stability of secondary structures. *Biopolymers*, **49**(2):145–165, 1999.
110. T. Xia, J. SantaLucia, Jr., M.E. Burkard, R. Kierzek, S.J. Schroeder, X. Jiao, C. Cox, and D.H. Turner. Thermodynamic parameters for an expanded nearest-neighbor model for formation of RNA duplexes with Watson–Crick base pairs. *Biochemistry*, **37**(42):14719–14735, 1998.
111. L. Zhang, L. Ding, T.H. Cheung, M.Q. Dong, J. Chen, A.K. Sewell, X. Liu, J.R. Yates 3rd, and M. Han. Systematic identification of C. elegans miRISC proteins, miRNAs, and mRNA targets by their interactions with GW182 proteins AIN-1 and AIN-2. *Mol Cell*, **28**(4):598–613, 2007.
112. J.J. Zhao and G. Lemke. Rules for ribozymes. *Mol Cell Neurosci*, **11**(1–2):92–97, 1998.
113. Y. Zhao, E. Samal, and D. Srivastava. Serum response factor regulates a muscle-specific microRNA that targets Hand2 during cardiogenesis. *Nature*, **436**(7048):214–220, 2005.
114. Y. Zhao, J.F. Ransom, A. Li, V. Vedantham, M. von Drehle, A.N. Muth, T. Tsuchihashi, M.T. McManus, R.J. Schwartz, and D. Srivastava. Dysregulation of cardiogenesis, cardiac conduction, and cell cycle in mice lacking miRNA-1-2. *Cell*, **129**(2):303–317, 2007.
115. M. Zuker. Calculating nucleic acid secondary structure. *Curr Opin Struct Biol*, **10**(3):303–310, 2000.
116. M. Zuker and P. Stiegler. Optimal computer folding of large RNA sequences using thermodynamics and auxiliary information. *Nucleic Acids Res*, **9**(1):133–148, 1981.

Chapter 3
Some Critical Data Quality Control Issues of Oligoarrays

Wenjiang J. Fu, Ming Li, Yalu Wen, and Likit Preeyanon

3.1 Introduction

The microarray technology has been widely used as a high-throughput tool in biological and biomedical research since its debut in 1990s. It has advanced rapidly during the past decade, from low-density arrays to high-density arrays, from gene expression arrays to single nucleotide polymorphism (SNP) arrays, tiling arrays, and mitochondrial arrays. It is anticipated that the copy number variants (CNV) arrays will become available very soon. On one hand, the array technology has been advancing at an unprecedented speed, especially the oligonucleotide arrays. On the other hand, a large number of computational methods have been introduced and advanced our knowledge in understanding and analyzing the array data. However, recent studies have shown that the oligoarrays have been widely applied but still poorly understood [27], which may imply that computational methods have not been able to capture the fast development of the array technologies. Interestingly, there has been a debate on the reliability and reproducibility of the microarray studies during the past five years [32, 33, 36]. The recent observation of the oligoarray design and data quality problems [27] raised further concerns over the array data quality control (QC), which becomes more and more severe and may take more time to resolve than previously expected.

Among the platforms of the commercial microarrays, the Affymetrix SNP arrays have received increasing attention for genome-wide association studies (GWASs) and CNV studies, partly due to the fact that the Affymetrix Inc has taken an open-source approach of making the detailed array information available on the internet, including the probe sequence structure and HapMap samples with the gold-standard HapMap annotation. Such an approach allows methodological research to fine-tune

W.J. Fu (✉) · M. Li · Y. Wen · L. Preeyanon
The Computational Genomics Lab, Department of Epidemiology, Michigan State University, East Lansing, MI 48824, USA
e-mail: fuw@epi.msu.edu

J. Feng et al. (eds.), *Frontiers in Computational and Systems Biology*, Computational Biology 15,
DOI 10.1007/978-1-84996-196-7_3,

the current computational methods and allows the optimization of future array design based on the current results and future research needs. Although more and more array design and data quality issues have been raised lately on the oligoarrays, this by no means implies that the oligoarrays are the only array platform that has data QC issues. The open-source policy has enabled research in the array design and data quality assessment, and the oligoarrays have received favorable attention with major improvement in array design, data QC, and computational methods for data analysis. In this chapter, we will study some critical issues in the oligoarrays and demonstrate that certain data issues inherent in the array design may be resolved through computational methods if the mechanism of the array design is properly incorporated into the mathematical models and computational methods.

Currently, there are a large number of computational methods available for microarray data analysis. Depending on the study outcome and the array design, these methods can be grouped into several categories. The gene expression data may be analyzed with single-probe intensity-based methods, e.g., SAM [38], PAM [37], or gene set methods [10, 35, 49] to identify differentially expressed genes or gene sets, pathway analysis and protein interaction in system biology [18]. The SNP array data can be analyzed for genetic association studies and haplotype analysis based on the SNP genotype calls, e.g., [6, 34, 50], or for GWASs based on the genome scan of the entire genome [46], or the most recent CNV studies [4].

The above computational methods have offered promise to deciphering the human genome by analyzing microarray data. While most of them use advanced statistical and computational tools, including data mining and machine learning, and have demonstrated to reveal successfully biological and biomedical discoveries, the mechanism of the array technology has not been fully incorporated into the data analysis. It is well known that the microarray has been designed through multidisciplinary research, and thus techniques used in the array design play a crucial role in the data analysis and should be incorporated through mathematical and statistical modeling.

Traditionally, the discipline of statistics was developed from the experimental sciences, such as agriculture. It thus has a deep root in biological experiments and emphasizes experimental design and data collection. This applies to studies not only at the population level but also at the molecular and genetic level. In particular, data from microarray studies are collected through biomedical experiments at both population level and molecular level through collaboration of multidisciplinary research team and thus deserve special interdisciplinary effort for data analysis. Ignorance or failure to recognize this uniqueness of microarray data may lead to inaccurate or biased results, or even lead to practice against rules in biology. In this paper, we will emphasize the importance of this approach and demonstrate with examples that incorporating the array mechanism in microarray data analysis will lead to considerable improvement, which not only provides much needed theoretical support from the related disciplines but also offers guidance to improved data analysis.

The outline of this chapter is as follows. We will summarize the microarray data quality control issues in Sect. 3.2. Section 3.3 discusses the modeling of physicochemical properties and their application to microarray data. Section 3.4 studies a special issue, the mismatch phenomenon. Section 3.5 discusses the relationship among the abundance of gene expression, DNA copy numbers and probe intensity, and presents a newly developed probe intensity composite representation (PICR) model for estimating the copy numbers with oligo SNP arrays. Section 3.6 provides a new mixed-effects model-based imputation method to repair bright spots on the arrays. Section 3.7 provides concluding remarks.

3.2 Quality Control in Microarray Data Analysis

So far, microarrays have been shown to boost biomedical research in many areas, from the early studies of rare diseases (breast cancer, leukemia) to common diseases (hypertension, diabetes), from infectious diseases (hepatitis, HIV) to neural disorders (autism, schizophrenia), etc. The array technology not only offers the much needed techniques to scan the human genome in searching for new genetic markers to achieve better understanding of the disease etiology but also provides fine-tuned techniques to pursue the goal of the translational research and customized medicine, particularly for drug development and therapeutic treatment through successful genetic profiling in clinical diagnosis. Although microarray studies have different designs and different scientific objectives to achieve, the fundamental principles remain the same, i.e., to identify the genetic difference and biological variabilities that are associated with or responsible for the diseases or phenotypes by studying the expression or abundance of genes (or gene sets) and the DNA structure (SNPs, CNVs) through mathematical modeling and statistical analysis. Successful examples have provided evidence that the microarray technology has assisted researchers to achieve their goals that otherwise would be impossible to achieve, including genetic profiling for the prognosis of breast cancer patient survival after surgery [41], classification of leukemia [12], accurate location of transcription regulatory regions and binding sites [42], recent large-scale GWAS study on tens of thousands of subjects [46], CNV studies [4, 45], etc.

One of the major issues in microarray studies is the array data quality control, or the so called low-level analysis, which has been emphasized by Professor Speed's group in a number of publications, including the pioneering work on microarray normalization [2]. It has been observed that microarray data, usually the probe intensity data, reflect not only biological variability between different groups of subjects, which is the primary objective of the investigations, but also wild noise and artifacts that are generated during the microarray experiments, including but not limited to variations in DNA or RNA tissue preservation, PCR amplification, experiment reagent, environmental and experimental conditions of the laboratories, etc. The data quality issues are present in various formats. Some are observable, such as systematic shift of probe intensity (array brightness), unequal variability of intensities across arrays (array probe contrast), and uneven reagent spray (array paintbrush). While others are more difficult to detect and require sophisticated analytic tools, such as array batch effect, population stratification [1], the recently discovered genomic wave [8, 22], etc.

The array quality has sparked a major debate about the microarray quality control: whether the array technology generates reliable data for biomedical research. The debate was initiated by observations that minimal consistency of differentially expressed genes was found in a series of microarray experiment studies on the same disease [36]. This observation makes the use of microarrays in biomedical research questionable and motivated a series of studies of the impact of the array data quality control in the literature. For example, it has been demonstrated that new criteria should be used in judging the consistency of findings of the microarray studies [32]. Others have shown that genotyping error on the SNP arrays for genetic association studies leads to increased type I and type II errors [17, 24], which further leads to increased false positives, false negatives, and decreased power for scientific discoveries.

Since our discussion focuses on the Affymetrix oligonucleotide arrays, some general concepts apply similarly to other array platforms, while others may require special treatment before the application.

The Affymetrix microarrays are based on the design of probes that have 25 nucleotide bases in each probe. The design uses probe sequence pairs of a perfect match (PM) probe and a mismatch (MM) probe to annotate the target sequences. The 25 nucleotide bases on the PM probes are perfectly complement to the target sequences, while those on the MM probes are complement nucleotides at all positions except for the center nucleotide (the 13th position on the probe), which is a mismatch to the nucleotide of the target sequence. Gene expression arrays have a number (about 10 to 20) of probe pairs within the domain of each gene to measure the abundance of gene expression, while the high-density SNP arrays have 20 and 12 pairs of probes for the GeneChip 100-K and 500-K arrays, respectively. The most recent Affymetrix GeneChip 6.0 SNP arrays use only 12 PM probes to annotate each SNP with 6 PM probes for one allele and other 6 PM probes for the other allele. The MM probes are not used in this new design.

The unique design of Affymetrix oligonucleotide arrays using PM and MM probes [23] requires special methods for the array data analysis. The rationale is that the mismatch nucleotide in the probe sequence usually induces lower intensity value than its corresponding PM probe, and the latter reflects the specific probe sequence binding by design, while the former reflects nonspecific probe binding and thus to some extent may represent the background level of the probe intensity. However, it has also been observed that a fairly large portion (about 30%) of MM probes in gene expression arrays have a larger intensity than their corresponding PM probes. In the SNP arrays, the portion is down to about 9%. This observation surely casts doubt on whether the MM probe intensity can be used for array background control, as they lead to negative intensity value if subtracted from their corresponding PM probe intensity as in general practice. It also raises further questions: Can microarray data analysis be conducted at a level different from the probe intensity? Can it be improved with other approaches given that the probe intensity level analysis has made major contributions in biomedical research.

The special observation of the large portion of MM $>$ PM implies that the occurrence of the MM phenomenon is not solely by chance but reflects some array mechanism, which is not fully understood. Thus it requires special analytical methods for data analysis. Successful examples include a statistical model by Li and Wong [19], in which they introduced a model-based method to analyze the array data by separately modeling the probe intensities of the PM probes from the MM probes. However, as pointed out in a recent paper, the oligoarrays have been widely applied but poorly understood [27], as there are far more unresolved issues than the ones we understand so far. We will discuss the MM phenomenon with more details in a later section.

It is well known that microarray data are subject to artifacts and wild noise at the probe intensity level. A series of normalization methods have been proposed to remove the artifacts [2, 9, 16, 28, 30, 53]. Although normalization procedures try to take the summary statistics of the probe intensities, such as the mean or median of the perfect match intensities (mismatch probes were deemed of no use and hence were recommended not to be used in a number of articles), the wild noise is only tapered to certain extent. For example, assuming that the probe intensity data has a measurement error with mean 0 and variance σ^2, the mean probe intensity will have the variance σ^2/m, where m is the number of PM probes annotating the same SNP. Although this approach may reduce the variability of the probe intensity data, it will retain the noise, possibly at a slightly smaller scale. Furthermore, in the gene expression data, the probe intensities are usually analyzed separately for each probe, and the probes within the same gene are not studied together, which may result in information loss. One may also ask: How to coherently combine the information from different probes together in data analysis? Although gene clustering has been studied in gene expression analysis, it does not take the advantage of modeling probes together within the same gene, which is meaningful since each probe intensity reflects part of the gene expression directly. There has been no satisfactory answer yet to the above question.

3.3 Physico-Chemical Properties in Sequence Duplex Hybridization

Although many methods have been studied for microarray data analysis, most of them are solely based on the probe intensity level data (e.g., SAM, PAM, etc.) after appropriate normalization procedure to remove the artifacts but do not incorporate the modeling of the probe sequence structure. Notice that microarray data are probe intensities generated from the array image based on fluorescence scan in the hybridization between probe sequence and target sequence. Hence, from the point of view of statistical experimental design, better understanding of the process in the microarray experiments and incorporating the array mechanism through the probe structure of the nucleotides into the model may improve the analysis.

So far, a series of papers have studied the array mechanism through the physico-chemical properties of the microarray duplex hybridization between probe and target sequences. In particular, the effect of the nucleotides and their positions on the probes have been studied with thermodynamic models of the sequence binding. Specifically, Zhang et al. studied the free energy of the probe sequences through a positional-dependent nearest-neighbor (PDNN) model by considering the nucleotide pairs in the probe sequence [51]. The free energy of each perfect match probe sequence was modeled by

$$E = \sum_{l=1}^{24} \omega_l \lambda(s_l, s_{l+1}),$$

where $\lambda(s_l, s_{l+1})$ is the stacking energy of the nucleotide pair at the position $(l, l+1)$ on the probe with $l = 1, \ldots, 24$, and ω_l is the effect of the lth position on the probe. They further studied the probe intensity and decomposed the probe intensity signal into gene-specific binding (GSB) and nonspecific binding (NSB). The signal of each probe intensity was modeled by

$$I = N_s \phi(E_s) + N_{ns} \phi(E_{ns}) + B,$$

where N_s is the concentration of the target that involves GSB, E_s is the free energy for the formation of the specific RNA–DNA or DNA–DNA duplex with the GSB target fragment. N_{ns} is the concentration of RNA or DNA target molecules that contributes to the NSB, and E_{ns} is the average free energy for NSB. B is the array baseline intensity. The function $\phi(x) = 1/(1 + e^x)$ yields the binding affinity based on the free energy x of the probe sequence for the GSB or the average free energy for the NSB. Fitting this probe intensity model together with the PDNN model for the free energy, Zhang et al. achieved a strikingly good fit to the probe intensity [51].

Furthermore, Zhang et al. also found that the effect of the nucleotide positions and the nucleotide pairs are not symmetric. They showed that the effect of nucleotides were higher at the central locations than at the locations far from the center, and the nucleotide pairs [AG], [CG], and [TG] had a much higher stacking energy than the pairs [GA], [GC], and [GT], respectively (Fig. 1c in [51]), while the pair [GG] had moderate stacking energy. This further implies that consideration of GC content of probe effect is important but not enough. The combinatorial pattern of the probe nucleotides also makes a major difference, and the structure of probe sequences needs to be modeled as well. In addition, in an experiment of very low concentration of target molecules, they found that the nucleotide triplets also present very different behavior, and the [CGC] triplet usually has a relatively low intensity ratio of PM/MM, while the triplet [CCC] has a relatively high intensity ratio of PM/MM. Their work has demonstrated that the study of the physico-chemical properties of the microarray data is crucial to the thorough understanding and analysis of the microarray data. Further study of the modeling of the probe intensities through the structure and binding of the sequence duplex may lead to improved modeling and analysis of the oligonucleotide microarray data.

The above Zhang affinity equation is similar to the Langmuir adsorption equation $\phi(x) = K/(1 + Kx)$ in studying the adsorption of gas molecules on a solid surface except that the target sequences are now in special solution and are usually at a relatively low concentration in the microarray experiments. Ono et al. further studied the DNA duplex binding through a more general finite hybridization (FH) thermodynamic model [26]. They have shown that the FH model unifies the Zhang gene-specific and Langmuir adsorption models. The Langmuir model can be expressed as

$$I^{Lm} = \alpha \frac{Kx}{1+Kx} + I^{bg},$$

where $K = \exp(-\Delta G/RT)$ gives the equilibrium constant of probe-target duplex formation, ΔG denotes the free energy, R is the gas constant, T is the temperature, x is the concentration of target molecules, and I^{bg} denotes the background intensity. The Zhang gene-specific model can be expressed as

$$I^{Zh} = \alpha' \left(\frac{x}{1+\exp(E_s)} + \frac{N}{1+\exp(E_{ns})} \right) + I^{bg},$$

where x is the same as in the Langmuir model, N is the population of the RNA molecules that contributes to the NSB. E_s and E_{ns} remain the same as above. The FH model takes the form

$$I^{FH} = \frac{C}{2} \left\{ \frac{1}{K^{sp}} + A + x - \sqrt{\left(\frac{1}{K^{sp}} + A + x \right)^2 - 4Ax} \right\} + I^{bg},$$

where C is the scale of intensity, and $A = [P^{\text{total}}]$ and $x = [T^{\text{total}}]$ are the probe and target concentrations, respectively. Ono et al. have shown that both Langmuir and Zhang models are limiting cases of the FH model. As the target concentration x is relatively high, i.e., $x \gg A$, the FH model can be approximated by the Langmuir model

$$I^{FH} \simeq AC \frac{xK^{sp}}{1+xK^{sp}} + I^{bg}.$$

As the target concentration is relatively low, i.e., $x \ll A$, the FH model can be approximated by the Zhang model

$$I^{FH} \simeq AC \frac{xK^{sp}}{1+AK^{sp}} + I^{bg}.$$

They further showed in their experiments of varying probe sequence length that the Langmuir model fits the intensity data well for high intensity values but not for low intensity values and that the Zhang model fits the intensity data well for low intensity values but not for high intensity values. In comparison, the FH model unifies both models and thus performs better than the other two for both high and low intensity values.

Both Zhang et al. and Ono et al. observed a considerable nonspecific binding effect, which may contribute up to 40% of the array probe signal through data analysis by the Zhang model (data not shown). This large amount of NSB may help to explain why microarray studies attained a low minimal consistency and further raises the concerns of the reliability and reproducibility of microarray studies.

To further study the nonspecific binding, Furusawa et al. [11] developed a thermodynamic model of NSB by duplex formation of probes and multiple hypothetical targets. They generalized the PDNN model to n-nearest neighbor model for free energy estimation and also considered the effect of secondary structure, i.e., the folding of duplex. Their new model improved the prediction of nonspecific signals.

Modeling the dissociation of molecules, Held et al. [13, 14] studied the wash-out effect of the microarray experiment and found that the residuals from the dissociation of the target fragments to the probes in the dynamics of the duplex hybridization contributes to the overall modeling of the expression. In a most recent work, Li et al. [21] studied a competitive thermodynamic model to characterize the dissociation between the probe and target sequences and achieved further improvement in fitting the array probe intensity data.

As pointed out in Pozhitkov et al. [27], "more systematic physico-chemical studies will be required to better understand the hybridization and dissociation behavior of oligonucleotides." The above work has demonstrated that proper modeling of the physico-chemical properties of the array duplex hybridization allows continuous improvement of the modeling of the array probe intensity data, which will eventually lead to improvement of data analysis through better understanding of the array mechanism.

3.4 The MM Phenomenon: MM > PM

It is known that MM probes are designed to tune the background intensity so that the true signal would be yielded by subtracting the MM intensity from the PM intensity, i.e., $I_{PM} - I_{MM}$ would provide the true signal at the probe. However, the MM phenomenon, where a large number of MM probes achieve higher intensity than their corresponding PM probes, makes the above adjustment method invalid through subtraction. This MM phenomenon implies that the array mechanism deserves serious investigation. It indicates that certain mismatch nucleotides may yield larger binding affinity than the complementary perfect match nucleotides, unexpected by our understanding that perfect match usually induces larger binding than the mismatch. This prompted a group of researchers to believe that MM probes are of no use and should be excluded from the array design [5, 29], which eventually led to the removal of the MM probes from the most recent Affymetrix 6.0 SNP arrays.

The MM phenomenon has been studied in a few articles. Initially, it was observed by Naef et al. [25] and also reported in Bolstad et al. [3] on gene expression arrays. It was then studied by Urakawa et al. [39], Zhang et al. [51], Wu et al. [47, 48], Wang et al. [44]. It was also studied on tiling arrays by Siringhaus et al. [31], who pointed out that the inclusion of MM probes is to differentiate specific from nonspecific hybridization. This practice was based on three assumptions: (1) Nonspecific bindings affect PM and MM probes equally; (2) The mismatch reduces the affinity of GSB to that of MM; and (3) fluorescence signal is identical for PM and MM probes. These assumptions imply that MM < PM. The observation of the MM phenomenon makes the above assumptions invalid.

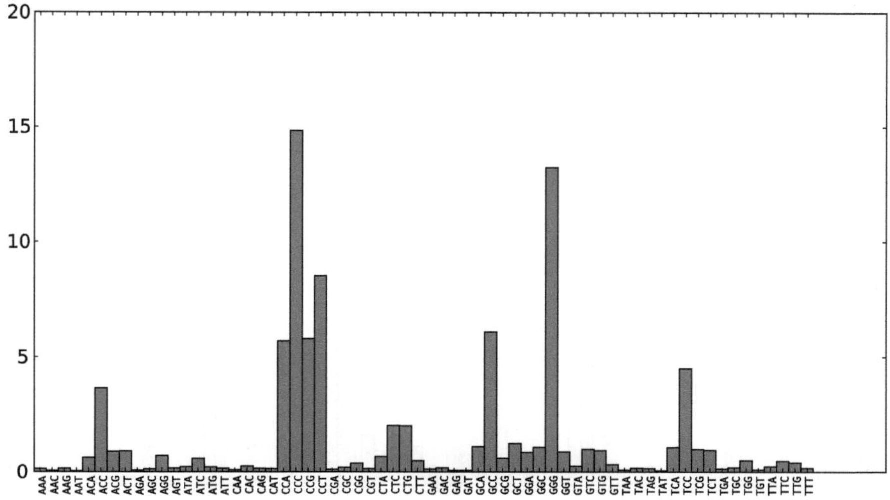

Fig. 3.1 Relative frequency of probe pairs showing MM > 2 PM in RNA U133 arrays of human breast tissue. The horizontal labels the nucleotide triplet in the center of the MM probe

Zhang et al. [51] studied the effect of nucleotide triplets through the PDNN model and found that at a very low concentration level where the GSB is negligible, MM > PM occurs when a Guanine (G) occupies the PM probe center position, and MM < PM occurs when a Cytosine (C) occupies the PM probe center position. Wu et al. [48] studied the positional effect of the nucleotide for GSB and NSB separately and studied the standard errors of the log probe intensity for PM and MM and the cost of cross-hybridization. Wu et al. [47] further studied the effect of G-stack—probes that contain multiple Guanines in a row—and concluded that the probes that contain a [GGGG] or [CCCC] may have different mechanism, which leads to their poor performance on the microarrays. Wang et al. [44] studied the MM phenomenon and characterized it using a special pattern of the center nucleotide of the probes. They found that among all probes on the U133 plus 2.0 RNA arrays of their study samples, the center nucleotide G had the highest percentage for MM > PM, while among the outlier of MM > PM, the center nucleotide C and G had the highest percentage with that of C slightly higher than G. However, their conclusion was based on the assumption that all four nucleotides have an equal 25% chance of occurrence, and their reported percentage was not adjusted by the distribution of the special probe classes.

To study the MM phenomenon, we used RNA arrays based on several different tissues, including the human mixed tissues of heart, teste, and cerebellum, the human tissue of the breast, and mouse tissues. We found a special pattern of the center nucleotide triplets. Different from the previous work [44], we studied the frequency pattern of center nucleotide triplets and adjusted with its distribution among the total number of probe pairs as follows. For each $c_1 = 2$, 4, or 6, the frequency (n^J_{raw}) of the probe pairs in which MM > c_1PM was calculated for each J among the center nucleotide class $J \in \{[\text{AAA}], [\text{AAC}], \ldots, [\text{TTG}], [\text{TTT}]\}$. The percent-

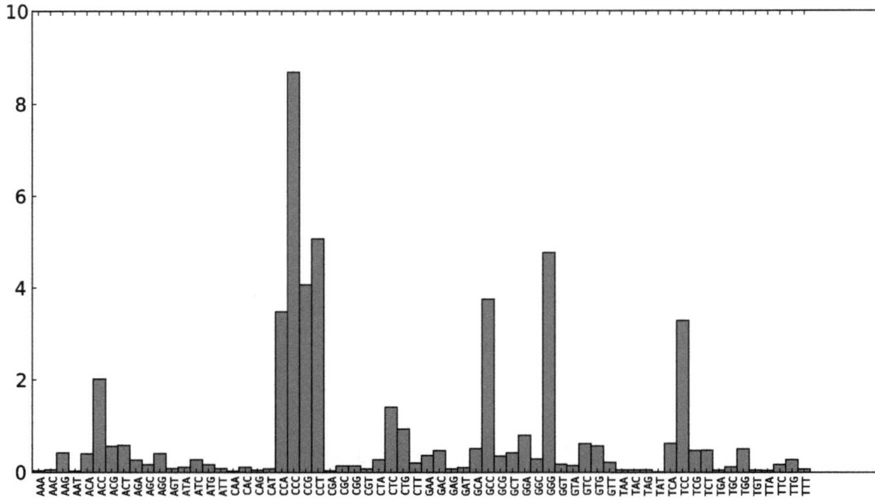

Fig. 3.2 Relative frequency of probe pairs showing MM > 2 PM in RNA U133 arrays of human heart, teste, and cerebellum tissues. The horizontal labels the nucleotide triplet in the center of the MM probe

age of the probe pairs (N^J) with the center nucleotide triplet J on the array out of the total number of probe pairs was also calculated. The adjusted relative frequency was calculated as $n_{\text{adj}}^J = n_{\text{raw}}^J / N^J$.

It is shown in Figs. 3.1, 3.2, 3.3, 3.4 that the adjusted relative frequency of the probe pairs in which MM > c_1PM with human breast tissue and human mixed tissues of the heart, teste, and cerebellum with different multiples $c_1 = 2$, 4, and 6. It is illustrated that the MM phenomenon has a higher relative frequency if the MM probe has a nucleotide pair [CC] in the center, extremely so if the center nucleotide triplet is [CCC]. While the center nucleotide triplet [GGG] has a high relative frequency in MM > 2PM but not in MM > 4PM and MM > 6PM. In contrast, such a special pattern disappears for PM > c_2MM, with $c_2 = 2$, 4, and 6 as shown in Fig. 3.5 for $c_2 = 2$. This observation is consistent with the one by Zhang et al. [51] for very low target concentration. We further found that similar pattern in the Affymetrix tiling arrays (data not shown), which implies that the MM phenomenon is not specific to RNA arrays but also may occur in other platforms of short oligoarrays.

Of note, the following inference can be made.

(1) There have been observations that some SNP genotype calling methods based solely on summary statistics of the PM probe intensities have higher error rate in annotating heterozygous SNPs. This is, at least partly, due to the fact that these SNP genotype calling methods use the largest probe intensity to make genotype calls assuming that perfectly complement nucleotides yield the largest probe intensity. For example, consider a (GC) SNP with flanking nucleotide G on both sides, i.e., the target nucleotide triplets at the SNP position are [GGG]

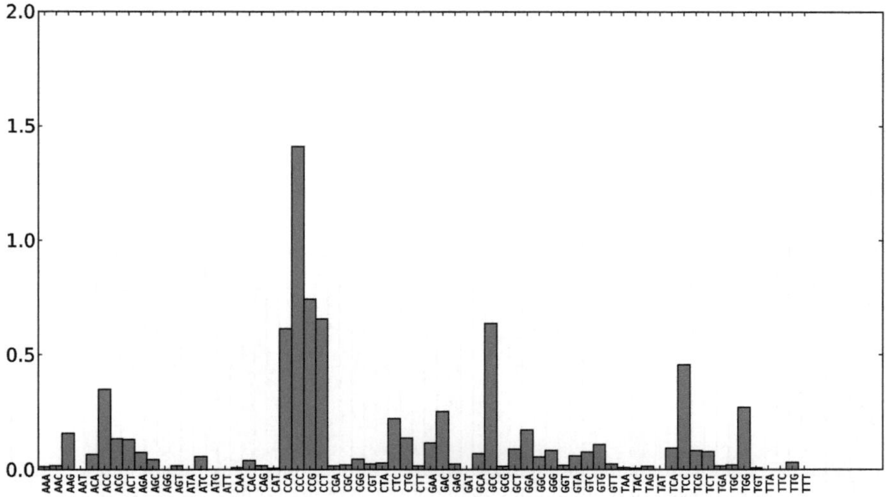

Fig. 3.3 Relative frequency of probe pairs showing MM > 4 PM in RNA U133 arrays of human heart, teste, and cerebellum tissues. The horizontal labels the nucleotide triplet in the center of the MM probe

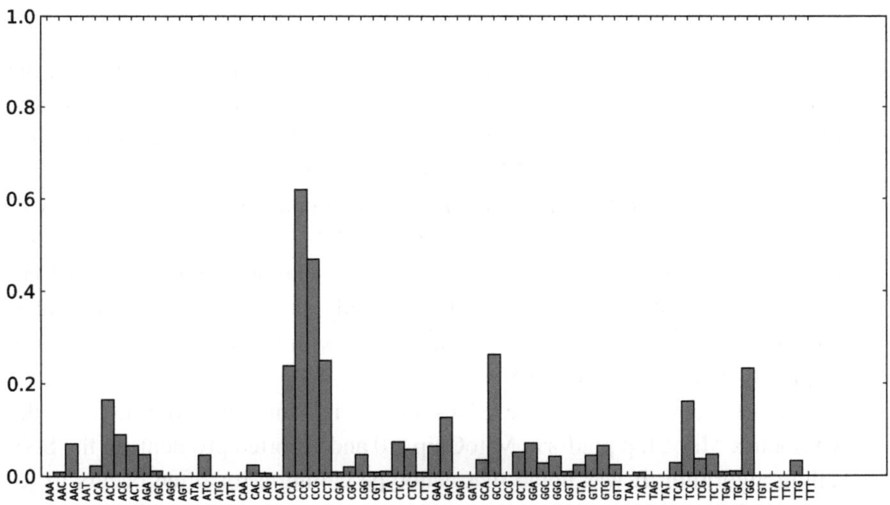

Fig. 3.4 Relative frequency of probe pairs showing MM > 6 PM in RNA U133 arrays of human heart, teste, and cerebellum tissues. The horizontal labels the nucleotide triplet in the center of the MM probe

or [GCG]. According to the SNP array design, the two PM probes annotating this SNP have the central nucleotide triplets: PA = [CCC] and PB = [CGC]. Notice that when annotating the target [GCG], PA probe has a mismatch in the center, while PB has a perfect match. Therefore, by the MM phenomenon, PA presents a larger intensity than PB, i.e., PA > PB. Hence, a nucleotide type G

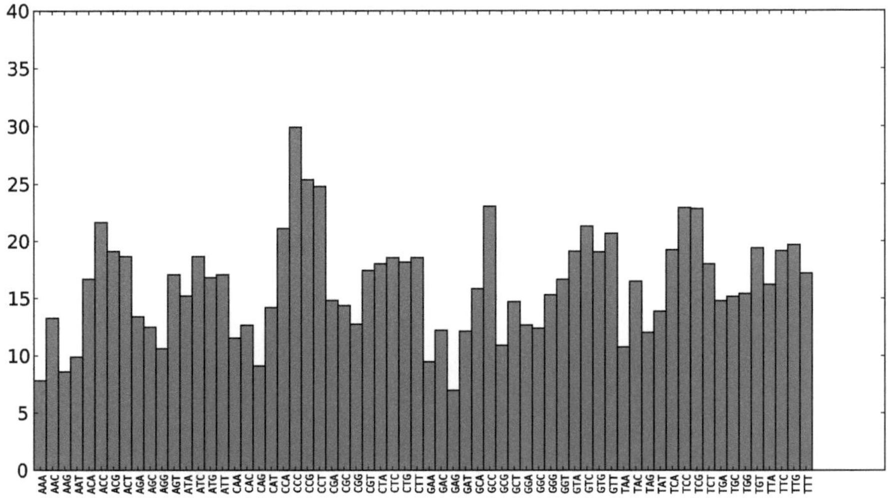

Fig. 3.5 Relative frequency of probe pairs showing PM > 2 MM in RNA U133 arrays of human heart, teste, and cerebellum tissues. The horizontal labels the nucleotide triplet in the center of the MM probe

will be called for this target. Similarly, when annotating the target [GGG], the PA probe has a perfect match, and the PB probe has a mismatch. In this case, the mismatch probe PB = [CGC] does not present the MM phenomenon, and thus the PA intensity is larger than PB intensity, i.e., PA > PB. Hence, a nucleotide type G will be called for this target. Therefore, a heterozygous (GC) SNP will be called incorrectly as a homozygous (GG) SNP. For the same reason, a homozygous (CC) SNP will be called incorrectly as a homozygous (GG) SNP, but a homozygous (GG) SNP will be called correctly. It is clear that this MM phenomenon leads to SNP genotype calling error, with heterozygous SNPs more frequently than homozygous SNPs.

(2) Vallone et al. [40] studied the SNPs of the human mitochondria with the Affymetrix MitoChip platform MitoChip 2.0 and reported problems in the SNP calling algorithm. It was noted that genotype calling error often occurs to result in a G-stack at the SNP position. This can be well explained by the above illustration assuming that the MM phenomenon occurs in MitoChip at a large chance as well, which may potentially be due to the same array mechanism of nucleotide binding.

From the above two implications of the MM phenomenon it can be concluded that mismatch probes are of particular importance so that they should not be removed from the oligonucleotide array design. If they are removed, SNP genotype calls should not be based on only simple summary statistics of the PM probe intensities. More sophisticated genotype calling methods must be used.

3.5 Abundance of Gene Expression: Copy Number Versus Probe Intensity

Microarray studies generate probe intensity data, and the analysis has thus far focused on the probe intensity data. Although PM probe intensity is known to be highly correlated with the copy numbers, as shown by experiments [15], and has been used as a surrogate of the copy numbers for the quantification of the abundance of gene expression, there exists major difference between the copy numbers and probe intensity in many aspects. Gene expression is a characterization of the abundance of gene activity and is presumably inherent in the tissues and is not subject to any measurement errors. However, probe intensity is an external measurement obtained through the high-throughput technology to quantify the energy involved in the DNA/DNA or DNA/RNA duplex hybridization and depends on many experimental factors, such as fluorescent quantity, array spray reagent, temperature, and the scanning of the array image. Any single factor may affect the probe intensity reading and contribute to the noise and artifacts in microarray data, where the latter refers to the factors that vary with experimental conditions but do not reflect biological variability of different individuals. Although the above artifacts have been known in microarray data analysis and a number of methods have been studied to address these issues, such as the normalization methods for microarray data [2], the fundamental issues of the difference between the two have not been fully recognized. Recognizing this difference will eventually lead to a search of the methods that focus on the genuine abundance of gene expression but do not vary with the experimental conditions, which may thus filter out the wild range of noise in the probe intensity data. Given the current experimental conditions and the microarray technologies, and assuming that the probe intensities are the data from the microarray studies, is it possible to find methods that quantify the abundance of gene expression and do not suffer from the current artifacts and wild range of noise?

To answer this question, it is crucial to separate true signal data that reflect the activities of genes from the false signal data that reflect the artifacts and the noise in the probe intensity experiments through decomposition of the probe intensity. It is believed that improvement of experimental conditions, such as following stringent protocols, will help to reduce the amount of artifacts but cannot completely remove it. Changing array design, such as using the Agilent long oligoarrays that have 60 nucleotide bases in each probe sequence rather than 25 nucleotide bases in the Affymetrix arrays, will help to reduce the noise by effectively reducing the number of short DNA sequence bindings of small homologue. However, these methods cannot completely remove the artifacts and thus are not what we are looking for.

Given the above problem in the experimental process, one needs to search for a resolution with computational approaches. Although this does sound unlikely, we will show that computational methods are promising in resolving this issue. In a series of studies, Zhang and colleagues [51, 52] developed a GSB method in modeling the RNA/DNA or DNA/DNA duplex hybridization, and the method has led to major improvement in microarray studies. Successful examples also include the CRLMM, an SNP array genotype calling method that accounts both probe sequence nucleotide effect and their positional effect as well as the length and GC content of the target sequence.

It is worthwhile to note that the DNA copy numbers at a single SNP locus have been estimated so far with the probe intensity ratio between an individual subject and a reference, such as a reference defined by the mean or median of the probe intensity of a normal control group. Since the microarray probe intensity data have many artifacts and are subject to wild range of noise due to experimental conditions, the copy numbers obtained through such a ratio method will surely inherit the noise even if array normalization procedures are cautiously taken as a preventive procedure.

In the following, we demonstrate that incorporation of the physico-chemical properties will lead to a new computational method for the estimation of copy numbers based on the probe intensities. In this work [43], we found that further study of the DNA/DNA or DNA/RNA duplex binding by modeling the positional and neighborhood effects will not only improve the SNP array data analysis for genotype calling and copy number estimation but also decompose the probe intensity data into signal data and nonsignal artifact data.

We first generalized the PDNN model by Zhang et al. [51] to a generalized PDNN (GPDNN) model by modeling the oligonucleotide binding with up to two mismatch nucleotides to estimate the probe sequence binding free energy. The free energy for binding with no mismatch nucleotide is modeled with the same PDNN model as before,

$$E = \sum_{l=1}^{24} \omega_l \lambda(s_l, s_{l+1}),$$

where $\lambda(s_l, s_{l+1})$ and ω_l remain the same. The free energy with one mismatch nucleotide at the shift position $(13 + j)$ is modeled by

$$E_1 = \sum_{l=1, l \neq 12+j, 13+j}^{24} \theta_l^j \lambda(s_l, s_{l+1}) + \kappa^j \delta\{(S_{12+j}^P, S_{13+j}^P, S_{14+j}^P), (S_{12+j}^T, S_{13+j}^T, S_{14+j}^T)\},$$

where θ_l^j is the positional factor, κ^j the positional factor of the mismatch nucleotide with shift j, and δ is the effect of the nucleotide triplet with mismatch nucleotide at position j. For two mismatch nucleotides, the free energy is modeled by

$$E_2 = E_1 + \xi^j\{(S_{12+j}^P, S_{13+j}^P, S_{14+j}^P), (S_{12+j}^T, S_{13+j}^T, S_{14+j}^T)\},$$

where E_1 is the free energy that the hybridization would have if there were only one mismatch nucleotide at the center position, ξ^j reflects the difference of the free energy due to the second mismatch nucleotide at position $13 + j$. This GPDNN model calculates the binding free energy for all oligonucleotide probe sequences.

We then decomposed each probe intensity into four terms for the high-density 100-K and 500-K SNP arrays by the following PICR model:

$$\begin{cases} \quad \vdots \\ I^{PA,ks} = N_A f(E^{PA,ks,A}) + N_B f(E_1^{PA,ks,B}) + B_{PA,ks} + \varepsilon_{PA}, \\ I^{MA,ks} = N_A f(E_1^{MA,ks,A}) + N_B f(E_{t_k}^{MA,ks,B}) + B_{MA,ks} + \varepsilon_{MA}, \\ I^{PB,ks} = N_A f(E_1^{PB,ks,A}) + N_B f(E^{PB,ks,B}) + B_{PB,ks} + \varepsilon_{PB}, \\ I^{MB,ks} = N_A f(E_{t_k}^{MB,ks,A}) + N_B f(E_1^{MB,ks,B}) + B_{MB,ks} + \varepsilon_{MB}, \\ \quad \vdots \end{cases}$$

where I's represent the probe intensities, N_A and N_B are allelic copy numbers for alleles A and B, respectively. E's are the free energies and can be calculated based on the GPDNN model with $t_k = 2$ for shift $k \neq 0$ and $t_k = 1$ for $k = 0$. $f(E)$ represents the binding affinity with the Zhang equation. The baseline B's are assumed to be identical within the same DNA strand with $s = 0$ for sense-strand and $s = 1$ for anti-sense strand. The measurement error terms ε's are independent and identically distributed with mean 0 and variance σ^2.

For given SNP of each array, the PICR model can be fitted to the probe intensities of the probe set annotating the same SNP, and the allelic copy numbers N_A and N_B can be estimated with a linear regression model with the precalculated binding affinities $f(E)$'s for different probes and strand. We have demonstrated that this model can be trained with only a single array to obtain all model parameters. Furthermore, we also developed an SNP genotype calling algorithm and demonstrated with the HapMap samples that this SNP genotyping method was robust across arrays, laboratories, and array platforms with high SNP genotype calling accuracy about 99.7% on 100-K SNP arrays and 99.2% on 500-K SNP arrays and outperformed other methods [43].

It was also pointed out [43] that the PICR model provides a novel approach by transforming the probe intensity level data, which are subject to wild range of noise and artifacts, to the copy number data, which are cleaned and biologically meaningful data. The robustness of the highly accurate genotype calling method based on the PICR model has demonstrated that this PICR model does yield cleaned copy number data with high accuracy. Recognizing the difference between the probe intensity data and the copy number data is of crucial importance and will lead to novel and powerful approaches for scientific discovery. While the copy numbers represent certain inborn characteristic of the tissue and are not subject to any measurement error or artifacts, the probe intensity data are subject to measurement errors and all kind of artifacts. Therefore this PICR model offers a means to transform the probe intensity data to the copy number data through a regression model and thus potentially offers a promise to revealing scientific discoveries with more powerful tests by modeling and analyzing the cleaned and biologically meaningful copy number data in subsequent analysis.

3.6 Array Image Quality and Repair Through an Imputation Method with a Mixed Effects Model

It is well known that microarray data may be subject to uneven spray of chemicals, which leads to bright or dark spots or paintbrush in the array image. Such defects of arrays, if not properly adjusted, will lead to large bias in the probe intensity data and will affect data analysis result. Although various normalization methods for microarray data analysis have been studied, most of them deal with systematic bias, such as the use of quantile normalization method to remove artifacts [2], few methods have been developed to deal with the bright spots and paintbrushes. We here present a newly developed imputation method based on a mixed-effects model that extends the PICR model from a single SNP to multiple SNPs on the same array to fix the bright spots [20].

Li et al. [20] developed a mixed-effects model for multiple SNPs on the same array:

$$\begin{cases} I_{ij} = \beta_{i0} + \beta_{i1} f_{ij1} + \beta_{i2} f_{ij2} + \varepsilon_{ij}, \\ \beta_{i0} = \beta_0 + \alpha_{i0}, \\ \beta_{i1} = \beta_1 + \alpha_{i1}, \\ \beta_{i2} = \beta_2 + \alpha_{i2}, \end{cases}$$

where I_{ij} is the intensity of the jth probe of SNP i for $i = 1, 2, \ldots, N$ and $j = 1, 2, \ldots, m$ with m being the number of probes for each SNP in a specific platform of arrays, e.g., $m = 40$ for Xba arrays. f_{ij1} and f_{ij2} are the binding affinities for two alleles A and B, respectively. Here, $\beta_0, \beta_1, \beta_2$ are fixed effects for baseline background, copy numbers for allele A and allele B, respectively. $\alpha_0, \alpha_1, \alpha_2$ are the corresponding random effects on SNPs and are independently normally distributed with $N(0, \sigma_0^2)$, $N(0, \sigma_1^2)$, and $N(0, \sigma_2^2)$, respectively. ε_{ij} is independent of the α's and is normally distributed with $N(0, \sigma^2)$.

Since the binding affinities f_{ij1} and f_{ij2} are calculated based on the PICR model that was trained with a single Xba array, they remain the same and are robust across different samples and different platforms of arrays [43]. This mixed-effects model is fitted to multiple SNPs on the same array and can provide prediction for the missing probe intensity. We apply this mixed-effects model to the arrays with damaged areas, in which the probe intensities are assumed missing to avoid taking biased intensity values. The predicted values of the mixed effects model over the damaged areas will provide probe intensity values for multiple imputation.

Figure 3.6 presents an Xba SNP array image obtained by the dChip program on one of the HapMap samples [7]. It is shown that there is one bright spot on the array due to uneven spray of chemical. This spot has about 2000 pairs of probe intensities and thus affects the intensity of about 2000 SNPs. To fix this problem, we took the mixed effects model approach with the following procedure.

Procedure to fix damaged area with the mixed-effects model

(1) Randomly select 100 SNPs among the unaffected SNPs by this spot.

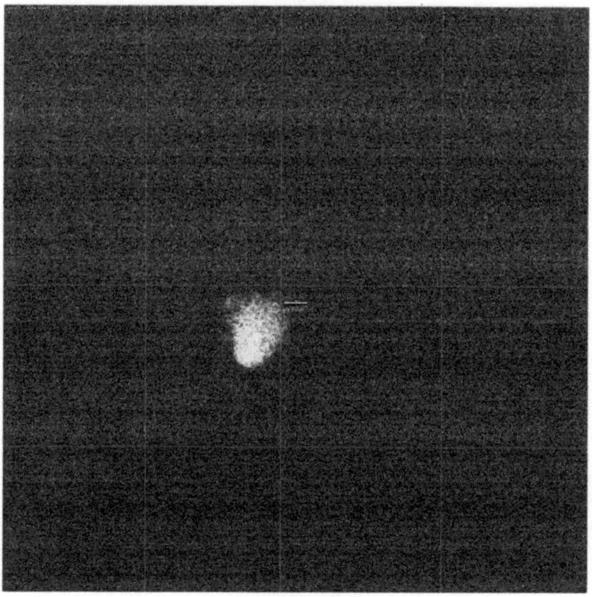

Fig. 3.6 A HapMap Xba array image illustrating a damaged area with *bright spot*

(2) Combine each affected SNP with the 100 selected SNPs to fit the above mixed-effects model.
(3) Predict the probe intensities in the damaged area by the mixed-effects model with the probe affinities in the area.

The imputation of the probe intensities by the above mixed-effects model removed the bright spots and yielded a smooth image as shown in Fig. 3.7. This imputation raised the SNP genotype calling accuracy from 98.96% to 99.74% by the PICR genotype calling method. Similar improvement was observed over other damaged arrays in the HapMap samples.

3.7 Concluding Remarks

The microarray technology has advanced rapidly during the last decade and has been demonstrated to be successful in assisting biological and biomedical research for scientific discoveries. It has gone through dramatic changes from the initial relatively uncomplicated design and understanding of the probe intensities with relatively simple computational methods to the current complex techniques and sophisticated computational methods. Although many advances and improvement have been made in the array technology and the related methodologies, there are still a number of aspects of the array mechanisms that are not thoroughly understood [27], such as the nonspecific binding of the arrays, the MM phenomenon, and the genomic wave [8, 22].

Fig. 3.7 Array image illustrating a damaged area fixed by imputation with the mixed-effects model

The recent development of the GWAS and CNV studies generated a huge amount of health data using the arrays [45, 46]. This important resource for biomedical research will keep the arrays a viable tool for biomedical studies. It is thus imperative that array data quality be ensured and novel analytical methods be developed to improve the current computational methods for higher accuracy and statistical power with high sensitivity and specificity and low false positive and false negative.

Acknowledgements The authors would like to thank Professor Robert C. Elston for his valuable comments and suggestions. This work was partly supported by an Oncology Summer Scholarship from the College of Human Medicine of Michigan State University.

References

1. C. Barnes, V. Plagnol, T. Fitzgerald, R. Redon, J. Marchini, D. Clayton, and M.E. Hurles. A robust statistical method for case-control association testing with copy number variation. *Nat Genet*, **40**(10):1245–1252, 2008.
2. B.M. Bolstad, R.A. Irizarry, M. Astrand, and T.P. Speed. A comparison of normalization methods for high density oligonucleotide array data based on variance and bias. *Bioinformatics*, **19**:185–193, 2003.
3. B.M. Bolstad, R.A. Irizarry, L. Gautier, and Z. Wu. Preprocessing high-density oligonucleotide arrays. In R. Gentleman, V. Carey, W. Huber, R.A. Irizarry, and S. Dudoit, editors, *Bioinformatics and Computational Biology Solutions Using R and Bioconductor*. Springer, New York, 2005.
4. N.P. Carter. Methods and strategies for analyzing copy number variation using DNA microarrays. *Nat Genet*, **39**:S16–21, 2007.

5. B. Carvalho, H. Bengtsson, T.P. Speed, and R.A. Irizarry. Exploration normalization, and genotype calls of high-density oligonucleotide SNP array data. *Biostatistics*, **8**(2):485–499, 2007.
6. N. Chatterjee and R.J. Carroll. Semiparametric maximum likelihood estimation in case-control studies of gene-environmental interactions. *Biometrika*, **92**:399–418, 2005.
7. dChip Software. http://biosun1.harvard.edu/complab/dchip/.
8. S.J. Diskin, M. Li, C. Hou, S. Yang, J. Glessner, H. Hakonarson, M. Bucan, J.M. Maris, and K. Wang. Adjustment of genomic waves in signal intensities from whole-genome SNP genotyping platforms. *Nucleic Acids Res*, **36**:e126, 2008.
9. D. Edwards. Nonlinear normalization and background correction in one-channel cDNA microarray studies. *Bioinformatics*, **19**:825–833, 2003.
10. B. Efron and R. Tibshirani. On testing the significance of sets of genes. *Ann Appl Stat*, **1**:107–129, 2007.
11. C. Furusawa, N. Ono, S. Suzuki, T. Agata, H. Shimizu, and T. Yomo. Model-based analysis of non-specific binding for background correction of high-density oligonucleotide microarrays. *Bioinformatics*, **25**(1):36–41, 2009.
12. T.R. Golub, D.K. Slonim, P. Tamayo, C. Huard, M. Gaasenbeek, J.P. Mesirov, H. Coller, et al. Molecular classification of cancer: class discovery and class prediction by gene expression. *Science*, **286**:531–537, 1999.
13. G.A. Held, G. Grinstein, and Y. Tu. Modeling of DNA microarray data by using physical properties of hybridization. *Proc Natl Acad Sci USA*, **100**:7575–7580, 2003.
14. G.A. Held, G. Grinstein, and Y. Tu. Relationship between gene expression and observed intensities in DNA microarrays—a modeling study. *Nucleic Acids Res*, **34**:e70, 2006.
15. J. Huang, W. Wei, J. Chen, J. Zhang, G. Liu, X. Di, R. Mei, S. Ishikawa, H. Aburatani, K.W. Jones, et al. CARAT: a novel method for allelic detection of DNA copy number changes using high density oligonucleotide arrays. *BMC Bioinform*, **7**:83, 2006.
16. R.A. Irizarry, B. Hobbs, F. Collin, Y.D. Beazer-Barclay, K.J. Antonellis, U. Scherf, and T.P. Speed. Exploration normalization, and summaries of high density oligonucleotide array probe level data. *Biostatistics*, **4**(2):249–264, 2003.
17. J.J.P. Lebrec, H. Putter, J.J. Houwing-Duistermaat, and H.C. van Houweliingen. Influence of genotyping error in linkage mapping for complex traits—an analytic study. *BMC Genetics* **9**(57), 2008. doi:10.1186/1471-2156-9-57.
18. H.J. Lee, M.H. Deng, F.Z. Sun, and T. Chen. An integrated approach to the prediction of domain–domain interactions. *BMC Bioinform*, **7**:269, 2006.
19. C. Li and W.H. Wong. Model-based analysis of oligonucleotide arrays: expression index computation and outlier detection. *Proc Natl Acad Sci USA*, **98**(1):31–36, 2001.
20. M. Li, Y. Wen, and W.J. Fu. A random effects model to repair oligoarray image through imputation. Technical Report, The Computational Genomics Lab, Department of Epidemiology, Michigan State University, MI, 2009.
21. S. Li, A. Pozhitkov, and M.A. Brouwer. Competitive hybridization model predicts probe signal intensity on high density DNA microarrays. *Nucleic Acids Res*, **36**(20):6585–6591, 2008.
22. J.C. Marioni, N.P. Thorne, A. Valsesia, T. Fitzgerald, R. Redon, H. Fiegler, T.D. Andrews, B.E. Stranger, A.G. Lynch, E.T. Dermitzakis, et al. Breaking the waves: improved detection of copy number variation from microarray-based comparative genomic hybridization. *Genome Biol*, **8**:R228, 2007.
23. R. Mei, P.C. Galipeau, C. Prass, A. Berno, G. Ghandour, N. Patil, R.K. Wolff, M.S. Chee, B.J. Reid, and D.J. Lockhart. Genome-wide detection of allelic imbalance using human SNPs and high-density DNA arrays. *Genome Res*, **10**(8):1126–1137, 2000.
24. V. Moskvina, N. Craddock, P. Holmans, M.J. Owen, and M.C. O'Donovan. Effects of differential genotyping error rate on the type I error probability of case-control studies. *Hum Hered*, **61**:55–64, 2006.
25. F. Naef, D.A. Lim, N. Patil, and M. Magnasco. DNA hybridization to mismatched templates: A chip study. *Phys Rev E* **65**(4), 2002.

26. N. Ono, S. Suzuki, C. Furasawa, T. Agata, A. Kashiwagi, H. Shimizu, and T. Yomo. An improved physico-chemical model of hybridization on high-density oligonucleotide microarrays. *Bioinformatics*, **24**(10):1278–1285, 2008.
27. A.E. Pozhitkov, D. Tautz, and P.A. Noble. Oligonucleotide microarrays: widely applied-poorly understood. *Brief Funct Genomics Proteomics*, **6**:141–148, 2007.
28. J. Quackenbush. Microarray datanormalization and transformation. *Nat Genet*, **32**(S2):496–501, 2002.
29. N. Rabbee and T.P. Speed. A genotype calling algorithm for Affymetrix SNP arrays. *Bioinformatics*, **22**:7–12, 2006.
30. E. Schadt, C. Li, B. Eliss, and W.H. Wong. Feature extraction and normalization algorithms for high-density oligonucleotide gene expression data. *J Cell Biochem*, **84**(S37):120–125, 2002.
31. M. Seringhaus, J. Rozowsky, T. Royce, U. Nagalakshmi, J. Jee, M. Snyder, and M. Gerstein. Mismatch oligonucleotides in human and yeast: guidelines for probe design on tiling microarrays. *BMC Genomics*, **9**:635, 2008. doi:10.1186/1471-2164-9-635.
32. L. Shi, L.H. Reid, W.D. Jones and M.A.Q.C. Consortium. The MicroArray Quality Control (MAQC) project shows inter- and intraplatform reproducibility of gene expression measurements. *Nat Biotechnol*, **24**:1151–1161, 2006.
33. L. Shi, W.D. Jones, R.V. Jensen, and S.C. Harris. The balance of reproducibility, sensitivity, and specificity of lists of differentially expressed genes in microarray studies. *BMC Bioinformatics*, **9**(9):S10, 2008. doi:10.1186/1471-2105-9-S9-S10.
34. C. Spinka, R.J. Carroll, and N. Chatterjee. Analysis of case-control studies of genetic and environmental factors with missing genetic information and haplotype-phase ambiguity. *Genet Epidemiol*, **29**:108–127, 2005.
35. A. Subramanian, P. Tamayo, V.K. Mootha, S. Mukherjee, B.L. Ebert, M.A. Gillette, A. Paulovich, S.L. Pomeroy, T.R. Golub, E.S. Lander, and J.P. Mesirov. Gene set enrichment analysis: A knowledge-based approach for interpreting genome-wide expression profiles. *Proc Natl Acad Sci USA*, **102**:15545–15550, 2005.
36. P.K. Tan, T.J. Downey, Jr., El. Spitznagel, P. Xu, D. Fu, D.S. Simitrov, R.A. Lempicki, B.M. Raaka, and M.C. Cam. Evaluation of gene expression measurements from commercial microarray platforms. *Nucleic Acid Res*, **31**(19):5676–5684, 2003.
37. R. Tibshirani, T. Hastie, B. Narasimhan, and G. Chu. Diagnosis of multiple cancer types by shrunken centroids of gene expression. *Proc Natl Acad Sci USA*, **99**(10):6567–6572, 2002.
38. V.G. Tusher, R. Tibshirani, and G. Chu. Significance analysis of microarrays applied to the ionizing radiation response. *Proc Natl Acad Sci USA*, **98**(9):5116–5121, 2001.
39. H. Urakawa, S.E. Fantroussi, H. Smidt, J.C. Smoot, E.H. Tribou, J.J. Kelly, P.A. Noble, and D.A. Stahl. Optimization of single-base-pair mismatch discrimination in oligonucleotide microarrays. *Appl Environ Microbiol*, **69**(5):2848–2856, 2003.
40. P.M. Vallon, J.P. Jakupciak, and M.D. Coble. Forensic application of the affymetrix human mitochondrial resequencing array. *Forensic Sci Int Genet*, **1**:196–198, 2007.
41. L.J. van't Veer, H. Dai, M.J. van de Vijver, Y.D. He, et al. Gene expression profiling predicts clinical outcome of breast cancer. *Nature*, **415**:530–536, 2002.
42. L. Wan, D. Li, D. Zhang, X. Liu, W.J. Fu, L. Zu, M. Deng, F. Sun, and M. Qian. Conservation and implications of eukaryote transcriptional regulatory regions across multiple species. *BMC Genomics*, **9**:623, 2008.
43. L. Wan, K. Sun, Q. Ding, Y.H. Cui, M. Li, Y. Wen, R.C. Elston, M. Qian, and W.J. Fu. Hybridization modeling of oligonucleotide SNP arrays for accurate DNA copy number estimation. *Nucl Acid Res*, 2009. doi:10.1093/nar/gkp559.
44. Y. Wang, Z.-H. Miao, Y. Pommier, E.S. Kawasaki, and A. Player. Characterization of mismatch and high-signal intensity probes associated with Affymetrix genechips, 2007. doi:10.1093/bioinformatics/btm306.
45. B.A. Weir, M.S. Woo, G. Getz, S. Perner, L. Ding, R. Beroukhim, W.M. Lin, M.A. Province, A. Kraja, L.A. Johnson, et al. Characterizing the cancer genome in lung adenocarcinoma. *Nature*, **450**:893–898, 2007.
46. Wellcome Trust Case Control Consortium. Genome-wide association study of 14,000 cases of seven common diseases and 3,000 shared controls. *Nature*, **447**:661–678, 2007.

47. C. Wu, H. Zhao, K. Baggerly, R. Carta, and L. Zhang. Short oligonucleotide probes containing G-stacks display abnormal binding affinity on Affymetrix microarrays. *Bioinformatics*, **23**(19):2566–2572, 2007.
48. C. Wu, R. Carta, and L. Zhang. Sequence dependence of cross-hybridization on short oligo microarrays. *Nucleic Acids Res*, **33**(9):e84, 2005.
49. X. Yan and F. Sun. Testing gene set enrichment for subset of genes: Sub-GSE. *BMC Bioinform*, **9**:362, 2008. doi:10.1186/1471-2105-9-362.
50. K. Zhang, M. Deng, T. Chen, T.S. Waterman, and F. Sun. A dynamical programming algorithm for haplotype block partitioning. *Proc Natl Acad Sci USA*, **99**:7335–7339, 2002.
51. L. Zhang, M.F. Miles, and K.D. Aldape. A model of molecular interactions on short oligonucleotide microarrays. *Nat Biotechnol*, **21**:818–821, 2003.
52. L. Zhang, C. Wu, R. Carta, and H. Zhao. Free energy of DNA duplex formation on short oligonucleotide microarrays. *Nucleic Acids Res*, **35**:e18, 2007.
53. Y. Zhao, M.-C. Li, and R. Simon. An adaptive method for cDNA microarray normalization. *BMC Bioinform*, **6**:28, 2005.

Chapter 4
Stochastic-Process Approach to Nonequilibrium Thermodynamics and Biological Signal Transduction

Hao Ge

4.1 Introduction

4.1.1 Nonequilibrium Thermodynamics

Equilibrium thermodynamics emerged when Carnot proposed the first theoretical treatise on mechanical work and efficiency in heat engines in the early nineteenth century. Over the course of that century, a complete physical theory on changes in heat, mechanical work, and internal energy of molecular systems was developed due to the elegant contributions by Clausius, Boltzmann, Helmholtz, Gibbs, and others.

The original objective of equilibrium thermodynamics is to describe the transformations of energy in all its forms. During its development, a number of key physical quantities were introduced, such as entropy, enthalpy, free energy, and so on. These quantities and their relations are essential for understanding biochemical systems from a physical point of view.

However, one hundred fifty years after its formulation, the second law of thermodynamics still appears more as a program than a well-defined theory and all the thermodynamic potentials could be well defined only in equilibrium states. This is one of the main reasons why the equilibrium thermodynamics could hardly be applied to real biochemical systems, because living cells must continually extract energy from their surroundings in order to sustain the characteristic features of life such as growth, cell division, intercellular communication, movement, and responsiveness to their environment.

Therefore, a central problem in physical chemistry arises: Does there exist a generalization of the Second Law which is valid away from equilibrium? This is an old question having its origin in Boltzmann's work in gas kinetics [3], and during

H. Ge
School of Mathematical Sciences, and Center for Computational Systems Biology,
Fudan University, Shanghai 200433, People's Republic of China
e-mail: gehao@fudan.edu.cn

J. Feng et al. (eds.), *Frontiers in Computational and Systems Biology*,
Computational Biology 15,
DOI 10.1007/978-1-84996-196-7_4,

the 20th century, its development has achieved great success, including two Nobel Prizes of Chemistry awarded to L. Onsager and I. Prigogine, respectively.

On the other hand, if one needs to study the nonequilibrium thermodynamics from a microscopic or mesoscopic point of view and wants to be less ambitious and get a rather satisfactory understanding, effective stochastic models may be the best approach to choose. It could help us break through the shackles of former equilibrium and near-equilibrium statistical mechanics and would accomplish a rather complete theory of nonequilibrium thermodynamics [12, 13].

4.1.2 Biological Signal Transduction

Biological signal transduction processes are increasingly understood in quantitative terms such that the switching of enzymes and proteins between phosphorylated and dephosphorylated states becomes a universal module [10, 35]. The biological activity of a target protein is often wakened by the phosphorylation reaction catalyzed by a specific kinase and restrained by the dephosphorylation reaction catalyzed by a specific phosphatase, which is quite similar to the turning on and off procedure of an ordinary switch.

One of the key concepts in Phosphorylation–dephosphorylation cycle (PdPC) signaling is the switching sensitivity: the sharpness of the activation of the substrate protein in response to the concentration of the kinase is basic in the perspective of metabolic control analysis, usually termed as Hill coefficient first proposed by Hill [24].

Actually, the research about the sensitivity of single-enzyme catalysis activity, also known as the allosteric cooperativity, has already been developed for about forty years, since the classic papers of Monod, Wyman, and Changeux [41] and Koshland, Nemethy, and Filmer [33]. However, in the case of multienzyme systems such as the phosphorylation–dephosphorylation module, the situation is quite different. In the early 1980s, Goldbeter and Koshland [20, 34] discovered the ultrasensitivity phenomenon of a PdPC switch in terms of the zeroth-order kinetics of kinase and phosphatase, where the Hill coefficient can be extremely high. Moreover, it has already been observed in experiments [27].

Most of the previous models [20, 23, 47, 57] built for the phosphorylation and dephosphorylation module were traditionally based on coupled nonlinear deterministic differential equations in terms of regulatory mechanisms and kinetic parameters, which are widely used in the field of computational biology [8, 42]. Nowadays, as there is a growing awareness of the basic character of noise in the study of the effects of noise in biological networks, it becomes more and more important to develop stochastic models with chemical master equations (CME) based on biochemical reaction stoichiometry, molecular numbers, and kinetic rate constants [18, 31, 40, 58]. Moreover, several recently interesting experimental results can only be explained by stochastic models [11].

We aim to thoroughly investigate temporal cooperativity [47] emerged in the signal transduction module of phosphorylation–dephosphorylation cycle (PdPC) and to compare it with allosteric cooperativity through stochastic models. The cooperativity in the cyclic reaction is temporal, with energy "stored" in time rather than in space as for allosteric cooperativity. This kind of cooperativity utilizes multiple kinetic cycles in time, in contrast to allosteric cooperativity that utilizes multiple subunits in a protein [16].

4.2 Stochastic-Process Approach: Examples

4.2.1 Mesoscopic Description of Biochemical Systems

4.2.1.1 Markov Chain (Master Equation)

Finite Markov chain is a jump process in the state space $S = \{1, 2, \dots, N\}$ with transition density matrix $Q(t) = \{q_{ij}(t)\}_{N \times N}$.

The starting point is a master equation formula of the system

$$\frac{d}{dt} p_i(t) = \sum_{j=1}^{N} \big(q_{ji}(t) p_j(t) - q_{ij} p_i(t) \big) \tag{4.1}$$

for the dynamical evolution of a probability distribution $p_i(t)$ over states $i = 1, 2, \dots, N$. The quantity $q_{ij}(t)$ is the transition density (probability per time) to state j from state i. It contains internal rate constants and external conditions imposed by the coupling to the reservoir systems.

Let us also mention that the number N need not to be finite, and the system could also be regarded as the stochastic model of coupled chemical reactions (chemical master equation) [40, 54].

4.2.1.2 Hill's Model of Muscle Contraction

T.L. Hill applied his general mesoscopic model for biochemical polymers to muscle contraction [25]. See Fig. 4.1 for the three-state model, where M is the myosin cross-bridge, A the actin cite, $A \cdot MDP_i$ and $A \cdot MD$ are both "attached" states of M and A, and the transition from $A \cdot MD$ to MDP_i will hydrolyze one molecule of $ATP (ATP \to ADP + P_i)$.

4.2.1.3 Stochastic Michaelis–Menten Kinetics

One considers a three-step mechanism of the Michaelis–Menten kinetics [46] in which the conversion of S into P in the catalytic site of the enzyme is represented

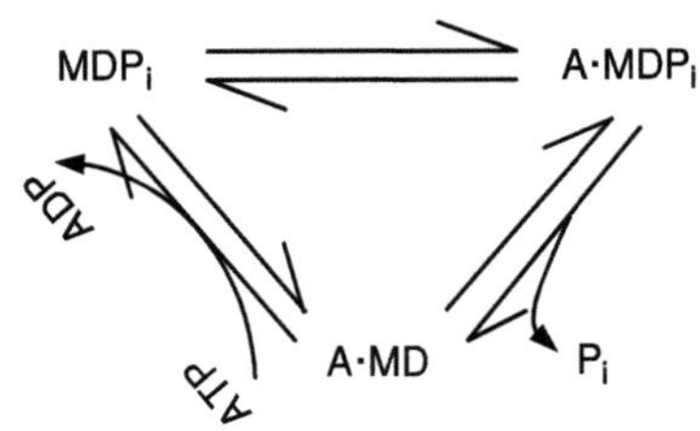

Fig. 4.1 Three-state ATPase attachment–detachment cycle used to illustrate the theoretical formalism for muscle contraction. Copied from [25]

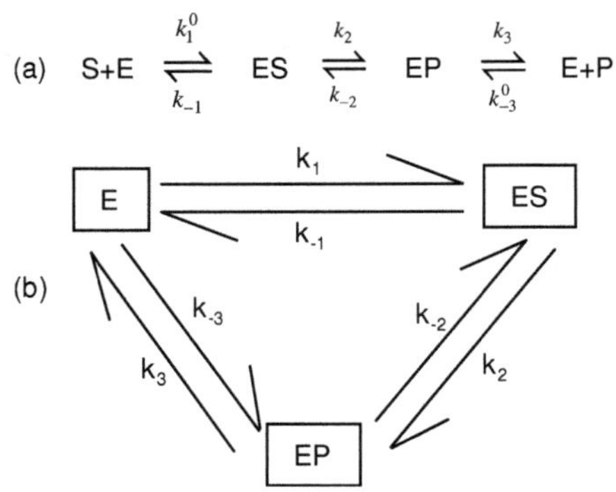

Fig. 4.2 Kinetic scheme of a simple reversible enzyme reaction (**a**) in which k_1^0 and k_{-3}^0 are second-order rate constants. From the perspective of a single enzyme molecule, the reaction is unimolecular and cyclic (**b**). The pseudo-first-order rate constants $k_1 = k_1^0 c_S$ and $k_{-3} = k_{-3}^0 c_P$ where c_S and c_P are the concentrations of substrate S and P in the steady state. Copied from [11]

as a process separate from release of P from the enzyme (Fig. 4.2(a)):

$$E + S \underset{k_{-1}}{\overset{k_1^0}{\rightleftharpoons}} ES \underset{k_{-2}}{\overset{k_2}{\rightleftharpoons}} EP \underset{k_{-3}^0}{\overset{k_3}{\rightleftharpoons}} E + P. \tag{4.2}$$

From the perspective of a single enzyme molecule, the reaction is unimolecular and cyclic, and the rate equation for the probabilities of the states is the master equation

$$\begin{aligned} \frac{dP_E(t)}{dt} &= -(k_1 + k_{-3})P_E(t) + k_{-1}P_{ES}(t) + k_3 P_{EP}(t), \\ \frac{dP_{ES}(t)}{dt} &= k_1 P_E(t) - (k_{-1} + k_2)P_{ES}(t) + k_{-2}P_{EP}(t), \\ \frac{dP_{EP}(t)}{dt} &= k_{-3}P_E(t) + k_2 P_{ES}(t) - (k_{-2} + k_3)P_{EP}(t). \end{aligned} \tag{4.3}$$

4.2.2 Langevin Systems

In statistical physics, Langevin dynamics is an approach to mechanics using simplified models and using stochastic differential equations describing Brownian motion in a potential to account for omitted degrees of freedom.

For a system of N particles with masses m and time-dependent coordinates $X = X(t)$, the Langevin equation is

$$m\ddot{X} = -\nabla V(X) - f + \sqrt{2m\gamma kT}\xi(t),$$

where $V(X)$ is the particle interaction potential such that $\nabla V(X)$ is the force calculated from the particle interaction potentials, $\dot{X}$ is the velocity and thus $f = \gamma m\dot{X}$ is the friction force, T is the temperature, k is Boltzmann's constant, and $\xi(t)$ is a delta-correlated stationary zero-mean Gaussian process satisfying

$$\langle \xi(t) \rangle = 0, \quad \langle \xi(t)\xi(t') \rangle = \delta(t - t'),$$

where δ is the Dirac delta.

If the friction coefficient γ is large enough, then $m\ddot{X} \approx 0$. Thus,

$$\gamma \dot{X} = -\frac{\nabla V(X)}{m} + \sqrt{\frac{2\gamma kT}{m}}\xi(t).$$

Generally speaking, Langevin dynamics is only one class of diffusion processes, and the latter could be also applied to model the stretching of single molecule [29] and the well-known Molecular motors [32, 52].

4.3 Stochastic Thermodynamics

If a system is gently driven away from equilibrium by a small time-dependent perturbation, then the response of the system to the perturbation can be described by linear response theory [39, 44]. However, if the system is driven far from equilibrium by a large perturbation, then linear response and other near-equilibrium approximations are generally not applicable.

The research on irreversible systems far from equilibrium began with the works by Haken [21] about laser and Prigogine et al. [19, 43] about oscillations of chemical reactions. Prigogine and his collaborators provided explicit expressions for entropy production in various situations and regarded a nonequilibrium steady state as a stationary open system with a positive entropy production rate [43].

Nonetheless, in contrast to equilibrium systems, with their elegant theoretical framework, the understanding of nonequilibrium steady-state systems is still primitive. What we really expect is a unified framework that describes both equilibrium and nonequilibrium phenomena, obtained by extending the Second Law not only to the steady states but also to any arbitrary nonequilibrium states. Many different kinds of approach have been put forward in the last several decades, and a recent comparative study of nonequilibrium thermodynamics can be found in [37].

In 1998, Oono and Paniconi [45] proposed a framework of steady-state thermodynamics. The steadily generated heat, which is generated even when the system remains in a single steady state, and the total heat are distinguished. They call the former the "housekeeping heat," which is equal to the entropy production in steady state and may come from the chemical driven force in biochemical systems [46, 48]. Subtracting the housekeeping heat from the total heat defines the excess heat, which reflects the time-dependent variation of the system. The key point of their work is that "if we can carefully remove the steadily produced heat due to housekeeping dissipation, then the state should not be very different from equilibrium." Moreover, they also put forward a phenomenological extended form of the Second Law: "A process converting work into excess heat is irreversible. And 'reversibility' is modulo house-keeping heat, which is produced anyway."

On the other hand, now stochastic models are widely used in physics, chemistry, biology, and even in economics. The Markov process could be applied to model chemical reactions, which are of special interest in biology, in relation with their coupling with active transport across membrane [8, 25] and also recent mechanisms of molecular motors [32]. Furthermore, in real biochemical systems, the external parameters such as the concentrations of external signal proteins always oscillate or remarkably fluctuates, which give rise to the necessity for the analysis of time-dependent processes. We believe that if one wants to comprehensively investigate the temporally phenomena of nonequilibrium state, then the stochastic processes, especially Markov chains and diffusions, would be the proper mathematical models to apply.

In 1953, Onsager and Machlup [39, 44] proposed the Onsager–Machlup principle, which is actually a functional formula about the probability density of a stochastic process close to equilibrium. Then, it was T.L. Hill who first successfully constructed a general mesoscopic stochastic model for the combination and transformation of biochemical polymers in vivid metabolic systems and investigated its thermodynamic properties far from equilibrium [25]. In the mean time, Schnakenberg developed an elegant network theory for the microscopic and macroscopic behavior of master-equation systems, namely, finite Markov chains in a mathematical language [54]. After that, a rather complete mathematical theory for nonequilibrium steady states has been developed for stochastic models [50, 51]. Here, we recommend a recent book [30] for systematic interpretations of this theory.

In recent years it has been realized that a trajectory perspective of stochastic processes might encode surprisingly more information than one might expect from traditional thermodynamic arguments, and a few interesting relations that describe the statistical dynamics of driven systems even far from equilibrium have been discovered. They include the fluctuation theorems of sample entropy production [7, 36, 38], Jarzynski's equality [28], Crooks' relations [5, 6], etc. Jarzynski [28, 29] considered nonequilibrium transitions between two equilibrium states, providing equilibrium Helmholtz free energy differences in terms of nonequilibrium measurements of the work required to switch from one ensemble to another.

Although the concept of Helmholtz free energy fails in nonequilibrium steady states (NESS), Hatano and Sasa [22] generalized Jarzynski's work relation to NESS, which is more relevant to motor proteins. Their work was inspired by Oono and Paniconi's framework [45], and they derived the first explicit expression for the extended form of the Second Law of Thermodynamics, namely $T\triangle S \geq Q_{\text{ex}}$, where S is the general entropy defined in their paper, and Q_{ex} is the excess heat.

It is indispensable to emphasize that the Jarzynski and Hatano–Sasa equalities become rather trivial and make no sense for time-independent processes. Hence recently, we put forward a unified rigorous proof of them [14, 15] for time-dependent[1] Markov chains and also diffusion processes and tried to gain deeper insights into these issues.

First of all, we accept the opinion that for the Second Law, in particular, a proper formulation and interpretation of entropy is more subtle. However, a nonequilibrium state needs for its description time-dependent variables, because of exchange of mass and energy between the system and its surroundings, and the problem of the definition of entropy is still open and actually not yet definitively solved. Fortunately, for stochastic processes, the so-called Gibbs entropy has already been widely accepted in statistical physics and information theory. Then it becomes our starting point.

The common definition of Gibbs entropy associated with any discrete probability distribution $\{p_i\}$ is

$$S\big[\{p_i\}\big] = -k \sum_i p_i \log p_i,$$

where k is the Boltzmann constant. In statistical mechanics, it gives the entropy for a canonical ensemble of a molecular system at a constant temperature and is a generalization of Boltzmann's formula to a situation with nonuniform probability distribution.

It is widely known that the entropy change dS can be decomposed into two terms [43, 46]: the first, $d_e S$ is the transfer of entropy across the boundaries of the system, and the second $d_i S$ is the entropy produced within the system.

Here, it is easy to derive that [46]

$$\frac{dS(t)}{dt} = d_i S + d_e S = epr(t) - hdr(t), \tag{4.4}$$

where $epr(t)$ is just the instantaneous entropy production rate [17], and $hdr(t)$ is due to the exchange of heat with the exterior, called the heat dissipation rate.

Although entropy may be considered as an ensemble property, Seifert [55] successfully developed a theory of entropy production along a stochastic trajectory enlightened by the newly developed fluctuation theorems [5, 7, 28] and showed that the entropy production defined along a single stochastic trajectory also can be divided into a medium part and a part of the particle (system) [55].

[1] Also called "inhomogeneous" in the language of Markov processes.

Meanwhile, the idea of decomposing the total heat into a "housekeeping" part and another "excess" part was put forward by Oono and Paniconi [45] and made explicit in Langevin systems by Hatano and Sasa [22]. It says that $Q_{\text{tot}} = Q_{\text{ex}} + Q_{\text{hk}}$, and more importantly, we found out that the housekeeping heat is always nonnegative, which implies the nonequilibrium essence of the system [12, 13].

For equilibrium system, Q_{ex} reduces to the total heat Q_{tot}, because in this case $Q_{\text{hk}} \equiv 0$ due to the detailed balance condition, and in time-independent steady state, $Q_{\text{ex}}(t) \equiv 0$, and hence the housekeeping heat Q_{hk} equals the work done by the external driven force, which is all dissipated [46, 48].

However, the situation is quite different for the time-dependent nonequilibrium system, in which the housekeeping heat still comes from the work done by some external driven force, but what is the origin of the excess heat? We will show that it is just the change of a thermodynamic quantity called *general internal energy*.

If the system satisfies the detailed balance conditions for all time, then the traditional concept of internal energy exists, and both of the excess heat and dissipative work contribute to its change, which is actually the First Law [15]. Therefore, we believe that the situation will not be essentially different even if detailed balance conditions fails.

Further, we find out that the integral of the dissipative work subtracting the excess heat does not depend on the particular "path" taking through the parameter space, namely, only depends upon the initial and final states. Thus, there exactly exists a "*general internal energy*," whose derivative is just the difference of the dissipative work and excess heat, i.e.,

$$\frac{dU(t)}{dt} = -Q_{\text{ex}}(t) + W(t). \tag{4.5}$$

It is just the ordinary internal energy for the equilibrium canonical ensemble according to the Maxwell–Boltzmann law. Hence (4.5) is just the generalized First Law of thermodynamics. From the trajectory view, we can also define the internal energy of the state $X(t)$. Then one has $U(X(t),t) = -Q_{\text{ex}}(X(t),t) + W(X(t),t)$, which implies that the First Law is also satisfied along every trajectory.

Despite the dissipative work, there also exists another kind of work done by the external driven force, denoted as $Edf(t)$, and we derived that $Edf(t) \equiv Q_{\text{hk}}(t)$, which is only known to be valid in steady state before [46, 48]. Now we understand that there exist two kinds of external works done on the system; one is the dissipative work $W(t)$, and the other, $Edf(t)$, is from the external driven force. They result in the change of general internal energy and the heat dissipation, respectively.

Based on the elementary definition of free energy in equilibrium thermodynamics $F = U - TS$, here we can define a *general free energy* in the same way: $F(t) = U(t) - TS(t)$. For an equilibrium system, it is just the Gibbs free energy in a spontaneously occurring chemical reaction at constant pressure p and temperature T, and also the Helmholtz free energy for systems at constant V and T [53]. Its change gives the maximum work, other than pV work. Therefore, it is called a "hybrid free energy" by Ross [53].

More important, Schnakenberg [54] has shown that it is just the Lyapunov function as well as Prigogine–Glansdorff criterion certificating the thermodynamic stability for the steady state of the time-independent master equation system. It has been revisited lately [12].

On the other hand,

$$\begin{aligned}\frac{dF(t)}{dt} &= \frac{dU(t)}{dt} - T\frac{dS(t)}{dt}\\ &= W(t) - \big(T \cdot epr(t) - Q_{\text{hk}}(t)\big).\end{aligned} \tag{4.6}$$

Here we introduce a new concept named *Free heat* $Q_f(t) = T \cdot epr(t) - Q_{\text{hk}}(t)$ identifying the free energy change in the form of heat, i.e.,

$$\frac{dF(t)}{dt} = W(t) - Q_f(t).$$

This concept will play the central role in the extended form of the Second Law of Thermodynamics below.

Regarding the Second Law of Thermodynamics, although all the thermodynamic quantities in the previous sections can be defined along the sample trajectory, the Clausius inequality and many other thermodynamic constrains related to the Second Law should be interpreted statistically through ensemble average.

Notice that every term in the expression of the entropy production rate $epr(t)$ is nonnegative; hence $epr(t) \geq 0$, and the equality holds if and only if the detailed balance condition holds. Then according to (4.4) and (4.6), we derive several inequalities of the differential forms:

$$T\frac{dS(t)}{dt} + Q_{\text{tot}}(t) = T \cdot epr(t) \geq 0, \tag{4.7a}$$

$$\frac{dF(t)}{dt} - W(t) - Q_{\text{hk}}(t) = -T \cdot epr(t) \leq 0. \tag{4.7b}$$

Equation (4.7a) is just the well-known Clausius inequality ($dS \geq -\frac{Q_{\text{tot}}}{T}$), which is rectified to obtain expressions for the entropy produced (dS) as the result of heat exchanges (Q_{tot}), and (4.7b) is a general version of the free energy inequality for the amount of work performed on the system, since the work values must then be consistent with the Kelvin–Planck statement and forbids the systematic conversion of heat to work.

More precise, the quantity $Q_{\text{hk}}(t)$ in (4.7b) vanishes when the detailed balance condition holds, and then it returns back to the traditional Helmholtz or Gibbs free energy inequalities of equilibrium thermodynamics depending on whether it is an NVT or NPT system [2]. In this case, $-dF \geq -W$, which implies that the decrease of free energy gives the maximum dissipative work done upon the external environment.

Also, their corresponding integral forms are

$$T\Delta S + \int Q_{\text{tot}}(t)\,dt \geq 0, \tag{4.8a}$$

$$\triangle F - \int W(t)\,dt - \int Q_{\mathrm{hk}}(t)\,dt \le 0. \tag{4.8b}$$

However, there still leaves a fundamental question: "what is precisely the reversible process that connects two different equilibrium states?" Then a new concept of "instantaneous reversible process" [17] with zero entropy production rate naturally emerges, which corresponds to the ideal reversible process involved in the classic theory of equilibrium thermodynamics, and it will imply that there does not exist any real reversible process connecting two different equilibrium states.

Furthermore, an extended quantitative form of Second Law of Thermodynamics will be developed built on the nonnegativity of the new concept '*free heat*" ($Q_f(t) = epr(t) - T Q_{\mathrm{hk}}(t) \ge 0$), which only appears during time-dependent processes [12, 13]. Then according to (4.4) and (4.6), we have another group of inequalities in the differential forms:

$$T\frac{dS(t)}{dt} + Q_{\mathrm{ex}}(t) = T \cdot epr(t) - Q_{\mathrm{hk}}(t) \ge 0, \tag{4.9a}$$

$$\frac{dF(t)}{dt} - W(t) = -Q_f(t) = -T \cdot epr(t) + Q_{\mathrm{hk}}(t) \le 0, \tag{4.9b}$$

followed by their corresponding integral forms

$$T\triangle S + \int Q_{\mathrm{ex}}(t)\,dt \ge 0, \tag{4.10a}$$

$$\triangle F - \int W(t)\,dt \le 0. \tag{4.10b}$$

Inequality (4.10a) is an extended form of the Clausius inequality during any nonequilibrium time-dependent process, whose special case is included in Hatano and Sasa's work [22], and inequality (4.10b) is a different general form of free energy inequality. It implies that the dissipative work value must be consistent with the Oono–Paniconi statement of the extended Second Law of thermodynamics [45], which forbids the systematic conversion of excess heat to work.

For the equilibrium case, $Q_{\mathrm{ex}} = Q_{\mathrm{tot}}$, and they both actually return to (4.7a). Moreover, if the system is in steady state, then $Q_f(t) \equiv 0$, and this form of the Second Law is eliminated.

4.4 Ultrasensitivity and Temporal Cooperativity of PdPC Module

4.4.1 Reversible Kinetic Model for Covalent Modification

Many references [2, 20, 47, 57] have considered the important phosphorylation–dephosphorylation cycle (PdPC) catalyzed by kinase E_1 and phosphatase E_2, respectively. The phosphorylation covalently modifies the protein W to become W^*:

$$W + E_1 + ATP \underset{d_1}{\overset{a_1^0}{\rightleftharpoons}} W \cdot E_1 \cdot ATP \underset{q_1^0}{\overset{k_1}{\rightleftharpoons}} W^* + E_1 + ADP;$$

$$W^* + E_2 \underset{d_2}{\overset{a_2}{\rightleftharpoons}} W^* E_2 \underset{q_2^0}{\overset{k_2}{\rightleftharpoons}} W + E_2 + Pi.$$

Then at constant concentrations for *ATP*, *ADP*, and *Pi*, introducing the pseudo-reaction orders $a_1 = a_1^0[ATP]$, $q_1 = q_1^0[ADP]$, and $q_2 = q_2^0[Pi]$, these reactions become

$$W + E_1 \underset{d_1}{\overset{a_1}{\rightleftharpoons}} WE_1; \qquad WE_1 \underset{q_1}{\overset{k_1}{\rightleftharpoons}} W^* + E_1;$$

$$W^* + E_2 \underset{d_2}{\overset{a_2}{\rightleftharpoons}} W^* E_2; \qquad W^* E_2 \underset{q_2}{\overset{k_2}{\rightleftharpoons}} W + E_2.$$

This biochemical scheme is also isomorphic to another important module in cellular signal transduction across the cell membrane, namely the GTPase system.

It is indispensable to note that the sustained high concentration of ATP (~1 mM) and low concentrations of adenosine diphosphate (ADP) (~10 μM) and Pi (orthophosphate) (~1 mM) give rise to an equilibrium constant of 4.9×10^5 M for ATP hydrolysis, and the phosphorylation potential in a normal cell is approximately 12 kcal mol^{-1} [26].

4.4.2 Reduced Models

It is always supposed that the total concentration of W and W^* is much larger than that of the kinase and phosphatase (i.e., $W_T \gg E_{1T} + E_{2T}$ or, equivalently, $W_T = [W] + [W^*]$) [20, 47]. Therefore, the dynamics of kinase and phosphatase can be considered separately:

$$\begin{aligned} &\text{(a):} \quad W + E_1 \underset{d_1}{\overset{a_1}{\rightleftharpoons}} WE_1 \underset{q_1}{\overset{k_1}{\rightleftharpoons}} W^* + E_1; \\ &\text{(b):} \quad W + E_2 \underset{k_2}{\overset{q_2}{\rightleftharpoons}} W^* E_2 \underset{a_2}{\overset{d_2}{\rightleftharpoons}} W^* + E_2. \end{aligned} \tag{4.11}$$

The steady states in the above Michaelis–Menten kinetics have been solved in the classic enzymology [4], and the fluxes from W to W^* and from W^* to W in reactions (a) and (b) of (4.11) are

$$v_1([W]) = \frac{\frac{V_1[W]}{K_1}}{1 + \frac{[W]}{K_1} + \frac{[W^*]}{K_1^*}}, \qquad v_1^*([W^*]) = \frac{\frac{V_1^*[W^*]}{K_1^*}}{1 + \frac{[W]}{K_1} + \frac{[W^*]}{K_1^*}}$$

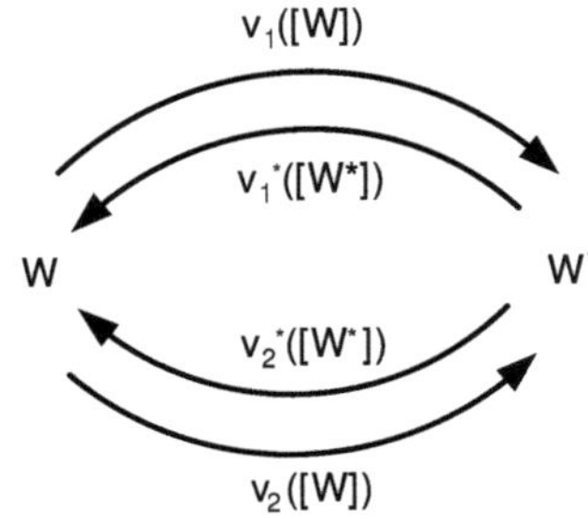

Fig. 4.3 The reduced model of PdPC switch

and

$$v_2([W]) = \frac{\frac{V_2[W]}{K_2}}{1 + \frac{[W]}{K_2} + \frac{[W^*]}{K_2^*}}, \qquad v_2^*([W^*]) = \frac{\frac{V_2^*[W^*]}{K_2^*}}{1 + \frac{[W]}{K_2} + \frac{[W^*]}{K_2^*}},$$

respectively, in which the parameters $V_1 = k_1 E_{1T}$, $V_1^* = d_1 E_{1T}$, $V_2 = d_2 E_{2T}$, and $V_2^* = k_2 E_{2T}$ are the maximal forward ($W \to W^*$) and backward ($W^* \to W$) fluxes of the reactions (a) and (b); and $K_1 = \frac{d_1+k_1}{a_1}$, $K_2^* = \frac{d_2+k_2}{a_2}$, $K_1^* = \frac{d_1+k_1}{q_1}$, and $K_2 = \frac{d_2+k_2}{q_2}$ are the corresponding Michaelis constants.

Hence our model is now reduced to the form of Fig. 4.3, which can be also found in the latest book [2] and further reduced to

$$W \underset{f_2([W^*])}{\overset{f_1([W])}{\rightleftharpoons}} W^*, \tag{4.12}$$

where $f_1 = v_1 + v_2$ is the total flux from W to W^*, $f_2 = v_1^* + v_2^*$ is the total flux from W^* to W, and $[W] + [W^*] = W_T$ (constant).

4.4.3 Deterministic Model

The ordinary differential equation of the model (4.12) is

$$\frac{d[W^*]}{dt} = f_1(W_T - [W^*]) - f_2([W^*]), \tag{4.13}$$

whose steady state $[W^*]^{ss}$ satisfies $f_1(W_T - [W^*]^{ss}) = f_2([W^*]^{ss})$ and $[W]^{ss} = W_T - [W^*]^{ss}$.

What we concern most is the steady-state fraction of phosphorylated protein W^*, i.e., $\phi = \frac{[W^*]^{ss}}{W_T}$. Beard and Qian [2] have written down the general equation for $\phi = \frac{[W^*]^{ss}}{W_T}$ in the deterministic model under the restrictions $W_T \gg E_{1T} + E_{2T}$ ($W_T = [W] + [W^*]$) and $q_1, q_2 \ll 1$:

$$\sigma \overset{def}{=} \frac{\theta K_1}{K_2^*} = \frac{V_1}{V_2^*} = \frac{\mu\gamma[\mu - (\mu+1)\phi](\phi - 1 - \frac{K_1}{W_T})}{[\mu\gamma - (\mu\gamma+1)\phi](\phi + \frac{K_2^*}{W_T})}. \tag{4.14}$$

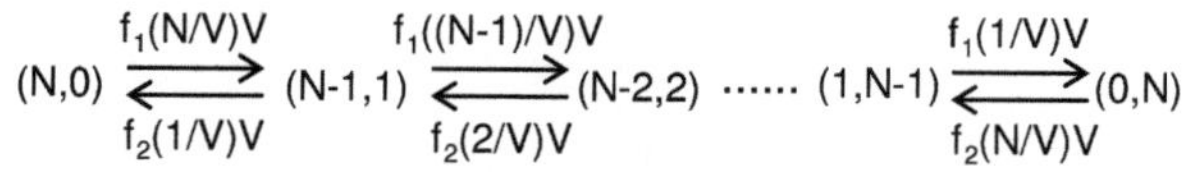

Fig. 4.4 The illustrated chemical master equation of the reduced model of the PdPC switch. The two-dimensional vector $(N-i,i)$ represents the random state that the molecule number of the species W is $(N-i)$ and the molecule number of the species W^* is i

Let $q_1 = q_2 = 0$; then one can get

$$\sigma = \frac{\phi(1-\phi+\frac{K_1}{W_T})}{(1-\phi)(\phi+\frac{K_2^*}{W_T})},$$

which is just the celebrated Goldbeter–Koshland equation [20] in their pioneer work on zero-order ultrasensitivity.

4.4.4 Stochastic Model: Chemical Master Equation

In order to illustrate the essence of temporal cooperativity, we should turn to the stochastic model, chemical master equation. Let V be the volume of the system; then the total molecule number of W and W^* is $N = W_T V$. Due to the existence of unavoidable fluctuations, one cannot determine the molecule numbers of each species at any arbitrary time t and, instead, can only determine the probability that the vector representing the molecule numbers of species W and W^* is $(N-i,i)$.

Denote the probability of the state $(N-i,i)$ at time t as $P(N-i,i;t)$; then it satisfies the **chemical master equation**

$$\begin{aligned}
\frac{dP(N,0;t)}{dt} &= f_2(1/V)VP(N-1,1;t) - f_1(N/V)VP(N,0;t);\\
\frac{dP(N-i,i;t)}{dt} &= f_1\big((N+1-i)/V\big)VP(N+1-i,i-1;t)\\
&\quad + f_2\big((i+1)/V\big)VP(N-1-i,i+1;t)\\
&\quad - \big[f_1\big((N-i)/V\big)\\
&\quad + f_2(i/V)\big]VP(N-i,i;t), \quad i=1,2,\ldots,N-1;\\
\frac{dP(0,N;t)}{dt} &= f_1(1/V)VP(1,N-1;t) - f_2(N/V)VP(0,N;t).
\end{aligned} \tag{4.15}$$

From (4.15), in the steady state, the ratio of the probabilities of the states $(N-i,i)$ and $(N,0)$ is $\Pi_{j=1}^{i}[\frac{f_1((N+1-j)/V)}{f_2(j/V)}]$; then the steady distribution of the state

$(N-i,i)$ is

$$P^{ss}(N-i,i)=\frac{\prod_{j=1}^{i}\frac{f_1((N+1-j)/V)}{f_2(j/V)}}{1+\sum_{i=1}^{N}\prod_{j=1}^{i}\frac{f_1((N+1-j)/V)}{f_2(j/V)}}. \tag{4.16}$$

Similar to the deterministic model, we introduce the ratio of the averaged molecule number $\langle W^*\rangle$ of phosphorylated protein molecules and the total molecule number N, i.e.,

$$\langle\phi\rangle \stackrel{def}{=} \frac{\langle W^*\rangle}{N}=\frac{\sum_{i=1}^{N} i\prod_{j=1}^{i}\frac{f_1((N+1-j)/V)}{f_2(j/V)}}{N(1+\sum_{i=1}^{N}\prod_{j=1}^{i}\frac{f_1((N+1-j)/V)}{f_2(j/V)})}. \tag{4.17}$$

Define the quantities $K_j=\frac{(N+1-j)f_2(j/V)}{jf_1((N+1-j)/V)}$ representing the "dissociation capability" of the jth molecule in the state $(N-j,j)$ transiting back from the activated species W^* to the inactivated one W that are called "dissociation constants," and their reciprocals are representing the "association capability" of the jth molecule transiting from the inactivated species W to the activated one W^* that can be called "association constants."

With these in our model, there exists the temporal cooperative phenomenon if the quantities $\{K_j,\ j=1,2,\dots,N\}$ successively decrease, which means the more number of molecules of W^* is, the larger the association constant of the next molecule transiting from the state W to W^* becomes. Furthermore, the cooperative phenomenon appears more and more distinct when the gradient of the decreasing quantities $\{K_j,\ j=1,2,\dots,N\}$ increases.

4.4.5 Simple PdPC Switch: First-Order Approximation

Suppose $W_T \ll K_1, K_2^* \ll K_1^*, K_2$ (nonsaturated); then $f_1([W])\approx\frac{V_1[W]}{K_1}+\frac{V_2[W]}{K_2}$ and $f_2([W^*])\approx\frac{V_2^*[W^*]}{K_2^*}+\frac{V_1^*[W^*]}{K_1^*}$ are both first-order, which is just the ordinary PdPC switch discussed in [47].

It is easy to derive that both the fractional saturation ϕ in the deterministic model and the $\langle\phi\rangle$ in the stochastic model are equal to $\frac{\alpha}{1+\alpha}$. Then since $\alpha=\frac{\frac{V_1}{K_1}+\frac{V_2}{K_2}}{\frac{V_2^*}{K_2^*}+\frac{V_1^*}{K_1^*}}$ is an increasing hyperbolic function of E_{1T}, $\langle\phi\rangle=\phi$ is also an increasing hyperbolic function of E_{1T} illustrating no cooperative effect either, which implies that the N molecules of W and W^* are all independent.

The dissociation constants $\{K_i\}$ of temporal cooperativity with different volumes. It is found that in such a simple PdPC switch, these dissociation constants are all very close to 1, regardless of the variety of volumes, reconfirming no obvious cooperative phenomenon.

4.4.6 Ultrasensitive PdPC Switch: Zero-Order Approximation

Supposing that $K_2, K_1^* \gg W_T \gg K_1, K_2^*$(saturated) and that $K_2^* \ll K_2$, $K_1 \ll K_1^*$, one can arrive at the limit case ($\frac{[W^*]}{K_1^*} \approx 0$ and $\frac{[W]}{K_2} \approx 0$) $f_1([W]) = v_1([W]) + v_2([W]) \approx V_1$ and $f_2([W^*]) = v_1^*([W^*]) + v_2^*([W^*]) \approx V_2^*$.

These are both in the zero-order case, which should be considered as nonlinear since $f_1(0) \neq 0$ and $f_2(0) \neq 0$. This is just the situation of ultrasensitive PdPC switch [47] and zero-order ultrasensitivity phenomenon put forward by Goldbeter and Koshland [20]. The Hill coefficient of the response curve can approach thousands and tens of thousands.

In the deterministic model of this limit case, we have $\phi = \delta_{\{V_1 > V_2^*\}}$, which is a step function with ideal infinite sensitivity. In the stochastic model, the steady distribution of the state $(N-i, i)$ is $\frac{\alpha^i}{N(1+\sum_{i=1}^N \alpha^i)}$(truncated geometric distribution), so

$$\langle\phi\rangle = \frac{\langle W^*\rangle}{N} = \frac{\sum_{i=1}^N i\alpha^i}{N(1+\sum_{i=1}^N \alpha^i)} = \begin{cases} \frac{N\alpha^{N+1} - \frac{\alpha^{N+1}-\alpha}{\alpha-1}}{N(\alpha^{N+1}-1)}, & \alpha \neq 1, \\ 1/2, & \alpha = 1, \end{cases} \tag{4.18}$$

where $\alpha = \frac{V_1}{V_2^*}$ is the ratio of the forward flux from W to W^* and the backward flux from W^* to W.

Obviously, $\langle\phi\rangle$ is an increasing function of α and, consequently, an increasing function of E_{1T}. Moreover, as $N \to \infty$, one has $\langle\phi\rangle \to 1$ if $\alpha > 1$ and $\langle\phi\rangle \to 0$ if $\alpha < 1$ (see Fig. 4.5). In this case, the classical Hill coefficient $n_H = 2\frac{d\log\langle\phi\rangle}{d\log\alpha}|_{\langle\phi\rangle=\frac{1}{2}} = \frac{1}{3}N + \frac{2}{3}$. Therefore, as the total molecule number N tends to infinity, the Hill coefficient can increase to an arbitrary value.

Hence, when the Michaelis constants K_1, K_2 are quite small, the ultrasensitive cooperative phenomenon emerges both in deterministic and stochastic models, although their sensitivities cannot be as high as in the limit case discussed above.

Figure 4.5 illustrates the curves of $\langle\phi\rangle$ with respect to E_{1T} at different volumes in the stochastic model of ultrasensitive PdPC switch under the zero-order approximation, in which it is found that the sensitivities of these curves are increasing with the volumes (molecule numbers) and finally approach the ideal jumping curve of ϕ with infinite sensitivity.

Figure 4.6 illustrates the curves of $\langle\phi\rangle$ with respect to E_{1T} at different volumes in the stochastic model (4.15) by formula (4.17) of the ultrasensitive PdPC switch without the zero-order approximation, in which it is found that the sensitivities of these curves are increasing with the volumes (molecule numbers).

Figure 4.7 represents the dissociation constants $\{K_i\}$ of cooperativity with different volumes. It is found that in the ultrasensitive PdPC switch, these dissociation constants clearly decrease, and the gradient increases with the total molecule numbers, suggesting more and more distinct cooperative phenomenon.

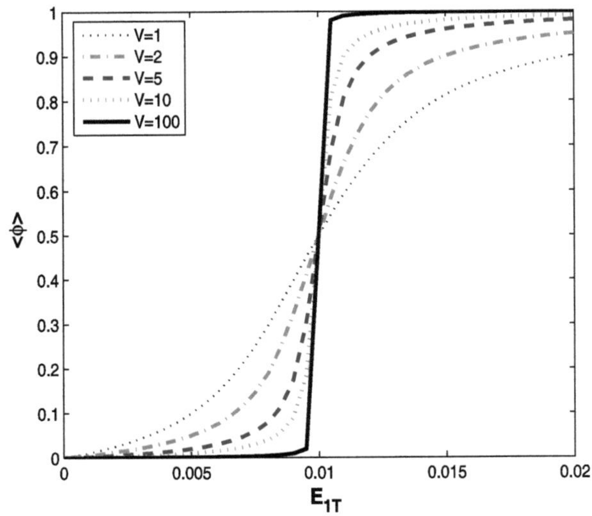

Fig. 4.5 The curve of $\langle\phi\rangle$ with respect to E_{1T} at different volumes in the stochastic model of ultrasensitive PdPC switch under the zero-order approximation, where the other parameters are the same as those in Fig. 4.7

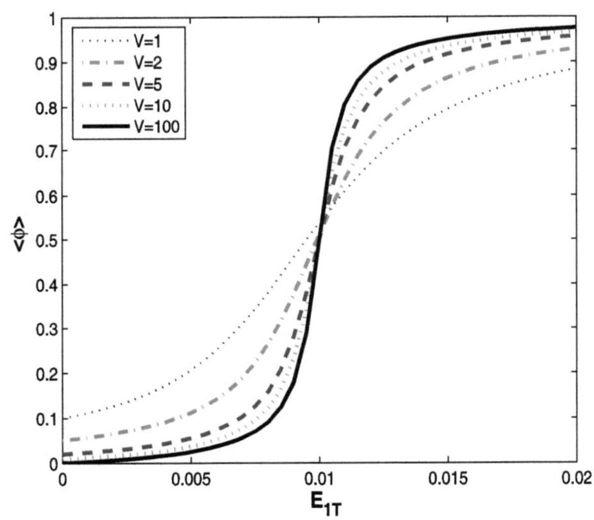

Fig. 4.6 The curve of $\langle\phi\rangle$ with respect to E_{1T} of different volumes in the stochastic model of the ultrasensitive PdPC switch without the zero-order approximation, where the other parameters are the same as those in Fig. 4.7

4.4.7 Mathematical Equivalence to Allosteric Cooperativity

Although the sharp activation in PdPC switches has always been compared to allosteric cooperative transitions [34], it has never been made very clear what the essential similarities and differences between them are. This significant question could date back to Fischer and Krebs [9, 10], who discovered protein phosphorylation as a regulatory mechanism for enzyme activity and won the Nobel Prize in 1992.

Cooperativity can be generally considered in relation to the Adair scheme, first proposed by Adair [1] in relation to the binding of oxygen to haemoglobin, and the

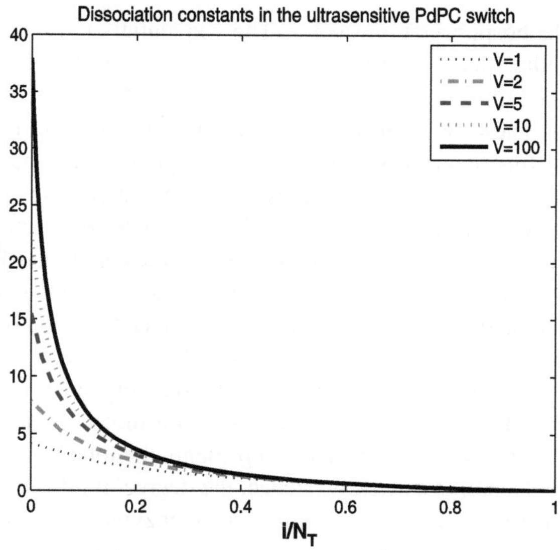

Fig. 4.7 The dissociation constants in the ultrasensitive PdPC switch, where $a_1 = 10$, $d_1 = 1$, $k_1 = 1.5$, $q_1 = 0.0001$, $E_{1T} = 0.01$, $a_2 = 10$, $d_2 = 1$, $k_2 = 1.5$, $q_2 = 0.0001$, $E_{2T} = 0.01$, $W_T = 10$, and $\alpha = V_1/V_2^*$. The volume V takes values 1, 2, 5, 10, and 100, and the molecule numbers $N = W_T V$ are 10, 20, 50, 100, and 1000, respectively

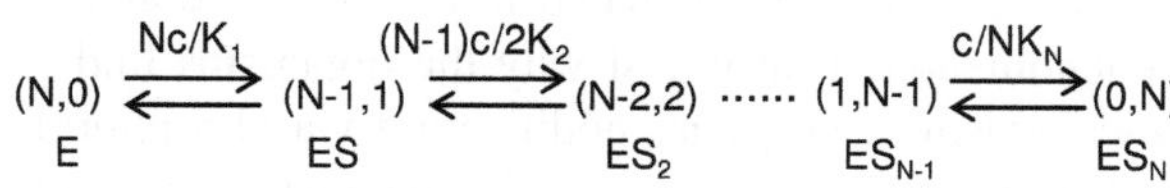

Fig. 4.8 General model of the allosteric cooperative phenomenon, where E is the enzyme, S is the substrate, and $c = [S]$

general form of the Adair equation is

$$\phi = \frac{\sum_{i=1}^{N} \frac{(N-1)!}{(i-1)!(N-i)!} \frac{c^i}{\prod_{j=1}^{i} K_j}}{1 + \sum_{i=1}^{N} \frac{N!}{i!(N-i)!} \frac{c^i}{\prod_{j=1}^{i} K_j}},$$

where $c = [S]$, $K_j = \frac{(N-j+1)c[ES_{j-1}]}{j[ES_j]}$ is the dissociation constant of the jth molecule of the substrate (regardless of site). Consequently, there is an important corollary that the Hill coefficient of the $[S]-\phi$ curve determined by the Adair equation cannot exceed the total number N of sites on a single enzyme, i.e., $n_H \leq N$.

Figure 4.8 is the general model of allosteric cooperative phenomenon including both the famous MWC and KNF models [33, 41]. In this model, the concentration of the substrate S is fixed, and the vector $(N-i, i)$ represents the state in which there are i sites occupied with substrates among the total N sites.

Therefore, we can show the equivalence of the underlying mathematics in temporal cooperativity and allosteric cooperativity, both of which can be expressed by "dissociation constants," which also raises the essential differences between the simple and ultrasensitive PdPC switches.

It is very important to point out that Fig. 4.8 is nearly the same as Fig. 4.4, where the temporal cooperativity is on the scale of the N sequential phosphorylation–

dephosphorylation cycles. The sequential states in Fig. 4.4 are adjacent in time rather than in space, which is the case in allosteric cooperativity. The model in Fig. 4.8 is a special case of the model in Fig. 4.4 where $\frac{f_1(N+1-i/V)}{f_2(i/V)} = \frac{(N+1-i)[S]}{iK_i}$.

These two kinds of cooperativity phenomena both come from the nonlinearity of functions f_1 and f_2 (i.e., the varying of K_i), but the former emerges from the complex chemical reactions, while the latter arises from the allosteric interactions between different sites. Actually, although there is no direct interaction between the substrate enzymes, the total N molecules of W and W^* are not really independent: they all compete for the single kinase and phosphatase, and hence there are implicit interactions between them. Because this interaction is not through space but, instead, is sequential in time, so Qian [47, 49] refer to it as temporal cooperativity.

Moreover, the meanings of the quantity N in Figs. 4.4 and 4.8 are totally different: the former represents the total molecule number in the temporal cooperativity model, and the latter represents the total number of sites on a single enzyme molecule, respectively. Hence, the degree of allosteric cooperativity is restricted by the total number of sites in a single enzyme molecule which cannot be freely regulated, while temporal cooperativity is only restricted by the total molecule number of the target protein which can be regulated in a wide range and gives rise to the ultrasensitivity phenomenon. That is just why the organisms find it advantageous to develop the mechanism of covalent modification via phosphorylation and *ATP* hydrolysis to control the biological activity of proteins rather than the mechanism of allosteric transitions.

Therefore, the improving of the total number of molecules of target protein cannot increase the degree of allosteric cooperativity, while it can obviously increase the degree of temporal cooperativity, indicated by the increasing gradients of the fractional saturation function $\langle\phi\rangle$ (Fig. 4.6) and the decreasing dissociation constants $\{K_j,\ j=1,2,\dots,N\}$ (Fig. 4.7)!

4.5 Conclusion and Discussion

We are now at the beginning of a new development in theoretical chemistry and physics, in which thermodynamic concepts may play an even more basic role. To investigate these concepts, stochastic thermodynamics has developed much further than other approaches during the last two decades [29, 56]. For stochastic systems, the central problem is around the extension of the Second Law, which originally describes the fundamental limitation on possible transitions between equilibrium states.

Based on stochastic processes, we put forward a rather unified theory of nonequilibrium thermodynamics [12, 13], which should be more convincing and rigorous than the previous phenomenological frameworks. In addition, it would be interesting to test experimentally all the quantities and relations, especially in nonharmonic time-dependent potentials, where the Gaussian distribution assumption should be violated.

On the other hand, quantitative understanding and mathematical modeling of biological systems presents a significant challenge and a unique opportunity for scientists of diverse disciplines, including both deterministic models and stochastic approaches. It is often thought that noise added to the biological models only provides moderate refinements to the behaviors otherwise predicted by the classical deterministic system description, while it is quite clear that the main result, namely the mathematical equivalence between temporal and allosteric cooperativity can only be explicitly expressed by the chemical master equation model (see Fig. 4.4). Also the concept of temporal cooperativity in terms of the random-walk model is not limited to PdPC and kinetically isomorphic GTPases but also applies to many other signaling processes [49].

Acknowledgements I would like to dedicate this work to Professor Minping Qian on the occasion of her 70th birthday. Her work on mathematical theory of nonequilibrium steady states has had major influence on my scientific career and will have long-lasting consequences in physical and mathematical biochemistry. Also, I am very grateful to Prof. Min Qian, Prof. Hong Qian, and Prof. Daquan Jiang for stimulating discussions.

References

1. G.S. Adair. The hemoglobin system. VI. The oxygen dissociation curve of hemoglobin. *J Biol Chem*, **63**:529–545, 1925.
2. D.A. Beard and H. Qian. *Chemical Biophysics: Quantitative Analysis of Cellular Systems*. Cambridge University Press, Cambridge, 2008.
3. L. Boltzmann. *Lectures on Gas Theory*. University of California Press, Berkeley, 1964. (Translated by S.G. Brush)
4. A. Cornish-Bowden. *Fundamentals of Enzyme Kinetics*, 3rd edition. Portland Press, London, 2004.
5. G.E. Crooks. Entropy production fluctuation theorem and the nonequilibrium work relation for free energy differences. *Phys Rev E*, **60**:2721–2726, 1999.
6. G.E. Crooks. Path-ensemble averages in systems driven far from equilibrium. *Phys Rev E*, **61**(3):2361–2366, 2000.
7. D.J. Evans, E.G.D. Cohen, and G.P. Morriss. Probability of Second Law violation in steady flows. *Phys Rev Lett*, **71**:2401–2404, 1993.
8. C.P. Fall, E.S. Marland, J.M. Wagner, and J.J. Tyson. *Computational Cell Biology*. Springer, New York, 2002.
9. E.H. Fischer and E.G. Krebs. Conversion of phosphorylase *b* to phosphorylase *a* in muscle extracts. *J Mol Chem*, **216**:121–133, 1955.
10. E.H. Fischer, L.M.G. Heilmeyer, and R.H. Haschke. Phosphorylation and the control of glycogen degradation. *Curr Top Cell Regul*, **4**:211–251, 1971.
11. H. Ge. Waiting cycle times and generalized Haldane equation in the steady-state cycle kinetics of single enzymes. *J Phys Chem B*, **112**:61–70, 2008.
12. H. Ge. Extended forms of the second law for general time-dependent stochastic processes. *Phys. Rev. E* **80**:021137, 2009.
13. H. Ge. Nonequilibrium thermodynamics of time-dependent Langevin system: Energy balance relation and the extended form of the Second Law. arXiv:0904.2059, 2009, preprint.
14. H. Ge and D.Q. Jiang. Generalized Jarzynski's equality of multidimensional inhomogeneous diffusion processes. *J Stat Phys*, **131**:675–689, 2008.
15. H. Ge and M. Qian. Generalized Jarzynski's equality in inhomogeneous Markov chains. *J Math Phys*, **48**:053302, 2007.

16. H. Ge and M. Qian. Sensitivity amplification in the phosphorylation-dephosphorylation cycle: nonequilibrium steady states, chemical master equation and temporal cooperativity. *J Chem Phys*, **129**:015104, 2008.
17. H. Ge, D.Q. Jiang, and M. Qian. Reversibility and entropy production of inhomogeneous Markov chains. *J Appl Prob*, **43**(4):1028–1043, 2006.
18. D.T. Gillespie. A general method for numerically simulating the stochastic time evolution of coupled chemical reactions. *J Comput Phys*, **22**:403, 1976.
19. P. Glansdorff and I. Prigogine. *Thermodynamic Theory of Structure, Stability and Fluctuations*. Wiley-Interscience, London, 1971.
20. A. Goldbeter and D.E. Koshland Jr. An amplified sensitivity arising from covalent modification in biological systems. *Proc Natl Acad Sci USA*, **78**:6840–6844, 1981.
21. H. Haken. *Synergetics: An Introduction: Nonequilibrium Phase Transitions and Self-Organization in Physics, Chemistry, and Biology*. Springer, Berlin, 1977.
22. T. Hatano and S. Sasa. Steady-states thermodynamics of Langevin systems. *Phys Rev Lett*, **86**:3463–3466, 2001.
23. R. Heinrich, B.G. Neel, and T.A. Rapoport. Mathematical models of protein kinase signal transduction. *Mol Cell*, **9**:957, 2002.
24. A.V. Hill. The possible effects of the aggregation of the molecules of haemoglobin on its dissociation curves. *J Phys*, **40**:iv–vii, 1910.
25. T.L. Hill. *Free Energy Transduction in Biology*. Academic Press, New York, 1977.
26. J. Howard. *Mechanics of Motor Proteins and the Cytoskeleton*. Sinauer, Sunderland, 2001.
27. C.F. Huang and J.E. Ferrell Jr. Ultrasensitivity in the mitogen-activated protein cascade. *Proc Natl Acad Sci USA*, **93**:10078, 1996.
28. C. Jarzynski. Nonequilibrium equality for free energy differences. *Phys Rev Lett*, **78**:2690–2693, 1997.
29. C. Jarzynski. Nonequilibrium work relations: foundations and applications. *Eur Phys J B*, **64**:331–340, 2008.
30. D.Q. Jiang, M. Qian, and M.P. Qian. *Mathematical Theory of Nonequilibrium Steady States—on the Frontier of Probability and Dynamical Systems*, volume 1833 of *Lect. Notes Math.* Springer, Berlin, 2004.
31. J. Keizer. Master equations, Langevin equations and the effect of diffusion on concentration fluctuations. *J Chem Phys*, **67**(4):1473–1476, 1977.
32. A.B. Kolomeisky and M.E. Fisher. Molecular motors: A theorist's perspective. *Annu Rev Phys Chem*, **58**:675–695, 2007.
33. D.E. Koshland Jr., G. Nemethy, and D. Filmer. Comparison of experimental binding data and theoretical models in proteins containing subunits. *Biochemistry*, **5**:365–385, 1966.
34. D.E. Koshland Jr., A. Goldbeter, and J.B. Stock. Amplification and adaptation in regulatory and sensory systems. *Science*, **217**:220, 1982.
35. E.G. Krebs. Phosphorylation and dephosphorylation of glycogen phosphorylase: a prototype for reversible covalent enzyme modification. *Curr Top Cell Regul*, **18**:401, 1980.
36. J. Kurchan. Fluctuation theorem for stochastic dynamics. *J Phys A, Math Gen*, **31**:3719–3729, 1998.
37. G. Lebon, D. Jou, and J. Casas-Vázquez. *Understanding Non-Equilibrium Thermodynamics*. Springer, Berlin, 2008.
38. J.L. Lebowitz and H. Spohn. A Gallavotti–Cohen-type symmetry in the large deviation functional for stochastic dynamics. *J Stat Phys*, **95**(1–2):333–365, 1999.
39. S. Machlup and L. Onsager. Fluctuations irreversible processes II. Systems with kinetic energy. *Phys Rev*, **91**:1512–1515, 1953.
40. D.A. McQuarrie. Stochastic approach to chemical kinetics. *J Appl Prob*, **4**(3):413–478, 1967.
41. J. Monod, J. Wyman, and J.P. Changeux. On the nature of allosteric transitions: a plausible model. *J Mol Biol*, **12**:88–118, 1965.
42. J.D. Murray. *Mathematical Biology*, 3rd edition. Springer, New York, 2002.
43. G. Nicolis and I. Prigogine. *Self-organization in Nonequilibrium Systems: From Dissipative Structures to Order Through Fluctuations*. Wiley, New York, 1977.

44. L. Onsager and S. Machlup. Fluctuations irreversible processes. *Phys Rev*, **91**:1505–1512, 1953.
45. Y. Oono and M. Paniconi. Steady state thermodynamics. *Prog Theor Phys Suppl*, **130**:29–44, 1998.
46. H. Qian. Cycle kinetics, steady state thermodynamics and motors-a paradigm for living matter physics. *J Phys Condens Matter*, **17**:S3783–S3794, 2005.
47. H. Qian. Phosphorylation energy hypothesis: Open chemical systems and their biological functions. *Annu Rev Phys Chem*, **58**:113–142, 2007.
48. H. Qian and D.A. Beard. Thermodynamics of stoichiometric biochemical networks in living systems far from equilibrium. *Biophys Chem*, **114**:213–220, 2005.
49. H. Qian and J.A. Cooper. Temporal cooperativity and sensitivity amplification in biological signal transduction. *Biophys J*, **47**:2211, 2008.
50. M.P. Qian and M. Qian. Circulation for recurrent Markov chains. *Z Wahrscheinlichkeitstheor Verw Geb*, **59**:203–210, 1982.
51. M.P. Qian, M. Qian, and G.L. Gong. The reversibility and the entropy production of Markov processes. *Contemp Math*, **118**:255–261, 1991.
52. P. Reimann. Brownian motors: noisy transport far from equilibrium. *Phys Rep*, **361**:57–265, 2002.
53. J. Ross. *Thermodynamics and Fluctuations Far from Equilibrium*. Springer, Berlin, 2008.
54. J. Schnakenberg. Network theory of microscopic and macroscopic behaviour of master equation systems. *Rev Mod Phys*, **48**(4):571–585, 1976.
55. U. Seifert. Entropy production along a stochastic trajectory and an integral fluctuation theorem. *Phys Rev Lett*, **95**:040602, 2005.
56. U. Seifert. Stochastic thermodynamics: principles and perspectives. *Eur Phys J B*, **64**:423–431, 2008.
57. E.R. Stadtman and P.B. Chock. Superiority of interconvertible enzyme cascades in metabolite regulation: analysis of multicyclic systems. *Proc Natl Acad Sci USA*, **74**:2761, 1977.
58. N.G. Van Kampen. *Stochastic Processes in Physics and Chemistry*. North-Holland, Amsterdam, 1981.

Chapter 5
Granger Causality: Theory and Applications

Shuixia Guo, Christophe Ladroue,
and Jianfeng Feng

5.1 Introduction

A question of great interest in systems biology is how to uncover complex network structures from experimental data [1, 3, 18, 38, 55]. With the rapid progress of experimental techniques, a crucial task is to develop methodologies that are both statistically sound and computationally feasible for analysing increasingly large datasets and reliably inferring biological interactions from them [16, 17, 22, 37, 40, 42]. The building block of such enterprise is to being able to detect relations (causal, statistical or functional) between nodes of the network. Over the past two decades, a number of approaches have been developed: information theory [4], control theory [16] or Bayesian statistics [35]. Here we will be focusing on another successful alternative approach: Granger causality. In recent Cell papers [7, 8], the authors have come to the conclusion that the ordinary differential equation approach outperforms the other reverse engineering approaches (Bayesian network and information theory) in building causal networks. We have demonstrated that the Granger causality achieves better results than the ordinary differential approach [54].

The basic idea of Granger causality can be traced back to Wiener [47] who conceived the notion that, if the prediction of one time series is improved by incorporating the knowledge of a second time series, then the latter is said to have a causal influence on the first. Granger [23, 24] later formalised Wiener's idea in the context of linear regression models. Specifically, two auto-regressive models are fitted to the first time series—with and without including the second time series—and the improvement of the prediction is measured by the ratio of the variance of the error

S. Guo · J. Feng (✉)
Mathematics and Computer Science College, Hunan Normal University, Changsha 410081, P.R. China
e-mail: jianfeng64@gmail.com

C. Ladroue · J. Feng
Department of Computer Science and Mathematics, Warwick University, Coventry CV4 7AL, UK

J. Feng et al. (eds.), *Frontiers in Computational and Systems Biology*, Computational Biology 15,
DOI 10.1007/978-1-84996-196-7_5,

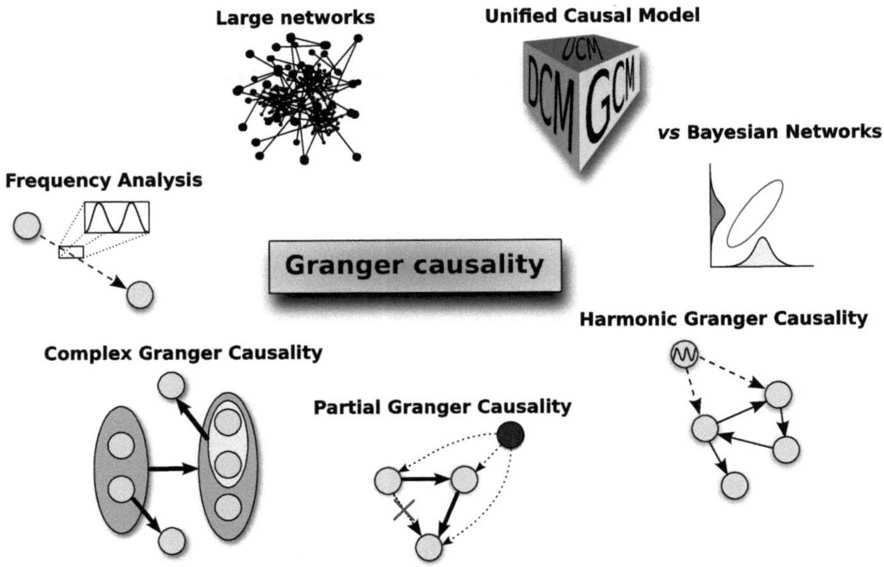

Fig. 5.1 The extensions of traditional Granger causality that will be discussed in this chapter

terms. A ratio larger than one signifies an improvement, hence a causal connection. At worst, the ratio is 1 and signifies causal independence from the second time series to the first. Geweke's decomposition of a vector autoregressive process [20, 21] led to a set of causality measures which have a spectral representation and make the interpretation more informative and useful by extending Granger causality to the frequency domain. In this chapter, we aim to present Granger causality and how its original formalism has been extended to address biological and computational issues, as summarised in Fig. 5.1.

Partial Granger Causality In its original conception, Granger causality is limited to the investigation of pairs of time series. If strong enough, indirect connections produce spurious relations between distant nodes. Conditional Granger causality [10, 12, 21] is able to deal with this situation by removing the influence of an external node and thus discarding what would be misleading connections. However, it requires the explicit knowledge of the influencing node; in other words, its applicability largely depends on the ability to measure all relevant variables, which is usually not possible in biological recordings. Both exogenous inputs and endogenous variables can confound accurate causal influences and thus degrade the credibility of the uncovered network structure. In order to eliminate the influences of exogenous inputs and latent variables, and inspired by the definition of partial correlation in statistics, we introduce a new definition of Granger causality: partial Granger causality, which is robust against perturbations due to common unseen variables [25, 26, 29, 50].

Frequency Analysis Thanks to Geweke's decomposition of a power spectrum [20, 21], Granger causality can be expressed in the frequency domain [26, 49] and thus provides more information: a spectrum indicating at which frequencies the connection between two time series occurs is produced, instead of a single number. This property is especially useful for studying phenomena in which frequencies play a major role, like electro-physiological recordings (e.g. multi-electrode array in the brain) or rhytmic behaviour like ciracadian rhythm [15]. The two representations (time and frequency domains) are consistent: integrating the spectrum over all frequencies is equal in practice to the time domain Granger causality.

Complex Granger Causality In complex systems of genes, proteins or neurons, elements often work cooperatively or competitively to achieve a task. If we want to understand biological process in details, it is of substantial importance to study the interactions among groups of nodes. Such group interactions are ubiquitous in biological processes: enzymes act on the production rate of metabolites [18, 28], information is passed on from one layer of neurons to the next, transcription factors form complexes which influence gene activity, etc. These interactions will be missed out with traditional Granger causality approaches. To tackle this problem, we extend traditional Granger causality to *complex* Granger causality, defined both in the time and frequency domains [15, 29]. Furthermore, we validate this approach with real biological data.

Harmonic Granger Causality Partial Granger causality was developed to eliminate the influence of common inputs, be they exogenous and endogenous. But full elimination is only possible if all common external inputs have equal influence on all measured variables, which is generally not realistic to expect in experimental recordings. For example, we know that certain genes are sensitive to local pH or temperature, while others are not. In this situation, we want to know which variables are impacted by environmental inputs, and the extent of this influence. In Sect. 5.5, Granger causality is modified to explicitly include a model of an oscillating external input [9, 11, 27, 44, 48–50]. Its influence on each of the observed variables can then be quantified, and an accurate network can be built. The method is applied to microarray experiments to study the circadian rhythm in a plant.

Granger Causality and Bayesian Networks Bayesian networks are a popular approach for investigating biological systems which has proved successful on many occasions [35, 39]. A natural question is to know whether one should choose Bayesian network or Granger causality when faced with data. In Sect. 5.6, we present a systematic and computationally intensive comparison between the two methods [52]. The experiment is done on simulated data, of which the true structure is known and for which we have total control on the parameters. It results that the length of the time series is a crucial factor for the contrasted performances of the two methods.

Unified Causal Model In contrast to many similar frameworks [17, 37], Granger causality does not use the concept of perturbation to define causality. In these frameworks, the states of the system are compared before and after some event (e.g. injection of different inhibitors in a cell in [39]). Instead, Granger causality relies on dependence over time to define causality. Unified Causal Model (UCM) is an attempt at including the notions of stimuli and modifying coupling to traditional Granger causality. Section 5.7 explains how UCM unifies the seemingly different approaches of Granger causality, phenomenological in nature, and the model-based Dynamic Causal Model (DCM, [16]).

Large Networks While partial Granger causality is efficient for uncovering small network structures, the method does not scale up for practical reasons: removing the influence of all other variables requires the fit of a linear model so large that it exhausts the number of observables. The system is underdetermined, and the resulting Granger causality unreliable. To address this issue, we developed an iterative procedure that builds a global network of possibly hundreds of nodes from local investigations. The idea is to gradually prune the network by removing indirect connections, considered one at a time. Applied on simulated networks of 200 nodes, the method shows very good performance. Section 5.8 shows its application on a large network of 800 proteins [53].

5.2 Partial Granger Causality

Traditional Granger causality is defined for two time series only. In order to build a network of interactions from a collection of time series, a simple approach is to apply it to all possible pairs of signals. However, indirect links can lead to spurious connections if single connections are strong enough: if $A \to B$ and $B \to C$, then very likely $A \to C$ will be picked up by Granger causality. The end-result is often a very densely connected network. *Conditional* Granger causality [12, 22] works by explicitly removing the influence of a third signal and thus avoids producing misleading links.

Critically, the ability of conditional Granger causality to deal with indirect interactions depends on being able to measure all relevant variables in a system. This is not possible most of the times, and both environmental inputs and unmeasured latent variables can confound real causal influences. For example, in experimental data recorded from the inferotemporal (IT) cortex, every measured neuron receives common exogenous inputs from the visual cortex and feedback from the prefrontal cortex. Moreover, even with advanced multielectrode array techniques, only a tiny subset of interacting neurons in a single area is recorded and there bound to be latent, unobserved variables. In this section, we introduce partial Granger causality [25], an extension of conditional Granger causality which addresses the issue of exogenous and latent variables.

5.2.1 Time Domain Formulation

For three time series X_t, Y_t and Z_t, define $\vec{y}_t = (X_t, Y_t, Z_t)$, where X_t and Y_t are one-dimensional time series, and Z_t is a set of time series of dimension m. A general form of an autoregressive model with zero mean and exogenous variable $\vec{\varepsilon}_t^E$ has the following vector autoregressive representation with the use of the lag operator[1] $\mathcal{L}$:

$$\mathbf{B}(\mathcal{L})\vec{y}_t = \vec{\varepsilon}_t^E + \vec{\varepsilon}_t \tag{5.1}$$

where $\mathbf{B}$ is a polynomial matrix of $\mathcal{L}$, $\mathbf{B}(0) = I_n$, the $n \times n$ identity matrix. The two random vectors $\vec{\varepsilon}^E$ and $\vec{\varepsilon}$ are independent. The exogenous variable $\vec{\varepsilon}^E$ represents the environmental drive and is typically present in any experimental setup.

As already mentioned, the confounding influence of latent variables is possibly even more disruptive than that due to exogenous inputs. To incorporate latent variables, assume that the ith network element receives unmeasured inputs of the form $\sum_{j=1}^N x_{ij}(t)/N$, where each x_{ij} is a stationary time series, and j is the latent index.

According to the Wold representation, any stationary variable $\xi(t)$ can be expressed as the summation of the form $\sum_k \psi_k \varepsilon(t-k)$, and we have

$$x_{ij}(t) = \sum_{k=1} \psi_{ij,k}\varepsilon_{ij}^L(t-k)$$

where ε_{ij}^L is the latent variable of indices (i, j). Therefore

$$\begin{aligned}\sum_{j=1}^N x_{ij}(t)/N &= \sum_{j=1}^N \sum_{k=1} \psi_{ij,k}\varepsilon_{ij}^L(t-k)/N \\ &= \sum_{k=1} \bar{\psi}_{i,k}\varepsilon_i^L(t-k)\end{aligned}$$

where $\psi_.$ are constants. In other words, each node i receives a latent input ε_i^L which depends on its history.

So the model (5.1) becomes

$$\mathbf{B}(\mathcal{L})\vec{y}_t = \vec{\varepsilon}_t^E + \vec{\varepsilon}_t + \mathbf{B}^{(1)}(\mathcal{L})\vec{\varepsilon}_t^L \tag{5.2}$$

where the random vectors $(\vec{\varepsilon}_t^E, \vec{\varepsilon}_t^L)$ and $\vec{\varepsilon}_t$ are independent, and $\mathbf{B}^{(1)}(\mathcal{L})$ is another polynomial matrix of $\mathcal{L}$ of appropriate size.

---

[1] A lag operator $\mathcal{L}$ is such that $\mathcal{L}X_t = X_{t-1}$. Thus, applying the operator k times yields: $\mathcal{L}^k X_t = X_{t-k}$. The auto-regressive model $X_t + A_1 X_{t-1} + A_2 X_{t-2} + \cdots = \varepsilon_t$ can be represented as $(Id + A_1\mathcal{L} + A_2\mathcal{L}^2 + \cdots)X_t = \varepsilon_t$.

Now consider two time series X_t and Z_t which admit a joint autoregressive representation of the form

$$\begin{cases} X_t = \sum_{i=1}^{\infty} a_{1i} X_{t-i} + \sum_{i=1}^{\infty} c_{1i} Z_{t-i} + \vec{\varepsilon}_{1t} + \vec{\varepsilon}_{1t}^{E} + \overrightarrow{B_1(\mathcal{L})\varepsilon_{1t}^{L}} \\ Z_t = \sum_{i=1}^{\infty} b_{1i} Z_{t-i} + \sum_{i=1}^{\infty} d_{1i} X_{t-i} + \vec{\varepsilon}_{2t} + \vec{\varepsilon}_{2t}^{E} + \overrightarrow{B_2(\mathcal{L})\varepsilon_{2t}^{L}} \end{cases} \tag{5.3}$$

For simplicity of notation, let us define

$$u_i(t) = \vec{\varepsilon}_{it} + \vec{\varepsilon}_{it}^{E} + \overrightarrow{B_i(\mathcal{L})\varepsilon_{it}^{L}}$$

The noise covariance matrix for the model can be represented as

$$S = \begin{bmatrix} \mathrm{var}(u_{1t}) & \mathrm{cov}(u_{1t}, u_{2t}) \\ \mathrm{cov}(u_{2t}, u_{1t}) & \mathrm{var}(u_{2t}) \end{bmatrix} = \begin{bmatrix} S_{xx} & S_{xz} \\ S_{zx} & S_{zz} \end{bmatrix}$$

In the same fashion, the vector autoregressive representation for a system involving three variables X_t, Y_t and Z_t can be written as follows:

$$\begin{cases} X_t = \sum_{i=1}^{\infty} a_{2i} X_{t-i} + \sum_{i=1}^{\infty} b_{2i} Y_{t-i} + \sum_{i=1}^{\infty} c_{2i} Z_{t-i} + \vec{\varepsilon}_{3t} + \vec{\varepsilon}_{3t}^{E} + \overrightarrow{B_3(\mathcal{L})\varepsilon_{3t}^{L}} \\ Y_t = \sum_{i=1}^{\infty} d_{2i} X_{t-i} + \sum_{i=1}^{\infty} e_{2i} Y_{t-i} + \sum_{i=1}^{\infty} f_{2i} Z_{t-i} + \vec{\varepsilon}_{4t} + \vec{\varepsilon}_{4t}^{E} + \overrightarrow{B_4(\mathcal{L})\varepsilon_{4t}^{L}} \\ Z_t = \sum_{i=1}^{\infty} g_{2i} X_{t-i} + \sum_{i=1}^{\infty} h_{2i} Y_{t-i} + \sum_{i=1}^{\infty} k_{2i} Z_{t-i} + \vec{\varepsilon}_{5t} + \vec{\varepsilon}_{5t}^{E} + \overrightarrow{B_5(\mathcal{L})\varepsilon_{5t}^{L}} \end{cases} \tag{5.4}$$

The noise covariance matrix for the model can be represented as

$$\Sigma = \begin{bmatrix} \mathrm{var}(u_{3t}) & \mathrm{cov}(u_{3t}, u_{4t}) & \mathrm{cov}(u_{3t}, u_{5t}) \\ \mathrm{cov}(u_{4t}, u_{3t}) & \mathrm{var}(u_{4t}) & \mathrm{cov}(u_{4t}, u_{5t}) \\ \mathrm{cov}(u_{5t}, u_{3t}) & \mathrm{cov}(u_{5t}, u_{4t}) & \mathrm{var}(u_{5t}) \end{bmatrix} = \begin{bmatrix} \Sigma_{xx} & \Sigma_{xy} & \Sigma_{xz} \\ \Sigma_{yx} & \Sigma_{yy} & \Sigma_{yz} \\ \Sigma_{zx} & \Sigma_{zy} & \Sigma_{zz} \end{bmatrix}$$

In order to consider the influence from Y to X while controlling for the effect of the exogenous input, we consider the variance of u_{1t} when we eliminate the influence of u_{2t}:

$$\mathrm{cov}(u_{1t}, u_{1t}) - \mathrm{cov}(u_{1t}, u_{2t})\mathrm{cov}(u_{2t}, u_{2t})^{-1}\mathrm{cov}(u_{2t}, u_{1t}) = S_{xx} - S_{xz}S_{zz}^{-1}S_{zx}$$

Similarly, the variance of u_{3t} while eliminating the influence of u_{5t} equals to

$$\mathrm{cov}(u_{3t}, u_{3t}) - \mathrm{cov}(u_{3t}, u_{5t})\mathrm{cov}(u_{5t}, u_{5t})^{-1}\mathrm{cov}(u_{5t}, u_{3t}) = \Sigma_{xx} - \Sigma_{xz}\Sigma_{zz}^{-1}\Sigma_{zx}$$

The value of $S_{xx} - S_{xz}S_{zz}^{-1}S_{zx}$, a scalar, measures the accuracy of the autoregressive prediction of X based on its previous values conditioned on Z and eliminating the influence of the latent variables, whereas the value of $\Sigma_{xx} - \Sigma_{xz}\Sigma_{zz}^{-1}\Sigma_{zx}$, also a scalar, represents the accuracy of predicting present value of X based on the previous history of both X and Y, conditioned on Z and eliminating the influence of latent variables. Granger causality defines the causality from one process to another by comparing the improvement in prediction when the first process is taken into account. Similarly we define this causal influence by

$$F = \ln\left(\frac{|S_{xx} - S_{xz}S_{zz}^{-1}S_{zx}|}{|\Sigma_{xx} - \Sigma_{xz}\Sigma_{zz}^{-1}\Sigma_{zx}|}\right) \tag{5.5}$$

We call F *partial* Granger causality. Note that conditional Granger causality is defined by $F = \ln(\frac{|S_{xx}|}{|\Sigma_{xx}|})$. The essential difference between them is that with conditional Granger causality, the effect of latent variables remains present both in the denominator $|\Sigma_{xx}|$ and in the numerator $|S_{xx}|$. In contrast, partial Granger causality uses the conditional variance in both the denominator $|\Sigma_{xx} - \Sigma_{xz}\Sigma_{zz}^{-1}\Sigma_{zx}|$ and numerator $|S_{xx} - S_{xz}S_{zz}^{-1}S_{zx}|$. As a result, the effects of the latent and exogenous variables are both taken into account.

5.2.2 Numerical Example

Example 1 We simulated a 5-node oscillatory network structurally connected with different delays. In order to illustrate the robustness of partial Granger causality, we add exogenous inputs and latent variables to the model:

$$\begin{cases} x_1(t) = 0.95\sqrt{2}x_1(t-1) - 0.9025x_1(t-2) + \varepsilon_1(t) + a_1\varepsilon_6(t) \\ \qquad + b_1\varepsilon_7(t-1) + c_1\varepsilon_7(t-2) \\ x_2(t) = 0.5x_1(t-2) + \varepsilon_2(t) + a_2\varepsilon_6(t) + b_2\varepsilon_7(t-1) + c_2\varepsilon_7(t-2) \\ x_3(t) = -0.4x_1(t-3) + \varepsilon_3(t) + a_3\varepsilon_6(t) + b_3\varepsilon_7(t-1) + c_3\varepsilon_7(t-2) \\ x_4(t) = -0.5x_1(t-2) + 0.25\sqrt{2}x_4(t-1) + 0.25\sqrt{2}x_5(t-1) + \varepsilon_4(t) \\ \qquad + a_4\varepsilon_6(t) + b_4\varepsilon_7(t-1) + c_4\varepsilon_7(t-2) \\ x_5(t) = -0.25\sqrt{2}x_4(t-1) + 0.25\sqrt{2}x_5(t-1) + \varepsilon_5(t) + a_5\varepsilon_6(t) \\ \qquad + b_5\varepsilon_7(t-1) + c_5\varepsilon_7(t-2) \end{cases}$$

where $\varepsilon_i(t), i = 1, 2, \ldots, 7$, are zero-mean uncorrelated processes with identical variances, $a_i\varepsilon_6$ is the exogenous input, and the term $b_i\varepsilon_7(t-1) + c_i\varepsilon_7(t-2)$ represents the influence of latent variables. From the model (depicted in Fig. 5.2(A)) one can see that $x_1(t)$ is a direct source to $x_2(t)$, $x_3(t)$, and $x_4(t)$, $x_4(t)$ and $x_5(t)$ share a feedback loop. There is no direct connection between $x_1(t)$ and $x_5(t)$. We perform a simulation of this system with $a_i \sim U[0, 1]$, $b_i = 2$, $c_i = 5$, $i = 1, \ldots, 5$, to generate a data set of 2000 data points with a sample rate of 200 Hz.

A bootstrap is used to calculate 95% confidence intervals. Figure 5.2 (B, upper panel) shows the values for both partial Granger causality (F_1) and conditional

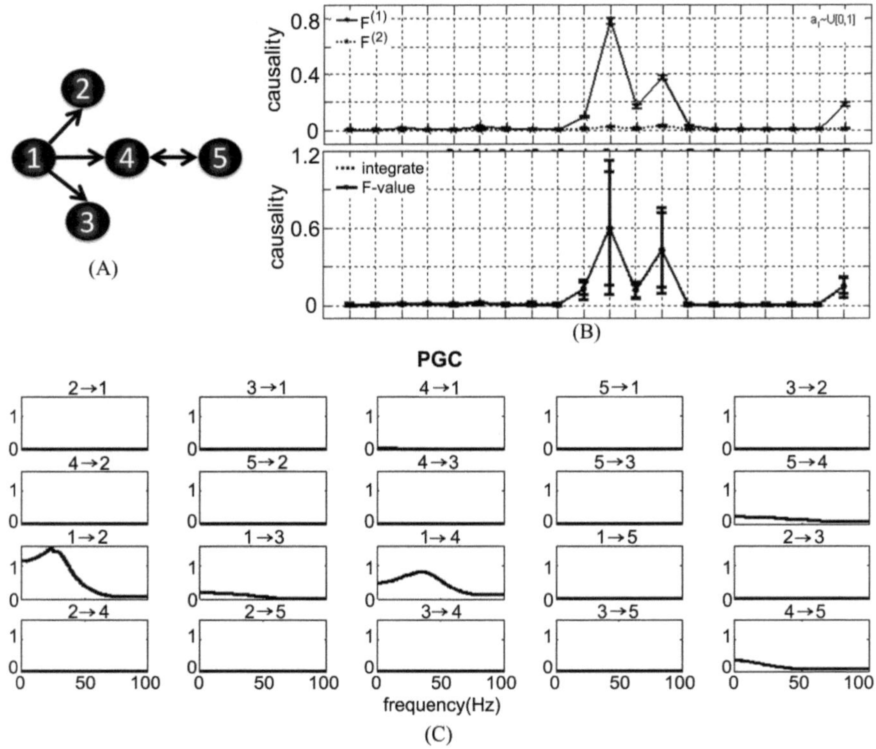

Fig. 5.2 Granger Causality applied to the system in Example 1. (**A**) The true network structure. (**B**) (*upper panel*) Comparison of the partial Granger causality F_1 and the conditional Granger causality F_2. F_2 fails to pick up any true connections, while the inferred links from F_1 are consistent with the correct structure (**A**). (*Bottom panel*) Comparison of the partial Granger causality in the time domain (*blue line*) and frequency domain (*red line*, the integral of the frequency domain formulation in the interval $[-\pi, \pi]$. (**C**) Results of the frequency domain decomposition of all 20 pairs of signals (for the color version, see Color Plates on p. 390)

Granger causality (F_2) when applied to the simulated data. Partial Granger causality outperforms the conditional Granger causality.

The values of the conditional Granger causality are all very small due to the dominating nuisance effect of latent variables and common inputs, while the partial Granger causality reveals the correct structure.

5.3 Frequency Analysis

Granger causality summarises the influence of one time series on another with a single nonnegative number. It is possible to extract more information about their connection by going to the frequency domain. Thanks to Geweke's decomposition

[20, 21], the power spectrum of the target signal can be written as a sum of easily interpretable quantities, leading to a natural definition of Granger causality in the frequency domain. Instead of a single number, a whole spectrum expliciting at which frequencies the signals interact is obtained. We present here a similar decomposition for the more recent partial Granger causality.

To derive the spectral decomposition of the time domain partial Granger causality, we first multiply the matrix

$$P_1 = \begin{pmatrix} 1 & -S_{xz}S_{zz}^{-1} \\ \mathbf{0} & I_m \end{pmatrix} \tag{5.6}$$

to both sides of (5.3). The normalised equations are represented as

$$\begin{pmatrix} D_{11}(L) & D_{12}(L) \\ D_{21}(L) & D_{22}(L) \end{pmatrix} \begin{pmatrix} X_t \\ Z_t \end{pmatrix} = \begin{pmatrix} X_t^* \\ Z_t^* \end{pmatrix} \tag{5.7}$$

with $D_{11}(0) = 1$, $D_{22}(0) = I_m$, $D_{21}(0) = \mathbf{0}$, $\operatorname{cov}(X_t^*, Z_t^*) = 0$, we note that $\operatorname{var}(X_t^*) = S_{xx} - S_{xz}S_{zz}^{-1}S_{zx}$, $\operatorname{var}(Z_t^*) = S_{zz}$. For (5.4), we also multiply the matrix

$$P = P_3 \cdot P_2 \tag{5.8}$$

where

$$P_2 = \begin{pmatrix} 1 & \mathbf{0} & -\Sigma_{xz}\Sigma_{zz}^{-1} \\ \mathbf{0} & 1 & -\Sigma_{yz}\Sigma_{zz}^{-1} \\ \mathbf{0} & \mathbf{0} & I_m \end{pmatrix} \tag{5.9}$$

and

$$P_3 = \begin{pmatrix} 1 & \mathbf{0} & \mathbf{0} \\ -(\Sigma_{xy} - \Sigma_{xz}\Sigma_{zz}^{-1}\Sigma_{zy})(\Sigma_{xx} - \Sigma_{xz}\Sigma_{zz}^{-1}\Sigma_{zx})^{-1} & 1 & \mathbf{0} \\ \mathbf{0} & \mathbf{0} & I_m \end{pmatrix} \tag{5.10}$$

to both sides of (5.4). The normalised equation of (5.4) becomes

$$\begin{pmatrix} B_{11}(L) & B_{12}(L) & B_{13}(L) \\ B_{21}(L) & B_{22}(L) & B_{23}(L) \\ B_{31}(L) & B_{32}(L) & B_{33}(L) \end{pmatrix} \begin{pmatrix} X_t \\ Y_t \\ Z_t \end{pmatrix} = \begin{pmatrix} \varepsilon_{xt} \\ \varepsilon_{yt} \\ \varepsilon_{zt} \end{pmatrix} \tag{5.11}$$

where $\varepsilon_{xt}, \varepsilon_{yt}, \varepsilon_{zt}$ are independent, with variances $\hat{\Sigma}_{xx}$, $\hat{\Sigma}_{yy}$ and $\hat{\Sigma}_{zz}$:

$$\begin{cases} \hat{\Sigma}_{zz} = \Sigma_{zz} \\ \hat{\Sigma}_{xx} = \Sigma_{xx} - \Sigma_{xz}\Sigma_{zz}^{-1}\Sigma_{zx} \\ \hat{\Sigma}_{yy} = \Sigma_{yy} - \Sigma_{yz}\Sigma_{zz}^{-1}\Sigma_{zy} - \dfrac{(\Sigma_{yx} - \Sigma_{yz}\Sigma_{zz}^{-1}\Sigma_{zx})(\Sigma_{xy} - \Sigma_{xz}\Sigma_{zz}^{-1}\Sigma_{zy})}{(\Sigma_{xx} - \Sigma_{xz}\Sigma_{zz}^{-1}\Sigma_{zx})} \end{cases}$$

After Fourier transforming (5.7) and (5.11), we can rewrite these two equations as the following expression:

$$\begin{pmatrix} X(\lambda) \\ Z(\lambda) \end{pmatrix} = \begin{pmatrix} G_{xx}(\lambda) & G_{xz}(\lambda) \\ G_{zx}(\lambda) & G_{zz}(\lambda) \end{pmatrix} \begin{pmatrix} X^*(\lambda) \\ Z^*(\lambda) \end{pmatrix} \tag{5.12}$$

and

$$\begin{pmatrix} X(\lambda) \\ Y(\lambda) \\ Z(\lambda) \end{pmatrix} = \begin{pmatrix} H_{xx}(\lambda) & H_{xy}(\lambda) & H_{xz}(\lambda) \\ H_{yx}(\lambda) & H_{yy}(\lambda) & H_{yz}(\lambda) \\ H_{zx}(\lambda) & H_{zy}(\lambda) & H_{zz}(\lambda) \end{pmatrix} \begin{pmatrix} E_x(\lambda) \\ E_y(\lambda) \\ E_z(\lambda) \end{pmatrix} \tag{5.13}$$

Note that $X(\lambda)$ and $Z(\lambda)$ from (5.12) are identical with those from (5.13), and we thus have

$$\begin{aligned} \begin{pmatrix} X^*(\lambda) \\ Y(\lambda) \\ Z^*(\lambda) \end{pmatrix} &= \begin{pmatrix} G_{xx}(\lambda) & 0 & G_{xz}(\lambda) \\ 0 & 1 & 0 \\ G_{zx}(\lambda) & 0 & G_{zz}(\lambda) \end{pmatrix}^{-1} \begin{pmatrix} H_{xx}(\lambda) & H_{xy}(\lambda) & H_{xz}(\lambda) \\ H_{yx}(\lambda) & H_{yy}(\lambda) & H_{yz}(\lambda) \\ H_{zx}(\lambda) & H_{zy}(\lambda) & H_{zz}(\lambda) \end{pmatrix} \\ &\quad \times \begin{pmatrix} E_x(\lambda) \\ E_y(\lambda) \\ E_z(\lambda) \end{pmatrix} \\ &= \begin{pmatrix} Q_{xx}(\lambda) & Q_{xy}(\lambda) & Q_{xz}(\lambda) \\ Q_{yx}(\lambda) & Q_{yy}(\lambda) & Q_{yz}(\lambda) \\ Q_{zx}(\lambda) & Q_{zy}(\lambda) & Q_{zz}(\lambda) \end{pmatrix} \begin{pmatrix} E_x(\lambda) \\ E_y(\lambda) \\ E_z(\lambda) \end{pmatrix} \end{aligned} \tag{5.14}$$

where $\mathbf{Q}(\lambda) = \mathbf{G}^{-1}(\lambda)\mathbf{H}(\lambda)$. Now the power spectrum of X^* is

$$S_{x^*x^*}(\lambda) = Q_{xx}(\lambda)\hat{\Sigma}_{xx}Q'_{xx}(\lambda) + Q_{xy}(\lambda)\hat{\Sigma}_{yy}Q'_{xy}(\lambda) + Q_{xz}(\lambda)\hat{\Sigma}_{zz}Q'_{xz}(\lambda) \tag{5.15}$$

where $'$ denotes the conjugate transpose. Note that $\hat{\Sigma}_{xx} = \Sigma_{xx} - \Sigma_{xz}\Sigma_{zz}^{-1}\Sigma_{zx}$; the first term of $S_{x^*x^*}$ can be thought of as the intrinsic power eliminating exogenous inputs and latent variables, while the remaining two terms as the combined causal influence from Y mediated by Z. This interpretation leads immediately to the definition

$$f_{Y\to X|Z}(\lambda) = \ln \frac{|S_{x^*x^*}(\lambda)|}{|Q_{xx}(\lambda)\hat{\Sigma}_{xx}Q'_{xx}(\lambda)|} \tag{5.16}$$

Note that according to (5.7), the variance of X^* equals $S_{xx} - S_{xz}S_{zz}^{-1}S_{zx}$. By the Kolmogorov formula [21] for spectral decompositions and under some mild conditions, the Granger causality in the frequency domain and in the time domain measures satisfy

$$F_{Y\to X|Z} = \frac{1}{2\pi}\int_{-\pi}^{\pi} f_{Y\to X|Z}(\lambda)\,d\lambda \tag{5.17}$$

Example 2 We apply the frequency analysis to the data presented in Example 1. Figure 5.2 (B, bottom panel) presents a comparison between the time domain partial Granger causality and the integrated frequency domain partial Granger causality (the summation over all frequencies). As expected from Kolmogorov formula, the decomposition in the frequency domain fits very well with the partial Granger causality in the time domain. Figure 5.2(C) shows the spectra for the partial Granger causality for all 20 pairs of signals. There are direct causal links from 1 to 2, 3 and 4, and a feedback between 5 and 4. Most importantly they are consistent with the results in the time domain.

This approach has been used with success on data where one would expect frequency to be important, e.g. electrophysiological experiments, but also on microarray or protein data [26, 53]. Section 5.5.4 presents an example on gene expression changes during circadian rhythm in a plant.

5.4 Group Interaction: Complex Granger Causality

So far, we have only considered Granger causality between two individual signals. This can be limiting for the study of biological systems, where cooperative and competitive actions are a frequent occurrence. For example, one would like to study the flow of information between brain regions rather than between individual neurons, or to elucidate transcription factor complexes (an AND-like combination of proteins) in the cell. In this section, we present *complex* Granger causality [29], a measure of causality between collections of time series. It can be considered the natural extension of partial Granger causality to the multidimensional case. However, we will see that cross-interaction between sources can now have an influence on the strength of the connection to a target group.

5.4.1 Time Domain Formulation

Consider three multiple stationary time series $\vec{X}_t$, $\vec{Y}_t$ and $\vec{Z}_t$ with k, l and m dimensions, respectively. We first consider the relationship from $\vec{Y}_t$ to $\vec{X}_t$ conditioned on $\vec{Z}_t$. The joint autoregressive representation for $\vec{X}_t$ and $\vec{Z}_t$ can be written as

$$\begin{cases} \vec{X}_t = \sum_{i=1}^{\infty} a_{1i}\vec{X}_{t-i} + \sum_{i=1}^{\infty} c_{1i}\vec{Z}_{t-i} + \vec{\varepsilon}_{1t} \\ \vec{Z}_t = \sum_{i=1}^{\infty} b_{1i}\vec{Z}_{t-i} + \sum_{i=1}^{\infty} d_{1i}\vec{X}_{t-i} + \vec{\varepsilon}_{2t} \end{cases} \tag{5.18}$$

where $\vec{\varepsilon}_t$ are vectors representing exogenous and endogenous inputs and noise. The noise covariance matrix for the system can be represented as

$$S = \begin{pmatrix} \mathrm{var}(\vec{\varepsilon}_{1t}) & \mathrm{cov}(\vec{\varepsilon}_{1t}, \vec{\varepsilon}_{2t}) \\ \mathrm{cov}(\vec{\varepsilon}_{2t}, \vec{\varepsilon}_{1t}) & \mathrm{var}(\vec{\varepsilon}_{2t}) \end{pmatrix} = \begin{pmatrix} S_{xx} & S_{xz} \\ S_{zx} & S_{zz} \end{pmatrix}$$

where var and cov represent variance and covariance, respectively. In the same manner, the vector autoregressive representation for the system involving the three time series $\vec{X}_t$, $\vec{Y}_t$ and $\vec{Z}_t$ can be written in the following way:

$$\begin{cases} \vec{X}_t = \sum_{i=1}^{\infty} a_{2i}\vec{X}_{t-i} + \sum_{i=1}^{\infty} b_{2i}\vec{Y}_{t-i} + \sum_{i=1}^{\infty} c_{2i}\vec{Z}_{t-i} + \vec{\varepsilon}_{3t} \\ \vec{Y}_t = \sum_{i=1}^{\infty} d_{2i}\vec{X}_{t-i} + \sum_{i=1}^{\infty} e_{2i}\vec{Y}_{t-i} + \sum_{i=1}^{\infty} f_{2i}\vec{Z}_{t-i} + \vec{\varepsilon}_{4t} \\ \vec{Z}_t = \sum_{i=1}^{\infty} g_{2i}\vec{X}_{t-i} + \sum_{i=1}^{\infty} h_{2i}\vec{Y}_{t-i} + \sum_{i=1}^{\infty} k_{2i}\vec{Z}_{t-i} + \vec{\varepsilon}_{5t} \end{cases} \tag{5.19}$$

The noise covariance matrix for the above system can be represented as

$$\Sigma = \begin{pmatrix} \mathrm{var}(\vec{\varepsilon}_{3t}) & \mathrm{cov}(\vec{\varepsilon}_{3t}, \vec{\varepsilon}_{4t}) & \mathrm{cov}(\vec{\varepsilon}_{3t}, \vec{\varepsilon}_{5t}) \\ \mathrm{cov}(\vec{\varepsilon}_{4t}, \vec{\varepsilon}_{3t}) & \mathrm{var}(\vec{\varepsilon}_{4t}) & \mathrm{cov}(\vec{\varepsilon}_{4t}, \vec{\varepsilon}_{5t}) \\ \mathrm{cov}(\vec{\varepsilon}_{5t}, \vec{\varepsilon}_{3t}) & \mathrm{cov}(\vec{\varepsilon}_{5t}, \vec{\varepsilon}_{4t}) & \mathrm{var}(\vec{\varepsilon}_{5t}) \end{pmatrix} = \begin{bmatrix} \Sigma_{xx} & \Sigma_{xy} & \Sigma_{xz} \\ \Sigma_{yx} & \Sigma_{yy} & \Sigma_{yz} \\ \Sigma_{zx} & \Sigma_{zy} & \Sigma_{zz} \end{bmatrix}$$

The conditional variance $S_{xx} - S_{xz}S_{zz}^{-1}S_{zx}$ measures the accuracy of the autoregressive prediction of $\vec{X}$ based on its previous values conditioned on $\vec{Z}$, whereas the conditional variance $\Sigma_{xx} - \Sigma_{xz}\Sigma_{zz}^{-1}\Sigma_{zx}$ measures the accuracy of the autoregressive prediction of $\vec{X}$ based on its previous values of both $\vec{X}$ and $\vec{Y}$ conditioned on $\vec{Z}$. Note that now both $S_{xx} - S_{xz}S_{zz}^{-1}S_{zx}$ and $\Sigma_{xx} - \Sigma_{xz}\Sigma_{zz}^{-1}\Sigma_{zx}$ are matrices and not scalars. We denote $T_{x|z}$ and $T_{xy|z}$ their respective traces, which we use to compare their relative size. Following the original concept of Granger causality, we define the partial complex Granger causality from group $\vec{Y}$ to group $\vec{X}$ conditioned on group $\vec{Z}$ to be

$$F_{\vec{Y}\to\vec{X}|\vec{Z}} = \ln\left(\frac{T_{x|z}}{T_{xy|z}}\right) \tag{5.20}$$

If Y and X are one-dimensional, the definition reduces to that of partial Granger causality. Partial Complex causality has the same property of removing the influence of explicit and unseen variables from the connection from Y to X.

5.4.2 Frequency Domain Formulation

The derivation of the frequency domain formulation follows the steps seen in Sect. 5.3, with the minor difference that the trace replaces the absolute value.

We first normalise (5.18) by multiplying the matrix

$$P_1 = \begin{pmatrix} I_k & -S_{xz}S_{zz}^{-1} \\ \mathbf{0} & I_m \end{pmatrix} \tag{5.21}$$

to both sides of it. The normalised equations are represented as

$$\begin{pmatrix} D_{11}(L) & D_{12}(L) \\ D_{21}(L) & D_{22}(L) \end{pmatrix} \begin{pmatrix} \vec{X}_t \\ \vec{Z}_t \end{pmatrix} = \begin{pmatrix} \vec{X}_t^* \\ \vec{Z}_t^* \end{pmatrix} \tag{5.22}$$

where $\mathrm{var}(\vec{X}_t^*) = S_{xx} - S_{xz}S_{zz}^{-1}S_{zx}$, $\mathrm{var}(\vec{Z}_t^*) = S_{zz}$. For (5.19), we also multiply the matrix

$$P = P_3 \cdot P_2 \tag{5.23}$$

where

$$P_2 = \begin{pmatrix} I_k & \mathbf{0} & -\Sigma_{xz}\Sigma_{zz}^{-1} \\ \mathbf{0} & I_l & -\Sigma_{yz}\Sigma_{zz}^{-1} \\ \mathbf{0} & \mathbf{0} & I_m \end{pmatrix} \tag{5.24}$$

and

$$P_3 = \begin{pmatrix} I_k & \mathbf{0} & \mathbf{0} \\ -(\Sigma_{yx} - \Sigma_{yz}\Sigma_{zz}^{-1}\Sigma_{zx})(\Sigma_{xx} - \Sigma_{xz}\Sigma_{zz}^{-1}\Sigma_{zx})^{-1} & I_l & \mathbf{0} \\ \mathbf{0} & \mathbf{0} & I_m \end{pmatrix} \tag{5.25}$$

to both sides of (5.19). The normalised equation of (5.19) becomes

$$\begin{pmatrix} B_{11}(L) & B_{12}(L) & B_{13}(L) \\ B_{21}(L) & B_{22}(L) & B_{23}(L) \\ B_{31}(L) & B_{32}(L) & B_{33}(L) \end{pmatrix} \begin{pmatrix} \vec{X}_t \\ \vec{Y}_t \\ \vec{Z}_t \end{pmatrix} = \begin{pmatrix} \vec{\varepsilon}_{xt} \\ \vec{\varepsilon}_{yt} \\ \vec{\varepsilon}_{zt} \end{pmatrix} \tag{5.26}$$

where $\vec{\varepsilon}_{xt}$, $\vec{\varepsilon}_{yt}$, $\vec{\varepsilon}_{zt}$ are independent, and their variances being $\hat{\Sigma}_{xx}$, $\hat{\Sigma}_{yy}$ and $\hat{\Sigma}_{zz}$ with

$$\begin{cases} \hat{\Sigma}_{zz} = \Sigma_{zz} \\ \hat{\Sigma}_{xx} = \Sigma_{xx} - \Sigma_{xz}\Sigma_{zz}^{-1}\Sigma_{zx} \\ \hat{\Sigma}_{yy} = \Sigma_{yy} - \Sigma_{yz}\Sigma_{zz}^{-1}\Sigma_{zy} - \dfrac{(\Sigma_{yx} - \Sigma_{yz}\Sigma_{zz}^{-1}\Sigma_{zx})(\Sigma_{xy} - \Sigma_{xz}\Sigma_{zz}^{-1}\Sigma_{zy})}{(\Sigma_{yy} - \Sigma_{yz}\Sigma_{zz}^{-1}\Sigma_{zy})} \end{cases}$$

As shown in Sect. 5.3, we can obtain the power spectrum of $\vec{X}_t^*$

$$S_{x^*x^*}(\omega) = Q_{xx}(\omega)\hat{\Sigma}_{xx}Q_{xx}^*(\omega) + Q_{xy}(\omega)\hat{\Sigma}_{yy}Q_{xy}^*(\omega) + Q_{xz}(\omega)\hat{\Sigma}_{zz}Q_{xz}^*(\omega) \tag{5.27}$$

Considering the traces of both sides of (5.27), we have:

$$\begin{aligned} \mathrm{tr}\big(S_{x^*x^*}(\omega)\big) &= \mathrm{tr}\big(Q_{xx}(\omega)\hat{\Sigma}_{xx}Q_{xx}^*(\omega)\big) + \mathrm{tr}\big(Q_{xy}(\omega)\hat{\Sigma}_{yy}Q_{xy}^*(\omega)\big) \\ &\quad + \mathrm{tr}\big(Q_{xz}(\omega)\hat{\Sigma}_{zz}Q_{xz}^*(\omega)\big) \end{aligned} \tag{5.28}$$

As before, we can think of the first term as the intrinsic power while eliminating exogenous inputs and latent variables, and the remaining two terms as the combined influence from $\vec{Y}$ mediated by $\vec{Z}$.

This interpretation leads immediately to the definition

$$f_{\vec{Y}\to\vec{X}|\vec{Z}}(\omega) = \ln \frac{\mathrm{tr}(S_{x^*x^*}(\omega))}{\mathrm{tr}(Q_{xx}(\omega)\hat{\Sigma}_{xx}Q^*_{xx}(\omega))} \tag{5.29}$$

Note that according to (5.20), the variance of X^* equals $\Sigma_{xx} - \Sigma_{xz}\Sigma_{zz}^{-1}\Sigma_{zx}$. By the Kolmogorov formula [21] for spectral decompositions and under the same mild conditions, the Granger causality in the frequency domain and in the time domain measures satisfy

$$F_{\vec{Y}\to\vec{X}|\vec{Z}} = \frac{1}{2\pi}\int_{-\pi}^{\pi} f_{\vec{Y}\to\vec{X}|\vec{Z}}(\omega)\, d\omega \tag{5.30}$$

5.4.3 Effect of Correlation Between Sources

The complex Granger causality between a group and a target signal can be affected by the source signals' cross-correlations. Let us consider a model where y_i, $i = 1, 2, \ldots, N$, are identical random processes. The Granger causality from $(y_i(t),\ i = 1, \ldots, N)$ to their weighted sum $y(t) := a\sum_{i=1}^{N} y_i(t) + \varepsilon_t$ is $\log(1 + a^2N(1 + \rho(N-1)))$ where ρ is the correlation coefficient between y_i's and ε_t is normally distributed. Figure 5.3 illustrates how the complex interaction depends on the correlation. If the original signals are not correlated (black dashed line), taken as group, they have increasingly higher interaction with y with the number of units. But this interaction is always higher the more positively cross-correlated they are. Conversely, negative cross-correlation reduces the interaction, all the way down to zero even though the target signal y is made up of each of these signals by construction. Collaborative activity enhances the interaction, but antagonistic activity reduces or even suppresses the interaction.

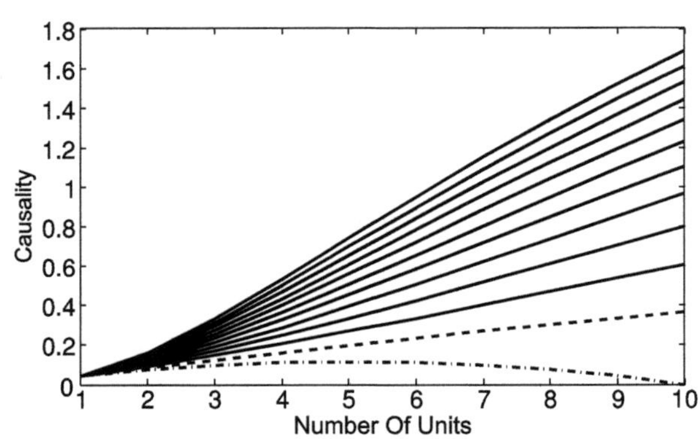

Fig. 5.3 The role of correlation in the complex interaction. The Granger causality vs. units (N) for different cross-correlation coefficients with $a = 0.022$. ($\rho = -1/9$ is the smallest possible value for the cross-correlations of 10 units)

5.5 Harmonic Granger Causality

With partial Granger causality, the external influence is assumed present but not known in explicit form. However, there are situations in which a model of an external factor is available, for example the oscillating amount of sunlight received by an organism. In this section, we consider the inclusion of a harmonic oscillator in the formulation of Granger causality. This new term represents a periodically changing environmental input (e.g. sunlight, pH, temperature), and since its contribution to the prediction is explicit, it is possible to elucidate its Granger influence on each of the observed time series.

This extension should be amenable to a more general class of models: any function can be written as a sum of periodic functions through Fourier analysis [27, 44]. Moreover, it should be possible to define it for more than two time series and use it like partial Granger causality. Both aspects are the subject of further study.

5.5.1 Time Domain Formulation

As we have seen in the previous sections, the core idea of Granger analysis is to measure how much the prediction of a target time series has improved with the knowledge of a possible source. The innovation of harmonic Granger causality is to have two possible sources of influences—another observed time series and the oscillator—which leads to two quantities of interest.

Consider a target time series X, a possible source Y and the periodic environmental factor E. As usual, we use an autoregressive model to predict X, but this time we include different combinations of contributions:

$$\begin{cases} X_t = \sum_{i=1}^{\infty} a_{1i} X_{t-i} + C_{1x} \cos(2\pi f_{1x} t + \phi_x) + \varepsilon_{1t} \\ X_t = \sum_{i=1}^{\infty} a_{2i} X_{t-i} + \sum_{i=1}^{\infty} b_{2i} Y_{t-i} + C_{2x} \cos(2\pi f_{2x} t + \phi_x) + \varepsilon_{3t} \\ X_t = \sum_{i=1}^{\infty} a_{3i} X_{t-i} + \sum_{i=1}^{\infty} b_{3i} Y_{t-i} + \varepsilon_{5t} \end{cases} \quad (5.31)$$

We denote Σ_i the variance of each error term ε_{it}. The only difference between the first and second equation is the inclusion of Y in the model. Therefore, we can naturally define the harmonic Granger causality from Y to X as

$$F_{Y \to X} = \ln \frac{\Sigma_1}{\Sigma_3}$$

Similarly, the only difference between the second and the third equation is the inclusion of the oscillator. We define the harmonic Granger causality from E to X

as

$$F_{E \to X} = \ln \frac{\Sigma_5}{\Sigma_3}$$

These newly defined quantities behave in the same manner as before: if Y (resp. E) does not contribute to X, $b_{2i} = 0$ and $\varepsilon_{3t} = \varepsilon_{1t}$, which implies $F_{Y \to X} = 0$. Otherwise, $\Sigma_1 > \Sigma_3$, since the fit has more degrees of freedom, and $F_{Y \to X} > 0$.

5.5.2 Two Remarks About Harmonic Granger Causality

1. The addition of the harmonic term in the causality analysis is motivated by the fact that signals from experimental data are often periodic. A closer model would include more than one oscillator, to take into account a larger number of background periodic influences. This Fourier-like method could lead to a more accurate picture of the network structure.
2. Harmonic Granger causality requires the fitting of three autoregressive models, whose parameters have to be estimated. When the model does not feature an harmonic term, a usual least-square fitting or classical more sophisticated techniques are sufficient and can easily be implemented [12, 33, 36]. When the autoregressive model includes an harmonic term, a two-step procedure can improve parameter estimation. First, we identify (manually or automatically) the dominating oscillation present in X by looking at its spectrum after Fourier transformation. The dominating frequency is filtered out to produce a new $\tilde{X}$ lacking this frequency, which is then fitted using usual methods [27, 44].

5.5.3 Frequency Domain Formulation

Harmonic Granger causality has an equivalent formulation in the frequency domain, whereby we can obtain the causality spectra showing the frequencies at which the influence of one node is exerted on another. Deriving the frequency domain formulation requires some mathematical manipulations in order to decompose the spectrum into clearly separated expressions.

To begin with, we rewrite the following auto-regressive model:

$$\begin{cases} X_t = \sum_{i=1}^{\infty} a_{2i} X_{t-i} + \sum_{i=1}^{\infty} b_{2i} Y_{t-i} + C_{2x} \cos(2\pi f_{2x} t + \phi_x) + \varepsilon_{3t} \\ Y_t = \sum_{i=1}^{\infty} c_{2i} Y_{t-i} + \sum_{i=1}^{\infty} d_{2i} Y_{t-i} + C_{2y} \cos(2\pi f_{2y} t + \phi_y) + \varepsilon_{4t} \end{cases} \tag{5.32}$$

in terms of the lag operator

$$\begin{pmatrix} D_{11}(L) & D_{12}(L) \\ D_{21}(L) & D_{22}(L) \end{pmatrix} \begin{pmatrix} X_t + O_t^x \\ Y_t + O_t^y \end{pmatrix} = \begin{pmatrix} \varepsilon_{3t} \\ \varepsilon_{4t} \end{pmatrix} \tag{5.33}$$

where O_t^x and O_t^y represent the oscillator added to X and Y, respectively. We first normalise (5.33) by multiplying the matrix

$$P = \begin{pmatrix} 1 & 0 \\ -\frac{\Sigma_{34}}{\Sigma_3} & 1 \end{pmatrix} \tag{5.34}$$

to both sides of it, where $\Sigma_{34} = \mathrm{cov}(\varepsilon_{3t}, \varepsilon_{4t})$, $\Sigma_3 = \mathrm{var}(\varepsilon_{3t})$, The normalised equations are represented as

$$\begin{pmatrix} \tilde{D}_{11}(L) & \tilde{D}_{12}(L) \\ \tilde{D}_{21}(L) & \tilde{D}_{22}(L) \end{pmatrix} \begin{pmatrix} X_t + O_t^x \\ Y_t + O_t^y \end{pmatrix} = \begin{pmatrix} \tilde{\varepsilon}_{3t} \\ \tilde{\varepsilon}_{4t} \end{pmatrix} \tag{5.35}$$

where $\tilde{\varepsilon}_{3t}$ and $\tilde{\varepsilon}_{4t}$ are now independent. Fourier-transforming both sides of (5.35) leads to

$$\begin{bmatrix} X(\omega) + O_x(\omega) \\ Y(\omega) + O_y(\omega) \end{bmatrix} = \begin{bmatrix} H_{xx}(\omega) & H_{xy}(\omega) \\ H_{yx}(\omega) & H_{yy}(\omega) \end{bmatrix} \begin{bmatrix} E_x(\omega) \\ E_y(\omega) \end{bmatrix} \tag{5.36}$$

where the transfer function is $H(\omega) = \tilde{\mathbf{D}}^{-1}(\omega)$. We can also rewrite (5.36) in the following expression:

$$\begin{bmatrix} X(\omega) \\ Y(\omega) \end{bmatrix} = \begin{bmatrix} H_{xx}(\omega) & H_{xy}(\omega) \\ H_{yx}(\omega) & H_{yy}(\omega) \end{bmatrix} \begin{bmatrix} E_x(\omega) \\ E_y(\omega) \end{bmatrix} + \begin{bmatrix} \tilde{O}_x(\omega) \\ \tilde{O}_y(\omega) \end{bmatrix} \tag{5.37}$$

where

$$\begin{bmatrix} \tilde{O}_x(\omega) \\ \tilde{O}_y(\omega) \end{bmatrix} = - \begin{bmatrix} O_x(\omega) \\ O_y(\omega) \end{bmatrix}$$

It can now be seen that X is defined as follows:

$$X(\omega) = H_{xx}(\omega)E_x(\omega) + H_{xy}(\omega)E_y(\omega) + \tilde{O}_x(\omega) \tag{5.38}$$

To obtain the frequency decomposition of the time domain causality, we look at the auto-spectrum of X_t:

$$\begin{aligned} S_{xx}(\omega) &= X(\omega)X^*(\omega) \\ &= H_{xx}E_xE_x^*H_{xx}^* + \tilde{O}_x\tilde{O}_x^* + H_{xx}E_x\tilde{O}_x^* + \tilde{O}_xE_x^*H_{xx}^* \\ &\quad + H_{xy}E_y\tilde{O}_x^* + H_{xy}E_yE_y^*H_{xy}^* + \tilde{O}_xE_y^*H_{xy}^* \\ &= S_1 + S_2 \end{aligned} \tag{5.39}$$

$S_1 = H_{xx}E_xE_x^*H_{xx}^* + \tilde{O}_x\tilde{O}_x^* + H_{xx}E_x\tilde{O}_x^* + \tilde{O}_xE_x^*H_{xx}^*$, viewed as intrinsic part, involves only the variance of ε_{3t}, which is the noise term from the model for X_t. $S_2 = H_{xy}E_y\tilde{O}_x^* + H_{xy}E_yE_y^*H_{xy}^* + \tilde{O}_xE_y^*H_{xy}^*$, viewed as the causal part, involves only the variance of ε_{4t}, which is the noise term from the model for Y_t.

Note that if the harmonic term O_x is not present, there are only two terms in the expression for S_{xx}; it is consistent with the frequency decomposition of the time domain pairwise causality [12].

Finally, we can define the causal influence from Y_t to X_t at frequency ω as

$$f_{Y \to X}(\omega) = \ln\left(\frac{S_{xx}(\omega)}{S_1(\omega)}\right) \tag{5.40}$$

5.5.4 A Circadian Circuit

Next we show a biological example to further confirm our partial complex Granger causality approach and harmonic Granger causality approach. We collected microarray data of Arabidopsis leaves of 32,448 genes that were observed over 11 days. The plants are grown in laboratory conditions, where they are subjected to 12 hours of artificial daylight followed by 12 hours of no light representing night time. Gene microarray data is collected at regular intervals (twice a day) throughout the experiment, so the data length is 22. A circadian circuit has been reported in the literature [13, 30, 31]. The circuit comprises of 8 genes: PRR5, PRR7, PRR9, ELF4, LHY, CCA1, TOC1 and GI. The time domain trace of the expression of these genes is shown in Fig. 5.5(A). Each of the genes with the exception of GI exhibits highly oscillatory behaviour with a time period of one day. This periodicity is attributed to the presence of incident sunlight during the day time and its absence during the night.

By calculating node-to-node partial Granger causality between single node, we find the gene circuit as plotted in Fig. 5.4(A). GI is an isolated gene in our structure, without having any interactions with other six genes. In fact, this also coincides with the experimental findings. On p. 4 of [30], it is mentioned that *The GI single mutant had a relatively weak phenotype, whereas our assays of the triple GI*; *LHY*; *CCA1 mutant demonstrate GI's importance*. We thus turn our attention to partial complex Granger causality. Figure 5.4(B) tells us that all single genes ELF4, TOC1, LHY, CCA1 and (LHY, CCA1) have very little influence on GI. However, ELF4, TOC1, LHY and CCA1 together exhibit a significant interaction with GI, which is in agreement with the experimental finding. We then analyse the interactions in the frequency domain, the detailed results are shown in Fig. 5.4(C). No surprisingly, most of the interactions show a periodic behaviour by exhibiting a peak at 11 day period.

Due to the strong oscillatory behaviour of the data shown in Fig. 5.5(A), in the next step, we use harmonic Granger causality to analyse it. The task regarding this data set is twofold. Firstly we wish to identify which of the genes are driven by the external oscillation, and secondly, we wish to determine how the genes are connected to form the network governing flowering of the plant [31, 41, 43]. The method to determine environmental input and network connectivity is as follows. There are 56 pairwise combinations possible with eight genes; for each of these 56 gene pairs, the parameters of three candidate models as (5.31) meaning with and without the oscillation term and with and without the causal term respectively are

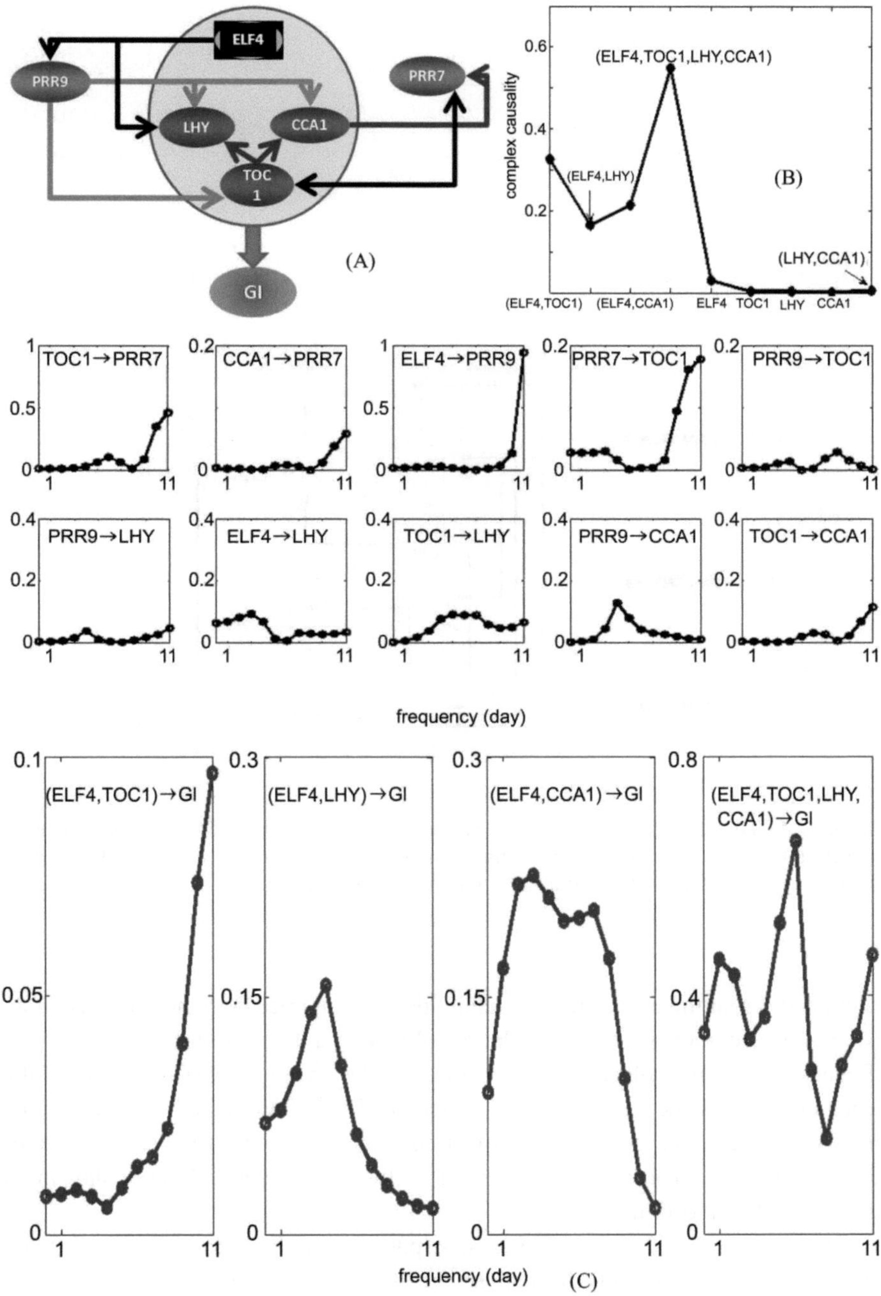

Fig. 5.4 Network of a circadian circuit in plant Arabidopsis leaf. (**A**) The gene circuit obtained in terms of partial Granger causality, GI has not any interactions with other six genes when the relationships between single genes are considered. However, ELF4, TOC1, LHY, CCA1 together exhibit a significant interaction with GI. (**B**) Complex interactions between different group of genes and GI. (**C**) Gene interactions in the frequency domain

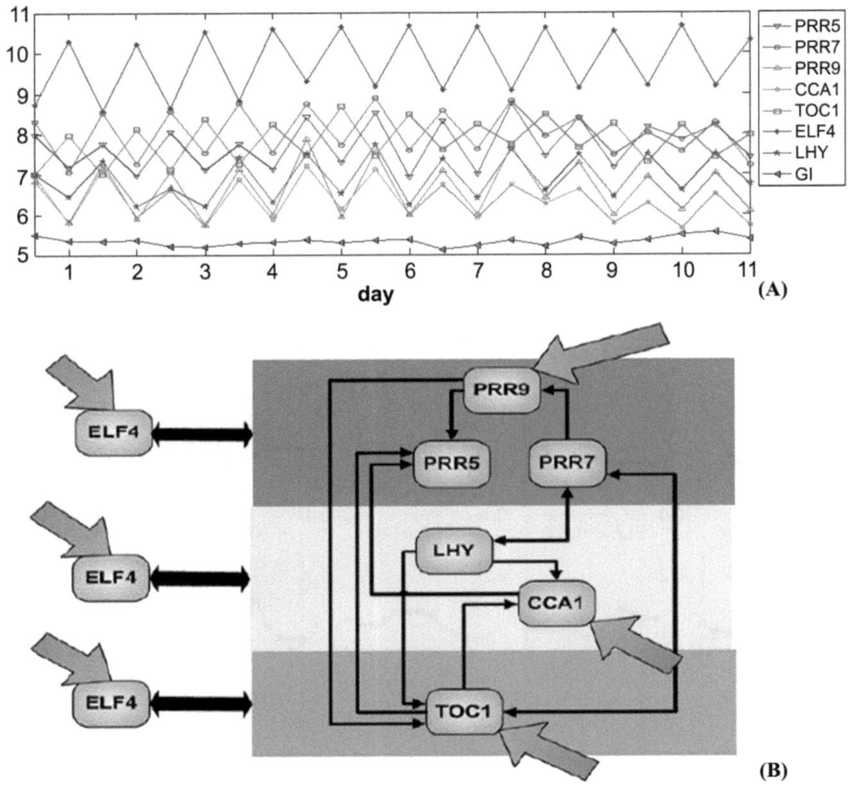

Fig. 5.5 (**A**) Time domain traces of gene expression of eight genes. (**B**) Network of a circadian circuit in plant Arabidopsis leaf using harmonic Granger causality approach. Four genes including PRR9, CCA1, TOC1 and ELF4 receive external inputs

calculated. From the error term of each model we can infer both the presence of an external environmental driver and the possibility of a connection between the pair of genes. The final results are shown in Fig. 5.5(B). The plot reveals that four of the genes in this network receive external inputs: PRR9, CCA1, TOC1 and ELF4. The first two of these four genes agree with Ueda's network [45, 46]. The structure of Ueda's network is also very close to the structure of our network, both showing a high level of connectivity.

Here we use two different approaches: partial complex Granger causality and harmonic Granger causality to analyse the same data, the two structures shown in Figs. 5.4(A) and 5.5(B) are also different. The reason is that the network in Fig. 5.4(A) is inferred from partial Granger causality, while network Fig. 5.5(B) is inferred from pairwise Granger causality. More reliable network would be constructed by combining both of these approaches, which is our further topic to be studied.

5.6 A Comparative Study Between Granger Causality and Bayesian Network

A popular approach for building a causal network is the use of Bayesian networks [6, 34, 35, 39]. Bayesian networks are based on the concept of conditional probability and are part of a class of probabilistic graphical models. While having sound theoretical foundations, they suffer some limitations, amongst which: a same set of probability distributions can have multiple graphical representations, and all networks are directed acyclic graphs, which precludes feedback loops. The latter can be addressed by Dynamic Bayesian Networks [51], which consider how the data changes over time. The usual approach for the estimation of a Bayesian network is to decide on a scoring function (the likelihood of observing the data given a network structure) and search for the best candidate in the space of possible graphs. Given the size of this space (exponential in the number of nodes), one usually relies on sampling methods like Markov-Chain Monte Carlo [5], a long and computer intensive process.

In this section, we compare the performances of Granger causality and Bayesian Network on a simulated dataset of known structure in order to investigate the relative merits of the two methods. As a benchmark, we use a modified version of the system from Example 1 where all exogenous inputs and latent variables have been removed:

Example 3

$$\begin{cases} x_1(t) = 0.95\sqrt{2}x_1(t-1) - 0.9025x_1(t-2) + \varepsilon_1(t) \\ x_2(t) = 0.5x_1(t-2) + \varepsilon_2(t) \\ x_3(t) = -0.4x_1(t-3) + \varepsilon_3(t) \\ x_4(t) = -0.5x_1(t-2) + 0.25\sqrt{2}x_4(t-1) + 0.25\sqrt{2}x_5(t-1) + \varepsilon_4(t) \\ x_5(t) = -0.25\sqrt{2}x_4(t-1) + 0.25\sqrt{2}x_5(t-1) + \varepsilon_5(t) \end{cases}$$

The experiment goes as follows: applied to the same data, both Granger causality and a dynamic Bayesian network produce a tentative network. The operation is done 100 times, and a final network from each approach is built with edges that appear at least 95% of the time. The sample size (or number of time points) varies from 1000 down to 20, in order to investigate its impact on accuracy. Figure 5.6 shows the performances of the two methods, with Bayesian networks on the right-hand side and Granger Causality on the left-hand side.

From this experiment we find that both approaches can reveal correct network structures for the data with a large sample size (1000 here). As one would expect, the accuracy decreases with the size of the data, with more and more links gone missing but the two methods perform identically. However, at $n = 20$, Granger

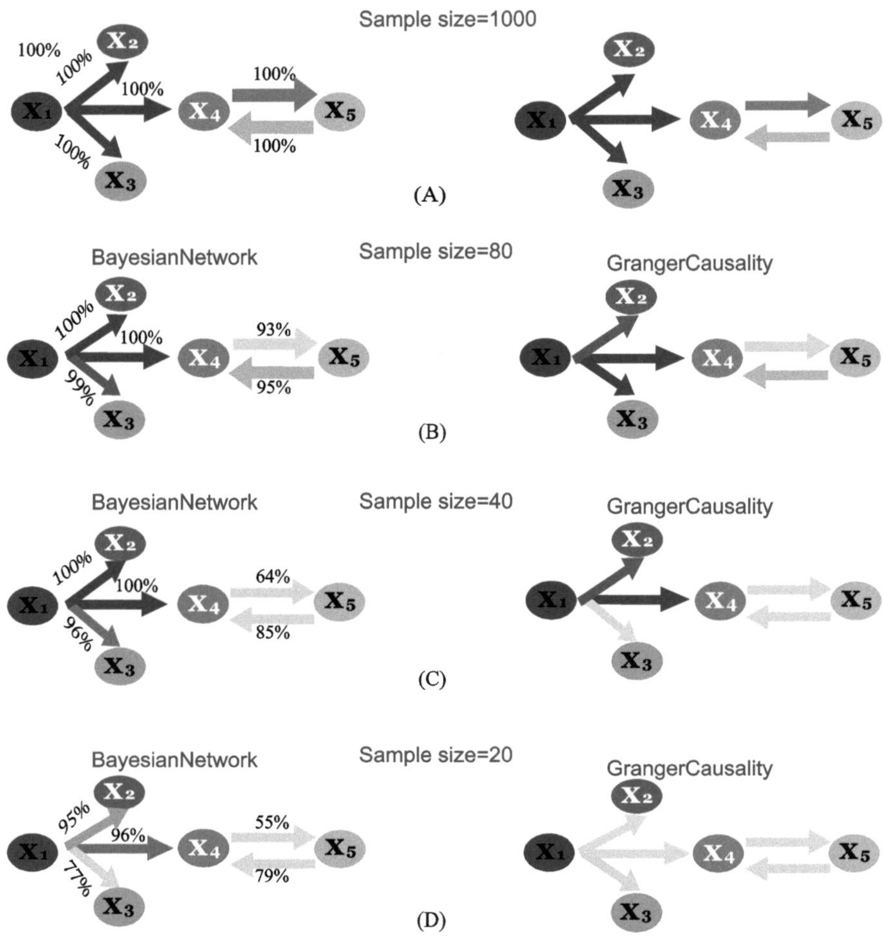

Fig. 5.6 Networks inferred with Bayesian Network and Granger causality for various sample sizes. *Grey edges* indicate undetected causalities (false negatives). For each sample size n, we simulated a data set of 100 realizations of n time points. High-confidence arcs, appearing in at least 95% of the 100 networks are shown. (**A**) The sample size is 1000. (**B**) The sample size is 80. (**C**) The sample size is 40. (**D**) The sample size is 20

causality lacks all correct links, while the Bayesian network still features true positives ($X_1 \rightarrow X_4$ and $X_1 \rightarrow X_2$). The reason for the poorer performance of Granger causality is that at small sample size the linear fit becomes less constrained, which makes the Granger coefficient unstable.

In conclusion, both Granger causality and Bayesian Networks are sensitive to sample size [52]. With long enough time series, they are as accurate as each other, and the choice for one or the other should be done on other parameters, like computational requirements (light for Granger causality, heavy for Bayesian networks) or inclusion of prior knowledge (natural in a Bayesian setting).

5.7 Unified Causal Model (UCM)

A successful approach for the study of brain region communication is the Dynamic Causal Model (DCM, [16, 17]). DCM uses a model of how the data were generated, equating the observed variables to the result of a convolution of a linear combination of latent variables. In a functional magnetic neuroimaging context, the convolution captures the varying delays on the blood flow (haemodynamics response). Moreover, the model includes the effect of a deterministic input—for example a stimulus—on the signal dynamics. A causal inference can be drawn from fitting the model (and other competing models) and estimating the likelihood of observing the data. Based on these two properties, DCM seems radically different from the Granger Causality Model (GCM). GCM does not impose a model but is purely based on the data statistics. And GCM does not natively incorporate the notion of input—although some extensions do, as seen in this chapter, e.g. Harmonic Granger causality, Sect. 5.5.

The Unified Causal Model is an attempt at reconciling the two approaches, whose mathematical formulations are actually quite close. We first recall the definition of DCM:

$$\begin{aligned} \frac{dX}{dt} &= (A + u_t B)X_t + u_t C \\ Y_t &= g(X) + \varepsilon_t \end{aligned} \tag{5.41}$$

where X represents brain region states and are unobserved. Matrices A, B and C model which interactions take place and to what extent and a deterministic input is represented by u_t. The function g is the convolution capturing the physiological effect of the blood flow. The observations Y_t are used to estimate the validity of the model. The traditional Granger causality has to be modified in such a way that it includes the deterministic input in a very similar fashion. We write the two modified auto-regressive models as

$$\begin{aligned} X_t &= \sum_{i=1} (a_{1i} + b_{1i} u_{t-i}) X_{t-i} + c_1 v_{t-1} + \varepsilon_{1t} \\ X_t &= \sum_{i=1} (a_{2i} + b_{2i} u_{t-i}) X_{t-i} + c_2 v_{t-1} + \sum_{i=1} (d_{2i} + e_{2i} u_{t-i}) Y_{t-i} + \varepsilon_{2t} \end{aligned} \tag{5.42}$$

where both X_t and Y_t are two observed signals. As usual, we write X_t first in terms of its past, then adding the knowledge of Y_t. The definition of the Granger causality is the measure of the prediction improvement:

$$F_{Y \to X} = \ln \frac{\mathrm{var}(\varepsilon_1)}{\mathrm{var}(\varepsilon_2)} \tag{5.43}$$

Thus modified, Granger causality is able to use known input signals in a way similar to that of DCM. It is also possible to derive a frequency domain formulation of the Granger causality. This method has been applied on local field potential recordings in sheep's brain and demonstrated learning induced changes in inter- and intra-hemispheric connectivity [19].

5.8 Large Networks

So far, the focus has been put on relatively small networks, with 20 nodes at most. The reason for this is purely numerical: with a limited number of time points, as is the case with experimental data, the linear auto-regressive model fit rapidly becomes under-constrained as the number of signals increases and the solution ceases to be unique. The resulting Granger causality, defined as a ratio, is unstable and unreliable. For example, a partial Granger causality between two signals, conditioned on n others, requires the estimation of $(2+n)^2 p$ parameters, where p is the order (maximum time delay) of the model; for a dataset of 10 signals and the minimum order $p = 1$, one needs at least 100 time points in order to calculate the Granger causality between pairs of nodes conditioned on the rest.

To bypass this difficulty, we proposed a procedure for building large networks of hundreds of nodes using partial Granger causality. The rationale is as follows: if the usual Granger causality from $Y \to X$ is large but significantly decreases when conditioned on a third signal Z ($F_{Y \to X|Z}$), then the connection $Y \to X$ is only indirect and should be discarded. We use this principle to find the direct ancestors (signals acting on a target X with no intermediate) of each nodes. At step 0, we search for all signals Y such that $F_{Y \to X}$ is large. We call Ω_0 this collection of candidate ancestors. At step 1, we filter this set further with keeping the signals $Y \in \Omega_0$ such that $F_{Y \to X|Z}$ is still large for all $Z \in \Omega_0$. We call Ω_1 this new set and carry on the procedure by conditioning on groups of 2, then 3, etc. signals until such an operation is not possible (the size of Ω_i decreases at each iteration). The result is a list of direct ancestors for each node, which we aggregate to produce the global network.

We test the validity of this approach on simulated data. We built an Erdös–Rényi random graph with $N = 200$ nodes and $M = N \ln N = 1060$ edges. We generate N time series with an auto-regressive model such that they follow the random network's structure: the transition matrix A is build from the transpose of network's adjacency matrix by replacing nonzero entries by a random value. The matrix A is then scaled to have a maximum eigenvalue less than 1, in order to make the system stable. Each time series is 200 time-points long and normal noise of unit variance is added throughout.

Figure 5.7(A) shows the resulting receiver operating characteristic (ROC) curve [14], that is the graph obtained by plotting the false positive rate against the true positive rate. A random guess is represented by the dashed line. The method shows a maximum true positive rate just over 0.5, which is not very high. However, the false positive rate is always very low: the method misses many ancestors, but its guesses are rarely wrong. This is crucial for biological applications: it means that the results can be used in further experiments, for example by indicating which protein/gene to manipulate. Figure 5.7(B) shows how the true positive rate varies with respect to the strength of the connection (i.e. the associated weight in the transition matrix). Weak connections are more likely missed out.

The procedure is easy to parallelise (each node can be processed separately) and easy to implement. It has been applied on a dataset of 812 proteomic time series [53] to produce a large and complete network.

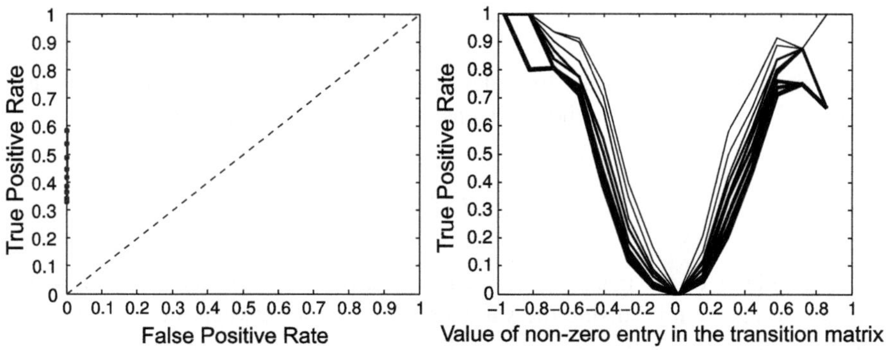

Fig. 5.7 Performance of the network building procedure. (**A**) Receiver operator characteristic curve. (**B**) Sensitivity of the procedure to the connection strength

5.9 Summary

In this chapter, we introduced several important extensions of Granger causality that have been devised for tackling issues specific to biological phenomena. We started with *partial Granger causality*, which is not only able to remove the influence of other observables but also to reduce the influence of external, unseen inputs. This is of considerable importance for biological data where typically only a small portion of the quantities of interest is actually observed. We showed that this extension (and the others) has its counterpart in the *frequency domain*, so that instead of summarizing the strength of the connection in one single number as is the case in the time domain, we can obtain a spectrum indicating at which frequency the interaction takes place. Partial Granger causality has then been further extended to *Complex Granger causality*. Complex Granger causality captures the effect of group action, a frequent occurrence in biology where the whole is often more significant than the parts. *Harmonic Granger causality* used an explicit form of possible external influence for modelling periodic inputs like sunlight. Since it is one of the most popular approaches for reverse-engineering biological systems, *Bayesian network* has been systematically compared with the much simpler Granger causality and it was shown that they both give similar results, provided that the number of time points is large enough. It was also shown that another popular approach for studying brain imaging data in particular called Dynamic Causal Model, and Granger causality both can be seen as a particular case of the more general *Unified Causal Model*. And finally, we presented a new method for building *large networks* of hundreds of nodes with limited data, a standing problem in systems biology.

Acknowledgements This work is supported by grants from EPRSC (UK, CARMEN EP/E002331/1) and EU grant (BION) and NSFC (China).

Appendix: Estimating the Error Covariance Matrix

The main quantity of interest for calculating Granger causality is the covariance matrix of the error term ε in the d-dimensional auto-regressive (AR) model

$$X_t = \sum_{i=1}^{p} A_i X_{t-i} + \varepsilon(t)$$

The coefficients of the matrices A_i need to be estimated from the data. Typically, they are found by minimising the variance of the error between the prediction X_t and the observation at the same time t. Morf's procedure [33] provides a fast and robust way of estimating A_i and the covariance matrix of ε via a recursive algorithm.

Unless prior knowledge informs us about the likely order p of the AR model (for example by knowing at which time-scale one should expect interactions to take place), it also has to be estimated from the data. The goodness of fit alone is not sufficient for selecting the optimal order: adding a new order implies adding d^2 more unknowns to the system, which considerably improves the fitting simply by increasing the degrees of freedom. The Akaike Information Criterion (AIC, [2]) provides a trade-off between the model complexity (a function of p) and the fit. For an AR model of dimension d, of order p and with T observations, this quantity can be written as

$$\mathtt{AIC}(p) = 2d^2 p + d(T-p)\log\left(2\pi \sum_{\substack{t=p+1\ldots T\\ i=1\ldots d}} \frac{\varepsilon_i^2(t)}{d(T-p)}\right)$$

The optimal order p is then defined as the one that minimises the AIC.

A point estimate of the Granger causality is often not sufficient for making a strong conclusion about the data and a confidence interval is required. In some cases ([25] for partial Granger causality), it is possible to derive the confidence interval in closed form. In general, it is estimated via a bootstrap procedure [32]—given the optimal AR model, an ensemble of signals are generated which each produce a value for the Granger causality. The final estimation of the Granger causality and its confidence interval are defined as the average and the standard error of this set of estimates.

References

1. R. Aebersold, L.E. Hood, and J.D. Watts. Equipping scientists for the new biology. *Nat Biotechnol*, **18**(4):359, 2000.
2. H. Akaike. A new look at the statistical model identification. *Autom Control, IEEE Trans*, **19**(6):716–723, 1974.
3. U. Alon. Biological networks: the tinkerer as an engineer. *Science*, **301**(5641):1866–1867, 2003.

4. D. Anastassiou. Computational analysis of the synergy among multiple interacting genes. *Mol Syst Biol*, **3**:83, 2007.
5. C. Andrieu, N. de Freitas, A. Doucet, and M.I. Jordan. An introduction to MCMC for machine learning. *Mach Learn*, **V50**(1):5–43, 2003.
6. W.L. Buntine. Operations for learning with graphical models. *J Artif Intell Res*, **2**:159, 1994.
7. D.M. Camacho and J.J. Collins. Systems biology strikes gold. *Cell*, **137**(1):24–26, 2009.
8. I. Cantone, L. Marucci, F. Iorio, M.A. Ricci, V. Belcastro, M. Bansal, S. Santini, M. di Bernardo, D. di Bernardo, and M.P. Cosma. A yeast synthetic network for in vivo assessment of reverse-engineering and modeling approaches. *Cell*, **137**(1):172–181, 2009.
9. B. Chance, R.W. Estabrook, and A. Ghosh. Damped sinusoidal oscillations of cytoplasmic reduced pyridine nucleotide in yeast cells. *Proc Natl Acad Sci*, **51**(6):1244–1251, 1964.
10. Y. Chen, S.L. Bressler, and M. Ding. Frequency decomposition of conditional Granger causality and application to multivariate neural field potential data. *J Neurosci Methods*, **150**(2):228–237, 2006.
11. J.J. Chrobak and G. Buzsaki. Gamma oscillations in the entorhinal cortex of the freely behaving rat. *J Neurosci*, **18**:388–398, 1998.
12. M. Ding, Y. Chen, and S.L. Bressler. Granger causality: Basic theory and application to neuroscience. In J. Timmer, B. Schelter, M. Winterhalder, editors, *Handbook of Time Series Analysis*, pages 451–474. Wiley-VCH, Weinheins, 2006.
13. M.R. Doyle, S.J. Davis, R.M. Bastow, H.G. McWatters, L. Kozma-Bognár, F. Nagy, A.J. Milla, and R.M. Amasino. The elf4 gene controls circadian rhythms and flowering time in arabidopsis thaliana. *Nature*, **1419**:74–77, 2002.
14. T. Fawcett. An introduction to ROC analysis. *Pattern Recogn Lett*, **27**(8):861–874, 2006.
15. J.F. Feng, D.Y. Yi, R. Krishna, S.X. Guo, and V. Buchanan-Wollaston. Listen to genes: dealing with microarray data in the frequency domain. *PLoS ONE*, **4**(4):e5098+, 2009.
16. K. Friston. Causal modelling and brain connectivity in functional magnetic resonance imaging. *PLoS Biol*, **7**(2):e1000033+, 2009.
17. K.J. Friston, L. Harrison, and W. Penny. Dynamic causal modelling. *NeuroImage*, **19**(4):1273–1302, 2003.
18. T.S. Gardner, D. di Bernardo, D. Lorenz, and J.J. Collins. Inferring genetic networks and identifying compound mode of action via expression profiling. *Science*, **301**(5629):102–105, 2003.
19. T. Ge, K.M. Kendrick, and J.F. Feng. A unified dynamic and granger causal model approach demonstrates brain hemispheric differences during face recognition learning. *PLoS Comput Biol*, 2009, submitted.
20. J.F. Geweke. Measurement of linear dependence and feedback between multiple time series. *J Am Stat Assoc*, **77**(378):304–313, 1982.
21. J.F. Geweke. Measures of conditional linear-dependence and feedback between time series. *J Am Stat Assoc*, **79**(388):907–915, 1984.
22. B. Gourévitch, R.L. Bouquin-Jeannès, and G. Faucon. Linear nonlinear causality between signals: methods, examples and neurophysiological applications. *Biol Cybern*, **95**(4):349–369, 2006.
23. C. Granger. Investigating causal relations by econometric models and cross-spectral methods. *Econometrica*, **37**:424–438, 1969.
24. C. Granger. Testing for causality: a personal viewpoint. *J Econ Dynam Control*, **2**:329–352, 1980.
25. S. Guo, A.K. Seth, K.M. Kendrick, C. Zhou, and J.F. Feng. Partial Granger causality–eliminating exogenous inputs and latent variables. *J Neurosci Methods*, **172**(1):79, 2008.
26. S. Guo, J. Wu, M. Ding, and J.F. Feng. Uncovering interactions in the frequency domain. *PLoS Comput Biol*, **4**(5):e1000087, 2008.
27. S. He. Estimation of the mixed AR and hidden periodic model. *Acta Math Appl Sin Engl Ser*, **13**(2):196–208, 1997.

28. E. Klipp, R. Herwig, A. Kowald, C. Wierling, and H. Lehrach. Systems biology in practice: concepts, implementation and application, 2005.
29. C. Ladroue, S.X. Guo, K. Kendrick, and J.F. Feng. Beyond element-wise interactions: identifying complex interactions in biological processes. *PLoS ONE*, **4**(9):e6899, 2009.
30. J.C. Locke, L. Kozma-Bognar, P.D. Gould, B. Feher, E. Kevei, F. Nagy, M.S. Turner, A. Hall, and A.J. Millar. Experimental validation of a predicted feedback loop in the multi-oscillator clock of arabidopsis thaliana. *Mol Syst Biol*, **2**:59, 2006.
31. H.G. McWatters, E. Kolmos, A. Hall, M.R. Doyle, R.M. Amasino, P. Gyula, F. Nagy, A.J. Millar, and S.J. Davis. ELF4 is required for oscillatory properties of the circadian clock. *Plant Physiol*, **144**(1):391, 2007.
32. D.S. Moore. *The Basic Practice of Statistics*. Freeman, New York, 2003.
33. M. Morf, A. Vieira, D.T.L. Lee, and T. Kailath. Recursive multichannel maximum entropy spectral estimation. *Geosci Electron IEEE Trans*, **16**(2):85–94, 1978.
34. S. Mukherjee and T.P. Speed. Network inference using informative priors. *Proc Natl Acad Sci*, **105**(38):14313–14318, 2008.
35. C.J. Needham, J.R. Bradford, A.J. Bulpitt, and D.R. Westhead. A primer on learning in Bayesian networks for computational biology. *PLoS Comput Biol*, **3**(8):e129, 2007.
36. A. Neumaier and T. Schneider. Estimation of parameters and eigenmodes of multivariate autoregressive models. *ACM Trans Math Softw*, **27**(1):27–57, 2001.
37. J. Pearl. *Causality: Models, Reasoning, and Inference*. Cambridge University Press, Cambridge, 2000.
38. J. Quackenbush. Computational analysis of microarray data. *Nat Rev Genet*, **2**(6):418–427, 2001.
39. K. Sachs, O. Perez, D. Pe'er, D.A. Lauffenburger, and G.P. Nolan. Causal protein-signaling networks derived from multiparameter single-cell data. *Science*, **308**(5721):523–529, 2005.
40. M. Schelter, B. an Winterhalderm, and J. Timmer. *Handbook of Time Series Analysis: Recent Theoretical Developments and Applications*. Wiley-VCH, Weinheim, 2006.
41. T.F. Schultz and S.A. Kay. Circadian clocks in daily and seasonal control of development. *Science*, **301**(5631):326–328, 2003.
42. T.P. Speed. *Statistical Analysis of Gene Expression Microarray Data*. CRC Press, Boca Raton, 2003.
43. A.N. Stepanova and J.M. Alonso. Arabidopsis ethylene signaling pathway. *Science*, **276**:1872–1874, 2005.
44. G.C. Tiao and M.R. Grupe. Hidden periodic autoregressive-moving average models in time series data. *Biometrika*, **67**(2):365–373, 1980.
45. H.R. Ueda. Systems biology flowering in the plant clock field. *Mol Syst Biol*, **2**:60, 2006.
46. H.R. Ueda, W.B. Chen, A. Adachi, H. Wakamatsu, S. Hayashi, T. Takasugi, M. Nagano, K. Nakahama, Y. Suzuki, S. Sugano, M. Iino, Y. Shigeyoshi, and S. Hashimoto. A transcription factor response element for gene expression during circadian night. *Nature*, **418**(6897):534–539, 2002.
47. N. Wiener. The theory of prediction. *Mod Math Eng Ser*, **1**:125–139, 1956.
48. J.H. Wu, K. Kendrick, and J.F. Feng. Detecting correlation changes in electrophysiological data. *J Neurosci Methods*, **161**(1):155–165, 2007.
49. J.H. Wu, X.G. Liu, and J.F. Feng. Detecting causality between different frequencies. *J Neurosci Methods*, **167**(2):367–375, 2008.
50. J.H. Wu, J.L. Sinfield, and J.F. Feng. Impact of environmental inputs on reverse-engineering approach to network structures. *BMC Systems Biology*, **3**:113, 2009.
51. J. Yu, A.V. Smith, P.P. Wang, and A.J. Hartemink. Advances to Bayesian network inference for generating causal networks from observational biological data. *Bioinformatics*, **20**(18):3594–3603, 2004.
52. C.L. Zou and J.F. Feng. Granger causality vs. dynamic bayesian network inference: a comparative study. *BMC Bioinform*, **10**(1):122, 2009.

53. C.L. Zou, C. Ladroue, S.X. Guo, and J.F. Feng. Identifying interactions in the time and frequency domains in local and global networks. *BMC Bioinform*, 2010, under revision.
54. C.L. Zou, K.M. Kendrick, and J.F. Feng. The fourth way: Granger causality is better than the three other reverse-engineering approaches. *Cell*, 2009. http://www.cell.com/comments/S0092-8674(09)00156-1.
55. M. Zylka, L. Shearman, J. Levine, X. Jin, D. Weaver, and S. Reppert. Molecular analysis of mammalian *timeless*. *Neuron*, **21**(5):1115–1122, 1998.

Chapter 6
Transcription Factor Binding Site Identification by Phylogenetic Footprinting

Haiyan Hu and Xiaoman Li

6.1 Introduction

Transcription factor binding sites (TFBSs) are 6–14 base pair (bp) long DNA segments that transcription factors (TFs) bind to. When TFs bind to their TFBSs, those genes near to the TFBSs can be turned on/off. The TFBSs bound by a TF are often similar to each other, and we call the common pattern of the TFBSs bound by a TF a motif. Motifs can be represented as consensus sequences, position weight matrices [1] and motif logos [2] (Fig. 6.1). Because the motif of a TF represents the common DNA pattern bound by a TF, in some sense, to identify TFBSs refers to the same thing as to identify motifs.

The identification of TFBSs is crucial for the understanding of gene regulation. When a TF binds to a TFBS alone or with other TFs binding to multiple TFBSs, the binding of TF–TFBS pairs can promote or repress the transcription of nearby genes. Put simply, the TFBSs are the switches of the gene expression. The identification of these switches and the understanding of the building principle of these switches are the first step to understand gene regulation and to control gene expression [3].

There are many methods available for TFBS identification. Traditionally, researchers were using DNase footprinting [4] and gel mobility shift assay [5, 6]. It is

H. Hu (✉)
School of Electrical Engineering and Computer Science, University of Central Florida, 4000 Central Florida Blvd, Orlando, FL 32816, USA
e-mail: haihu@cs.ucf.edu

X. Li
Biomolecular Science Center, University of Central Florida, 4000 Central Florida Blvd, Orlando, FL 32816, USA
e-mail: xiaoman@mail.ucf.edu

H. Hu · X. Li
Harris Corporation and Engineering Center, University of Central Florida, 4000 Central Florida Blvd, Orlando, FL 32816, USA

J. Feng et al. (eds.), *Frontiers in Computational and Systems Biology*, Computational Biology 15,
DOI 10.1007/978-1-84996-196-7_6,

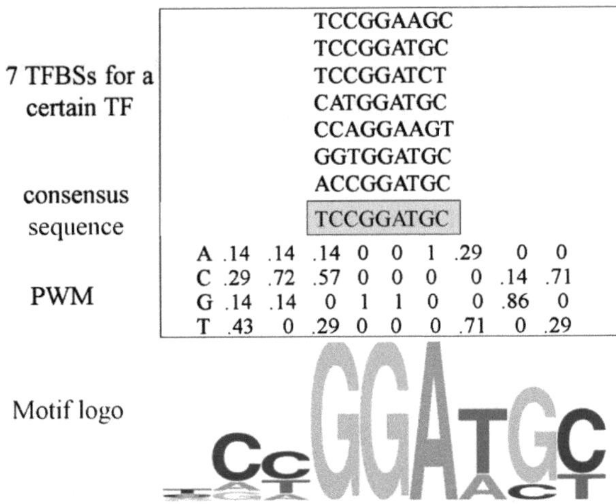

Fig. 6.1 Transcription factor binding sites, motif, and their representation. Here the PWM is a frequency matrix. Researchers often use the relative entropy instead of frequency in PWMs

through these gene-by-gene approaches that we gain some basic understanding of the gene regulation and the nature of TFBSs. With the advent of genome sequencing technology, more and more genome sequences become available. It is necessary to develop computational approaches to deal with the enormous amount of sequence data. Since the 1980s, a large number of computational methods have been developed for such purpose [7–35, 39]. Computational methods are capable of predicting TFBSs from a group of sequences and/or on the whole genome scale, which greatly accelerate the speed of identifying functional TFBSs.

The available computational methods often utilize one or more of the following three types of information to identify motifs and TFBSs (Fig. 6.2). The first type is the over-representation property of the motif [1, 7, 19, 22]. As we know, a TF can bind to many similar short DNA segments. That is, a TF can affect the expression of many genes, alone or together with other TFs. Therefore, if we have a group of genes that are regulated by a TF, we expect the TFBSs of the TF to occur in the noncoding sequences of many genes in this group, presumably in the upstream one kilobase (kb) long sequences (Fig. 6.2a). We can obtain such a group of coregulated genes from prior biological knowledge, microarray experiments, ChIP-chip experiments [36] (chromatin immunoprecipitation followed by microarray experiments), ChIP-seq experiments [37, 38] (chromatin immunoprecipitation followed by high-throughput sequencing experiments), and so on. The second type of information is the conservation property of motifs [9, 24–26, 28, 34, 39]. A TFBS is functional under certain experimental conditions. Such functional DNA segments are often more conserved compared with nonfunctional DNA segments. By comparing a sequence and its orthologous sequences, we can identify more conserved regions in these sequences and then identify TFBSs in these more conserved regions. Wasserman et al. [40] have found that 98% (74/75) of experimentally defined sequence-specific binding sites of skeletal-muscle-specific TFs are confined to the 19% of human sequences that are most conserved in the orthologous rodent sequences. Therefore, the identification of motifs and TFBSs in conserved regions is a feasible and reli-

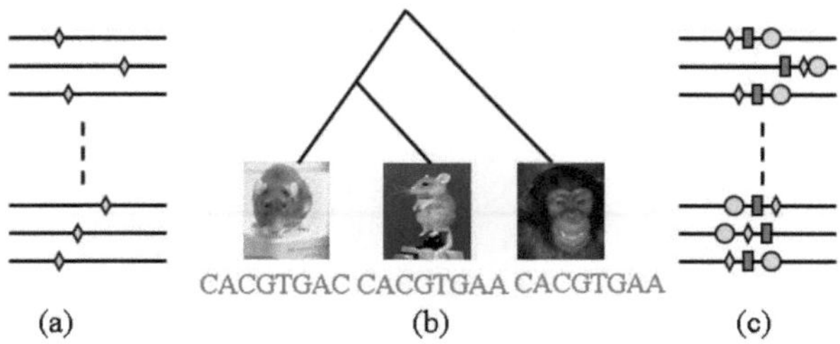

Fig. 6.2 Three types of information that are used for TFBS identification. (**a**) Over-representation. The *horizontal lines* are the sequences from coregulated genes in one species. The *small boxes* on the *line* are the TFBSs. (**b**) Conservation. The TFBSs in one group of orthologous chimpanzee, mouse and rat genes are similar. That is, the TFBSs in this gene are conserved across three species. (**c**) Clustering. Three different TFBSs often occur together in short regions in the input sequences. Such short regions are often called cis-regulatory modules (CRMs) (for the color version, see Color Plates on p. 390)

able strategy. The third type of information is the clustering property of motifs [14, 41, 42] (Fig. 6.2c). Early experimental studies in Sea urchin and fly have shown that many TFBSs from the same TF or from different TFs often occur in short regions of a few hundred bp [3]. We call such short regions containing multiple TFBSs a cis-regulatory module (CRM). The chance of finding a CRM in the noncoding sequences is much smaller than that of finding an individual TFBS in the noncoding sequences.

In this chapter, we are focusing on the computational methods that utilize the conservation property of motifs. Such methods are often called phylogenetic footprinting methods. In biology, phylogenetics is the study of evolutionary relatedness among various groups of organisms (e.g., species, populations), which is discovered through molecular sequencing data and morphological data matrices. Bona fide TFBSs are functional and are shared by multiple species if these species are properly chosen. These shared TFBSs by orthologous sequences are likely the footprints in the sequences that inform how the functional elements evolve. Therefore, the methods based on finding such footprints are called phylogenetic footprinting methods. Note that, although the conservation property is an indispensable part in these methods, these methods also often utilize other properties of motifs.

An essential component of the phylogenetic footprinting methods is the comparison of orthologous sequences to identify conserved footprints. To compare orthologous sequences, an alignment approaches is often used. An alignment of two sequences occurs when the nucleotides from both sequences match in order with each other.That is, assume a_i and a_k are the nucleotides at any two positions i and k of the first sequence, respectively. Assume b_j and b_m are the nucleotides at any two positions j and m of the second sequence that match with a_i and a_k, respectively. If $i < j$, then $k < m$ (Fig. 6.3). In the alignment of two sequences, a nucleotide from one sequence can either match with one nucleotide in the other sequence and an indel that represents a deletion in the other sequence. Similarly, an alignment of multiple sequences is considered as a match of nucleotides from multiple sequences in order. Alignments can be classified into two large categories: global alignments

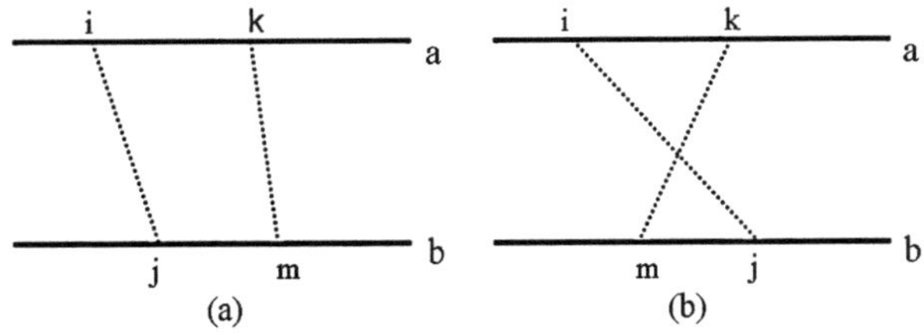

Fig. 6.3 Alignment of two sequences *a* and *b*. (**a**) Sequence alignments require the aligned order of the nucleotides in the sequences must be kept. (**b**) An illegal alignment where the order of the aligned nucleotides are changed

and local alignments. The former are to identify the match between or among the entire sequence regions, while the latter are to identify the match between or among any subregions in the input sequences (Fig. 6.4).

One obvious drawback of using alignments for phylogenetic footprinting is that the TFBSs may not be aligned with their orthologous counterparts in another species in the alignments, which is particular true for global alignments. This is because global alignments may neglect the optimal match of short regions while TFBSs are in general short. For instance, "ACCCTGA" in the two sequences in Fig. 6.4 are not aligned in the global alignment. Local alignments tend to produce many good alignment candidates for TFBS identification; however, in practice only a handful of optimal local alignments can be kept for TFBS analysis (Fig. 6.4). To compare sequences without alignments, one may either enumerate all k-mers (DNA segments of k bp long) and their derivatives or apply statistical methods such as expectation maximization (EM) algorithms and Gibbs sampling methods to compare DNA segments.

In this chapter, we will first describe current TFBS identification methods based on alignments, followed by the introduction of current TFBS identification methods without alignments. All these methods use multiple genome sequences. Then we will point out the future direction of TFBS identification by phylogenetic footprinting.

6.2 Current TFBS Identification Based on Alignments

As we mentioned earlier, a routine way for researchers to identify TFBSs is to look for conserved over-represented segments in the alignments of orthologous sequences. Orthologous sequences are the corresponding sequences in different species. In high eukaryotes, such as human, it is much easier to identify orthologous sequences of coding regions rather than noncoding regions. A common practice to obtain orthologous sequences of noncoding regions is to use the aligned sequences in multiple genome alignments as orthologous sequences. Researchers also use the sequences around transcription start sites (TSSs) of orthologous genes as orthologous sequences.

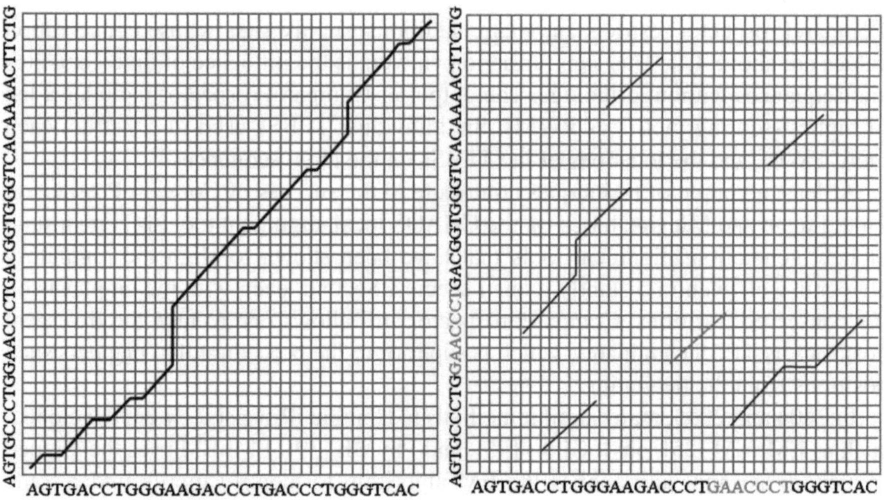

Fig. 6.4 (Color online) Global and local alignments. (**a**) A global alignment of two sequences. The curve shows the optimal match between the two sequences. (**b**) A local alignment of two sequences. The short curves are the optimal local alignments. For example, the two *red segments* are optimal matches in a local alignment, represented by the *red straight line*. The figure is from Michael Brudno

One of the best alignment-based methods is CompareProspector [23]. CompareProspector takes as input a list of sequences that are assumed to share motifs from one anchor species. Such sequences can be obtained from high-throughput genomics techniques such as gene expression profile clustering, ChIP-chip target regions and ChIP-seq peak regions. To utilize the multiple genome sequences, CompareProspector uses Lagan (Mlagan) to align two (multiple) orthologous sequences [43] and generates a list of percent identity values representing the cross-species conservation of each nucleotide in the anchor species. Then CompareProspector applies a similar Gibbs sampling approach as was implemented in BioProspector [22].

In detail, in the Gibbs sampling iterations, CompareProspector biases the motif finding towards sequences conserved across species. First, the user can specify two thresholds, T_{ch} (high conservation threshold) and T_{cl} (low conservation threshold). In BioProspector, a site score Ax is calculated for every site x in the input sequence as the ratio of the probability of generating x from the motif model over the probability of generating x from the background distribution. A new site is sampled with probability proportional to Ax. In CompareProspector, during initial iterations of Gibbs sampling, only positions whose conservation values are above T_{ch} are sampled. Subsequently, the conservation cutoff is gradually decreased from T_{ch} to T_{cl} to allow sampling of less conserved positions. The new site score $A'x$ is weighted by sequence conservation to favor sampling of more conserved sequences. Sequences without orthologs are assigned T_{cl} as the conservation value for all x, so they only participate in sampling in later iterations. Finally, in the original BioProspector, sites with a high enough score Ax are automatically added to the motif without sampling.

CompareProspector restricts automatic additions to only sites whose conservation values are above T_{ch}. This step further down weighs the influence of divergent sites and sequences without orthologs. The output of CompareProspector includes a list of highest-scoring motifs as PWMs, the individual TFBSs used to construct each motif PWM, and the locations of the TFBSs on the input sequences.

At least a couple of techniques contribute to the success of CompareProspector. First, Lagan and Mlagan, based on the CHAOS local alignment tool [44], are a suite of new generation alignment tools that have been shown to align the regulatory elements well even in distantly related species. Second, even with the emphasis on the conserved positions, CompareProspector still allows one to take the divergent positions as TFBS candidates, which is superior to the procedure that neglects such positions from the beginning.

In 2004 and 2005, many similar methods to CompareProspector were developed, such as EMnEM [25], phyloGibbs [45], and PhyME [39]. EMnEM uses the aligned sequences as input and applies the EM algorithm to identify motifs under some evolution models. PhyloGibbs uses the aligned sequences by the Dialign program [46] and applies the Gibbs sampling methods to identify motifs under evolution models. Other difference between EMnEM and PhyloGibbs includes evolution tree difference, how to deal with the divergent sequences in the multiple alignments, and so on. PhyloGibbs considers a star phylogenetic tree while EMnEM considers a more suitable tree like a species tree. PhyloGibbs focuses on the well-conserved regions in the multiple alignments first and considers the unaligned TFBS candidates while EMnEM uses the entire aligned sequences simultaneously. In terms of these two aspects, PhyME basically uses the phylogenetic trees similarly as the EMnEM and utilizes the aligned sequences in similar fashion as the PhyloGibbs. Note that PhyME is also using the EM algorithm to identify motifs and TFBSs.

Kellis et al. developed an approach that is different from the above methods [47]. They have applied the method in yeast genomes and have produced impressive results. They start from the genome alignments of noncoding sequences of orthologous genes and try to enumerate gapped 6-mer motifs. In detail, they first identify conserved "mini-motifs," which are then used to construct full motifs. Mini-motifs are sequences of the form XYZn(0–21)UVW, consisting of two triplets of specified bases interrupted by a fixed number (from 0 to 21) of unspecified bases. For instance, TAGGAT and ATAnnGGC are two mini-motifs. If reverse complements are grouped together, the total number of distinct mini-motifs is 45760. Conserved mini-motifs are then defined according to three conservation criteria. In each case, conservation rates are normalized to appropriate random controls. The conservation criteria are: (1) intergenic conservation, the mini-motif shows a significantly high conservation rate in intergenic regions; (2) intergenic–genic conservation, the mini-motif shows significantly higher conservation in intergenic regions than in genic regions; (3) upstream–downstream conservation, the mini-motif shows significantly different conservation rates when it occurs upstream compared with downstream of a gene. The conserved mini-motifs are then used to construct full motifs. The mini-motifs are first extended by searching for nearby sequence positions showing significant correlation with a mini-motif. The extended motifs are then clustered, merging those with substantially overlapping sequences and those that tend to occur in the same intergenic regions. Finally, a full motif is created by deriving a consensus sequence (which may be degenerate). Each full motif is assessed for genome-wide conservation by calculating its MCS (motif conservation score), and those motifs with MCS > 4 are retained. Here the MCS of a motif is defined based on the conservation rate of the motif in intergenic regions and is measured in standard deviations above the rate for comparable control motifs. Each full motif was also tested for enrichment in upstream compared with down-stream regions, by comparing its conservation rate in divergent versus convergent intergenic regions. The motif analysis automatically identified 72 genome-wide elements, including most known regulatory motifs in yeast.

The success of the method from Kellis et al. lies on the comparison of motif identification with random sequences. In general, researchers define background sequences as those input segments by removing the TFBSs they predict. This makes sense in many cases, as was shown in the success of the software of MEME [7], CompareProspector [24], and others. However, whether a method works, we really need to test on the independent random sequences, which will greatly filter false positives.

Although successful, the method developed by Kellis et al. will miss many bona fide motifs and TFBSs. By using the whole genome sequences of four yeast species, Kellis et al. only identified 72 motifs, which is much less than the 300 hypothesized motifs in *Saccharomyces cerevisiae*. Therefore, even in the yeast, it is not a simple task to identify all the motifs.

6.3 Methods Independent of Alignments

6.3.1 Defects of Alignments Especially Genome Alignments

Alignment methods may presumably align many TFBSs and their counterparts in orthologous sequences. However, many TFBSs may not be aligned with their orthologous TFBSs. This is because the typical length of the sequences input for alignments is around 1 kb and the typical TFBSs are 6–14 bp long. If the species under consideration are not divergent enough, the alignments cannot really show the evolution constraints. On the other hand, if the species are divergent, it is likely that the noise of the diverged nonfunctional background will overcome the short conserved signal. The result is that the alignments may not align the short regulatory elements properly. If one tends to use the local alignments instead of the global alignments, the true TFBSs may be aligned properly with their counterpart across species. Given the short length of the TFBSs, we will have numerous aligned short segments from local alignments if the species under consideration diverged not that long. We can only afford to consider some of these local alignments and may miss many other possible TFBSs. If the species are divergent, we may rarely be able to find good candidates since the alignment methods in general do not take the divergent time into account.

Moreover, the current alignment methods are still not perfect. For instance, although we use the genome alignments on the daily basis, the current genome alignments from different methods are not consistent. Recently, Margulies et al. [48] compared four genome alignment methods, Mlagan [43], TBA [49], Mavid [50], and Pecan, on orthologous sequences in the 44 ENCODE regions (The ENCODE Project Consortium 2004) from 28 vertebrates. For 14 mammals, a total of 206 megabase (Mb) of sequence was obtained from mapped bacterial artificial chromosomes and finished to "comparative grade" standards [51] specifically for these studies. For another 14 species, 340 Mb of sequence were obtained from genome-wide sequencing efforts at varying levels of completeness and quality. The details of the difference of the alignments are based on the four methods are as follows.

The comparison here was made at the nucleotide level, at which many downstream applications operate. The authors [48] found that the level of agreement between alignments varies significantly between species, with agreement much higher when comparing alignments of primates versus those of more distant species. In general, agreement between the different alignments is influenced significantly by the total coverage; for example, MAVID aligns 27.4% of human bases to an armadillo nucleotide, versus 42.4%, 41.2%, and 40.1% for Mlagan, PECAN, and TBA, respectively; and thus the maximum possible agreement between all the alignments is 27.4%. They found that 17.5% of all human nucleotides are aligned to the same armadillo nucleotide by all four alignments, and 66.1% of all human bases are identically aligned if they considered gapped columns (i.e., columns in which a human nucleotide is predicted do not have an orthologous nucleotide in the armadillo sequence). Their conclusions show that there are substantial variations between the nucleotide-level orthology predictions made by the four alignments, even though a significant majority of all human nucleotides is aligned identically between human and a given nonhuman sequence. As a surrogate for sensitivity, they also determined the coverage of annotated protein-coding sequences in each of the alignments. Since coding exons are regions of the human genome that are largely ancient and likely to be shared among all of the lineages analyzed here, these represent a set of nucleotides heavily enriched for "true positive" (i.e., actually orthologous) positions. The authors defined the alignment "coverage" as the number of human coding bases aligning to a given nonhuman species. It was found that coverage of coding exons varies considerably among the different alignments, especially when analyzing alignments between humans and more distant species. When counting the number of coding exons with at least one base pair aligned to a base in the mouse genome, for example, coverage ranges from 55% in MAVID to 72% in Mlagan, with TBA and PECAN showing intermediate values. Alternatively, when looking at only those coding exons that are fully covered (i.e., no gaps), these values range from 29% in MAVID to 38% in PECAN. PECAN and Mlagan exhibit the highest values by these measures and are similar for most species.

From the study by Margulies et al. it is clear that researchers should be cautious when they aim to identify TFBSs and motifs from those alignments. Kellis et al.'s results compared with PhyloNet below show that maybe only those obvious TFBSs and motifs can be identified from the whole genome alignments. In the following, we will introduce several methods that do not use any alignment tools.

6.3.2 TFBS Identification Without Sequence Alignments

One of the early methods that do not use the alignment to identify conserved TFBSs is the footprinter developed by Blanchette and Tompa [9, 10]. This algorithm is guaranteed to report all sets of motifs with the lowest parsimony scores, calculated with respect to the phylogenetic tree relating the input species. It is a deterministic method, using dynamic programming to obtain the motifs. This method is more about identifying conserved motifs in one gene across multiple species, instead of conserved motifs in multiple genes across multiple species. Thus, it does not require a group of coregulated genes as input. Although it is linear in terms of sequence length, it cannot be applied to long DNA sequences because TFBSs can be conserved by chance in long DNA sequences.

An early TFBS identification method without alignment that also considers over-representation property of motifs is the PhyloCon algorithm [34]. PhyloCon consists of three components: initial profile generation, profile comparison, and a greedy approach to combine common regions in different profiles. PhyloCon generates initial multiple aligned sequences using Wconsensus [52]. Wconsensus gives many ungapped suboptimal aligned segments. If the real TFBSs are correctly positioned in any of these aligned segments, they will emerge during subsequent profile comparisons. Initial multiple aligned sequences generated by Wconsensus are transformed into profiles, or position-specific scoring matrices. Each column is a vector of four elements, representing either counts or observed frequencies of different nucleotides at a position in the aligned segments. Each profile represents a conserved region in the initial orthologous sequences. A conserved region can be represented by more than one profile based on suboptimal aligned segments. A profile is treated as a sequence of columns, so the alignment between profiles is analogous to the alignment between sequences. Assuming position independence and a suitable scoring scheme, the score of an alignment between two profiles is the sum of the scores from comparing corresponding columns in the two profiles. PhyloCon uses a new statistic called Average Log Likelihood Ratio (ALLR) to compare two columns in different profiles. ALLR can be used to distinguish probability distributions from each other, as well as from the background. It measures the joint probability of observing the data generated by one distribution given the likelihood ratio of the other distribution over the background distribution. Given this scoring statistic and the assumption of position independence, the score of aligning local regions of two profiles is simply the sum of comparison scores of position pairs. A dynamic programming algorithm is implemented to identify high similarity regions in profiles using the ALLR statistic. PhyloCon then uses a greedy algorithm to combine profile comparison results.

As one of the first methods combining the over-representation property and the conservation property of motifs, PhyloCon is fast and works better than the Gibbs sampling methods based on over-representation property of motifs only [34]. However, in some cases, it is difficult to select motifs from Phylo-Con output, due to the fact that PhyloCon outputs the same motifs repeatedly, from the longer versions to the shorter versions.

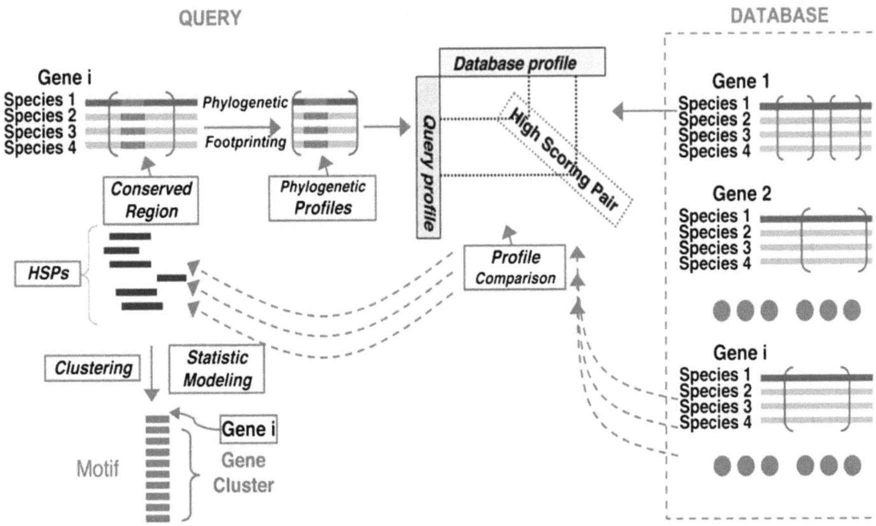

Fig. 6.5 An illustration of PhyloNet. Copied from the PhyloNet paper

The PHYLONET algorithm [35] is a much faster algorithm compared with PhyloCon (Fig. 6.5). Briefly, the algorithm enables "motif-BLAST" by integrating comparative genomics information and regulatory network topology and exploring a phylogenetic profile space. Similar to PhyloCon, PhyloNet generates conserved short regions as profiles. Each promoter is represented by its phylogenetic profiles and queried against a database of phylogenetic profiles of all promoters in a genome. Statistical significance of motifs is determined by Karlin–Altschul statistics that are modified for profile searches.

Note that PhyloNet identifies TFBSs and motifs on a whole genome scale, while PhyloCon is considering only a group of coregulated genes. To identify TFBSs on a genome scale, the authors of PhyloNet take an approximation approach when they compare profiles. They establish a partition of the profile space and develop a BLAST-like algorithm that has linear time and memory complexity. In detail, they partition the profile space into 15 subspaces by supervised learning and elect a consensus letter based on the weighted sum of all profiles in the subspace to represent a subspace. This partition coincides with common degenerate DNA representations, but the actual boundaries are optimized based on all pairwise profile comparisons. The new alphabet of 15 letters (A, C, G, T, a, c, g, t, W, S, R, Y, M, K, and N) replaces the original DNA sequences with additional conservation information. Based on the weighted similarity measurements by the ALLR statistic among profiles of any two subspaces, a substitution scoring matrix is constructed. The ALLR matrix is a log-odds scoring system that satisfies the restrictions needed to apply Karlin–Altschul statistics. With the use of the expansion of Karlin-Altschul statistics, good estimation of the statistical significance of scores of local profile alignments can be made. The letter representation of the profiles allowed the authors to use an efficient BLAST-like search engine to identify all similar profiles and identify the shared motifs among these similar profiles. This design allows a profile comparison to be 1,000 times faster than a pairwise comparison of all profiles by dynamic programming implemented in PhyloCon, with a minimum loss of sensitivity.

The advantage of methods without alignments can be seen from the comparisons of PhyloNet and Kellis et al.'s methods. PhyloNet successfully predicted 296 nonredundant motifs by using 3,524 *Saccharomyces cerevisiae* promoter sequences with orthologous counterparts *Saccharomyces mikatae*, *Saccharomyces kudriavzevii*, and *Saccharomyces bayanus*. From the analysis of four to six yeast genomes Kellis et al. [47] and Cliften et al. [53] predicted 71 and 92 regulatory motifs, respectively. Both collections identified many known TF motifs and many predicted motifs. However, the two collections overlap by 50%, demonstrating that neither collection reached saturation: 30 (42%) motifs in the Kellis set match a motif in the Cliften set, whereas 43 (47%) in the Cliften set match the Kellis prediction. Despite using the sequences from only four yeast species from Cliften et al., PHYLONET not only identified over twice as many predictions as either previous study, it also identified 86% ($n = 61$) of the Kellis motifs and 92% ($n = 85$) of the Cliften motifs, including all motifs supported by both studies. These comparisons highlight PHYLONET's ability to extract substantially more information from comparative analysis than previous methods. It also shows that the alignment- based methods may miss many motifs and TFBSs.

Tree Gibbs Sampler (TGS) [20] is another method that does not use the alignments to identify TFBSs. Similar to footprinter [10], TGS does not depend on alignments of orthologous sequences and takes the evolution of DNA sequences into account. Different from footprinter, TGS uses a PWM to represent a motif and applies a Gibbs sampling method, instead of deterministic methods to identify TFBSs. Such statistical approach avoids the arbitrary alignment score cutoffs to define the candidate functional sites, such as those often encountered in current methods. Without alignments, TGS has the flexibility to find similar motif instances in orthologous sequences, even if those TFBSs are inverted, translocated, or mutated. Moreover, TGS fully uses the phylogenetic information and tracks the trace of the functional sites during evolution, i.e., the motifs are allowed to evolve, although at a slower rate than the background. Furthermore, the method simultaneously finds similar motif instances not only in many genes but also in orthologous sequences, which enables it to find weak but conserved motifs. Finally, the method automatically chooses the width for motifs, which has the potential to identify all motifs that are enriched in the input sequences.

TGS uses two evolution models to describe nucleotide evolution. TGS assumes the TFBSs evolve more slowly than the nonfunctional background sites. Therefore, two 4×4 substitution matrices are used to describe the evolution along every branch of the phylogenetic tree for the TFBSs and the background sites, respectively. Each row of a matrix gives the probabilities of one type of nucleotide in the ancestor evolving into A, C, G, and T in the descendant, in the order of A, C, G, and T. For instance, the number at the entry (3, 2) in the matrix in Table 1 tells that the probability for a nucleotide G in the ancestor to evolve into the nucleotide C in the descendant is 0.0347.

Table 6.1 An example of the substitution matrices for the branch from the common ancestor of *S. cerevisiae* and *S. mikatae* to *S. cerevisiae* in the yeast phylogenetic tree. See TGS paper for details

	S. cerevisiae				
		A	C	G	T
Ancestor	A	.7743	.0347	.1329	.0581
	C	.0583	.6791	.0334	.2292
	G	.2320	.0347	.6752	.0581
	T	.0583	.1369	.0334	.7714

For background evolution, regions in the upstream region of orthologous genes are aligned, and the background nucleotide distribution and the branch lengths in the species tree are inferred from the alignments by using maximal-likelihood estimation. Then the background substitution matrix for every branch is obtained from the estimated background nucleotide distribution and the branch lengths. Note that in TGS, the species tree is used as the phylogenetic tree. To define the motif substitution matrix for a branch, TGS simply decreases the branch length estimated above by a fixed proportion, say 50%, and then constructs the motif substitution matrix for the branch from the decreased branch length. This is a primitive way to model the slower evolution of the functional motifs as compared with the background. With more experimentally verified motif sites available, motif substitution matrices may be constructed from the experimentally verified sites.

With the two evolution substitution matrices on every branch of the phylogenetic tree defined, TGS implements a Gibbs sampling method to infer the model parameters and motif instances under the assumption that there is at most one motif instance for every gene. See Fig. 6.6 for the illustrations. A novel strategy used by TGS is the up-down fashion to identify motifs. That is, TGS will first identify the ancestral TFBS for each group of orthologous sequences, and then it will identify child TFBSs of these ancestral TFBSs until we have predicted the TFBSs for all the sequences in all the species. It is because of this up-down fashion of sampling, TGS avoids the alignment procedure to find the TFBSs and their counterparts. For instance, in Fig. 6.6, for one group of orthologous sequences, if the ancestral TFBS at the root of the tree is known, to identify the TFBSs in the sequences of the current species is just to look at each position of the sequences and see whether there is some segment "similar" to its parent TFBSs. Here "similar" takes the divergent time into account, and it is not necessary that the TFBSs must be physically similar to their parent TFBSs. Because of this up-down fashion across many orthologous sequence groups, TGS can pinpoint the true motifs fast. For instance, if TGS starts from a non-TFBS segment at the ancestral level, it is difficult to find similar segments in many species and/or in many genes. On the other hand, if TGS starts from a TFBS-like segment at the ancestral level, it is more likely that many similar TFBSs can be found in the input sequences.

TGS has shown several advantages over other methods, such as PhyloCon and CompareProspector. The authors have shown that TGS has both better sensitivity and better specificity when applied to the yeast ChIP-chip datasets [54]. The advantage can also be shown in the distant species comparisons. By applying TGS on the 63 ribosomal protein gene pairs from two insect species, fruit fly (*Drosophila*

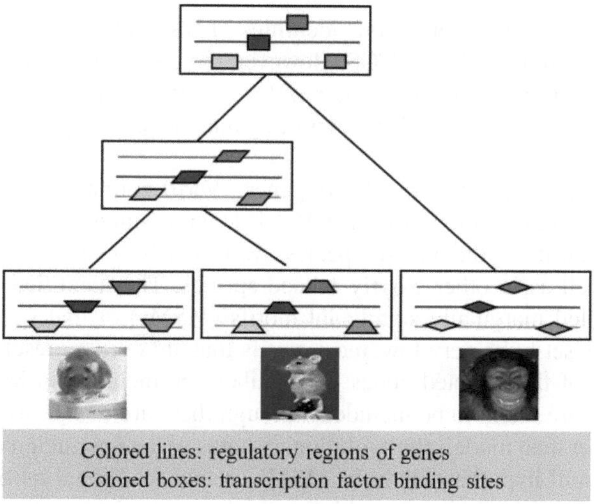

Fig. 6.6 A cartoon illustration of TGS. Each *rectangular box* represents one species. The *colored lines* are the regulatory regions of coregulated genes. The *small boxes* are the TFBSs in each sequence. TGS assumes that there is at most one TFBS in each sequence. The motifs may be different in different species although they evolve from the same ancestral motifs (for the color version, see Color Plates on p. 391)

melanogaster) and mosquito (*Anopheles gambiae*), the authors predicted a pair of motifs. Note that the two species diverged more than 250 million years ago, and methods that rely on alignment typically give poor predictions in this situation. The two motifs predicted by TGS have the same consensus, ACAGCTGTCAAAA. Moreover, TGS found TFBSs in all 63 gene pairs. If MEME [7] is applied on genes in individual species, GCGGTCACACT (fly) and CAGCTGTCAAACGG (mosquito) are identified in 41 and 44 genes, respectively. Although the underlined parts in the motif consensus identified by MEME look similar, the instances corresponding to the two motifs in the orthologous gene pairs rarely share >5 nucleotides in the underlined 9-mer parts. The TFBSs found by TGS share a median 8 bp of 13 positions. Moreover, the motif ACAGCTGTCAAAA identified by TGS is similar to an experimentally verified motif CAGTCACA, which was found to regulate 14 ribosomal protein genes in *Schizosaccharomyces pombe* [55]. Although phylocon outputs a motif ACCAGCTGTCAAAGGGG, which contains the one identified by our method, only 7 orthologous pairs are found by PhyloCon to contain the motif instances. Moreover, the p-value of this motif is not significant compared with those of other motifs output by PhyloCon. As to CompareProspector, none is found in the top 15 output motifs similar to the one identified by TGS. Note that 63 TFBS pairs found by TGS share 8 bp of 13 positions on average. This shows that using prealigned sequences as input to find motifs will likely miss many TFBSs for distant species, because those alignments in general cannot take the evolutionary distance into account, and many "well" conserved instances are missed.

Verifier [21] is another method that identifies TFBSs without alignments. Verifier uses similar evolution models as TGS. However, the goal is very different from that of TGS. TGS is trying to identify true motifs, and presumably, the true motifs are among the top predictions. Verifier is expecting that the predicted top motifs are true motifs.

In detail, for a set of coregulated genes, Verifier first uses a motif over-representation based method, e.g., MEME [7], with a nonstringent threshold to find potential motifs in the anchor species, from which the coregulated genes are obtained, as well as in other closely related species. The identified motifs in any species are called marginally significant motifs (MSMs) of that species. Because the threshold is set to be very low, most motifs that are over-represented in the genomic regions of the selected genes would likely be included in MSMs; i.e., the genuine motifs are likely to be included although there are many false positive predictions. Verifier then models the evolutionary paths of the neutral intergenic regions and poses the null hypothesis that the MSMs are not functional motifs, and therefore their TFBSs evolved like neutral intergenic regions. Verifier then tests whether there are MSMs that are much more conserved in the multiple species than what are expected under the null hypothesis. Verifier performs the tests by enumerating all of the groupings' of MSMs and calculating the probability that the current grouping of MSMs evolved from the same common ancestral motif under the null model. In the end, Verifier reports the significantly conserved MSMs as putative TF binding motifs and ranks these motifs according to their significance.

Again, Verifier performs better than PhyloCon and CompareProspector. It has better sensitivity and specificity as well. Especially the specificity, in 30 out of 35 predictions, verifier has ranked the bona fide motif as the top one motif.

6.4 Future Direction of Phylogenetic Footprinting

Although computational TFBS identification can be traced back to 1984 or earlier and we may have several hundred computational methods already, computational TFBS identification is far from mature. We will mention some challenges in the field of cis-regulatory analysis in the following.

One challenge is how to select one or several TFBS identification methods for a group of input sequences [56]. With many TFBS identification methods and software available, biologists and even computational biologists have difficulty in choosing the best tools for their TFBS finding endeavor. Tompa et al. [57] have shown that it has been a challenging task to conduct studies on performance comparisons of TFBS finding tools. The difficulty in performance assessment of these tools stems from several sources. The tools have been developed based on varied and complex motif models, and therefore, individual tools may do better on one type of data but do worse on other types of data. In addition, our incomplete understanding of the biology of regulatory mechanism does not always provide adequate evaluation of underlying algorithms over motif models. Tompa et al. have made a benchmark dataset for testing computational methods that are based on sequences from one species. Most of these datasets only contain a few sequences. Benchmark datasets with more sequences might be better, because in general TFBSs are identified in many sequences, not just 4 or 5 sequences in one species. Moreover, great effort is needed for a benchmark dataset for comparing the phylogenetic methods as well. It is also important to develop strategies to integrate these available methods for better prediction of TFBSs.

Another challenge is how to identify TFBSs in the entire noncoding sequences in high eukaryotes, especially in human. It is relatively simple to identify TFBSs in the yeast genome and current methods such as PhyloNet work well in yeast. However, to simply apply such methods to the human genome, it is most likely to generate more false positives than true positives. This is because the noncoding region of a human gene is several million bp long on average, as compared to the several hundred bp long noncoding regions of yeast genes. Very few of the current methods can identify TFBSs in the entire noncoding regions of the human genome [58]. Without considering the entire noncoding region of a gene, many conclusions about the expression pattern of a gene must be misleading and incomplete. We doubt that computational methods alone can achieve such a goal without considering the high-throughput experimental data. Currently, there is a large amount of data available concerning DNA methylation, histone methylation, DNase I hypersensitive sites, and so on. To investigate the relationship between TFBSs and these chromatin parameters will narrow the TFBS searching space. With more and more understanding of the relationship of TF binding and chromatin structure changes, we may develop computational methods to identify CRMs in the entire noncoding regions.

Another challenge is how to understand the dynamics of the TF binding. Purely sequence-based approaches for binding site identification do not capture the cellular state and thus do not reflect the dynamic nature of transcriptional regulation [59]. Many computational approaches that have taken mRNA levels of genes into account can already infer the interaction of TFs and tissue specific TF bindings. Unfortunately, these approaches still cannot describe the dynamics of the gene regulatory networks on a larger scale. A highly relevant attribute of the cellular state is the chromatin structure and epigenetic state of the genome. Recently, computational models have been proposed to predict DNase I hypersensitive regions [60], nucleosome positioning [61], and unmethylated CpG islands [62]. Incorporating these attributes should enhance binding site and its dynamics prediction.

Hannenhalli [59] proposed a new challenge in TFBS analysis: how to represent TFBSs more accurately. He pointed out that previous approaches to improve binding site predictions have either attempted to develop enhanced, more informative motif representations or models, or tried to exploit additional genomic or transcriptomic attributes. Given several experimentally determined binding sites for a TF, he suggested that an ideal representation is one that strikes an optimal balance between sensitivity and specificity by extracting maximal information. While PWM representation assumes independence among positions within a binding site, a full dependence model, on the other extreme, requires estimating an exponentially large joint distribution based on a small number of exemplars. Mixture models [63] represent a reasonable tradeoff. However, the functional relevance of multiple motif subtypes is not always clear. The optimal choice among these possibilities may vary among TFs, and a detailed evaluation of these choices needs to be done. Moreover, a significant portion of known binding sites have been determined using in vitro approaches, such as SELEX or DNA arrays, which may be different in vivo because of additional factors such as chromatin structure, epigenetic state, and the availability of other TFs. An unbiased and comprehensive evaluation of the differences between binding sites recognized in vivo and in vitro needs to be done. Also, posttranslational modification states of TF proteins can alter, directly or indirectly, the TF–DNA interaction [64]. The high-throughput technology to identify posttranslational modifications is limited to certain types of modifications and our understanding of how these modifications affect TF–DNA interaction is not sufficiently detailed. Ultimately, although we have PWMs for some TFs, such PWMs may contain sub-PWMs and should be divided into different subtypes, or these PWMs are only for in vitro models, and so on. To have more accurate PWMs is another challenge.

There are many other challenges, such as how to validate the computational predictions? How to incorporate CRM prediction with the phylogenetic footprinting? How to model chromatin structure and its dynamics in order to understand gene regulation? Currently, there are some methods developed for such purposes. Such efforts should be thought as the first step in the direction of these endeavors. We need to address these questions comprehensively in the future.

Acknowledgement The authors are supported by an NHGRI grant R01HG004359.

References

1. G.D. Stormo and G.W. Hartzell 3rd. Identifying protein-binding sites from unaligned DNA fragments. *Proc Natl Acad Sci USA*, **86**(4):1183–1187, 1989.
2. G.E. Crooks, G. Hon, J.M. Chandonia, et al. WebLogo: a sequence logo generator. *Genome Res*, **14**(6):1188–1190, 2004.
3. C.H. Yuh and E.H. Davidson. Modular cis-regulatory organization of Endo16, a gut-specific gene of the sea urchin embryo. *Development*, **122**(4):1069–1082, 1996.
4. D.J. Galas and A. Schmitz. DNAse footprinting: a simple method for the detection of protein-DNA binding specificity. *Nucleic Acids Res*, **5**(9):3157–3170, 1978.
5. M.M. Garner and A. Revzin. A gel electrophoresis method for quantifying the binding of proteins to specific DNA regions: application to components of the Escherichia coli lactose operon regulatory system. *Nucleic Acids Res*, **9**(13):3047–3060, 1981.
6. M. Fried and D.M. Crothers. Equilibria and kinetics of lac repressor-operator interactions by polyacrylamide gel electrophoresis. *Nucleic Acids Res*, **9**(23):6505–6525, 1981.
7. T.L. Bailey and C. Elkan. Fitting a mixture model by expectation maximization to discover motifs in biopolymers. *Proc Int Conf Intell Syst Mol Biol*, **2**:28–36, 1994.
8. B.P. Berman, B.D. Pfeiffer, T.R. Laverty, et al. Computational identification of developmental enhancers: conservation and function of transcription factor binding-site clusters in Drosophila melanogaster and Drosophila pseudoobscura. *Genome Biol*, **5**(9):R61, 2004.
9. M. Blanchette, B. Schwikowski, and M. Tompa. Algorithms for phylogenetic footprinting. *J Comput Biol*, **9**(2):211–223, 2002.
10. M. Blanchette and M. Tompa. Discovery of regulatory elements by a computational method for phylogenetic footprinting. *Genome Res*, **12**(5):739–748, 2002.

11. J. Buhler and M. Tompa. Finding motifs using random projections. *J Comput Biol*, **9**(2):225–242, 2002.
12. E.M. Conlon, X.S. Liu, J.D. Lieb, et al. Integrating regulatory motif discovery and genome-wide expression analysis. *Proc Natl Acad Sci USA*, **100**(6):3339–3344, 2003.
13. E. Eskin and P.A. Pevzner. Finding composite regulatory patterns in DNA sequences. *Bioinformatics*, **18**(1):S354–S363, 2002.
14. M. Gupta and J.S. Liu. De novo cis-regulatory module elicitation for eukaryotic genomes. *Proc Natl Acad Sci USA*, **102**(20):7079–7084, 2005.
15. S.T. Jensen and J.S. Liu. BioOptimizer: a Bayesian scoring function approach to motif discovery. *Bioinformatics*, **20**(10):1557–1564, 2004.
16. N.C. Jones and P.A. Pevzner. Comparative genomics reveals unusually long motifs in mammalian genomes. *Bioinformatics*, **22**(14):e236–e242, 2006.
17. U. Keich and P.A. Pevzner. Finding motifs in the twilight zone. *Bioinformatics*, **18**(10):1374–1381, 2002.
18. U. Keich and P.A. Pevzner. Subtle motifs: defining the limits of motif finding algorithms. *Bioinformatics*, **18**(10):1382–1390, 2002.
19. C.E. Lawrence and A.A. Reilly. An expectation maximization (EM) algorithm for the identification and characterization of common sites in unaligned biopolymer sequences. *Proteins*, **7**(1):41–51, 1990.
20. X. Li and W.H. Wong. Sampling motifs on phylogenetic trees. *Proc Natl Acad Sci USA*, **102**(27):9481–9486, 2005.
21. X. Li, S. Zhong, and W.H. Wong. Reliable prediction of transcription factor binding sites by phylogenetic verification. *Proc Natl Acad Sci USA*, **102**(47):16945–16950, 2005.
22. X. Liu, D.L. Brutlag, and J.S. Liu. BioProspector: discovering conserved DNA motifs in upstream regulatory regions of co-expressed genes. *Pac Symp Biocomput*:127–38, 2001.
23. X.S. Liu, D.L. Brutlag, and J.S. Liu. An algorithm for finding protein-DNA binding sites with applications to chromatin-immunoprecipitation microarray experiments. *Nat Biotechnol*, **20**(8):835–839, 2002.
24. Y. Liu, X.S. Liu, L. Wei, et al. Eukaryotic regulatory element conservation analysis and identification using comparative genomics. *Genome Res*, **14**(3):451–458, 2004.
25. A.M. Moses, D.Y. Chiang, and M.B. Eisen. Phylogenetic motif detection by expectation-maximization on evolutionary mixtures. *Pac Symp Biocomput*:324–35, 2004.
26. A.M. Moses, D.Y. Chiang, D.A. Pollard, et al. MONKEY: identifying conserved transcription-factor binding sites in multiple alignments using a binding site-specific evolutionary model. *Genome Biol*, **5**(12):R98, 2004.
27. A.F. Neuwald, J.S. Liu, and C.E. Lawrence. Gibbs motif sampling: detection of bacterial outer membrane protein repeats. *Protein Sci*, **4**(8):1618–1632, 1995.
28. A. Prakash, M. Blanchette, S. Sinha, et al. Motif discovery in heterogeneous sequence data. *Pac Symp Biocomput*, 348–359, 2004.
29. A. Price, S. Ramabhadran, and P.A. Pevzner. Finding subtle motifs by branching from sample strings. *Bioinformatics*, **19**(2):ii149–ii155, 2003.
30. Z.S. Qin, L.A. McCue, W. Thompson, et al. Identification of co-regulated genes through Bayesian clustering of predicted regulatory binding sites. *Nat Biotechnol*, **21**(4):435–439, 2003.
31. S. Sinha and M. Tompa. A statistical method for finding transcription factor binding sites. *Proc Int Conf Intell Syst Mol Biol*, **8**:344–354, 2000.
32. S. Sinha, M. Blanchette, and M. Tompa. PhyME: a probabilistic algorithm for finding motifs in sets of orthologous sequences. *BMC Bioinform*, **5**:170, 2004.
33. M. Tompa. An exact method for finding short motifs in sequences, with application to the ribosome binding site problem. *Proc Int Conf Intell Syst Mol Biol*, 262–271, 1999.
34. T. Wang and G.D. Stormo. Combining phylogenetic data with co-regulated genes to identify regulatory motifs. *Bioinformatics*, **19**(18):2369–2380, 2003.
35. T. Wang and G.D. Stormo. Identifying the conserved network of cis-regulatory sites of a eukaryotic genome. *Proc Natl Acad Sci USA*, **102**(48):17400–17405, 2005.

36. B. Ren, F. Robert, J.J. Wyrick, et al. Genome-wide location and function of DNA binding proteins. *Science*, **290**(5500):2306–2309, 2000.
37. D.S. Johnson, A. Mortazavi, R.M. Myers, et al. Genome-wide mapping of in vivo protein-DNA interactions. *Science*, **316**(5830):1497–1502, 2007.
38. G. Robertson, M. Hirst, M. Bainbridge, et al. Genome-wide profiles of STAT1 DNA association using chromatin immunoprecipitation and massively parallel sequencing. *Nat Methods*, **4**(8):651–657, 2007.
39. S. Sinha, M. Blanchette, and M. Tompa. PhyME: a probabilistic algorithm for finding motifs in sets of orthologous sequences. *BMC Bioinform*, **5**:170, 2004.
40. W.W. Wasserman, M. Palumbo, W. Thompson, et al. Human-mouse genome comparisons to locate regulatory sites. *Nat Genet*, **26**(2):225–228, 2000.
41. M.C. Frith, U. Hansen, and Z. Weng. Detection of cis-element clusters in higher eukaryotic DNA. *Bioinformatics*, **17**(10):878–889, 2001.
42. Q. Zhou and W.H. Wong. CisModule: de novo discovery of cis-regulatory modules by hierarchical mixture modeling. *Proc Natl Acad Sci USA*, **101**(33):12114–12119, 2004.
43. M. Brudno, C.B. Do, G.M. Cooper, et al. LAGAN and Multi-LAGAN: efficient tools for large-scale multiple alignment of genomic DNA. *Genome Res*, **13**(4):721–731, 2003.
44. M. Brudno, M. Chapman, B. Gottgens, et al. Fast and sensitive multiple alignment of large genomic sequences. *BMC Bioinformatics*, **4**:66, 2003.
45. R. Siddharthan, E.D. Siggia, and E. van Nimwegen. PhyloGibbs: a Gibbs sampling motif finder that incorporates phylogeny. *PLoS Comput Biol*, **1**(7):e67, 2005.
46. A.R. Subramanian, M. Kaufmann, and B. Morgenstern. DIALIGN-TX: greedy and progressive approaches for segment-based multiple sequence alignment. *Algorithms Mol Biol*, **3**:6, 2008.
47. M. Kellis, N. Patterson, M. Endrizzi, et al. Sequencing and comparison of yeast species to identify genes and regulatory elements. *Nature*, **423**(6937):241–254, 2003.
48. E.H. Margulies, G.M. Cooper, G. Asimenos, et al. Analyses of deep mammalian sequence alignments and constraint predictions for 1% of the human genome. *Genome Res*, **17**(6):760–774, 2007.
49. M. Blanchette, W.J. Kent, C. Riemer, et al. Aligning multiple genomic sequences with the threaded blockset aligner. *Genome Res*, **14**(4):708–715, 2004.
50. N. Bray and L. Pachter. MAVID: constrained ancestral alignment of multiple sequences. *Genome Res*, **14**(4):693–699, 2004.
51. R.W. Blakesley, N.F. Hansen, J.C. Mullikin, et al. An intermediate grade of finished genomic sequence suitable for comparative analyses. *Genome Res*, **14**(11):2235–2244, 2004.
52. G.Z. Hertz and G.D. Stormo. Identifying DNA and protein patterns with statistically significant alignments of multiple sequences. *Bioinformatics*, **15**(7–8):563–577, 1999.
53. P. Cliften, P. Sudarsanam, A. Desikan, et al. Finding functional features in Saccharomyces genomes by phylogenetic footprinting. *Science*, **301**(5629):71–76, 2003.
54. C.T. Harbison, D.B. Gordon, T.I. Lee, et al. Transcriptional regulatory code of a eukaryotic genome. *Nature*, **431**(7004):99–104, 2004.
55. I. Witt, N. Straub, N.F. Kaufer, et al. The CAGTCACA box in the fission yeast Schizosaccharomyces pombe functions like a TATA element and binds a novel factor. *Embo J*, **12**(3):1201–1208, 1993.
56. M.K. Das and H.K. Dai. A survey of DNA motif finding algorithms. *BMC Bioinform*, **8**(7):S21, 2007.
57. M. Tompa, N. Li, T.L. Bailey, et al. Assessing computational tools for the discovery of transcription factor binding sites. *Nat Biotechnol*, **23**(1):137–144, 2005.
58. M. Blanchette, A.R. Bataille, X. Chen, et al. Genome-wide computational prediction of transcriptional regulatory modules reveals new insights into human gene expression. *Genome Res*, **16**(5):656–668, 2006.
59. S. Hannenhalli. Eukaryotic transcription factor binding sites—modeling and integrative search methods. *Bioinformatics*, **24**(11):1325–1331, 2008.
60. W.S. Noble, S. Kuehn, R. Thurman, et al. Predicting the in vivo signature of human gene regulatory sequences. *Bioinformatics*, **21**(1):i338–i343, 2005.

61. E. Segal, Y. Fondufe-Mittendorf, L. Chen, et al. A genomic code for nucleosome positioning. *Nature*, **442**(7104):772–778, 2006.
62. F. Fang, S. Fan, X. Zhang, et al. Predicting methylation status of CpG islands in the human brain. *Bioinformatics*, **22**(18):2204–2209, 2006.
63. S. Hannenhalli and L.S. Wang. Enhanced position weight matrices using mixture models. *Bioinformatics*, **21**(1):i204–i212, 2005.
64. M. Neumann and M. Naumann. Beyond IkappaBs: alternative regulation of NF-kappaB activity. *Faseb J*, **21**(11):2642–2654, 2007.

Chapter 7
Learning Network from High-Dimensional Array Data

Li Hsu, Jie Peng, and Pei Wang

7.1 Introduction

The study of interactions among biological components helps to shed light on the functional interconnections among the regulatory genes and their signaling components, consequently resulting in a better understanding of disease pathologies. In many recent studies, correlations (or other essentially equivalent statistics, such as regression coefficients) derived from high-throughput array data have been used to infer interactions among molecular activities, for it is believed that strong interactions among various functional components often result in significant correlations among genes or clones measured in the experiments [17, 25, 29, 41]. The most straightforward method is to build a *relevance network* by declaring an edge between two genes if the absolute correlation of their molecular activity measurements exceed a threshold [3]. Such high correlations can be due to direct interactions with each other (e.g., one gene is regulated by another gene) or indirect interactions through intermediate genes (e.g., coregulated by a third gene). A relevance network cannot distinguish between these two types of interactions. To further investigate relationships among genes, *Gaussian Graphical Models* (*GGMs*) have been adopted (see, for example, [19, 22, 27, 33, 47]). In a GGM, each vertex represents a gene, and an edge will be drawn between two genes only if their corresponding molecular activity measurements are conditionally dependent given the activity measurements of all other genes. There is a rich literature on fitting GGMs (see, for example, [5–8, 45], and references therein). However, in array data where the

L. Hsu (✉) · P. Wang
Division of Public Health Sciences, Fred Hutchinson Cancer Research Center, Seattle, WA, USA
e-mail: lih@fhcrc.org

P. Wang (✉)
e-mail: pwang@fhcrc.org

J. Peng (✉)
Department of Statistics, University of California, Davis, CA, USA
e-mail: jie@wald.ucdavis.edu

J. Feng et al. (eds.), *Frontiers in Computational and Systems Biology,* Computational Biology 15,
DOI 10.1007/978-1-84996-196-7_7,

number of genes p is typically much larger than the number of samples n, the classical methods do not work any more. To tackle this challenge, pioneer work has been done for high-dimensional *continuous* variables, including likelihood-based approach [11, 19, 31, 47], regression-based approach [22], and many others (see, for example, [21, 33]). In a more recent work [27], we develop a new method **space**—Sparse Partial Correlation Estimation—for selecting nonzero partial correlations under the high-dimension-low-sample-size setting. This method employs sparse regression techniques by imposing a sparsity constraint on the network as a whole, which is particulary effective in identifying hub nodes of a network.

When the molecular activities are measured or summarized in *binary* variables, tools developed for high-dimensional *continuous* variables cannot be directly applied. As a parallel development to the continuous case, we propose a new algorithm **LogitNet** [43] for inferring the conditional dependence between pairs of *binary* variables given all others. Assuming a tree topology for the genetic network, we establish a connection between joint probability distribution of the p binary variables and p logistic regression models with symmetric coefficient matrix. As in `space`, we impose a sparsity constraint on the network as a whole.

The idea of utilizing conditional dependency can be generalized to build networks based on multiple types of data. Motivated by studying the influence of DNA copy number alterations on RNA transcript levels, we develop **remMap**—REgularized Multivariate regression for identifying MAster Predictors—for fitting multivariate regression models with high-dimensional responses and predictors [28]. `RemMap` employs the *MAP penalty*, which consists of an ℓ_1 norm part for controlling the overall sparsity of the network and an ℓ_2 norm part for encouraging the detection of hub nodes in the network. This combined regularization takes into account both model interpretability and computational tractability.

A coherent theme among the three methods is to utilize proper sparse regularization schemes under regression frameworks to deal with high dimensionality while incorporating desired network structures. In the rest of this chapter, we will discuss each of these three methods in detail (Sects. 7.2–7.4). We will also describe a real application (Sect. 7.5) and then conclude the chapter with a short summary (Sect. 7.6).

7.2 `space`

7.2.1 Model

In [27], we proposed a novel method for constructing GGMs for a high-dimensional vector based on i.i.d. samples. This method is referred to as `space`. Since high-throughput genomic experiments usually involve thousands of genes for tens or at most hundreds of samples, the challenge is on how to deal with the high-dimension-low-sample-size. While this seems to be a prohibitively difficult problem, it is widely believed that genetic regulatory relationships are intrinsically sparse. Thus one can reasonably assume that for most gene pairs, their corresponding measurements are conditionally independent. Moreover, for genetic regulatory networks (GRNs), there often exist well-connected hub nodes. Since these nodes play important roles in shaping network functionality, it is particularly interesting to identify them. To tackle these challenges, `space` utilizes sparse regression techniques to induce model sparsity while encouraging the selection of hub nodes.

Suppose that $Y = (y_1, \dots, y_p)^T$, where the superscript T is a transpose, has a joint distribution with mean 0 and covariance Σ. The *partial correlation* ρ^{ij} between y_i and y_j is defined as $\rho^{ij} := \mathrm{Corr}(\epsilon_i, \epsilon_j)$, where ϵ_i and ϵ_j are the residual errors of the best linear predictors of y_i and y_j based on $y_{-(i,j)} = \{y_k : 1 \le k \ne i, j \le p\}$, respectively. Note that, under the normality assumption, partial correlations equal to conditional correlations and thus a zero partial correlation means conditional independency. The inverse of the covariance matrix Σ is called the *concentration matrix*. Denote $\Sigma^{-1} = (\sigma^{ij})_{p \times p}$ and let $y_{-i} := \{y_k : 1 \le k \ne i \le p\}$. The following well-known result relates partial correlations to regression coefficients [27]: *for* $1 \le i \le p$, y_i *is expressed as* $y_i = \sum_{j \ne i} \beta_{ij} y_j + \epsilon_i$, *such that* ϵ_i *is uncorrelated with* y_{-i} *if and only if* $\beta_{ij} = -\frac{\sigma^{ij}}{\sigma^{ii}} = \rho^{ij}\sqrt{\frac{\sigma^{jj}}{\sigma^{ii}}}$. *Moreover, for such defined* β_{ij}, $\mathrm{Var}(\epsilon_i) = \frac{1}{\sigma^{ii}}$. Since $\rho^{ij} = \mathrm{sign}(\beta_{ij})\sqrt{\beta_{ij}\beta_{ji}}$, the search for nonzero partial correlations can be viewed as a model selection problem under a regression framework.

Suppose $\mathbf{Y}^k = (y_1^k, \dots, y_p^k)^T$ are i.i.d. observations from $(0, \Sigma)$ for $k = 1, \dots, n$. Denote the sample of the ith variable as $\mathbf{Y}_i = (y_i^1, \dots, y_i^n)^T$. The `space` method estimates the partial correlations $\boldsymbol{\theta} = (\rho^{12}, \dots, \rho^{(p-1)p})^T$ by minimizing the following *joint penalized loss function*:

$$\mathcal{L}_n(\boldsymbol{\theta}, \boldsymbol{\sigma}, \mathbf{Y}) = L_n(\boldsymbol{\theta}, \boldsymbol{\sigma}, \mathbf{Y}) + \mathcal{J}(\boldsymbol{\theta}). \tag{7.1}$$

In the above, $L_n(\boldsymbol{\theta}, \boldsymbol{\sigma}, \mathbf{Y})$ is the weighted sum of the ℓ_2 loss over all p regressions,

$$\begin{aligned} L_n(\boldsymbol{\theta}, \boldsymbol{\sigma}, \mathbf{Y}) &:= \frac{1}{2}\left(\sum_{i=1}^{p} w_i \left\| \mathbf{Y}_i - \sum_{j \ne i} \beta_{ij} \mathbf{Y}_j \right\|^2\right) \\ &= \frac{1}{2}\left(\sum_{i=1}^{p} w_i \left\| \mathbf{Y}_i - \sum_{j \ne i} \rho^{ij} \sqrt{\frac{\sigma^{jj}}{\sigma^{ii}}} \mathbf{Y}_j \right\|^2\right), \end{aligned} \tag{7.2}$$

where $\boldsymbol{\sigma} = \{\sigma^{ii}\}_{i=1}^p$, $\mathbf{Y} = \{\mathbf{Y}^k\}_{k=1}^n$, and $\boldsymbol{w} = \{w_i\}_{i=1}^p$ are nonnegative weights, and the penalty term $\mathcal{J}(\boldsymbol{\theta})$ aims to control the overall sparsity of the final estimation of $\boldsymbol{\theta}$. In [27], we focus on the ℓ_1 penalty (also a.k.a. *lasso* penalty [37]):

$$\mathcal{J}(\boldsymbol{\theta}) = \lambda \|\boldsymbol{\theta}\|_1 = \lambda \sum_{1 \le i < j \le p} |\rho^{ij}|. \tag{7.3}$$

In `space`, sparsity is imposed on the partial correlations $\boldsymbol{\theta}$ as a whole. This "joint" modeling helps to efficiently utilize data and also encourages the selection of hub nodes. This method also preserves the intrinsic symmetry of the partial correlation matrix, i.e., $\rho^{ij} = \rho^{ji}$ by estimating them directly. Moreover, `space` is flexible in incorporating prior knowledge. One may assign different weights to different nodes according to their "importance." For example, as in weighted least squares, the weight can be set as the (estimated) residual variance. Or one can let the weight be proportional to the (estimated) degree of each variable. This would result in a *preferential attachment* effect which explains the cumulative advantage phenomena observed in many real life networks including GRNs [23].

`Space` estimates $\boldsymbol{\theta}$ and $\boldsymbol{\sigma}$ by a two-step iterative procedure. Given an initial estimate $\boldsymbol{\sigma}^{(0)}$ of $\boldsymbol{\sigma}$, $\boldsymbol{\theta}$ is estimated by minimizing the penalized loss function (7.1), whose implementation is discussed below. Then given the current estimates $\boldsymbol{\theta}^{(c)}$ and $\boldsymbol{\sigma}^{(c)}$, $\boldsymbol{\sigma}$ is updated by $1/\hat{\sigma}^{ii} = \frac{1}{n}\|\mathbf{Y}_i - \sum_{j\neq i}\hat{\beta}_{ij}^{(c)}\mathbf{Y}_j\|^2$, where $\hat{\beta}_{ij}^{(c)} = (\rho^{ij})^{(c)}\sqrt{\frac{(\sigma^{jj})^{(c)}}{(\sigma^{ii})^{(c)}}}$. These two steps are then iterated until convergence is reached for all parameters. Based on our experience, it usually takes no more than three cycles for this procedure to stabilize.

Given $\boldsymbol{\sigma}$ and positive weights $\boldsymbol{w}$, let $\mathcal{Y} = (\tilde{\mathbf{Y}}_1^T, \ldots, \tilde{\mathbf{Y}}_p^T)^T$ be an $np \times 1$ column vector, where $\tilde{\mathbf{Y}}_i = \sqrt{w_i}\,\mathbf{Y}_i$, and let $\mathcal{X} = (\tilde{\mathcal{X}}_{(1,2)}, \ldots, \tilde{\mathcal{X}}_{(p-1,p)})$ be an np by $p(p-1)/2$ matrix, with

$$\tilde{\mathcal{X}}_{(i,j)} = \Bigg(\underset{\phantom{i\text{th block}}}{0, \ldots, 0}, \underset{\substack{\uparrow \\ i\text{th block}}}{\sqrt{\frac{\tilde{\sigma}^{jj}}{\tilde{\sigma}^{ii}}}\tilde{\mathbf{Y}}_j^T}, 0, \ldots, 0, \underset{\substack{\uparrow \\ j\text{th block}}}{\sqrt{\frac{\tilde{\sigma}^{ii}}{\tilde{\sigma}^{jj}}}\tilde{\mathbf{Y}}_i^T}, 0, \ldots, 0\Bigg)^T,$$

where $\tilde{\sigma}^{ii} = \sigma^{ii}/w_i$. Then it is easy to see that the loss function (7.2) equals to $\frac{1}{2}\|\mathcal{Y} - \mathcal{X}\boldsymbol{\theta}\|_2^2$, and the minimization with respect to $\boldsymbol{\theta}$ is equivalent to

$$\min_{\theta} \frac{1}{2}\|\mathcal{Y} - \mathcal{X}\boldsymbol{\theta}\|_2^2 + \lambda\|\boldsymbol{\theta}\|_1.$$

Note that the current dimensions $\tilde{n} = np$ and $\tilde{p} = p(p-1)/2$ are of much higher order than the original n and p. Fortunately, $\mathcal{X}$ is a block matrix with many zero blocks. Thus, algorithms for lasso regression can be efficiently implemented by taking into consideration this structure (see [27] for more details). To further improve the convergence speed, `space` implements an `active-shooting` algorithm, modifying the `shooting` algorithm, which is first proposed by Fu in [13] and then extended by many others including [10, 12, 14]. It solves ℓ_1 penalized regression problems by updating each coordinate iteratively until convergence. Since under a sparse model, only a small subset of variables have nonzero coefficients, a faster convergence rate can be achieved by focusing on the set of variables that are more likely to be in the model (referred to as the *active set*). Suppose that the goal is

$$\min_{\beta\in\mathcal{R}^p} \frac{1}{2}\|Y - X\beta\|_2^2 + \lambda\sum_{j=1}^{p}|\beta_j|.$$

The `active-shooting` algorithm proceeds as follows:

1. Initial step: previously obtained estimates or estimates obtained from univariate soft shrinkage,

$$\beta_j^{(0)} = \text{sign}(Y^T X_j)\big(|Y^T X_j / X_j^T X_j| - \lambda / X_j^T X_j\big)_+.$$

2. Define the current active set $\Lambda = \{k : \text{current } \beta_k \neq 0\}$.
   (1) For each $k \in \Lambda$, update β_k with all other coefficients fixed at the current values.
   (2) Repeat (1) until convergence is achieved on the active set.
3. For $i = 1$ to p, update β_i with all other coefficients fixed at the current value. If no β_i changes during this process, stop; otherwise, go back to step 2.

A simulation study in [27] shows that `active-shooting` greatly improves computational efficiency over `shooting` for sparse models. In particular, it takes less than 30 seconds (on average) to fit the `space` model by `active-shooting` (implemented in `c` code) for cases with 1000 variables and 200 samples and when the resulting model has around 1000 nonzero partial correlations on a server with two Dual/Core, CPU 3 GHz, and 4 GB RAM.

7.2.2 Simulation

In [27], simulation studies were performed to compare `space` and two alternative methods. The first one is the *neighborhood selection* approach proposed by Meinshausen and Buhlmann in [22] (referred to as `MB` hereafter), where p lasso regressions are performed separately. By comparing `space` with `MB`, we illustrated the advantages of the joint modeling approach. The second method is the penalized maximum likelihood approach proposed by [47] and efficiently implemented by [11] (referred to as `glasso` hereafter). This method assumes a multivariate normal distribution and imposes the ℓ_1 norm of the concentration matrix as penalty on the negative log-likelihood function. As suggested by the simulation results, the relatively simpler loss function used by `space` (which is quadratic in the unknown parameters) appears to pay off under the high-dimension-low-sample-size setting [27].

For `space`, the initial value of σ^{ii} was set to be one. The initial weights were also set to be one (i.e., equal weights). In each subsequent iteration, new weight w_i was set to be proportional to the estimated degree of y_i, i.e., $\#\{j : \hat{\rho}^{ij} \neq 0, j \neq i\}$. The corresponding method is referred to as **`space.dew`** (degree-based weight). For `glasso`, the diagonal of the concentration matrix was not penalized.

A network consisting of five disjointed modules was simulated, with each module having 100 nodes. This is aimed to mimic real-life large networks which often exhibit a modular structure comprised of many disjointed or loosely connected components of relatively small size. Moreover, each module consisted of three hubs with degrees around 15, and the other nodes with degrees in between one to four. In total, there were 500 nodes and 568 edges. (This network was referred to as the Hub network.) Nonzero partial correlations were assumed to fall

Table 7.1 Power (sensitivity) of `space.dew`, `MB`, and `glasso` in identifying correct edges when FDR is controlled at 0.05

p	n	`space.dew`	`MB`	`glasso`
500	250	0.844	0.784	0.655
1000	200	0.707	0.656	0.559
	300	0.856	0.790	0.690
	500	0.963	0.894	0.826

into $(-0.67,-0.1] \cup [0.1,0.67)$, with two modes around -0.28 and 0.28. Finally $n = 250$ i.i.d. samples were drawn from the corresponding multivariate normal distribution.

Each method was evaluated at a series of different values of the tuning parameter. (The tuning parameter controls the total number of edges in the estimated model.) Figure 7.1(a) shows the number of correctly detected edges (N_c) vs. the number of total detected edges (N_t) averaged over 50 independent data sets. It can be seen that `space.dew` performed the best among all methods. Specifically, when $N_t = 568$ (which is the number of true edges), `space.dew` detected 501 correct edges on average with a standard deviation 4.5 edges. On the other hand, `MB` and `glasso` detected 472 and 480 correct edges on average, respectively. In terms of hub detection, for a given N_t, a rank was assigned to each variable y_i based on its estimated degree (the larger the estimated degree, the smaller the rank value). The average rank of the 15 true hub nodes was then calculated for each method. The results are shown in Fig. 7.1(b). This average rank would achieve the minimum value 8 (indicated by the grey horizontal line) if the 15 true hubs had larger estimated degrees than all other nonhub nodes. As it can be seen from Fig. 7.1(b), the average rank curves (as a function of N_t) for `space.dew` was very close to the optimal minimum value 8 for a large range of N_t. This suggests that `space.dew` can successfully identify most of the true hubs. On the other hand, both `MB` and `glasso` identified far fewer hub nodes, as their corresponding average rank curves were much higher than the grey horizontal line.

To investigate the impact of dimensionality p and sample size n, simulations were conducted for a larger dimension with $p = 1000$ and various sample sizes with $n = 200$, 300, and 500. The simulated network included ten disjointed modules of size 100 each and had 1163 edges in total. Nonzero partial correlations formed a similar distribution as that of the $p = 500$ network discussed above. When false discovery rate (= 1-specificity) is controlled at 0.05, the power (= sensitivity) for detecting correct edges is given in Table 7.1. The sample size appeared to have a big impact on the performance of all methods. For $p = 1000$, when the sample size increased from 200 to 300, the power of `space.dew` increased more than 20%; when the sample size was 500, `space.dew` achieved an impressive power of 96%.

These methods were also applied to various networks without hubs. `space.dew` performed similarly as, if not better than, the other two methods, although its advantages became smaller compared to the results on the networks with hubs. Nonnormal distributions, in particular, multivariate t-distribution with degrees of freedom

3, 6, 10, were also considered. As expected, the performances of all methods deteriorated compared to the normal case; however, the relative performance of these methods remained essentially unchanged.

In summary, `space` utilizes a joint sparse regression model for selecting nonzero partial correlations under the high-dimension-low-sample-size setting. By controlling the overall sparsity of the partial correlation matrix, it is able to automatically adjust for different neighborhood sizes and thus utilizes data effectively, especially in terms of hub nodes identification. This method is implemented through a fast algorithm `active-shooting`, which can be readily extended to solve many other penalized optimization problems. See Sects. 7.3 and 7.4 for examples of such extension.

7.3 LogitNet

7.3.1 Model

This section describes a sparse regression-based approach to infer the conditional dependencies for binary variables [43]. This is a parallel development to the `space` method for continuous variables. Binary variables arise frequently in genomic data, for example, genomic aberration status or mutation status at marker loci in patient samples. Understanding the interactions of these aberration events will provide us insights into biological mechanism of disease process.

Let $X^T = (X_1, \dots, X_p)$ be a $p \times 1$ vector of binary variables. The pattern of conditional dependencies between these binary variables can be described by an undirected graph $\mathcal{G} = (V, E)$, where V is a finite set of vertices, $(1, \dots, p)$, that are associated with binary variables $(X_1, \dots, X_p)$; and E is a set of pairs of vertices such that each pair in E is conditionally dependent given the rest of binary variables. Assume that the edge set E does not contain cliques more than 2 and the joint probability distribution $\Pr(X)$ can be represented as a product of functions of pairs of binary variables. Furthermore, if $\Pr(X)$ is strictly positive for all values of X, then $\Pr(X)$ leads to the well-known *quadratic exponential model*

$$\Pr(X = x) = \Delta^{-1} \exp\bigl(x^T\theta + z^T\kappa\bigr),$$

where $z^T = (x_1x_2, x_1x_3, \dots, x_{p-1}x_p)$, $\theta^T = (\theta_1, \dots, \theta_p)$, $\kappa^T = (\kappa_{12}, \kappa_{13}, \dots, \kappa_{(p-1)p})$, and Δ is a normalization constant such that $\Delta = \sum_{x_1=0}^{1} \cdots \sum_{x_p=0}^{1} \exp(x^T\theta + z^T\kappa)$.

There are a couple of appealing properties of this model. First, the zero values in κ are equivalent to the conditional independence for the corresponding binary variables. This implies that to infer the edges in E, one only needs to examine whether the corresponding κ parameter values are zero or not. Second, κ can also be interpreted as a conditional odds ratio between paired binary variables given all

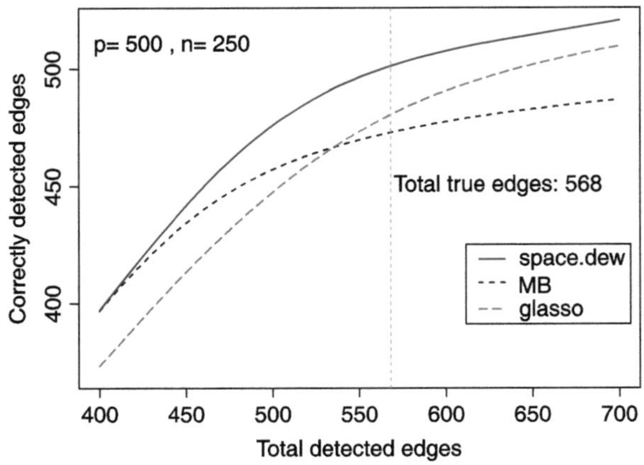

(a) *x-axis*: the number of total detected edges(i.e., the total number of pairs (i, j) with $\widehat{\rho}^{ij} \neq 0$); *y-axis*: the number of correctly identified edges. The vertical grey line corresponds to the number of true edges.

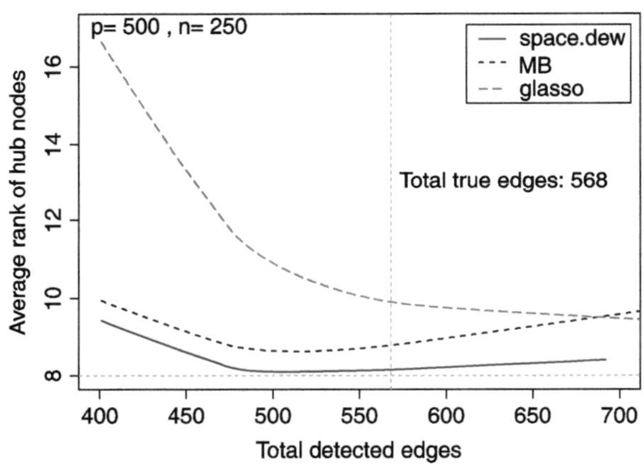

(b) *x-axis*: the number of total detected edges; *y-axis*: the average rank of the estimated degrees of the 15 true hub nodes.

Fig. 7.1 Simulation results for Hub network

others and obtained by p logistic regression models with each binary variable as an outcome and the rest of the binary variables as covariates. This can be written as

$$\begin{cases} \text{logit}\{\Pr(x_1 = 1|x_2, \ldots, x_p)\} = \theta_1 + \kappa_{12}x_2 + \cdots + \kappa_{1p}x_p, \\ \qquad\qquad\qquad\qquad \vdots \\ \text{logit}\{\Pr(x_p = 1|x_1, \ldots, x_{p-1})\} = \kappa_{1p}x_1 + \cdots + \kappa_{(p-1)p}x_{p-1} + \theta_p. \end{cases} \tag{7.4}$$

Let $\mathcal{B}$ be the matrix of the row combined regression coefficients from p logistic regression models, and β_{rs} the matrix element for the rth row and the sth column. It is easy to see that the $\mathcal{B}$ matrix is symmetric, i.e., $\beta_{rs} = \beta_{sr} = \kappa_{rs}, i \neq j$. Interestingly, the symmetry of $\mathcal{B}$ also ensures the compatibility of the p logistic conditional distributions, and the resulting joint distribution is the quadratic exponential model [15].

Therefore, to infer the edge set E of the graph, one can use the regression coefficient estimates obtained by fitting the p logistic regression models in (7.4) with symmetric $\mathcal{B}$. The log pseudo-likelihood function for (7.4) is then $l(\mathcal{B}) = -\sum_{r=1}^{p} l_p(\mathcal{B})$, where $l_p(\mathcal{B})$ is the log-likelihood function for the rth logistic regression model, $r = 1, \ldots, p$.

An advantage for using regression models to infer graphs is one can make use of sparse regression techniques to handle high-dimension-low-sample-size problems when the true model is sparse relative to the dimensionality of the data. The basic idea is to constrain the number of nonzero coefficients or shrink the estimates of coefficients so that the model is not overfit in the case of $p \gg n$. One particularly popular choice for constraining the parameter estimation is by using ℓ_1 norm or the `lasso` penalty [37]. Adding this penalty function to the negative of log pseudo-likelihood function (7.4) gives a regularized or penalized loss function

$$l_\lambda^{\text{lasso}}(\mathcal{B}) = -l(\mathcal{B}) + \lambda \sum_{r=1}^{p} \sum_{s=r+1}^{p} |\beta_{rs}|, \tag{7.5}$$

and $\hat{\mathcal{B}}(\lambda) := \arg\min_{\mathcal{B}} l_\lambda^{\text{lasso}}(\mathcal{B})$. The model defined in (7.5) is referred to as `LogitNet` model, and $\hat{\mathcal{B}}(\lambda)$ is the `LogitNet` estimator with $\widehat{\beta}_{rs}(\lambda)$ as the rsth element of $\hat{\mathcal{B}}(\lambda)$.

To minimize the loss function (7.5), the gradient descent algorithm [14] needs to be extended to enforce the symmetry of $\mathcal{B}$. Since there is no closed form solution to minimizing (7.5), a one-step Newton–Raphson algorithm is used to update the parameter estimates one at a time in the same spirit as the shooting algorithm [10, 13] for solving general linear `lasso` regression. Care is taken when the current estimate is 0 or the update of an estimate crosses 0, as ℓ_1 norm is not differentiable at 0. The algorithm also takes other steps to ensure a stable numerical procedure, for example, limiting the update size and setting the upper bound for the Hessian matrix [49]. This is the basic algorithm. To improve the convergence speed, the algorithm also adopts the `active-shooting idea` used in the `space` algorithm (Sect. 7.2), which focuses on the convergence of currently nonzero regression coefficients (active set) while looping the whole set of regression coefficients. Further details of the algorithm can be found in [43].

In the penalized regression, the penalty term shrinks the estimates towards zero by the amount determined by the penalty parameter λ. However, it is worth noting that each parameter is not penalized by the same amount. This can be seen from the update for $\widehat{\beta}_{rs}(\lambda)$, $r = 1, \ldots, p$ and $s = (r+1), \ldots, p$,

$$\Delta\beta_{rs}^{\text{lasso}} = \Delta\beta_{rs} - \frac{\lambda}{\ddot{l}(\beta_{rs})}\text{sgn}(\beta_{rs}), \tag{7.6}$$

where $\text{sgn}(\beta_{rs}) = 1$ if β_{rs} is positive and -1 if β_{rs} is negative; and $\ddot{l}(\beta_{rs})^{-1}$ is the variance of $\widehat{\beta}_{rs}$. When $\lambda = 0$, i.e., no penalty, the update for penalized lasso

$\Delta\beta_{rs}^{\text{lasso}}$ is same as $\Delta\beta_{rs}$. When $\lambda > 0$, the penalty is weighted by the variance of the estimate, i.e., estimates with larger variance are penalized more than estimates with smaller variances. It turns out that this type of penalization is very useful, as it offers ways to account for specific features of a particular data type. In the next section on the application of `LogitNet` to genomic instability data, it is shown a proposal for how to modify the weighted penalty parameter to account for specific features of the data.

Two common procedures can be used for selecting the penalty parameter λ: cross validation (CV) and Bayesian Information Criterion (BIC). For both procedures, the penalized log-likelihood function (7.5) is used for selecting nonzero coefficients. After the nonzero coefficients are determined, they are reestimated without penalty, and the reestimated ones are then plugged in the log-likelihood function, because when $p \gg n$, the estimates from the penalized likelihood function can be considerably shrunk and do not reflect the true values any more. The reestimation is a common practice when applying CV for ℓ_1 penalized regression under high-dimensional setting [9]. To further control the false positive rate, a `cv.vote` procedure from [28] is applied, which retains only those variables being consistently selected by many, for example, more than half of, cross validation folds in the final model.

7.3.2 Application to Genomic Instability Data

This section uses genomic instability data as an example to illustrate how `LogitNet` can be modified to accommodate a specific feature in the data. Specifically, genomic instability, referring to as the propensity of aberrations in chromosomes, plays a critical role in the development of many diseases. In this type of data, spatial correlation of aberrations is very common. This can pose problem in inferring nonzero coefficients, because loci that are spatially closest to the target are likely the strongest predictors in the model and would explain away most of the variation in the target locus. This leaves little variation for loci at other locations even if they are correlated with the target locus. Obviously this result is not desirable because the objective is to identify the network among all of these loci, particularly those that are not close spatially.

One approach to account for this undesirable spatial effect is to downweight the effect of the neighboring loci of the target locus, say, X_r when regressing X_r on the rest of the loci. Recall that the penalty term in (7.6) is weighted by the variance of the parameter estimates. Following the same idea, one can achieve the downweighting of neighboring loci by letting the penalty term be proportional to the strength of their correlations with X_r. This way one can shrink the effects of the neighboring loci with strong spatial correlation more than those that have less or no spatial correlation. Specifically, the update for the parameter estimate β_{rs} in (7.6) can be written as

$$\Delta\beta_{rs}^{\text{lasso}} = \Delta\beta_{rs} - w_{rs}\frac{\lambda}{\ddot{l}(\beta_{rs})}\text{sgn}(\beta_{rs}),$$

where w_{rs} is the weight for the spatial correlation. It is natural to assume that spatial correlation only exists within the same chromosome. Thus, for events on different chromosomes, the corresponding weight w_{rs} can be simply set to 1; while for events on the same chromosome, the corresponding w_{rs} is calculated based on the strength of the spatial correlation between locus r and locus s. One may first calculate the log odds ratio between the target locus (the outcome) and each of the rest of the loci, and then smooth the log odds ratios along the genome with a window size of say, 10 loci. The smoothed curve is set to zero when the curve starting from the target locus hits 0, and the weight may be set as exponential of the smoothed curve.

Interestingly, the above weighting scheme together with the enforcement of the symmetry of $\mathcal{B}$ in `LogitNet` encourages a group selection effect, i.e., highly correlated predictors tend to be in or out of the model simultaneously. This point can be illustrated with a simple example system of three variables X_1, X_2, and X_3. Suppose that X_2 and X_3 are very close on the genome and highly correlated and that X_1 is associated with X_2 and X_3 but sits on a different chromosome. Under the proposed weighting scheme, the weight matrix w is 1 for all entries except $w_{23} = w_{32} = a$, which is a large value because of the strong spatial correlation between X_2 and X_3. Then, for `LogitNet`, the joint logistic regression model

$$\texttt{logit}(X_1) \sim \beta_{11} + \beta_{12}X_2 + \beta_{13}X_3, \tag{7.7}$$

$$\texttt{logit}(X_2) \sim \beta_{12}X_1 + \beta_{22} + \beta_{23}X_3, \tag{7.8}$$

$$\texttt{logit}(X_3) \sim \beta_{13}X_1 + \beta_{23}X_2 + \beta_{33}, \tag{7.9}$$

is subject to the constraint $|\beta_{12}| + |\beta_{13}| + a|\beta_{23}| < s$. Because of the large value of a, β_{23} will likely be shrunk to zero, which ensures β_{12} and β_{13} to be nonzero in (7.8) and (7.9), respectively. With the symmetry constraint imposed on $\mathcal{B}$ matrix, both β_{12} and β_{13} are forced to be selected in (7.7). This grouping effect would not happen if only the model (7.7) were fit, for which only one of β_{12} and β_{13} would likely be selected [51], nor would it happen if one did not have a large value of a because β_{23} would have been the dominant coefficient in models (7.8) and (7.9). Indeed, the group selection effect of `LogitNet` is clearly observed in the simulation studies shown in the next section.

7.3.3 Simulation

The `LogitNet` model is closely related to the work by [30], which fits p `lasso` logistic regressions separately (hereafter referred to as `SepLogit`), whereas `LogitNet` fits p logistic regression models jointly while enforcing symmetry of regression coefficients. In [43], the performance of `LogitNet` was evaluated and compared with `SepLogit`. Since the `SepLogit` method did not ensure symmetry, there would be cases where $\beta_{rs} = 0$ but $\beta_{sr} \neq 0$ or vice versa. In these cases the result may be interpreted using the "OR" rule: X_r and X_s are deemed to be conditionally dependent if either β_{rs} or β_{sr} is nonzero. The "AND" rule was also used,

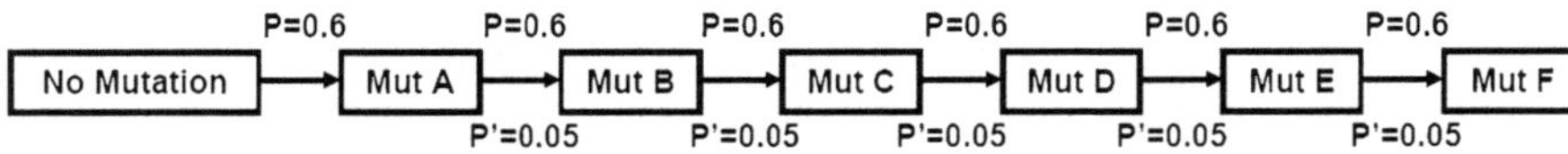

Fig. 7.2 Oncogenic pathway of a chain shape. For each adjacent mutation pair (M_1, M_2) with M_1 on the *left side* of the *arrow* and M_2 on the *right side* of the *arrow*, the number above (or below) the *arrow* gives the conditional probability of $\Pr(M_2 = 1|M_1 = 1)$ (or $\Pr(M_2 = 1|M_1 = 0)$)

however, it always yielded very high false negative rate. Due to space limitation, the results for the "AND" rule were omitted.

Data were generated mimicking genomic instability data. Background aberration events with spatial correlation were generated using a homogenous Bernoulli Markov model. It was part of the instability-selection model [24], which hypothesized that the genetic structure of a progenitor cell was subject to chromosomal instability that causes random aberrations. Then aberration events that followed a particular oncogenic pathway were generated with spatial correlation and superimposed on the background aberration events.

A total of $n = 200$ samples each with $p = 600$ markers were generated. The 600 marker loci were uniformly distributed across six different chromosomes with 100 loci on each chromosome. A simple oncogenic pathway model, a chain shape model (Fig. 7.2), was considered. The model contained six aberration events: A, B, C, D, E, F. Without loss of generality, these six aberrations were located in the middle of each chromosome. The true conditionally dependent pairs in this model were $\{(A, B), (B, C), (C, D), (D, E), (E, F)\}$.

The performance of the methods was evaluated by two metrics: the false positive rate (FPR) and the false negative rate (FNR) of edge detection. A nonzero $\hat{\beta}_{rs}$ was considered a false detection if its genome location indices (r, s) is far from the indices of any true edge (Manhattan distance > 30 loci), where 30 was the maximum aberration size around the target locus in the simulation set up. For example, in Figs. 7.3(b) and (c) red dots that do not fall into any grey diamond are considered false detection. Similarly, a conditionally dependent pair was considered missed if there was no nonzero β falling in the grey diamond. FPR is then the number of false detections divided by the total number of nonzero $\hat{\beta}_{rs}$, $r < s$; and FNR is the number of missed divided by the number of truly dependent pairs.

A total of 50 independent data sets were generated. Both `LogitNet` and `SepLogit` were applied to each simulated data set for a series of different values of λ. Figure 7.3(a) shows the total error rate of the two methods as a function of λ. Clearly `LogitNet` outperformed `SepLogit`. For `LogitNet`, the average optimal total error rate (FPR+FNR) across the 50 independent data sets was 0.014 (s.d. = 0.029), while the average optimal total error rate for `SepLogit` was 0.211 (s.d. = 0.203). The two coefficient matrices $\hat{\mathcal{B}}$ for a simulated data set are illustrated in Figs. 7.3(b) and (c). As one can see, there was a large degree of asymmetry in the result of `SepLogit`: 435 out of the 476 nonzero $\hat{\beta}_{rs}$ had inconsistent transpose elements, $\hat{\beta}_{sr} = 0$. On the contrary, by enforcing symmetry the `LogitNet` approach correctly identified all five true conditionally dependent pairs in the chain model. Moreover, the nonzero $\hat{\beta}_{rs}$'s plotted by red dots tended to be clustered within the

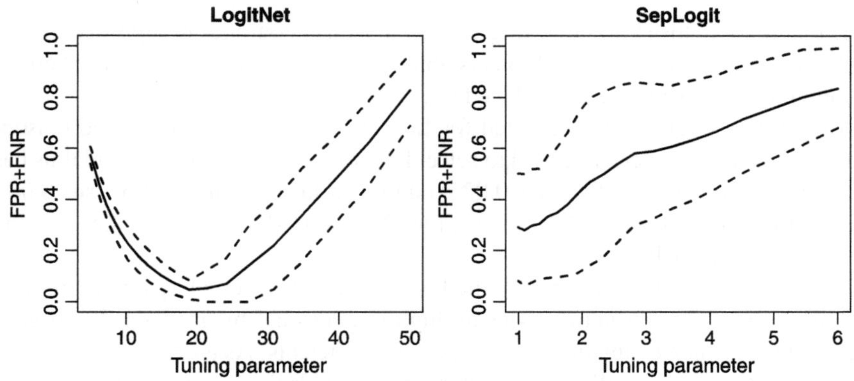

(a) Total error rate (FPR+FNR).

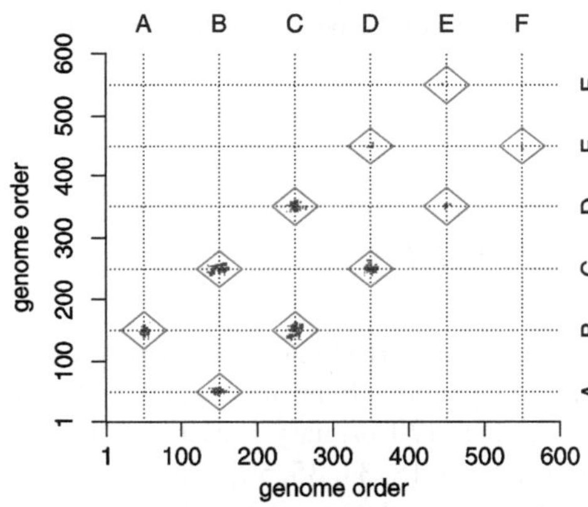

(b) Estimated $\mathscr{B}$ by `LogitNet` at the optimal λ

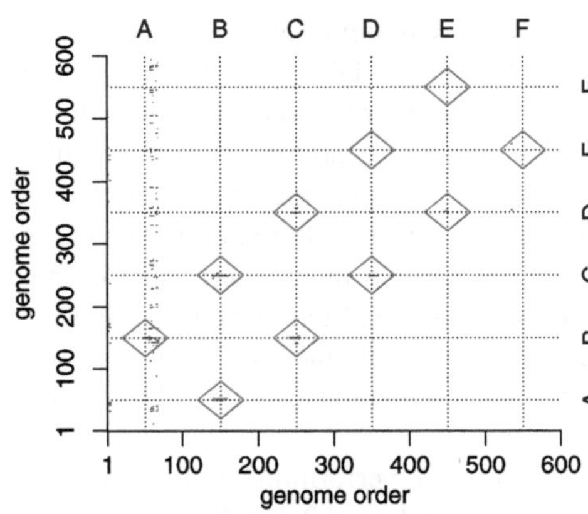

(c) Estimated $\mathscr{B}$ by `SepLogit` at the optimal λ

Fig. 7.3 (Color online) **(a)** Total error rates of `LogitNet` and `SepLogit` for the chain model in Fig. 7.2. The *solid line* (*dashed lines*) is the mean curve (mean ± one s.d.). **(b, c)** An example of the coefficient matrix $\mathcal{B}$ by `LogitNet` and `SepLogit`, respectively. *Red dots* represent nonzero β_{rs}. *Points* in the *grey diamond* are deemed as correct detections. The *dashed blue lines* indicate the locations of aberration $A - F$

grey diamonds. This shows that LogitNet indeed encourages group selection for highly correlated predictors and thus is able to make good use of the spatial correlation in the data when inferring the edges.

Both CV and BIC were evaluated for LogitNet, and they performed reasonably well. The CV criterion tended to select larger models than the BIC and thus has more false positives (0.079 versus 0.025) and fewer false negatives (0 versus 0.06). The average total error rate (FPR+FNR) for CV was 0.079, slightly smaller than the total error rate for BIC, 0.084.

Taken together, the LogitNet method [43] makes uses of sparse regression techniques for learning networks using high-dimensional binary data. It enforces the symmetry when estimating the regression coefficients. As a result, the LogitNet estimates are more efficient and more interpretable than approaches that do not enforce the symmetry. Here a lasso penalty function is imposed. Other penalty functions can easily be extended in the penalized loss function (7.5).

7.4 remMap

7.4.1 Motivation and Model

In a few recent breast cancer cohort studies, microarray expression experiments and array-based comparative genomic hybridization (array CGH) experiments have been conducted for more than 170 primary breast tumor specimens collected at multiple cancer centers [1, 2, 16, 18, 34, 35, 50]. One goal of these studies is to reveal the subtle and complicated regulatory relationships among DNA copy numbers (from CGH experiments) and RNA transcript levels (from microarray expression experiments). Such information will help shed light on cancer mechanisms.

The dependence of RNA levels on DNA copy numbers can be modeled through a straightforward multivariate linear regression model with the RNA levels as responses and the DNA copy numbers as predictors. While multivariate linear regression is well studied in statistical literature, the current problem bears new challenges due to (i) high-dimensionality in terms of both predictors and responses; (ii) the interest in identifying *master regulators* in genetic regulatory networks; and (iii) the complicated relationships among correlated response variables. Thus, the naive approach of regressing each response onto the predictors separately is unlikely to produce satisfactory results, as such methods often lead to high variability and overfitting.

Some work has been done for performing multivariate linear regression with high-dimensional predictors, including [20, 39, 48]. In [28], we proposed a method remMap—Regularized Multivariate regression for identifying MAster Predictors—for fitting multivariate regression with not only high-dimensional predictors but also high-dimension responses. RemMap uses a penalty which is designed to induce sparsity in the model and at the same time to encourage the selection of *master predictors*—predictors which affect (relatively) many responses.

Consider a multivariate regression with Q response variables $y_1, \dots, y_Q$ and P prediction variables $x_1, \dots, x_P$:

$$y_q = \sum_{p=1}^{P} x_p \beta_{pq} + \epsilon_q, \quad q = 1, \dots, Q, \tag{7.10}$$

where the error terms $\epsilon_1, \dots, \epsilon_Q$ have a joint distribution with mean 0 and covariance Σ_ϵ. All the response and prediction variables are standardized to have mean zero, and thus there is no intercept term in (7.10). The goal is to identify nonzero entries in the $P \times Q$ coefficient matrix $\mathbf{B} = (\beta_{pq})$ based on N i.i.d. samples. Under the normality assumption, the coefficients β_{pq} can be interpreted as proportional to the conditional correlation $\mathrm{Cor}(y_q, x_p | x_{-(p)})$, where $x_{-(p)} := \{x_{p'} : 1 \le p' \ne p \le P\}$. Let $Y_q = (y_q^1, \dots, y_q^N)^T$ and $X_p = (x_p^1, \dots, x_p^N)^T$ denote the sample of the qth response variable and that of the pth prediction variable, respectively. Also let $\mathbf{Y} = (Y_1 : \dots : Y_Q)$ denote the $N \times Q$ response matrix, and $\mathbf{X} = (X_1 : \dots : X_P)$ denote the $N \times P$ prediction matrix.

`RemMap` aims to fit the multivariate regression model (7.10) when both Q and P are larger than the sample size N. For example, in the breast cancer study mentioned above, the sample size is 172, while the number of genes and the number of chromosomal regions are on the order of a couple of hundred (after prescreening). When the dimensions of both predictors and responses are large, it is reasonable to assume that (i) only a subset of predictors enter the model, and (ii) a predictor may affect only some but not all responses. To take into account these aspects, `remMap` uses a combined regularization with a ℓ_1 norm penalty on the coefficient matrix $\mathbf{B}$ to control the overall sparsity of the model and a ℓ_2 norm penalty for the group of coefficients corresponding to the same predictor (one row of $\mathbf{B}$) to control the total number of predictors entering the model. Consequently, a predictor will not be selected into the model if the corresponding ℓ_2 norm is too small. Note that, this penalty favors the selection of the so-called master predictors. The ℓ_2 penalty (or its analogies) has been used for other purposes in the literature, see [46, 48] for examples.

Specifically, `remMap` utilizes the following penalized loss function:

$$L(\mathbf{B}; \lambda_1, \lambda_2) = \frac{1}{2} \left\| \mathbf{Y} - \sum_{p=1}^{P} X_p B_p \right\|_F^2 + \lambda_1 \sum_{p=1}^{P} \| C_p \cdot B_p \|_1 + \lambda_2 \sum_{p=1}^{P} \| C_p \cdot B_p \|_2, \tag{7.11}$$

where C_p is the pth row of the indicator matrix $\mathbf{C} = (c_{pq}) = (C_1^T : \dots : C_P^T)^T$, which is a prespecified $P \times Q$ 0–1 matrix indicating which coefficients should be penalized; B_p is the pth row of $\mathbf{B}$; $\| \cdot \|_F$ denotes the Frobenius norm of matrices; $\| \cdot \|_1$ and $\| \cdot \|_2$ are the ℓ_1 and ℓ_2 norms for vectors, respectively; and "$\cdot$" stands for Hadamard product (that is, entry-wise multiplication). The indicator matrix $\mathbf{C}$ is prespecified based on prior knowledge: if one knows in advance that predictor x_p affects response y_q, then the corresponding regression coefficient β_{pq} will not be penalized, i.e., $c_{pq} = 0$ (see Sect. 7.5 for an example). When there is no such information available, $\mathbf{C}$ can be simply set to be a constant matrix $c_{pq} \equiv 1$. The remMap estimate of the coefficient matrix $\mathbf{B}$ is defined as $\hat{\mathbf{B}}(\lambda_1, \lambda_2) := \arg\min_{\mathbf{B}} L(\mathbf{B}; \lambda_1, \lambda_2)$. The combined penalty in (7.11) is referred to as the MAP (MAster Predictor) penalty. Here, we want to point out a difference between the MAP penalty and the Elastic-Net penalty proposed by Zou et al. [51], which combines the ℓ_1 norm penalty with the *squared* ℓ_2 norm penalty under the multiple regression setting. The squared ℓ_2 norm itself does not induce sparsity and thus is intrinsically different from the ℓ_2 norm penalty discussed above.

The remMap estimator $\hat{\mathbf{B}}(\lambda_1, \lambda_2)$ can be obtained by an iterative algorithm. We first describe how to update one row of $\mathbf{B}$ (say the p_0^{th} row), when all other rows are fixed.

Theorem 7.1 *Given* $\{B_p\}_{p \neq p_0}$ *in* (7.11), *the solution for* $\min_{B_{p_0}} L(\mathbf{B}; \lambda_1, \lambda_2)$ *is given by* $\hat{B}_{p_0} = (\hat{\beta}_{p_0,1}, \ldots, \hat{\beta}_{p_0,Q})$ *which satisfies: for* $1 \leq q \leq Q$,

(i) *If* $c_{p_0,q} = 0$, $\hat{\beta}_{p_0,q} = X_{p_0}^T \tilde{Y}_q / \|X_{p_0}\|_2^2$ *(OLS), where* $\tilde{Y}_q = Y_q - \sum_{p \neq p_0} X_p \beta_{pq}$.
(ii) *If* $c_{p_0,q} = 1$,

$$\hat{\beta}_{p_0,q} = \begin{cases} 0 & \text{if } \|\hat{B}_{p_0}^{\text{lasso}}\|_{2,C} = 0, \\ \left(1 - \frac{\lambda_2}{\|\hat{\mathbf{B}}_{p_0}^{\text{lasso}}\|_{2,C} \cdot \|X_{p_0}\|_2^2}\right)_+ \hat{\beta}_{p_0,q}^{\text{lasso}} & \text{otherwise}, \end{cases} \tag{7.12}$$

where $\|\hat{B}_{p_0}^{\text{lasso}}\|_{2,C} := \{\sum_{t=1}^{Q} c_{p_0,t} (\hat{\beta}_{p_0,t}^{\text{lasso}})^2\}^{1/2}$, *and*

$$\hat{\beta}_{p_0,q}^{\text{lasso}} = \begin{cases} X_{p_0}^T \tilde{Y}_q / \|X_{p_0}\|_2^2 & \text{if } c_{p_0,q} = 0, \\ (|X_{p_0}^T \tilde{Y}_q| - \lambda_1)_+ \frac{\text{sign}(X_{p_0}^T \tilde{Y}_q)}{\|X_{p_0}\|_2^2} & \text{if } c_{p_0,q} = 1. \end{cases} \tag{7.13}$$

The proof of Theorem 7.1 is given in the supplementary material of [28]. Theorem 7.1 basically shows that the remMap procedure amounts to two folds of shrinkage: a lasso shrinkage of the OLS solution followed by a group shrinkage of the lasso solution. Theorem 7.1 naturally leads to an algorithm which updates the rows of $\mathbf{B}$ iteratively until convergence. The same active-shooting idea described earlier can also be adopted here to improve the convergence speed.

In [28], we select the tuning parameters (λ_1, λ_2) by v-fold cross validation (CV). Specifically, as described in Sect. 7.3.1, OLS estimates based on the selected model rather than the shrunken estimates are used in calculating the cross validation scores to avoid overfitting. In addition, in [28], we develop the cv.vote procedure mentioned earlier to further control the false positive rates. A BIC criterion, which is computationally cheaper but requires much more assumptions, is also considered.

7.4.2 Simulation

In [28], we compared the `remMap` method with two alternatives: (i) the `joint` method which only utilizes the ℓ_1 penalty, that is, $\lambda_2 = 0$ in (7.11); (ii) the `sep` method which performs Q separate lasso regressions. For each method, we considered three tuning strategies, which results in nine methods in total:

1. `remMap.cv`, `joint.cv`, `sep.cv`: The tuning parameters are selected through 10-fold cross validation.
2. `remMap.cv.vote`, `joint.cv.vote`, `sep.cv.vote`: The `cv.vote` procedure is applied that retains only variables selected by more than five cross validation folds.
3. `remMap.bic`, `joint.bic`, `sep.bic`: The tuning parameters are selected by a BIC criterion.

In the simulation, the sample size N was fixed at 200, and varying dimensions $P = Q = 400, 600, 800$ were considered. For a given set of (N, P, Q), predictors $(x_1, \ldots, x_P)^T$ were first generated according to $\text{Normal}_P(0, \Sigma_X)$, where Σ_X is the predictor covariance matrix ($\Sigma_X(p, p') := 0.4^{|p-p'|}$). Next, a $P \times Q$ 0–1 adjacency matrix $\mathbf{A}$ was simulated, which specified the topology of the network between predictors and responses. The networks were generated with five master predictors (hubs), each influencing 20–40 responses. In addition, the diagonals of $\mathbf{A}$ equaled one. This was aimed to mimic `cis-regulations` of DNA copy number alternations on its own expression levels. The $P \times Q$ regression coefficient matrix $\mathbf{B} = (\beta_{pq})$ were then simulated by setting $\beta_{pq} = 0$ if $\mathbf{A}(p, q) = 0$ and $\beta_{pq} \sim \text{Uniform}([-5, -1] \cup [1, 5])$ if $\mathbf{A}(p, q) = 1$. After that, the residuals $(\epsilon_1, \ldots, \epsilon_Q)^T$ were generated according to $\text{Normal}_Q(0, \sigma_\epsilon^2 I_Q)$. The residual variance σ_ϵ^2 was chosen such that the average signal-to-noise ratio equals 0.25. Finally, the responses $(y_1, \ldots, y_Q)^T$ were generated according to model (7.10). Each data set consisted of $N = 200$ i.i.d. samples of such generated predictors and responses. For all methods, predictors and responses were standardized to have (sample) mean zero and standard deviation one before model fitting. In addition, $\mathbf{C} = (c_{pq})$ was taken to be $c_{pq} = 0$ if $p = q$ (meaning that the cis-regulations were viewed as prior information) and $c_{pq} = 1$ otherwise. The primary goal was to identify the `trans-edges`, the predictor-response pairs (x_p, y_q) with $\mathbf{A}(p, q) = 1$ and $\mathbf{C}(p, q) = 1$, i.e., the edges that were not prespecified by the indicator matrix $\mathbf{C}$. The total number of `tran-edges` was 132 in the above network.

Results on `trans-edge` detection averaged across 25 independent data sets were summarized in Fig. 7.4 in terms of the number of false positive detections of `trans-edges` (FP) and the number of false negative detections of `trans-edges` (FN). It is clear that `remMap.cv` and `remMap.cv.vote` performed the best in terms of the total number of false detections (FP+FN), followed by `remMap.bic`. The three `sep` methods resulted in too many false positives (especially `sep.cv`). This is expected since there were a total of Q tuning parameters selected separately, and the relations among the responses were not utilized at all.

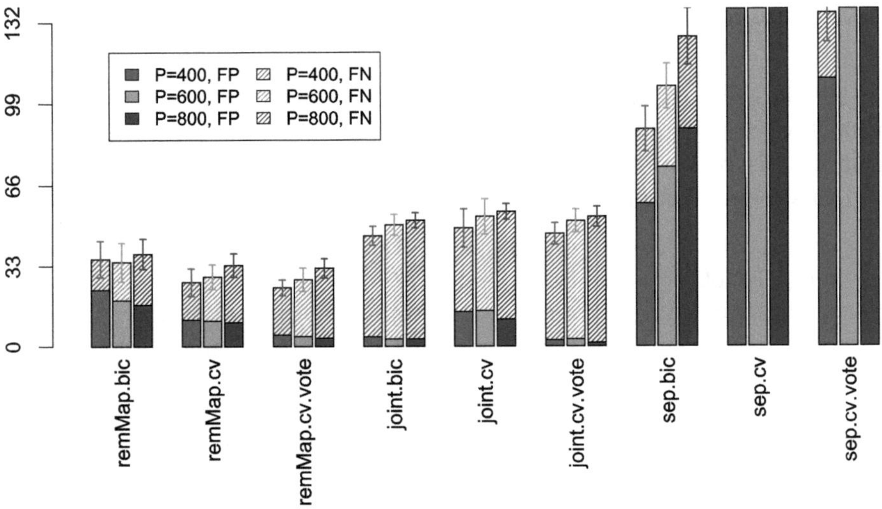

Fig. 7.4 Impact of dimensionality. Heights of *solid bars* represent numbers of false positive detections of `trans-edges` (FP); heights of *shaded bars* represent numbers of false negative detections of `trans-edges` (FN). All bars are truncated at height = 132 [28] (for the color version, see Color Plates on p. 391)

The three `joint` methods performed reasonably well, though they had considerably larger number of false negative detections compared to `remMap` methods. This is because the `joint` methods incorporated less information about the relations among the responses caused by the same predictors. Finally, comparing `cv.vote` to `cv`, the `cv.vote` procedure decreased the false positive detections while only inflated the false negative counts slightly. As to the impact of dimensionality (P, Q), the larger the dimension, the more false negative detections (Fig. 7.4). Moreover, `remMap` performed much better than `joint` and `sep` on master predictor selection, especially in terms of the number of false positive `trans-predictors` (results not shown). This is because the ℓ_2 norm penalty was more effective than the ℓ_1 norm penalty in excluding irrelevant predictors.

In summary, `remMap` is developed for fitting multivariate regression models under high-dimensionality and is particularly useful for identifying predictors that affect many responses. The `remMap` method can be applied to many genomic studies involving two types of biological measurements. A notable example is the e-QTL mapping, with the SNP status being the predictor variables and the expression levels used as response variables.

7.5 Real Application

In [28], `space` and `remMap` were applied to the breast cancer study discussed in Sect. 7.4.1. The goal was to search for genome regions whose copy number alterations have significant impact on RNA expression levels, especially on those of the unlinked genes, i.e., genes not falling into the same genome region as the copy number alterations.

Data Preprocessing A total of 172 tumor samples had both cDNA expression microarray and CGH array data. The experiments were described in [1, 2, 16, 18, 34, 35, 50]. In what follows, we outline the data preprocessing steps. More details can be found in [28].

Each CGH array contains measurements ($\log_2$ ratios) on about 17 K mapped human genes. A positive (negative) measurement suggests a possible copy number gain (loss). After proper normalization, `cghFLasso` [38] was used to estimate the underlying DNA copy numbers based on array outputs. *Copy number alteration intervals* (CNAIs) were then derived by employing the Fixed-Order Clustering (FOC) method [42], where CNAIs are defined as basic CNA units (genome regions) in which all genes tend to be amplified or deleted simultaneously in a sample. For each CNAI in each sample, the mean value of the estimated copy numbers of the genes falling into this CNAI was calculated. This resulted in a 172 (samples) by 384 (CNAIs) numeric matrix.

Each expression array contains measurements for about 18 K mapped human genes. After global normalization for each array, each gene was standardized across 172 samples to have median = 0 and MAD (median absolute deviation) = 1. Then we focused on a set of 654 breast cancer related genes, which was derived based on seven published breast cancer gene lists [4, 26, 32, 35, 36, 40, 44]. This resulted in a 172 (samples) by 654 (genes) numeric matrix.

Since different tumor subtypes might confound the detection of associations between CNAIs and gene expressions, a set of subtype indicator variables based on the expression data were derived. Specifically, following [35], 172 patients were grouped into five distinct groups based on their expression patterns. The corresponding subtype indicator variables were then used as additional predictors in the `remMap` model.

Interactions Among RNA When the copy number change of one CNAI affects the RNA level of an unlinked gene, there are two possibilities: (i) the copy number change directly affects the RNA level of the unlinked gene; (ii) the copy number change first affects the RNA level of an intermediate gene (either linked or unlinked), and then the RNA level of this intermediate gene affects that of the unlinked gene. Figure 7.5 gives an illustration of these two scenarios. In [28], the main interest was to find the first-type relationships. To achieve this, the interactions among RNA levels were first characterized and then accounted for in `remMap` in order to obtain the direct interactions. For this purpose, the `space` method was applied to search for associated RNA pairs through identifying nonzero partial correlations [27]. The estimated network (referred to as *Exp.Net.664* hereafter) has in total 664 edges—664 pairs of genes whose RNA levels are significantly correlated with each other after accounting for the expression levels of other genes (Fig. 7.6(a)).

Interactions Between CNAIs and RNA Expressions `remMap` was then applied to study the interactions between CNAIs and RNA transcript levels. For each of the

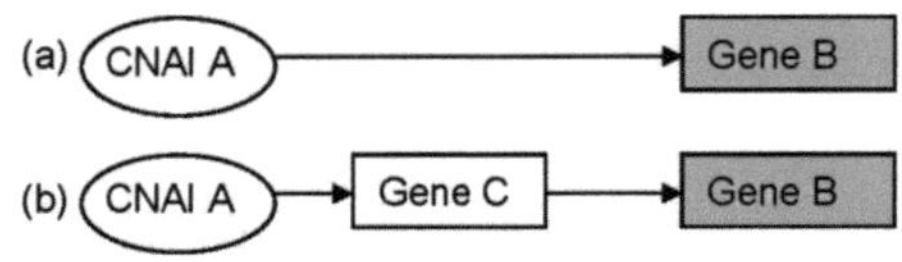

Fig. 7.5 (**a**) Direct interaction between CNAI A and the expression of gene B; (**b**) indirect interaction between CNAI A and the expression of Gene B through the intermediate Gene C [28]

654 breast cancer genes, the expression level was regressed on three sets of predictors: (i) expression levels of other genes that are connected to the target gene (the current response variable) in *Exp.Net.664*; (ii) the subtype indicator variables derived in the previous section; and (iii) the copy numbers of all 384 CNAIs. It is of interest whether any unlinked CNAIs are selected into this regression model, i.e., the corresponding regression coefficients are nonzero. This suggests potential trans-regulations (`trans-edges`) between the selected CNAIs and the target gene expression. The coefficients of the linked CNAI of the target gene were not included in the `MAP` penalty (this corresponds to $c_{pq} = 0$, see Sect. 7.4.1 for details). This is because the DNA copy number changes of one gene often influence its own expression level, and it is of less interest in this study. Furthermore, no penalties were imposed on the expression levels of connected genes either. Another view of this is that the cis-regulations between CNAIs and their linked expression levels, as well as the inferred RNA interaction network were considered as "prior knowledge" in this study.

Note that, different response variables (gene expressions) now have different sets of predictors, as their neighborhoods in `Exp.Net.664` are different. However, the `remMap` model can still be fitted with a slight modification. The idea is to treat all CNAI (384 in total), all gene expressions (654 in total), and subtype indicators as predictors. Then, for each target gene, the coefficients of those gene expressions that were not linked in *Exp.Net.664* were forced to be zero and not updated throughout the iterative fitting procedure.

We applied `remMap.cv.vote` on the data and identified 43 `trans-edges`, which correspond to three contiguous CNAIs on chromosome 17 and 31 distinct (unlinked) RNAs. Figure 7.6(b) illustrates the topology of the estimated regulatory relationships. The three CNAIs being identified as transregulators sit closely on chromosome 17, spanning from 34811630*bp* to 35699243*bp* and falling into cytoband 17q12-q21.2. This region (referred to as CNAI-17q12 hereafter) contains 24 known genes, including the famous breast cancer oncogene ERBB2 and the growth factor receptor-bound protein 7 (GRB7). As suggested by the result of the `remMap` model, the amplification of CNAI-17q12 also influences the expression levels of 31 unlinked genes/clones. This implies that CNAI-17q12 may harbor transcriptional factors whose activities closely relate to breast cancer. Indeed, there are four transcription factors (NEUROD2, IKZF3, THRA, NR1D1) and two transcriptional coactivators (MED1, MED24) in CNAI-17q12. It is possible that the amplification of CNAI-17q12 results in the over expression of one or more transcription factors/co-activators in this region, which then influence the expressions of the unlinked 31 genes/clones. In addition, none of the subtype indicator variables was

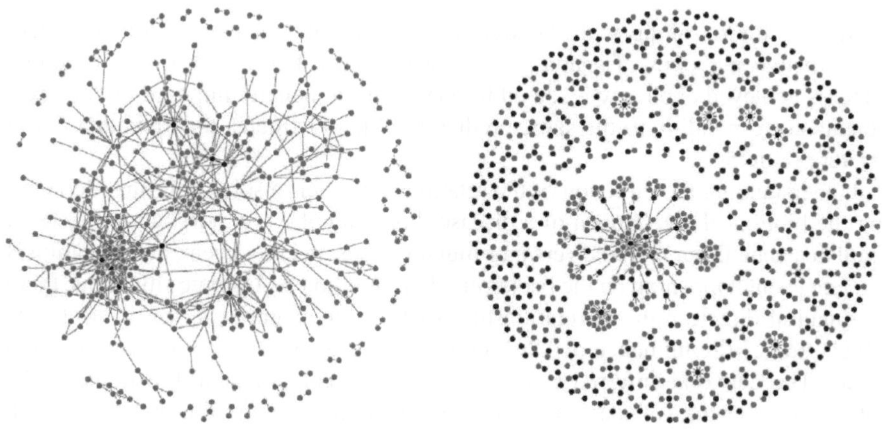

Fig. 7.6 (**a**) Inferred network for the 654 breast cancer related genes (based on their expression levels) by space. Nodes with degrees greater than ten are drawn in *blue*. (**b**) Network of the estimated regulatory relationships between the copy numbers of the 384 CNAIs and the expressions of the 654 breast cancer related genes. Each *blue node* stands for one CNAI, and each *green node* stands for one gene. *Red edges* represent inferred transregulations (43 in total). *Gray edges* represent cis-regulations [28] (for the color version, see Color Plates on p. 392)

selected into the final model. Even when we forced them to be in the model (by setting the corresponding $c_{pq} = 0$), the resulting hub CNAIs remained unchanged. These imply that the three hub CNAIs are unlikely confounded by tumor subtypes. However, besides RNA interactions and subtype stratification, there could be other unaccounted confounding factors. Therefore, caution must be taken when interpreting these results.

7.6 Concluding Remarks

In this chapter, three sparse regression models, space, LogitNet and remMap, for inferring networks from high-dimensional array data are presented. Space is designed for continuous variables, whereas LogitNet is designed for binary variables, and both aim to infer conditional dependency relationships among the variables. remMap infers networks relating two different types of high-dimensional data, with one set of variables naturally serving as predictors and the other serving as responses. The three methods have some common features. All methods model gene relationships jointly instead of looking at one gene at a time. This joint modeling approach not only improves the model efficiency by reducing the total error rates, but also allows the incorporation of specific features of the model such as the symmetry constraint in space and LigitNet. Furthermore, all three methods have produced grouping effects. In space, the overall sparsity constraint and the choice of weights encourage identification of hubs. In LogitNet, the symmetric coefficient matrix along with the weight function encourages conditional dependency between groups of spatially correlated clones. For `remMap`, the grouping effect is achieved explicitly by an additional ℓ_2 norm penalty imposed on the coefficients corresponding to the same predictor, which facilitates the identification of master predictors.

Understanding the direction of the relationship is very useful in further delineating the biological mechanism of a disease. For example, it is possible that gene A regulates gene B but not vice versa; or mutation A is an early event, and mutation B is a late event happening on the background of mutation A. In some situations, there is a natural direction in the relationship, such as the example shown in Sect. 7.5, which studies the influence of DNA copy number alterations on RNA expression levels. The networks inferred from `remMap` often contain such directions, as the nature of two types of data may suggest a natural ordering. On the other hand, `space` and `LogitNet` infer undirected graphs, as the conditional dependency relationship between the two events has no direction. To infer directed graphs, prior information regarding pathways (for example, KEGG database) or clonal evolution will be useful. Research along this line is under way and will be communicated in future works.

References

1. A. Bergamaschi, Y.H. Kim, P. Wang, T. Sorlie, T. Hernandez-Boussard, P.E. Lonning, R. Tibshirani, A.L. Borresen-Dale, and J.R. Pollack. Distinct patterns of DNA copy number alteration are associated with different clinicopathological features and gene-expression subtypes of breast cancer. *Genes Chromosomes Cancer*, **45**:1033–1040, 2006.
2. A. Bergamaschi, Y.H. Kim, K.A. Kwei, Y.L. Choi, M. Bocanegra, A. Langerod, W. Han, D.Y. Noh, D.G. Huntsman, S.S. Jeffrey, A.L. Borresen-Dale, and J.R. Pollack. CAMK1D amplification implicated in epithelial-mesenchymal transition in basal-like breast cancer. *Mol Oncol*, 2008, in press.
3. A.J. Butte, P. Tamayo, D. Slonim, T.R. Golub, and I.S. Kohane. Discovering functional relationships between RNA expression and chemotherapeutic susceptibility using relevance networks. *Proc Natl Acad Sci USA*, **97**(22):12182–12186, 2000.
4. H.Y. Chang, J.B. Sneddon, A.A. Alizadeh, R. Sood, R.B. West, et al Gene expression signature of fibroblast serum response predicts human cancer progression: Similarities between tumors and wounds. *PLoS Biol*, **2**(2):e7, 2004.
5. A.P. Dawid and S.L. Lauritzen. Hyper-Markov laws in the statistical analysis of decomposable graphical models. *Ann Stat*, **21**(3):1272–1317, 1993.
6. A. Dempster. Covariance selection. *Biometrics*, **28**(1):157–175, 1972.
7. M. Drton and M.D. Perlman. Model selection for Gaussian concentration graphs. *Biometrika*, **91**(3):591–602, 2004.
8. D. Edward. *Introduction to Graphical Modelling*, 2nd edition. Springer, New York, 2000.
9. B. Efron, T. Hastie, I. Johnstone, and R. Tibshirani. Least angle regression. *Ann Stat*, **32**:407–499, 2004.
10. J. Friedman, T. Hastie, and R. Tibshirani. Pathwise coordinate optimization. *Ann Appl Stat*, **1**(2):302–332, 2007.
11. J. Friedman, T. Hastie, and R. Tibshirani. Sparse inverse covariance estimation with the graphical lasso. *Biostatistics*, 2007. doi:10.1093/biostatistics/kxm045.
12. J. Friedman, T. Hastie, and R. Tibshirani. Regularized paths for generalized linear models via coordinate descent. Technical Report, Department of Statistics, Stanford University, 2009.

13. W. Fu. Penalized regressions: the bridge vs the lasso. *J Comput Graph Stat*, **7**(3):417–433, 1998.
14. A. Genkin, D.D. Lewis, and D. Madigan. Large-scale Bayesian logistic regression for text categorization. *Technometrics*, **49**:291–304, 2007.
15. H. Joe and Y. Liu. A model for a multivariate binary response with covariates based on compatible conditionally specified logistic regression. *Stat Probab Lett*, **31**:113–120, 1996.
16. A.V. Kapp, S.S. Jeffrey, A. Langerod, A.L. Borresen-Dale, W. Han, D.Y. Noh, I.R. Bukholm, M. Nicolau, P.O. Brown, and R. Tibshirani. Discovery and validation of breast cancer subtypes. *BMC Genomics*, **7**:231, 2006.
17. Y. Kim, L. Girard, C. Giacomini, P. Wang, T. Hernandez-Boussard, R. Tibshirani, J. Minna, and J. Pollack. Combined microarray analysis of small cell lung cancer reveals altered apoptotic balance and distinct expression signatures of MYC family gene amplification. *Oncogene*, **25**(1):130–138, 2006.
18. A. Langerod, H. Zhao, O. Borgan, J.M. Nesland, I.R. Bukholm, T. Ikdahl, R. Karesen, A.L. Borresen-Dale, and S.S. Jeffrey. TP53 mutation status and gene expression profiles are powerful prognostic markers of breast cancer. *Breast Cancer Res*, **9**:R30, 2007.
19. H. Li and J. Gui. Gradient directed regularization for sparse Gaussian concentration graphs, with applications to inference of genetic networks. *Biostatistics*, **7**(2):302–317, 2006.
20. R. Lutz and P. Bühlmann. Boosting for high-multivariate responses in high-dimensional linear regression. *Stat Sin*, **16**:471–494, 2006.
21. D. Madigan and J. York. Bayesian graphical models for discrete data. *Int Stat Rev*, **63**:215–232, 1995.
22. N. Meinshausen and P. Buhlmann. High dimensional graphs and variable selection with the lasso. *Ann Stat*, **34**:1436–1462, 2006.
23. M. Newman. The structure and function of complex networks. *Soc Ind Appl Math*, **45**(2):167–256, 2003.
24. M.A. Newton, M.N. Gould, C.A. Reznikoff, and J.D. Haag. On the statistical analysis of allelic-loss data. *Stat Med*, **17**:1425–1445, 1998.
25. L. Nie, G. Wu, and W. Zhang. Correlation between mRNA and protein abundance in Desulfovibrio vulgaris: a multiple regression to identify sources of variations. *Biochem Biophys Res Commun*, **339**(2):603–610, 2006.
26. S. Paik, S. Shak, G. Tang, C. Kim, J. Baker, et al A multigene assay to predict recurrence of tamoxifen-treated, node-negative breast cancer. *N Engl J Med*, **351**(27):2817–2826, 2004.
27. J. Peng, P. Wang, N. Zhou, and J. Zhu. Partial correlation estimation by joint sparse regression models. *J Am Stat Assoc*, **104**(486):735–746, 2009.
28. J. Peng, J. Zhu, A. Bergamaschi, W. Han, D.Y. Noh, J.R. Pollack, and P. Wang. Regularized multivariate regression for identifying master predictors with application to integrative genomics study of breast cancer. *Ann Appl Stat*, **4**(1):53–77, 2010.
29. J. Pollack, T. Srlie, C. Perou, C. Rees, S. Jeffrey, P. Lonning, R. Tibshirani, D. Botstein, A. Brresen-Dale, and P. Brown. Microarray analysis reveals a major direct role of DNA copy number alteration in the transcriptional program of human breast tumors. *Proc Natl Acad Sci*, **99**(20):12963–12968, 2002.
30. P. Ravikumar, M. Wainwright, and J. Lafferty. High-dimensional Ising model selection using l_1-regularized logistic regression. *Ann Stat*, **38**:1287–1319, 2010.
31. A.J. Rothman, P.J. Bickel, E. Levina, and J. Zhu. Sparse permutation invariant covariance estimation. *Electron J Stat*, **2**:494–515, 2008.
32. L.H. Saal, P. Johansson, K. Holm, S.K. Gruvberger-Saal , Q.B. She, et al Poor prognosis in carcinoma is associated with a gene expression signature of aberrant PTEN tumor suppressor pathway activity. *Proc Natl Acad Sci USA*, **104**(18):7564–7569, 2007.
33. J. Schafer and K. Strimmer. An empirical Bayes approach to inferring large-scale gene association networks. *Bioinformatics*, **21**(6):754–764, 2005.
34. T. Sorlie, C.M. Perou, R. Tibshirani, T. Aas, S. Geisler, H. Johnsen, T. Hastie, M.B. Eisen, M. van de Rijn, S.S. Jeffrey, T. Thorsen, H. Quist, J.C. Matese, P.O. Brown, D. Botstein, P. Eystein Lonning, and A.L. Borresen-Dale. Gene expression patterns of breast carcinomas

distinguish tumor subclasses with clinical implications. *Proc Natl Acad Sci USA*, **98**:10869–10874, 2001.
35. T. Sorlie, R. Tibshirani, J. Parker, T. Hastie, J.S. Marron, A. Nobel, S. Deng, H. Johnsen, R. Pesich, S. Geisler, J. Demeter, C.M. Perou, P.E. Lnning, P.O. Brown, A.L. Brresen-Dale, and D. Botstein. Repeated observation of breast tumor subtypes in independent gene expression data sets. *Proc Natl Acad Sci USA*, **100**:8418–8423, 2003.
36. C. Sotiriou, P. Wirapati, S. Loi, A. Harris, S. Fox, et al Gene expression profiling in breast cancer: Understanding the molecular basis of histologic grade to improve prognosis. *J Natl Cancer Inst*, **98**(4):262–272, 2006.
37. R. Tibshirani. Regression shrinkage and selection via the lasso. *J R Stat Soc B*, **58**:267–288, 1996.
38. R. Tibshirani and P. Wang. Spatial smoothing and hot spot detection for CGH data using the fused lasso. *Biostatistics*, **9**(1):18–29, 2008.
39. B. Turlach, W. Venables, and S. Wright. Simultaneous variable selection. *Technometrics*, **47**:349–363, 2005.
40. M.J. van de Vijver, Y.D. He, L.J. van't Veer, H. Dai, A.A. Hart, D.W. Voskuil, G.J. Schreiber, J.L. Peterse, C. Roberts, M.J. Marton, M. Parrish, D. Atsma, A. Witteveen, A. Glas, L. Delahaye, T. van der Velde, H. Bartelink, S. Rodenhuis, E.T. Rutgers, S.H. Friend, and R. Bernards. A gene-expression signature as a predictor of survival in breast cancer. *N Engl J Med*, **347**(25):1999–2009, 2002.
41. S. Varambally, J. Yu, B. Laxman, D. Rhodes, R. Mehra, S. Tomlins, R. Shah, U. Chandran, F. Monzon, M. Becich, J. Wei, K. Pienta, D. Ghosh, M. Rubin, and A. Chinnaiyan. Integrative genomic and proteomic analysis of prostate cancer reveals signatures of metastatic progression. *Cancer Cell*, **8**(5):393–406, 2005.
42. P. Wang. Statistical methods for CGH array analysis. Ph.D. Thesis, Stanford University, 2004.
43. P. Wang, D.L. Chao, and L. Hsu. Learning networks from high dimensional binary data: An application to genomic instability data. *Biometrics*, 2010, to appear. arXiv:0908.3882v1 [stat.ME].
44. Y. Wang, J.G. Klijn, Y. Zhang, A.M. Sieuwerts, M.P. Look, et al Gene-expression profiles to predict distant metastasis of lymph-node-negative primary breast cancer. *Lancet*, **365**(9460):671–679, 2005.
45. J. Whittaker. *Graphical Models in Applied Mathematical Multivariate Statistics*. Wiley, New York, 1990.
46. M. Yuan and Y. Lin. Model selection and estimation in regression with grouped variables. *J R Stat Soc, Ser B*, **68**(1):49–67, 2006.
47. M. Yuan and Y. Lin. Model selection and estimation in the Gaussian graphical model. *Biometrika*, **94**(1):19–35, 2007.
48. M. Yuan, A. Ekici, Z. Lu, and R. Monterio. Dimension reduction and coefficient estimation in multivariate linear regression. *J R Stat Soc B*, **69**(3):329–346, 2007.
49. T. Zhang and F. Oles. Text categorization based on regularized linear classifiers. *Inf Retr*, **4**:5–31, 2001.
50. H. Zhao, A. Langerod, Y. Ji, K.W. Nowels, J.M. Nesland, R. Tibshirani, I.K. Bukholm, R. Karesen, D. Botstein, A.L. Borresen-Dale, and S.S. Jeffrey. Different gene expression patterns in invasive lobular and ductal carcinomas of the breast. *Mol Biol Cell*, **15**:2523–2536, 2004.
51. H. Zou and T. Trevor. Regularization and variable selection via the elastic net. *J R Stat Soc, Ser B*, **67**(2):301–320, 2005.

Chapter 8
Computational Methods for Predicting Domain–Domain Interactions

Hyunju Lee, Ting Chen, and Fengzhu Sun

8.1 Introduction

In recent years, high-throughput technologies such as yeast two-hybrid (Y2H) assays have produced large-scale protein–protein interaction data sets at a genome scale from several organisms. Using these protein–protein interaction data sets, researchers have been able to study proteins in the context of their functional networks and to address important biological questions such as detecting signalling pathways, elucidating protein complexes, inferring protein functions, and predicting disease-related genes.

A domain is a functional unit of a protein, and different combinations of domains result in diverse range of proteins. Therefore, protein interactions are generally caused by domain interactions. It is essential to understand protein interactions at the domain level. In Eukaryotes, many proteins are composed of more than one domain, and these domains are conserved across the organisms. For example, domain G-γ (PF00631) is present in 259 protein sequences of 57 species. As another example, domain WD40 (PF00400) is present in 45,685 sequences of 608 species (as of April 2009). Domains on the surface of the proteins usually physically interact with specific domains in other proteins in order to perform their functions. Domain–domain interaction is the subunit of the protein–protein interaction and conserved across organisms. For example, it is reported from protein data bank (PDB) [2] that domains G-γ and WD40 interact with each other [PDB:1GP2] [25].

Domain–domain interaction is very useful for studying protein functions, protein interactions, and the gene regulatory network. As interacting domains and in-

H. Lee
Department of Information and Communications, Gwangju Institute of Science and Technology (GIST), Gwangju, Republic of Korea

T. Chen · F. Sun (✉)
Molecular and Computational Biology Program, Department of Biological Sciences, University of Southern California, Los Angeles, CA, USA
e-mail: fsun@college.usc.edu

J. Feng et al. (eds.), *Frontiers in Computational and Systems Biology,* Computational Biology 15,
DOI 10.1007/978-1-84996-196-7_8,

teracting proteins are highly likely to share functions or involve in the same cellular process, inferred interactions may give clues to determine the function of unknown proteins [3]. The mutation of specific domains of a protein can cause disease by disrupting the interaction with other proteins. Hence, the study of domain interactions can reveal the pathways related to disease [12]. In addition, as domain–domain interactions are conserved across several organisms, it is possible to use the predicted domain interactions from one organism in order to infer functional interaction network in the other organisms. The study of domain interaction conservation also provides insight about the evolution of proteins [15].

Domain interactions can be determined by the crystal structures of proteins using experimental methods such as X-ray crystallography or nuclear magnetic resonance (NMR). However, the experimental determination of new domain interactions is slow as these techniques are generally labor intensive. On the other hand, computational methods have been shown to be promising to infer domain interactions by integrating large-scale data sets such as protein–protein interactions, domain functions, and protein sequences. When we infer domain interactions from protein interactions, (1) what models should we use to link domain interactions with protein interactions? (2) how should we deal with false positive and false negative interactions? When we integrate multiple data sources to infer domain interactions, (3) what is the relationship between domain interaction and each data source such as the domain function and domain fusion? (4) what integrative methods should we use to improve the prediction accuracy? In this chapter, we review several computational methods inferring domain–domain interactions.

1. **Using protein–protein interaction data**: We introduce the following methods: an association-based method, a maximum-likelihood estimation (MLE)-based method, a Bayesian method, a domain pair exclusion analysis-based method, and a parsimony linear programming optimization-based method, for predicting domain interactions.
2. **Integrating multiple biological data sources**: We introduce an extension of the likelihood-based method based on protein interaction data from one species to multiple species and a naive Bayesian method for integrating many different data sources.

This chapter is organized as follows. We first introduce data sources used in the reviewed papers. Then, we describe various ways to assess the accuracy of predicted domain interactions. Next, we explain methods for predicting domain interactions using protein–protein interactions, approaches for combining protein interaction data sets from multiple organisms, and methods for integrating multiple biological data sources. Finally, we discuss future research questions.

We note that a large number of computational methods have been developed to predict domain interactions and it is difficult to have an exhaustive review of the field at this stage. The computational methods that we review in this chapter represent only a small fraction of such methods and certainly biased toward those that we have studied over the years. We apologize to many excellent researchers whose work is not represented in this review.

8.1.1 Data Sources

Domain databases include Interpro [1] and Pfam [6]. In Pfam, the conserved functional units are discovered through multiple sequence alignments and hidden Markov models (HMMs). The current release of Pfam (22.0) contains 9,314 protein families.

Protein structure database includes Protein Data Bank (PDB) [2]. It contains 57,013 structures as of April 2009. Domain interactions can be detected by the experimental approaches based on the crystal structures of proteins.

Domain–domain interaction data sets can be constructed using protein structure data set. If two domains are close enough to interact, the binding sites are inferred. iPfam [5] contains 2,580 domain interactions inferred from PDB (July 2004 version).

Protein interactions have traditionally been studied by genetic, biochemical, and biophysical techniques. The interactions detected by these methods are generally considered reliable but produce only small-scale data sets. The MIPS [19] database collected protein interactions from these small-scale experiments for yeast. Recently, several high-throughput methods such as yeast two-hybrid assay for the detection of protein interactions have been developed. These methods are used to generate protein interactions for yeast [13, 26], fruitfly [11], etc. DIP database [22] collects protein interactions from many species such as yeast, worm, fruitfly, and humans. It contains 57,000 proteins from over 270 organisms (as of May 2009). Other large-scale interaction databases include Biomolecular INteraction Network Database (BIND) and General Repository of Interaction Datasets (BioGRID). For humans, the Human Protein Reference Database (HPRD) [20] collects human protein–protein interactions from individual small-scale experiments published in the literature.

Domain functions can be obtained using the mapping table from Pfam to GO in the Gene Ontology webpage (http://www.geneontology.org) [7, 8].

Domain fusion data set can be obtained from protein–domain information in Pfam-A [6] to identify pairs of domains coexisting in one protein. The method is referred as domain fusion in the rest of the paper.

8.1.2 Assessing the Accuracy of Predicted Domain–Domain Interactions

One of the important issues after predicting domain–domain interactions is how to assess the accuracy of inferred interactions. The direct way is to compare the predicted domain interactions with experimentally determined domain interactions in databases such as PDB [2] or iPfam [5]. However, it is generally difficult to discover new protein structures and to analyze domain interactions from protein structures. Even though iPfam provides the analyzed domain interactions from PDB, the coverage of the iPfam is relatively small compared to the number of potential domain interactions.

Since the number of known domain interactions is relatively small compared to the entire domain interactions, indirect methods for evaluating the predicted domain interactions have been developed. One of the approaches is to use the predicted domain interactions to predict protein interactions in a new set of proteins. The accuracy of these inferred protein interactions is assessed by comparing with gold standard protein interactions such as MIPS in yeast. The basic idea is that if one or multiple domain pairs (one from each protein of the protein pair) interact, then the pair of proteins interact.

Another indirect method is to compare the coexpression patterns of genes whose products contain interacting domain pairs. It is assumed that protein pairs with interacting domain pairs are more likely to have similar expression patterns. The assessment is measured by calculating the statistical significance of the correlation between the expression pattern of protein pairs inferred as interacting compared to those from random protein pairs.

8.2 Computational Methods for Predicting Domain Interactions Using Protein–Protein Interactions

Sprinzak and Margalit [23] predicted domain interactions from protein interactions by introducing the idea of over-represented domain pairs in interacting protein pairs. This idea was later extended using the maximum likelihood estimation (MLE) method and an expectation and maximization (EM) algorithm by Deng et al. [3] and Riley et al. [21]. Guimarães et al. [12] addressed this problem using a linear programming optimization approach based on the maximum parsimony idea.

8.2.1 Predicting Domain Interactions Based on Over-represented Domain Pairs

The basic idea of the method based on over-represented domain pairs is that domain pairs which occur more frequently in interacting protein pairs than expected assuming random association of domain pairs are more likely to interact with each other [23]. Let I_{mn} be the observed frequency of interacting protein pairs with one protein containing domain D_m and the other protein containing domain D_n. Let I_m and I_n be the frequencies of proteins containing domains D_m and D_n in all the proteins, respectively. Then, the likelihood ratio defined as $A_{mn} = I_{mn}/(I_m I_n)$ was used to measure the strength of association between domains D_m and D_n. Throughout this chapter, we refer this method as the association method.

In Fig. 8.1, we give an artificial example where there are five proteins P_1, $P_2, \dots, P_5$ and six domains $D_1, D_2, \dots, D_6$. The adjacency matrix between the pro-

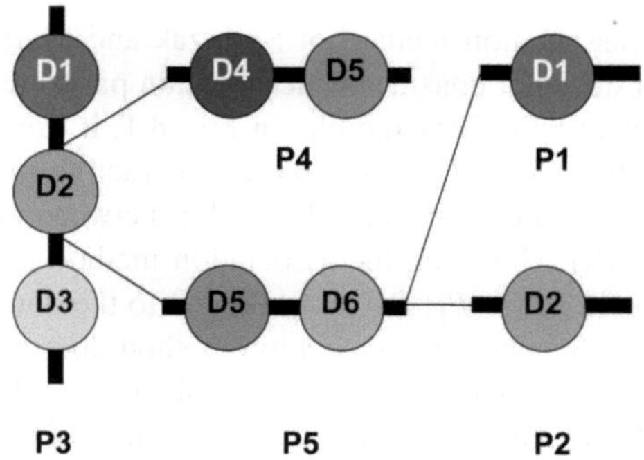

Fig. 8.1 An artificial example of five proteins ($P_1, P_2, \ldots, P_5$) and their domains ($D_1, D_2, \ldots, D_6$). The edges between the nodes indicate interactions between the proteins

Table 8.1 (Upper) The adjacency matrix between the proteins and the domains for the artificial example in Fig. 8.1. (Lower) The frequency of interacting protein pairs with one protein containing one domain and the other protein containing the other domain (upper triangle), and the calculated association scores A_{mn} for any domain pairs (lower triangle)

Proteins	Domains					
	D_1	D_2	D_3	D_4	D_5	D_6
P_1	1	0	0	0	0	0
P_2	0	1	0	0	0	0
P_3	1	1	1	0	0	0
P_4	0	0	0	1	1	0
P_5	0	0	0	0	1	1
I	2	2	1	1	2	1

Domains	Domains					
	D_1	D_2	D_3	D_4	D_5	D_6
D_1	–	0	0	1	3	1
D_2	0	–	0	1	3	1
D_3	0	0	–	1	2	1
D_4	0.5	0.5	1.0	–	0	0
D_5	0.75	0.75	1.0	0	–	0
D_6	0.5	0.5	1.0	0	0	–

teins and domains is given in Table 8.1 (upper). The upper part above the diagonal of Table 8.1 (lower) gives the values of I_{mn}, $m \neq n$, and the lower part below the diagonal gives the value of A_{mn}.

Using the association score defined above, Sprinzak and Margalit [23] identified several interesting interactions among domains which the authors referred as signatures. This original study raised significant interests in the computational biology community, and many more advanced methods have been developed to infer domain interactions from protein interactions.

The association method of Sprinzak and Margalit [23] has several limitations. First, it does not consider which domain pairs are responsible for a given interacting protein pair. For example, in Fig. 8.1, it can be inferred that D_1 may interact with either D_5 or D_6 based on the interacting protein pair (P_1, P_5). On the other hand, we cannot infer the relationship between (D_1, D_2, D_3) and (D_5, D_6) based on (P_3, P_5). However, the association method treats the contributions of interacting protein pairs (P_1, P_5) and (P_3, P_5) to the domain pairs (D_1, D_5) and (D_1, D_6) the same. Second, the association method does not take the noninteracting protein pairs into consideration. In the example given in Fig. 8.1, since P_1 does not interact with P_4, we can infer that domain D_1 does not interact with D_4 or D_5. Combining the above two reasons, it is more likely that D_1 interacts with D_6. To infer domain–domain interactions, global approaches considering all proteins and all domains are needed. Third, the association methods cannot efficiently deal with false positive and false negative interactions between the proteins.

8.2.2 Maximum Likelihood Estimation (MLE) Method

To overcome the limitations of the association method, Deng et al. [3] developed a probabilistic model linking domain interactions with protein interactions. This model considered both interacting protein pairs and noninteracting protein pairs, and incorporated the false-positive and false-negative protein interactions from high-throughput experiments. A maximum likelihood estimating (MLE) method was developed to estimate the probability of interaction for any pair of domains. An expectation maximization (EM) algorithm was used to obtain the MLE. This method made two key assumptions:

1. Pairs of domains interact with each other independent of other domain pairs.
2. A pair of proteins interact if and only if at least one pair of domains, one from each protein, interact with each other.

The conditions were relaxed in other later studies [21, 24]. The model can be briefly described as follows. We use the following notation.

- $P_1, P_2, \ldots, P_N$: the set of N proteins.
- P_{ij} indicates protein pair (P_i, P_j), and $P_{ij} = 1$ if protein P_i interacts with protein P_j.
- $D_1, D_2, \ldots, D_M$: the set of M domains.
- D_{mn} indicates domain pair (D_m, D_n), and $D_{mn} = 1$ if domain D_m interacts with domain D_n.
- $D_{mn}^{(ij)}$: interaction status of domains D_m and D_n within the protein pair (P_i, P_j).
- N_{mn}: the number of protein pairs containing domain pairs (D_m, D_n).
- λ_{mn}: probability that domain D_m interacts with domain D_n.
- $\mathcal{P}_{ij}$: the set of domain pairs, one domain from each of the proteins P_i and P_j.
- O_{ij}: the interaction status between protein P_i and protein P_j, with $O_{ij} = 1$ if proteins P_i and P_j are observed to interact and $O_{ij} = 0$ otherwise.

- f_n and f_p: false negative and false positive rates defined as $f_n = P(O_{ij} = 0|P_{ij} = 1)$ and $f_p = P(O_{ij} = 1|P_{ij} = 0)$.

Based on the two assumptions and the notation given above, we have

$$Pr(P_{ij} = 1) = 1.0 - \prod_{D_{mn} \in \mathcal{P}_{ij}} (1 - \lambda_{mn}) \tag{8.1}$$

and

$$Pr(O_{ij} = 1) = Pr(P_{ij} = 1)(1 - f_n) + \big(1 - Pr(P_{ij} = 1)\big) f_p. \tag{8.2}$$

Finally, the likelihood for the observed interaction data is

$$L = \prod_{1 \le i < j \le N} \big(Pr(O_{ij} = 1)\big)^{o_{ij}} \big(1 - Pr(O_{ij} = 1)\big)^{1-o_{ij}}, \tag{8.3}$$

where o_{ij} is the observed value of O_{ij} for proteins P_i and P_j. The objective is to estimate λ_{mn} so that L is maximized based on the observed protein interactions and the relationship between domains and proteins. In Deng et al. [3], they fixed the values of f_n and f_p, and approximate methods for determining their values were given. For convenience, we let $\theta = (\lambda_{mn}, 1 \le m < n \le M)$. To achieve this objective, an EM algorithm was designed as follows.

Assume that at the tth step of the EM algorithm, the current value of θ is $\theta^{(t-1)}$. In the E-step,

$$E\big(D_{mn}^{(ij)} = 1 | O_{kl} = o_{kl}, \forall k, l, \theta^{(t-1)}\big) = \frac{\lambda_{mn}^{(t-1)} (1 - f_n)^{o_{ij}} f_n^{1-o_{ij}}}{Pr(O_{ij} = o_{ij} | \theta^{(t-1)})}, \tag{8.4}$$

and in the M-step,

$$\lambda_{mn}^{(t)} = \frac{\lambda_{mn}^{(t-1)}}{N_{mn}} \sum_{i \in A_m, j \in A_n} \frac{(1 - f_n)^{o_{ij}} f_n^{1-o_{ij}}}{Pr(O_{ij} = o_{ij} | \theta^{(t-1)})}. \tag{8.5}$$

By repeating the E-step and M-step many times until the likelihood function in (8.3) does not change significantly, we can estimate the parameters.

The MLE approach is assessed using protein–protein interaction data sets. First, the probability of interaction between two domains is calculated using the high-throughput data sets such as Uetz and Ito protein interaction data. Second, probabilities of domain–domain interactions are used to infer probabilities of protein–protein interactions using (8.1). Third, the protein interactions with probability larger than a certain threshold is considered as predicted protein interactions, and these are compared with MIPS protein interaction data set. MIPS data set is usually considered as containing reliable protein interactions. When the predicted protein interaction belongs to the MIPS data set, it is called correctly predicted protein interaction. This process is repeated for both the association method and the MLE method, and it has been consistently shown that the MLE method outperforms the association method. For a given specificity (a fraction of correctly predicted protein interactions over all MIPS protein interactions), the sensitivity (a fraction of correctly predicted protein interactions over all predicted protein interactions) is always higher for the MLE approach than the association method, showing that the MLE approach outperforms the association method. This observation has been consistently shown in other studies [17, 21].

8.2.3 A Bayesian Method for Predicting Domain Interactions

The number of parameters λ_{mn} is usually large, and the number of observed interactions is comparatively small. Therefore, the accuracy of the estimated parameters can be low. To overcome this problem, Kim et al. [14] developed a Bayesian method to estimate the parameters. In addition, they also treated f_n and f_p as random variables and assumed their prior distributions. Some other modifications, including the consideration of the number of domains in each protein and protein interactions across several different organisms, were also introduced. Here we review the basic setup of the Bayesian model.

The basic model linking domain interactions to protein interactions as described in Sect. 8.2.2 was used in the Bayesian approach. Instead of treating (λ_{mn}, f_n, f_p) as deterministic parameters, the Bayesian model considers them as random variables. The prior distributions for f_n and f_p are assumed to be uniform on intervals $[u_n, v_n]$ and $[u_p, v_p]$, respectively. The prior distribution of λ_{mn} was assumed to be the beta distribution with parameters (α, β), i.e., $\lambda_{mn} \sim Beta(\alpha, \beta)$. Under the above assumptions,

$$f_n|\text{rest} \propto \prod_{1 \le i < j \le N} \left(Pr(O_{ij}=1)\right)^{o_{ij}} \left(1 - Pr(O_{ij}=1)\right)^{1-o_{ij}} g(f_n|f_p, \theta),$$

$$f_p|\text{rest} \propto \prod_{1 \le i < j \le N} \left(Pr(O_{ij}=1)\right)^{o_{ij}} \left(1 - Pr(O_{ij}=1)\right)^{1-o_{ij}} g(f_p|f_n, \theta),$$

$$\lambda_{mn}|\text{rest} \propto \prod_{1 \le i < j \le N} \left(Pr(O_{ij}=1)\right)^{o_{ij}} \left(1 - Pr(O_{ij}=1)\right)^{1-o_{ij}} g(\lambda_{mn}|rest),$$

where $g(\cdot)$ is the prior distribution for the parameters. Kim et al. [14] showed that all the three posterior distributions are log-concave functions. They thus used adaptive rejection sampling method [9] and the adaptive rejection Metropolis sampling method [10] to obtain the posterior distributions of the parameters.

8.2.4 A Likelihood-Ratio-Based Method: Domain Pair Exclusion Analysis (DPEA)

The MLE method of Deng et al. [3] ranked the domain–domain interactions based on the estimated values of λ_{mn} without specifically considering the number of protein pairs N_{mn} containing the domain pairs (D_m, D_n). When multiple domain proteins interact with each other, it is not clear which domain pairs are responsible for the protein interactions. Due to the relative small number of observed interactions, the estimated values of λ_{mn} may not be accurate making the ranking of domain interactions based on λ_{mn} less reliable. In statistics, one commonly used principle for hypothesis testing is the likelihood ratio test, which compares the likelihood of the observed data under the null hypothesis versus the likelihood of the data under the alternative hypothesis. For the current problem of evaluating the significance of domain interactions between domains D_m and D_n, the null hypothesis is $H_0 : \lambda_{mn} = 0$, and the alternative hypothesis is $H_1 : \lambda_{mn} > 0$.

Based on these considerations, Riley et al. [21] introduced a domain pair exclusion analysis (DPEA) approach to predict domain interactions. First, DPEA calculates the λ_{mn}, the probability that domain D_m and domain D_n interact using the MLE method of Deng et al. [3]. Similar method can be used to estimate the MLE of $\overline{\theta} = (\overline{\lambda}_{kl}^{(mn)}, 1 \leq k < l \leq N)$ by forcing $\lambda_{mn} = 0$. Second, let L_0 be the likelihood of the observed interaction data under $\overline{\theta}$. Then the E score is the $\log(L/L_0)$. In Riley et al. [21], the investigators used the highly reliable interactions in MIPS and assumed that $f_p = 0$. Preliminary studies also showed that f_n tends to be small and can be approximately by 0. Under all these assumptions,

$$E_{mn} = \sum_{i,j,D_{mn} \in \mathcal{P}_{ij}} \log \frac{Pr(O_{ij} = 1|m,n \text{ can interact})}{Pr(O_{ij} = 1|m,n \text{ do not interact})} \tag{8.6}$$

$$= \sum_{i,j,D_{mn} \in \mathcal{P}_{ij}} \log \frac{1 - \prod_{D_{kl} \in P_{ij}} (1 - \lambda_{kl})}{1 - \prod_{D_{kl} \in P_{ij}} (1 - \overline{\lambda}_{kl}^{(mn)})}, \tag{8.7}$$

where $\overline{\lambda}_{kl}^{(mn)}$ represents the probability of interaction between domains D_k and D_l using the MLE method when the probability of interaction between domains D_m and D_n is set to 0. E-score is the summation of the likelihood ratio between the probability that protein i and protein j interact when domain D_m and domain D_n interact, and the probability that protein i and protein j interact when domain D_m and domain D_n does not interact for all protein pairs with the domain pair D_m and D_n. Intuitively, this E-score calculates the contribution of domains D_m and D_n for interacting protein pairs with domains D_m and D_n.

An important difference between the MLE method and the E-score is that the more there are interacting protein pairs with domains D_m and D_n, the higher E_{mn} score. This is because more interacting proteins with domain pairs D_m and D_n have a chance to increase the E-score by the summation of the likelihood ratio. On the other hand, the score by the MLE method largely depends on the fraction of interacting protein pairs compared to all the pairs of proteins with domains D_m and D_n. Hence, it does not depend on the number of interacting protein pairs with domains D_m and D_n.

The prediction accuracy of DPEA was assessed using the PDB domain interaction data set [21]. If a method is reasonable, the set of highly ranked domain pairs should contain more true domain interactions from PDB. The authors compared DPEA with the MLE-based method of Deng et al. [3] and the association method of [23] using the following approach. For any given k, domain pairs with the top k highest scores according to each of the three methods were chosen and compared with true domain interactions in PDB. It was shown that the fraction of true domain interactions among domain pairs with the top k highest scores based on DPEA is the highest, followed by that of the MLE-based method, and the corresponding fraction based on the association method is the lowest for all the values of k. At least for the data the authors analyzed, the DPEA method outperforms the other two approaches. However, independent studies based on other protein interaction data sources are still needed.

8.2.5 Maximum-Parsimony-Based Method–Linear Programming Optimization

Although the MLE, Bayesian, and the likelihood-ratio-based methods described above performed well in predicting domain interactions, they also have several drawbacks. First, they are all model based and need several assumptions. However, these assumptions may not hold in reality. The effects of the misspecifications of the assumptions on the accuracy of the predicted domain interactions are not known. The assumption of independence of domain interactions among protein pairs certainly does not hold in reality. If one domain D_m already interacts with another domain D_n, D_m cannot participate in interacting with other domains. Second, it is difficult to specify a range for f_n and f_p in the model. To overcome these problems, model-free-based methods have been developed to predict domain interactions.

Guimarães et al. [12] assumed that interactions between proteins have evolved in a parsimonious way so that the minimal number of domain interactions can explain the observed protein interactions. They considered the problem of predicting interacting domain pairs as an optimization problem. The objective is to minimize the number of domain–domain interactions necessary to justify the underlying protein–protein interaction network. This approach is called the "Parsimonious Explanation (PE)" method. They formulated the problem using a linear programming. Let variable x_{mn} represent the score of the potential interaction between domains D_m and D_n. For a given interacting protein pair of P_i and P_j, we have a set of potential domain interaction pairs $\mathcal{P}_{ij}$. Using linear programming, the goal is to minimize the objective function $\sum_{m,n} x_{mn}$ subject to the set of constraints, which require that $\sum_{(D_m,D_n)\in\mathcal{P}_{ij}} x_{mn} \geq 1$ for all interacting protein pairs. The estimated value of x_{mn} is called an LP (linear programming)-score in [12].

The PE method assigns a high score to frequently occurring domain pairs in order to minimize the number of domain pairs in explaining observed protein interactions. To avoid over-prediction for frequently occurring domain pairs, a promiscuity versus witnesses (pw)-score is calculated for every predicted domain–domain interaction. The pw-score considers both statistical significance of the LP-score and false positive protein interactions. First, p-value(m,n) estimates statistical significance of the frequency of appearance of the domain pair in its LP-score. One thousand random networks are generated by permutating the randomly selected edges, and the LP-score is calculated for each random network. P-value(m, n) is the fraction of random network experiments that return the LP-score equal to or larger than the LP-score obtained by observed protein–protein interaction network. Second, let $w(m, n)$ be the number of observed protein interactions for a given domain pair (m, n), and let r be the reliability of the observed interacting protein network. The $(1 - r)^{w(m,n)}$ is the probability that all edges in the network that correspond to an interaction with domain pair (m, n) are false positive. The pw-score is the minimum between $(1 - r)^{w(m,n)}$ and p-value(m, n).

$$\text{pw-score}(m, n) = \min\bigl(\text{p-value}(m, n), (1 - r)^{w(m,n)}\bigr). \tag{8.8}$$

By selecting domain–domain interactions lower than pw-score cutoffs, the PE method can control the prediction accuracy. Using the same DIP protein interaction data and PDB domain interaction data from Riley et al. [21], the authors showed that the prediction accuracy increases with LP-score and decreases with pw-score.

The PE method does not penalize domain pairs in noninteracting protein pairs and adds weight for frequent domain pairs in the interacting proteins, allowing detection of high-specificity and low-promiscuity domain interactions. At the same time, the pw-score controls frequently occurring domain pairs which can be falsely considered as interacting independent of the underlying protein interaction structure.

8.3 Integrated Approaches for Predicting Domain Interactions

With the completion of genome sequences of many species, comparative analysis of many species became very important to understand the function of genes and their products. Comparative analysis of protein domains from more than 50 species revealed the consensus and differences of domain organizations between species [27]. The analysis of protein–protein interactions from several species allows the detection of conserved domain–domain interactions, and these conserved interactions can be used to predict high-confidence domain interactions [16, 17].

In addition to protein interaction data sets from multiple species, other data sources including domain fusion and domain function also contain information for domain interactions. We next review integrative approaches for predicting domain interactions [16, 17, 24].

8.3.1 An Extended Likelihood Approach for Predicting Domain Interactions Based on Protein Interactions from Multiple Species

Liu et al. [17] first proposed the idea of integrating protein interactions from multiple species for predicting domain interactions. They used the interaction data from yeast, worm, and drosophila to predict domain–domain interactions by extending the MLE and the EM algorithm of Deng et al. [3]. The likelihood function for the observed protein interaction data across all three species is defined as

$$L = \prod Pr(O_{ijk} = 1)^{o_{ijk}} \big[1 - Pr(O_{ijk} = 1)\big]^{1-o_{ijk}}, \tag{8.9}$$

where k indicates the species. The likelihood function is represented by $\theta = (\lambda_{mn}, f_{pk}, f_{nk})$. Liu et al. [17] assumed that the probability of domain interactions λ_{mn} does not depend on the species. However, the false positive and false negative rates are species specific. To obtain the maximum likelihood estimates of the parameters, they modified the EM algorithm of Deng et al. [3]. The complete data include all the domain interactions for protein pairs i and j in the three species, denoted by $D^{(ij)}_{mnk}$. The E-step is

$$E\big(D^{(ij)}_{mnk} | O_{ijk} = o_{ijk}, \lambda^{(t-1)}_{mn}\big) = \frac{\lambda^{(t-1)}_{mn} (1 - f_{nk})^{o_{ijk}} f_{nk}^{1-o_{ijk}}}{Pr(O_{ijk} = o_{ijk} | \lambda^{(t-1)}_{mn})}. \tag{8.10}$$

In the M-step, the probability of interactions between domains m and n, λ_{mn}, is calculated as

$$\lambda^{(t)}_{mn} = \frac{\lambda^{(t-1)}_{mn}}{N_{mn}} \sum_{i,j,k} \frac{(1 - f_{nk})^{o_{ijk}} f_{nk}^{1-o_{ijk}}}{Pr(O_{ijk} = o_{ijk} | \lambda^{(t-1)}_{mn})}, \tag{8.11}$$

where N_{mn} is the total number of all protein pairs with domains m and n across the three species. These two equations are the extension of (8.4) and (8.5) for integrating protein interactions from single species.

The extended likelihood approach allows the integration of protein interactions from diverse species. The accuracy of domain–domain interactions predicted from this integrated information outperforms that from only one species yeast. It is demonstrated by comparing a gold standard MIPS protein interactions with protein–protein interactions inferred from predicted domain–domain interactions using the extended likelihood approach. The Bayesian method has also been extended to multiple species [14].

Table 8.2 Domain fusion and domains with similar biological processes are more likely to interact. "Fraction" indicates the fraction of domain interactions in iPfam in a given set. "Fold" indicates the ratio of the fraction over expected value (0.17%)

Evidence	# protein pairs	# Overlap iPfam	Fraction	Fold
Random domain pairs	1,539,135	2,580	0.17%	–
Domain fusion	9,615	1,141	11.8%	69
Same GO terms	57,907	1,302	0.8%	13

8.3.2 Predicting Domain Interactions from Multiple Data Sources

In addition to protein interactions from multiple species, other information sources, such as domain fusion and domain function, can also be used to predict domain interactions.

Using Protein Interactions from Multiple Species Instead of using the extended likelihood approach of [17], Lee et al. [16] proposed another approach to combine protein interaction data from multiple species. For each species, instead of directly using λ_{mn} to measure the possibility of domain interactions, Lee et al. [16] proposed to use the expected number of domain interactions among the interacting protein pairs to score domain interactions, i.e.,

$$E(\#D_{mn}) = N_{mn} Pr(D_{mn} = 1), \tag{8.12}$$

where N_{mn} is the number of protein pairs having domains D_m and D_n. This expected number considers the biological intuition that if a pair of domains is observed in multiple protein interactions, this pair of domains is more likely to interact.

Using Domain Fusion Enright et al. [4] and Marcotte et al. [18] showed that two proteins are more likely to interact if they are fused into one protein in another species. Lee et al. [16] extended this idea to domains and examined whether two domains are more likely to interact with each other if they are fused into one protein in any species. A total of 9,615 pfam-A domain pairs which coexist in one protein in any species were collected. Among them, 11.8% of domain pairs overlap with iPfam. It is 69-fold higher than random domain pairs with 0.17% overlap with iPfam. Biological process of gene ontology is studied to see if domain pairs with similar functions are more likely to interact with each other. A total of 57,907 domain pairs with the same GO terms were collected, and 0.8% of domain pairs overlapped with iPfam. This ratio is 13-fold higher than random pairs as seen in Table 8.2. These preliminary studies proved our initial hypotheses.

Based on these observations, we defined CE(D_{mn}), where CE stands for Co-Existence, as the number of proteins that contain both domains D_m and D_n. The values of CE(D_{mn}) were binned into nine categories. For each category, we calculated the likelihood ratio corresponding to domain fusion by the fraction of interacting domain pairs in the category over that among random domain pairs. The likelihood ratio score increases with CE. For details, see [16].

Table 8.3 The probability of true domain interactions increases with the overall likelihood ratio score. The likelihood ratio values of predicted domain pairs, the number of predicted domain pairs, and the number of overlaps with iPfam

Likelihood ratio values	Interactions	Overlap with iPfam	Fraction	Fold
Random domain pairs	1,539,135	2,580	0.17%	–
>0	25,352	2,080	8.2%	48
≥1	6,386	1,641	25.7%	151
≥4	2,391	1,241	51.9%	305
≥6	2,044	1,142	55.9%	329
≥11	1,683	1,011	60.1%	353
≥21	886	634	71.6%	421
≥51	420	336	80.0%	471

Using Gene Ontology The gene ontology has a hierarchical structure, where parents represent the general function, and offsprings represent more specific terms. Two domains with more specific common terms are more functionally related with each other than two domains with more general common terms. A more specific function generally covers a smaller number of domains. Assume that domain D_m and domain D_n have the same function F_f. A score SG(D_{mn}), where SG stands for the Same Gene ontology, was defined as the number of domains having the function F_f. Values of SG(D_{mn}) were binned into six intervals, and then the likelihood ratio corresponding to SG was similarly calculated as for CE. The likelihood ratio decreases as SG(D_{mn}) increases.

Predicting Domain Interactions by Integrating Protein Interactions from Multiple Species, Domain Fusion, and Gene Ontology All information including protein interactions from four species, CE(D_{mn}), and SG(D_{mn}) were then combined to obtain a joint score by multiplying the likelihood ratio scores for all the six data sources. As shown in Table 8.3, as the total likelihood increases, the fraction of overlap with iPfam increases with the overall likelihood score. Figure 8.2 shows the accuracy of predicted domain interactions based on the ROC curve when domain fusion, domain function, and combination of all of the six evidences are used. Integration of all of the six evidences outperforms the other cases. This integrative approach has been formalized recently by Wang et al. [24].

8.4 Discussion

In this chapter, we review several computational methods to predict domain interactions from multiple biological data sets. First, we describe the methods using protein interactions. One of first systematic methods incorporating the global protein interaction network is the maximum likelihood estimation approach. This method has

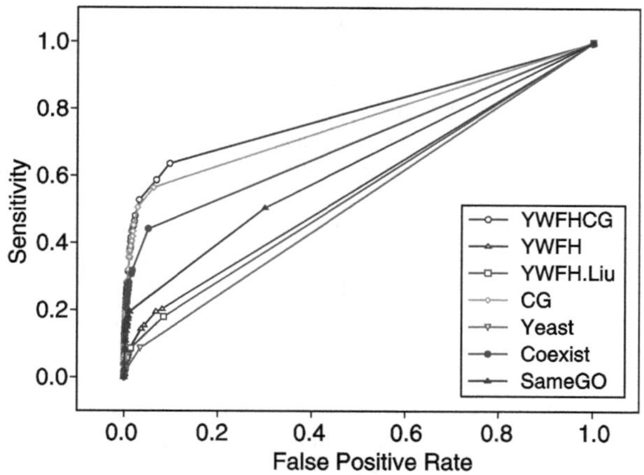

Fig. 8.2 The comparison of prediction accuracies by integrating multiple biological data sets using the naive Bayesian method. The letters Y, W, F, H, C, and G indicate domain interactions based on yeast, worm, fruitfly, humans, co-existence, and same GO function, respectively. YWFH. Liu shows the result of predicted domain interactions using the extended MLE method defined in Liu et al. [17] with protein interactions of yeast, worm, fruitfly, and humans. (This figure is excerpted from Lee et al. [16]) (for the color version, see Color Plates on p. 392)

been extended by several groups for more accurate prediction of domain interactions.

Second, we describe the integrative approach to use protein interactions from multiple species, domain function, and domain sequences. The fact that domains are conserved across species gives a great opportunity to use these data sets for comprehensive understanding of cellular activities at a domain level.

As it is hard to determine domain interactions using experiments, it is becoming more important to develop computational approaches to achieve this objective. We are able to infer more biological and medical knowledge from domain interactions if the accuracy of predicted domain interactions increases. For example, revealing how mutations on proteins are propagated through domain interaction network helps to predict disease-related proteins in the network level.

In summary, we show the power of integrating multiple biological data sources for predicting domain interactions. Still, significant questions such as incompleteness of data sets and difficulty of assessment of predictions are remained to be addressed. As new biological data sets are rapidly generated, it is important to study the relationship between these new data sets and domain interactions.

References

1. R. Apweiler, T.K. Attwood, A. Bairoch, A. Bateman, E. Birney, M. Biswas, et al. The InterPro database, an integrated documentation resource for protein families, domains and functional sites. *Nucleic Acids Res*, **29**(1):37–40, 2001.

2. H.M. Berman, J. Westbrook, Z. Feng, G. Gilliland, T.N. Bhat, H. Weissig, I.N. Shindyalov, and P.E. Bourne. The protein data bank. *Nucleic Acids Res*, **28**:235–242, 2000.
3. M. Deng, F. Sun, and T. Chen. Inferring domain–domain interactions from protein–protein interactions. *Genome Res*, **12**:1540–1548, 2002.
4. A.J. Enright, I. Iliopoulos, N.C. Kyrpides, and C.A. Ouzounis. Protein interaction maps for complete genomes based on gene fusion events. *Proc Natl Acad Sci USA*, **402**(6757):86–90, 1999.
5. R.D. Finn, M. Marshall, and A. Bateman. Visualisation of protein–protein interactions at domains and amino acid resolutions. *Bioinformatics*, **21**(3):410–412, 2005.
6. R.D. Finn, J. Tate, J. Mistry, P.C. Coggill, J.S. Sammut, H.R. Hotz, G. Ceric, K. Forslund, S.R. Eddy, E.L. Sonnhammer, and A. Bateman. The Pfam protein families database. *Nucleic Acids Res*, **36**:D281–D288, 2008.
7. The Gene Ontology (GO) Consortium. Gene Ontology: tool for the unification of biology. *Nat Genet*, **25**:25–29, 2000.
8. The Gene Ontology (GO) Consortium. Creating the Gene Ontology resource: design and implementation. *Genome Res*, **11**:1425–1433, 2001.
9. W.R. Gilks and P. Wild. Adaptive rejection sampling for Gibbs sampling. *Appl Stat*, **41**:337–348, 1992.
10. W.R. Gilks, N.G. Best, and K.K.C. Tan. Adaptive rejection Metropolis sampling. *Appl Stat*, **44**:455–472, 1995.
11. L. Giot, J.S. Bader, C. Brouwer, and A. Chaudhuri. A protein interaction map of Drosophila melanogaster. *Science*, **302**(5651):1727–1736, 2003.
12. K.S. Guimarães, R. Jothi, E. Zotenko, and T.M. Przytycka. Predicting domain–domain interactions using a parsimony approach. *Genome Biol*, **7**(11):R104, 2006.
13. T. Ito, T. Chiba, R. Ozawa, M. Yoshida, M. Hattori, and Y. Sakaki. A comprehensive two-hybrid analysis to explore the yeast protein interactome. *Proc Natl Acad Sci USA*, **98**:4569–4574, 2001.
14. I. Kim, Y. Liu, and H.Y. Zhao. Bayesian methods for predicting interacting protein pairs using domain information. *Biometrics*, **63**:824–833, 2007.
15. P.M. Kim, L.J. Lu, Y. Xia, and M.B. Gerstein. Relating three dimensional structures to protein networks provides evolutionary insights. *Science*, **314**:1938–1941, 2006.
16. H.J. Lee, M.H. Deng, F.Z. Sun, and T. Chen. An integrated approach to the prediction of domain–domain interactions. *BMC Bioinf*, **7**:269, 2006.
17. Y. Liu, N. Liu, and H. Zhao. Inferring protein–protein interactions through high-throughput interaction data from diverse organisms. *Bioinformatics*, **21**(15):3279–3285, 2005.
18. E.M. Marcotte, M. Pellegrini, H.L. Ng, D.W. Rice, T.O. Yeates, and D. Eisenberg. Detecting protein function and protein–protein interactions from genome sequences. *Science*, **285**:751–753, 1999.
19. H.W. Mewes, D. Frishman, U. Guldener, G. Mannhaupt, K. Mayer, M. Mokrejs, B. Morgenstern, M. Munsterkotter, S. Rudd, and B. Weil. MIPS: a database for genomes and protein sequences. *Nucleic Acids Res*, **30**:31–34, 2002.
20. S. Peri, J. Navarro, R. Amanchy, T. Kristiansen, C. Jonnalagadda, et al. Development of human protein reference database as an initial platform for approaching systems biology in humans. *Genome Res*, **13**:2363–2371, 2003.
21. R. Riley, C. Lee, C. Sabatti, and D. Eisenberg. Inferring protein domain interactions from databases of interacting proteins. *Genome Biol*, **6**(10):R89, 2005.
22. L. Salwinski, C.S. Miller, A.J. Smith, F.K. Pettit, J.U. Bowie, and D. Eisenberg. The database of interacting proteins: 2004 update. *Nucleic Acids Res*, **32**:D449–D451, 2004.
23. E. Sprinzak and H. Margalit. Correlated sequence-signatures as markers of protein–protein interaction. *J Mol Biol*, **311**:681–692, 2001.
24. H. Wang, E. Segal, A. Ben-Hur, Q.R. Li, M. Vidal, and D. Koller. InSite: a computational method for identifying protein–protein interaction binding sites on a proteome-wide scale. *Genome Biol*, **8**:R192, 2007.

25. M.A. Wall, D.E. Coleman, E. Lee, J.A. Iniguez-Lluhi, B.A. Posner, A.G. Gilman, and S.R. Sprang. The structure of the G protein heterotrimer Gi alpha 1 beta 1 gamma 2. *Cell*, **83**:1047–1058, 1995.
26. P. Uetz, L. Giot, G. Cagney, T.A. Mansfield, R.S. Judson, J.R. Knight, D. Lockshon, V. Narayan, M. Srinivasan, and P. Pochart. A comprehensive analysis of protein–protein interactions in Saccharomyces cerevisiae. *Nature*, **403**:623–627, 2000.
27. Y. Ye and A. Godzik. Comparative analysis of protein domain organization. *Genome Res*, **14**:343–353, 2004.

Chapter 9
Irreversible Stochastic Processes, Coupled Diffusions and Systems Biochemistry

Pei-Zhe Shi and Hong Qian

9.1 Introduction

One of the earliest research areas of Professor Min-Ping Qian is the irreversibility of stochastic processes, more specifically Markov chains [1, 2]. These days, part of her main interest is stochastic modeling of cellular processes at a systems biochemistry level [3]. The relation between these two topics might not be so obvious at first sight. But if one takes a more fundamental perspective of "what is life" as that of a physicist [4], then the philosophical connection is obvious: Since cellular processes are stochastic [5] and living systems are in nonequilibrium states [6]; the mathematical representation of living cellular processes must be stochastic processes that reflect irreversibility.

While the philosophical connection is now obvious, this paper focuses on the specific role played by irreversible Markov processes in modeling cellular biochemical systems [7]. In Sect. 9.2 we start with the simple Michaelis–Menten enzyme kinetics from a purely stochastic perspective, i.e., that of a single enzyme driven by the nonequilibrium chemical reaction of substrate to product. This provides the motivation for irreversible Markov processes (also known as nonsymmetric Markov processes following Kolmogorov). Then in Sect. 9.3, we discuss three biochemical models, each one of them leads to an irreversible Markov process called *coupled diffusion*. In the past, the mathematical theory of coupled diffusions has been extensively studied by the probability group at Peking University [8–10]. In Sect. 9.4, two limit cases of the coupled diffusion in the fast jump processes and the fast diffusion processes are analyzed. In particular, we study the nonequilibrium steady-state flux and entropy production of the system. In Sect. 9.5, we show a bifurcation, saddle-node or pitchfork, which occurs in certain coupled diffusion systems while decreasing the rates of jump processes. Section 9.6 discusses some recently developed numerical methods for solving the steady-state coupled diffusion equations.

P.-Z. Shi (✉) · H. Qian
Department of Applied Mathematics, University of Washington, Seattle, WA 98195-2420, USA
e-mail: ship@uw.edu

J. Feng et al. (eds.), *Frontiers in Computational and Systems Biology*, Computational Biology 15,
DOI 10.1007/978-1-84996-196-7_9,

Section 9.7 gives some discussions, and all the mathematical details are collected in Sect. 9.8.

9.2 Single-Molecule Michaelis–Menten Enzyme Kinetics and Irreversible Markov Processes

Let us consider the widely studied enzymatic reaction

$$S + E \underset{k_{-1}}{\overset{k_1^o}{\rightleftharpoons}} SE \underset{k_{-2}}{\overset{k_2}{\rightleftharpoons}} PE \underset{k_{-3}^o}{\overset{k_3}{\rightleftharpoons}} P + E, \tag{9.1}$$

in which k_1^o and k_{-3}^o are second-order rate constants, with dimension $[\text{time}]^{-1} \times [\text{concentration}]^{-1}$, while all the other k's are first-order rate constants with dimension $[\text{time}]^{-1}$.

The classical reversible Michaelis–Menten kinetics analysis focuses on the turnover of substrate S to product P. With relatively low concentration of the enzyme, the net turnover rate of the enzyme-catalyzed reaction is characterized nicely by Briggs and Haldane [11] under the quasi-steady-state assumption, in terms of $V_{\max}$s and K_ms. Furthermore, the *Haldane relationship* shows that the concentrations of substrate and product at equilibrium is determined by $V_{\max}$s and K_ms [12].

However, in another viewpoint, one can focus on the states of the enzyme. If we consider only a single enzyme molecule, as that in single-molecule enzymology [13], then the enzyme has three possible states: E, SE, and PE. The dynamics of the single-enzyme kinetics is a continuous-time three-state Markov chain (Q-process) with infinitesimal transition rates $k_{\pm i}$ ($i = 1, 2, 3$), where $k_1 = k_1^o c_S$ and $k_{-3} = k_{-3}^o c_P$. Here, instead of assuming a constant concentration of enzyme-substrate complex under the quasi-steady state, we assume that the concentration of substrate c_S and the concentration of product c_P are constant, independent of time. This is certainly a valid assumption for a wide range of biochemical reactions inside a living cell under homeostasis. The basis of this assumption is that living cells are open systems, which finally go to a nonequilibrium steady state; while the Haldane relationship in classical Michaelis–Menten kinetics considers a closed system, which reaches an equilibrium at the end.

The Michaelis–Menten kinetics of a single molecule is

$$\frac{dp_E}{dt} = k_{-1}p_{SE} - (k_1 + k_{-3})p_E + k_3 p_{PE} = J_1^- + J_3^+ - J_1^+ - J_3^-, \tag{9.2a}$$

$$\frac{dp_{SE}}{dt} = k_1 p_E - (k_{-1} + k_2)p_{SE} + k_{-2}p_{PE} = J_1^+ + J_2^- - J_1^- - J_2^+, \tag{9.2b}$$

$$\frac{dp_{PE}}{dt} = k_2 p_{SE} - (k_{-2} + k_3)p_{PE} + k_{-3}p_E = J_2^+ + J_3^- - J_2^- - J_3^+, \tag{9.2c}$$

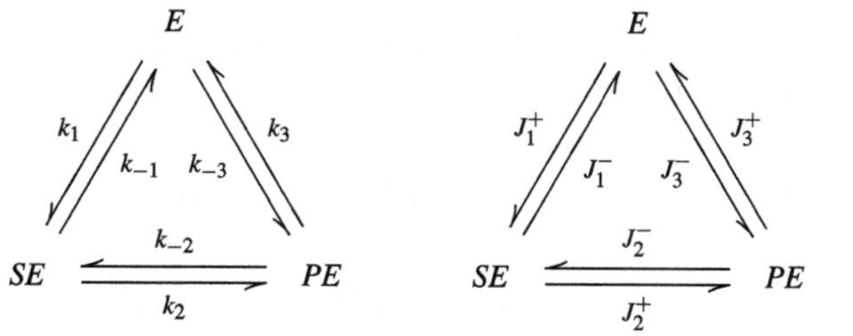

Fig. 9.1 A diagramic representation of enzyme catalyzed reaction (9.1). (**a**) A three-state continuous-time Markov chain with transition rates $k_{\pm i}$. (**b**) The probability flux between states, defined as the transition rate times the probability of the single enzyme being at the origin state of the flux

where p_X means the probability of being at state X ($X = E, ES, EP$), $J^{\pm}_{1,2,3}$ are the forward and backward flux of the reactions.

If a chemical reaction system reaches its equilibrium, the condition of detailed balance is satisfied, i.e., the net probability mass transition between any pair of reactants being zero: $J_i^+ - J_i^- \equiv 0$, $i = 1, 2, 3$. In terms of reaction constants, that is,

$$k_1 p_E = k_{-1} p_{ES}, \qquad k_2 p_{SE} = k_{-2} p_{PE}, \qquad k_3 p_{PE} = k_{-3} p_E.$$

This lead to the relation of the reaction rate constants

$$k_1 k_2 k_3 = k_{-1} k_{-2} k_{-3}, \tag{9.3}$$

which is known as the Kolmogorov cycle condition. Noticing that $k_1 = k_1^0 c_S$ and $k_{-3} = k_{-3}^0 c_P$, the condition (9.3) puts a constrain on the concentrations of S and P in chemical equilibrium.

However, for general values of c_S and c_P, the detailed balance condition will not be satisfied. In fact the steady state of the system is not an equilibrium. To investigate the irreversibility of this chemical process, we analyze the system in the viewpoint of energy [14]. Define the chemical potential (Gibbs free energy) of X as

$$\mu_X(t) = \mu_X^0 + k_B T \ln p_X(t), \qquad \mu_X^0 = h_X^0 - T s_X^0,$$

where $X = E, SE, PE$, μ_X^0, h_X^0, s_X^0 are the standard state free energy, enthalpy, and entropy, k_B is the Boltzmann constant, and T is temperature in Kelvin. The chemical potentials of S and P are as normally defined,

$$\mu_S = \mu_S^0 + k_B T \ln c_S, \qquad \mu_P = \mu_P^0 + k_B T \ln c_P.$$

For chemical reaction $E + S \rightleftharpoons SE$ at equilibrium, $\mu_E + \mu_S = \mu_{SE}$, so

$$\mu_E^0 + \mu_S^0 - \mu_{SE}^0 = k_B T \ln \frac{p_{SE}^{eq}}{c_S^{eq} p_E^{eq}} = k_B T \ln \frac{k_1^0}{k_{-1}}.$$

The second equation is due to the detailed balance at equilibrium $p_E^{eq} c_S^{eq} k_1^0 = p_{SE}^{eq} k_{-1}$. Then at any time

$$\mu_E + \mu_S - \mu_{SE} = \mu_E^0 + \mu_S^0 - \mu_{SE}^0 - k_B T \ln \frac{p_E c_S}{p_{ES}}$$

$$= k_B T \ln \frac{k_1^0 p_E c_S}{k_{-1} p_{ES}} = k_B T \ln \frac{J_1^+}{J_1^-}. \tag{9.4a}$$

Similar equations are set up for the other two reactions:

$$\mu_{SE} - \mu_{PE} = k_B T \ln \frac{J_2^+}{J_2^-}, \tag{9.4b}$$

$$\mu_{PE} - (\mu_E + \mu_P) = k_B T \ln \frac{J_3^+}{J_3^-}. \tag{9.4c}$$

Therefore, summing up both sides of (9.4a)–(9.4c), we get

$$\mu_S - \mu_P = k_B T \ln \frac{J_1^+ J_2^+ J_3 +}{J_1^- J_2^- J_3^-} = k_B T \ln \frac{k_1 k_2 k_3}{k_{-1} k_{-2} k_{-3}}. \tag{9.5}$$

If the detailed balance condition (9.3) holds, $\mu_S = \mu_P$, which means the chemical potential of substrate S and product P are equal, and the system reaches equilibrium. However, if the chemical potential difference between S and P is not zero, we have

$$\Delta G_{SP} = \mu_S - \mu_P = k_B T \ln \frac{k_1 k_2 k_3}{k_{-1} k_{-2} k_{-3}}. \tag{9.6}$$

So

$$\frac{k_1 k_2 k_3}{k_{-1} k_{-2} k_{-3}} = e^{\Delta G_{SP}/k_B T}. \tag{9.7}$$

After a sufficient long time, the system will settle down to a nonequilibrium steady state (NESS), while the mass flux is going from high energy (source) to low energy (sink) constantly. The chemical potential difference ΔG_{SP} is the driving force of the flux J^{ss}, which has the expression

$$J^{ss} = \frac{k_1 k_2 k_3 - k_{-1} k_{-2} k_{-3}}{k_1 k_2 + k_1 k_3 + k_1 k_{-2} + k_2 k_3 + k_2 k_{-3} + k_{-1} k_{-2} + k_{-1} k_3 + k_{-1} k_{-3} + k_{-2} k_{-3}}. \tag{9.8}$$

The rate of heat dissipation at NESS is

$$hdr = J^{ss} \cdot \Delta G_{SP}.$$

Since $J^{ss} = J_i^+ - J_i^-$, $i = 1, 2, 3$, we see

$$hdr = \sum_{i=1}^{3} (J_i^+ - J_i^-) k_B T \ln \frac{J_i^+}{J_i^-} \geq 0.$$

This is exactly the statement of The Second Law of Thermodynamics: with only a single temperature bath T, one can only continuously convert chemical work to heat, but not in reverse. The condition for the equality is $J^{ss} = 0$, which is equivalent to the condition of detailed balance. And in general, when $J^{ss} \neq 0$, the system has a positive heat dissipation rate, thus is irreversible.

To summarize, the reversibility of the chemical reaction system (9.1) in a steady state is determined by the pseudo-first-order reaction constants $k_{\pm i}$. The term pseudo refers to the implicitly included concentrations of S and P in the constants, i.e., $k_1 = k_1^0 c_S$, $k_{-3} = k_{-3}^0 c_P$. If

$$\frac{k_1 k_2 k_3}{k_{-1} k_{-2} k_{-3}} = 1,$$

then the system will reach equilibrium, and the process will be reversible. If

$$\frac{k_1 k_2 k_3}{k_{-1} k_{-2} k_{-3}} \neq 1,$$

then it reaches an NESS with positive heat dissipation rate, the process turns out to be irreversible. In NESS, there is a nonzero steady-state cyclic flux J^{ss} in the system, and a nonzero free-energy difference between S and P as the driving force of the system. This force comes from the external energy supply, as one needs to put S in and take P out continuously.

In the classical enzyme kinetics, Briggs and Haldane studied "reversible" enzyme reaction (9.1) [11]. With the assumption of quasi steady-state relationship

$$\frac{d[SE]}{dt} = \frac{d[PE]}{dt} = 0,$$

one can compute the net turnover rate of the enzyme-catalyzed reaction (9.1) [12]

$$v_{\text{net}} = \frac{V_{\max_f} \frac{[S]}{K_{m_S}} - V_{\max_r} \frac{[P]}{K_{m_P}}}{1 + \frac{[S]}{K_{m_S}} + \frac{[P]}{K_{m_P}}}, \tag{9.9}$$

where

$$K_{m_S} = \frac{k_{-1}k_3 + k_{-1}k_{-2} + k_2 k_3}{k_1^o (k_2 + k_{-2} + k_3)}, \qquad K_{m_P} = \frac{k_{-1}k_3 + k_{-1}k_{-2} + k_2 k_3}{k_{-3}^o (k_{-1} + k_2 + k_{-2})},$$

$$V_{\max_f} = \frac{k_2 k_3 [E]_t}{k_2 + k_{-2} + k_3}, \qquad V_{\max_r} = \frac{k_{-1} k_{-2} [E]_t}{k_{-1} + k_2 + k_{-2}}.$$

The celebrated Haldane equation between kinetic constants and equilibrium constant is

$$K_{eq} = \frac{[P]_{eq}}{[S]_{eq}} = \frac{k_1^o k_2 k_3}{k_{-1} k_{-2} k_{-3}^o} = \frac{V_{\max_f} K_{mP}}{V_{\max_r} K_{mS}}, \tag{9.10}$$

so at equilibrium

$$k_1^o [S]_{eq} k_2 k_3 = k_{-1} k_{-2} k_{-3}^o [P]_{eq}.$$

Now realizing that $k_1 = k_1^o[S]$ and $k_{-3} = k_{-3}^o[P]$, we see that the Haldanes equation is precisely our (9.3). In fact, the v_{net} in (9.9) is exactly the J^{ss} in (9.8) if we set $[E]_t = 1$. The Michaelis–Menten kinetics can be best understood in terms of single-enzyme steady-state flux [15].

From the mathematical stand point, the continuous-time discrete-state Markov process has the infinitesimal transition rate matrix (Q-matrix)

$$Q = \begin{bmatrix} -k_1 - k_{-3} & k_1 & k_{-3} \\ k_{-1} & -k_{-1} - k_2 & k_2 \\ k_3 & k_{-2} & -k_3 - k_{-2} \end{bmatrix}. \tag{9.11}$$

The stationary distribution $\pi = (p_1, p_2, p_3)$ satisfies $\pi = \pi Q$. By the theory of Markov processes [16], its stationary process is reversible if and only if

$$p_1 k_1 = p_2 k_{-1}, \qquad p_2 k_2 = p_3 k_{-2}, \qquad p_3 k_3 = p_1 k_{-3}, \tag{9.12}$$

which is equivalent to

$$k_1 k_2 k_3 = k_{-1} k_{-2} k_{-3}.$$

One can see that the Michaelis–Menten kinetics (9.2a)–(9.2c) is equivalent to the Q-matrix transition description, and the irreversibility of the chemical reaction system is exactly the same as the irreversibility of the continuous-time three-state Markov process.

9.3 Coupled Diffusion

Modeling the previous simple chemical reaction network as a Markov process is rather straightforward. One may ask how to apply the above theory to more complex biochemical processes? Here we show a general model of a large class of processes in cellular molecular biology, the coupled diffusion model, which intrinsically is an irreversible Markov process.

We use three biologically different examples to show how they fit into the same coupled diffusion model. The biological examples are conformational fluctuating enzymes, motor proteins, and self-regulating genes. The term coupled diffusion refers to a two-dimensional space, where in one dimension (generally continuous), the system follows a diffusion process, and in the other dimension (generally discrete), it follows a jump process. Then the probability distribution satisfies a Fokker–Planck equation (FPE) in the diffusion direction and satisfies a chemical master equation (CME) in the jumping direction. The coupled diffusion as a whole is a special form of differential Chapman–Kolmogorov equation, which describes a general continuous-time Markov process. FPE characterizes the detail in a single reaction, including information about potential surface and multiple time scales in diffusion and reaction. On the other hand, CME, which ignores the fluctuation within a single state, handles the intermolecular dynamics of a multimolecular system. The coupled diffusion model provides a flexible way of modeling a small biological system with both intramolecular details and intermolecular dynamics.

9.3.1 Fluctuating Enzymes

In biochemistry, an interesting and important phenomenon is cooperativity. In the traditional theory of allosteric cooperativity, the cooperativity in enzyme kinetics comes from the cooperative binding of substrates, where the binding of one enzymatic subunit enhances the binding of the rest of the subunits. Such cooperative mechanism often leads to a sigmoidal curve of the rate of production versus the concentration of substrate. According to classical Michaelis–Menten theory, positive cooperativity occurs only when the enzyme has multiple binding sites. In recent years, however, positive cooperativity of monomeric enzyme, glucokinase as an example [17], is reported, which marks the break down of Michaelis–Menten theory.

On another hand, the conformational fluctuation of enzyme molecules has been discovered and studied since 1970s [18]. The recent development in single-molecule enzymology has provided a wealth of information on fluctuating enzymes. Statistically significant narrowing of the conformational distribution is also observed [19]. Under this background, a unified coupled diffusion model with an additional conformational dimension arise, and under certain conditions, the model exhibits positive cooperativity for monosite enzymes (Fig. 9.2), even if the enzyme and enzyme-substrate complex have the same one macroscopic conformational state [20].

The enzyme-catalyzed reaction can be written in the simpler form

$$E \underset{k_{-1}}{\overset{k_1^0[S]}{\rightleftharpoons}} ES \overset{k_2}{\rightharpoonup} E, \tag{9.13}$$

where the second reaction is irreversible because product is constantly removed from the system. In the perspective of single enzyme molecule, the enzyme molecule has two states, unbound form E and bound form ES. Let $p_E(x,t)$ and $p_{ES}(x,t)$ denote the probabilities of the enzyme molecule being in the states E and ES with conformational coordinate x, respectively. By Michaelis–Menten kinetics, the probabilities of being at E and ES satisfy

$$\frac{dp_E}{dt} = -k_1^0[S]p_E + (k_{-1} + k_2)p_{ES},$$

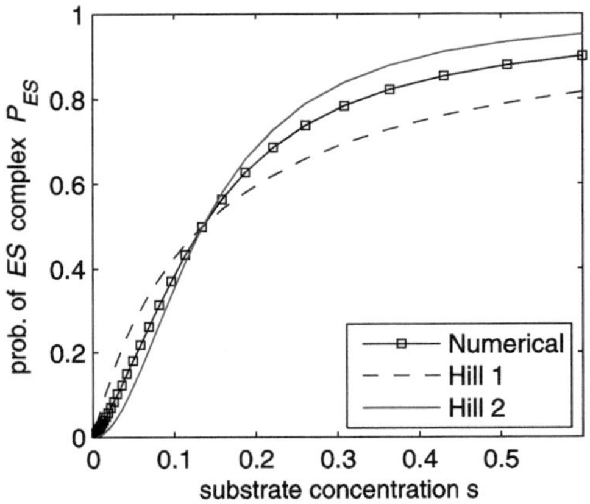

Fig. 9.2 Sigmoidal curve (*squares*): The probability of the enzyme molecule being *ES* at steady state as a function of substrate concentration s. *Dash line* is for Michaelis–Menten relation $p_{ES} = s/(s + K_m)$; *Solid line* is for cooperative binding with Hill coefficient 2: $p_{ES} = s^2/(s^2 + K_m')$. The curves are adjusted to have the same half-rate substrate concentration to compare the sigmoidal shape

$$\frac{dp_{ES}}{dt} = k_1^0[S]p_E - (k_{-1} + k_2)p_{ES}.$$

Meanwhile, the fluctuation follows a one-dimensional diffusion process on the potential landscape $u_{1,2}(x)$. Then, the kinetics of the enzyme with conformational fluctuation can be described by a set of coupled partial differential equations,

$$\frac{\partial p_E(x,t)}{\partial t} = \frac{k_BT}{\eta_1}\frac{\partial^2 p_E}{\partial x^2} + \frac{1}{\eta_1}\frac{\partial}{\partial x}\left(\frac{du_1(x)}{dx}p_E\right) - k_1^o(x)[S]p_E + \big(k_{-1}(x) + k_2(x)\big)p_{ES}, \tag{9.14a}$$

$$\frac{\partial p_{ES}(x,t)}{\partial t} = \frac{k_BT}{\eta_2}\frac{\partial^2 p_{ES}}{\partial x^2} + \frac{1}{\eta_2}\frac{\partial}{\partial x}\left(\frac{du_2(x)}{dx}p_{ES}\right) + k_1^o(x)[S]p_E - \big(k_{-1}(x) + k_2(x)\big)p_{ES}, \tag{9.14b}$$

with nonflux boundary conditions

$$J_E|_{\pm\infty} = -\left[\frac{k_BT}{\eta_1}\frac{\partial p_E}{\partial x} + \frac{1}{\eta_1}\frac{du_1(x)}{dx}p_E\right]_{\pm\infty} = 0, \tag{9.14c}$$

$$J_{ES}|_{\pm\infty} = -\left[\frac{k_BT}{\eta_2}\frac{\partial p_{ES}}{\partial x} + \frac{1}{\eta_2}\frac{du_2(x)}{dx}p_{ES}\right]_{\pm\infty} = 0. \tag{9.14d}$$

9.3.2 Motor Proteins

Motor proteins are molecules that can move along a periodic molecular track in one direction. A typical example of motor protein is myosin in the muscle cells. The track of myosin is the actin filament, with a periodic structure of period ~36 nm. Generally, we consider the movement of motor proteins on a periodic domain.

The dynamics of a single molecule with the presence of a periodic energy potential can be modeled by the Smoluchowski equation

$$\frac{\partial p(x,t)}{\partial t} = -\frac{\partial J(x,t)}{\partial x} = D\frac{\partial^2 p(x,t)}{\partial x^2} - \frac{\partial}{\partial x}\left(\frac{F(x)}{\beta}p(x,t)\right),$$

where x is the position variable, D and β are diffusion and frictional coefficients, satisfying the Einstein relation $\beta D = k_B T$, and $F(x) = -u'(x)$ is the force on the protein due to energy potential function $u(x)$. Without thermal agitation, the movement of the protein molecule will follow an overdamped Newtonian motion: $-\beta\dot{x} + F(x) = 0$. This Smoluchowski equation is equivalent to the stochastic equation

$$\frac{dx(t)}{dt} = \frac{F(x)}{\beta} + \sqrt{2D}w(t).$$

If the protein has only one macroscopic state with a unique periodic energy landscape, one can show that the mean velocity is 0 [21]. Therefore, to generate unidirectional movement, the motor protein needs more than one macroscopic state as well as external energy supply.

Consider the simple case where the motor protein has two states, with energy potential functions $u_1(x)$ and $u_2(x)$, respectively. The driving force for the transition between states comes from the ATP hydrolysis

$$\mathrm{ATP} + \mathrm{H_2O} \underset{g}{\overset{f}{\rightleftharpoons}} \mathrm{ADP} + \mathrm{Pi}.$$

As an example, myosin molecule can bind an ATP or ADP molecule in its head domain, and the transition between these two states is represented by the ATP hydrolysis. Note that in a real biological process like myosin movement, the motor protein generally undergoes many states, and ATP hydrolysis comes into play in various forms. Suppose that the two states of the motor protein are distinguished by the bound ATP or ADP molecule and the transition between the states of the motor protein is characterized by

$$\begin{aligned}\frac{dp_1}{dt} &= -fp_1 + gp_2,\\ \frac{dp_2}{dt} &= fp_1 - gp_2.\end{aligned} \tag{9.15}$$

Combining the Markov kinetics (9.15) between macroscopic states and the previously established Brownian dynamics of each single state, we have the coupled diffusion equation

$$\frac{\partial p_1(x,t)}{\partial t} = \frac{k_B T}{\beta_1}\frac{\partial^2 p_1}{\partial x^2} + \frac{1}{\beta_1}\frac{\partial}{\partial x}\big(u_1'(x)p_1\big) - f(x)p_1 + g(x)p_2, \quad (9.16a)$$

$$\frac{\partial p_2(x,t)}{\partial t} = \frac{k_B T}{\beta_2}\frac{\partial^2 p_2}{\partial x^2} + \frac{1}{\beta_2}\frac{\partial}{\partial x}\big(u_2'(x)p_2\big) + f(x)p_1 - g(x)p_2, \quad (9.16b)$$

where p_1, p_2 are the probability density functions of the motor protein being at position x and state 1 or 2. $f(x)$, $g(x)$ are the reaction rates between the two states, where the concentration of ATP, ADP, and Pi are included implicitly.

9.3.3 Self-regulating Genes

One of the amazing features of biological systems is that they can regulate their protein levels to accommodate the environment. This regulation is mostly accomplished by an entire genetic network. However, even a single gene can regulate its expression by itself. One famous example is Lac operon [22], whose expression products can inactivate the inhibitor in the environment and in turn enhance the expression level.

Now we consider a simplified self-regulating gene in a DNA molecule inside a cell, of which the only expression product is the repressor of the gene itself. If no repressor is bound on the regulatory site, the gene is in "ON" state and can produce the repressor protein at a constant rate of g_α; while if a repressor binds up and turns the gene "OFF", then the rate of producing repressor becomes a lower value g_β. Meanwhile, the repressor protein has a constant rate of degradation k. For a single repressor protein, the binding rate is h, and the releasing rate is f. Considering the total number of repressor proteins in the system, n, and distinguish the two states (ON/OFF) of the gene, we will have the following probability transition diagram:

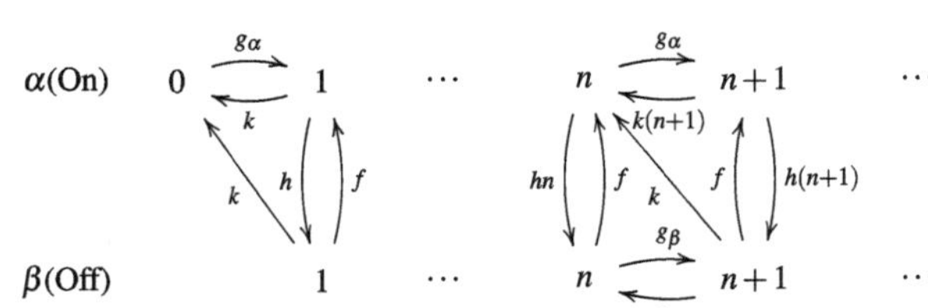

Let $p_\alpha(n,t)$, $p_\beta(n,t)$ be the probabilities that there are n repressor proteins in the system while the gene is "ON" and "OFF", respectively. Then for the above scheme, the master equations are

$$\frac{dp_\alpha(n)}{dt} = g_\alpha p_\alpha(n-1) - (g_\alpha + kn)p_\alpha(n) + k(n+1)p_\alpha(n+1)$$
$$+ kp_\beta(n+1) - hnp_\alpha(n) + fp_\beta(n), \quad (9.17a)$$

$$\frac{dp_\beta(n)}{dt} = g_\beta p_\beta(n-1) - (g_\beta + kn)p_\beta(n) + knp_\beta(n+1) + hnp_\alpha(n) - fp_\beta(n), \tag{9.17b}$$

with one-sided boundary conditions $p_\beta(0) = 0$ and $p_\alpha(-1) = 0$.

If we define the forward and backward difference operators

$$\delta^+ f(n) = f(n+1) - f(n), \qquad \delta^- f(n) = f(n) - f(n-1),$$

then the chemical master equation can be written as a coupled diffusion form. For the bound repressor degradable case,

$$\begin{aligned} &g_\alpha p_\alpha(n-1) - (g_\alpha + kn)p_\alpha(n) + k(n+1)p_\alpha(n+1) \\ &\quad = g_\alpha\big[p_\alpha(n-1) - 2p_\alpha(n) + p_\alpha(n+1)\big] + \big[k(n+1) - g_\alpha\big]p_\alpha(n+1) \\ &\qquad - [kn - g_\alpha]p_\alpha \\ &\quad = g_\alpha \delta^+\delta^- p_\alpha(n) + \delta^+\big[(kn - g_\alpha)p_\alpha(n)\big]. \end{aligned}$$

So

$$\frac{dp_\alpha(n)}{dt} = g_\alpha \delta^+\delta^- p_\alpha(n) + \delta^+\big[(kn - g_\alpha)p_\alpha(n)\big] + kp_\beta(n+1) - hnp_\alpha(n) + fp_\beta(n), \tag{9.18a}$$

$$\frac{dp_\beta(n)}{dt} = g_\beta \delta^+\delta^- p_\beta(n) + \delta^+\big[(kn - g_\beta)p_\beta(n)\big] - kp_\beta(n+1) + hnp_\alpha(n) - fp_\beta(n). \tag{9.18b}$$

9.3.4 General Form

Although these three biological processes are very different, the mathematical models have the same form of equations

$$\frac{\partial}{\partial t}\begin{pmatrix} p_1 \\ p_2 \end{pmatrix} = \begin{pmatrix} \mathcal{L}_1 & 0 \\ 0 & \mathcal{L}_2 \end{pmatrix}\begin{pmatrix} p_1 \\ p_2 \end{pmatrix} + \begin{pmatrix} -\alpha & \beta \\ \alpha & -\beta \end{pmatrix}\begin{pmatrix} p_1 \\ p_2 \end{pmatrix}, \tag{9.19}$$

where $p_{1,2}$ are probability density functions in a single molecule perspective, $\mathcal{L}_{1,2}$ are second-order differential or difference operators, and α, β are exchange rates. Particularly in continuous space,

$$\mathcal{L}_i = \partial_x(D_i \partial_x + u_i'), \quad i = 1, 2, \tag{9.20}$$

$D_{1,2}$ are diffusion constants, and $u_{1,2}$ are potential functions. This coupled diffusion is actually a differential Chapman–Kolmogorov equation, where the diffusion part is restricted within each of the two states, and jump process occurs only in between the states.

Now we restrict our discussion on continuous spacial coordinates equations. At steady state, the time-independent coupled diffusion equation can be written in the Sturm–Liouville form

$$\begin{aligned} 0 = \partial_x & \begin{pmatrix} e^{u_1(x)}\partial_x & 0 \\ 0 & e^{u_2(x)}\partial_x \end{pmatrix} \begin{pmatrix} p_1 \\ p_2 \end{pmatrix} \\ & + \begin{pmatrix} e^{u_1(x)}[u_1''(x) - \alpha] & e^{u_1(x)}\beta \\ e^{u_2(x)}\alpha & e^{u_2(x)}[u_2''(x) - \beta] \end{pmatrix} \begin{pmatrix} p_1 \\ p_2 \end{pmatrix}. \end{aligned} \tag{9.21}$$

Consider the Hilbert space $\mathcal{H} = L^2 \times L^2$ with inner product of $\mathbf{f} = (f_1, f_2)^T$ defined as

$$\langle \mathbf{f}, \mathbf{g} \rangle = \int_{-\infty}^{\infty} \mathbf{f} \cdot \mathbf{g} \, dx = \int_{-\infty}^{\infty} f_1(x) g_1(x) + f_2(x) g_2(x) \, dx.$$

The linear operator on $\mathcal{H}$

$$\mathcal{L} = \partial_x \begin{pmatrix} e^{u_1}\partial_x & 0 \\ 0 & e^{u_2}\partial_x \end{pmatrix} + \begin{pmatrix} e^{u_1(x)}[u_1''(x) - \alpha] & e^{u_1(x)}\beta \\ e^{u_2(x)}\alpha & e^{u_2(x)}[u_2''(x) - \beta] \end{pmatrix}$$

with reflecting or periodic boundary conditions is symmetric if and only if

$$e^{u_1}\beta = e^{u_2}\alpha. \tag{9.22}$$

See Sect. 9.8.1 for a proof. Equation (9.22) represents the Kolmogorov cycle condition for the coupled diffusion process.

It has been proved mathematically that, with a symmetric elliptic operator $\mathcal{L}$, the stationary coupled diffusion process (steady-state solution) is time reversible and has zero entropy production rate [23]. As one expected, (9.22) shows that, in the time-reversible process, detailed balance holds in between the two diffusion coordinates, since the solutions to the separated diffusion processes satisfy the coupled equation

$$p_1(x) = Ce^{-u_1(x)}, \qquad p_2(x) = Ce^{-u_2(x)},$$

where C is a scaling constant. Notice that $u_1(x), u_2(x)$ can vary by a constant without affecting the equation. Therefore, if $e^{u_1}\beta$ and $e^{u_2}\alpha$ are linearly dependent, the steady state of the coupled diffusion process is actually an equilibrium, and the process is time reversible.

In the models of biochemical processes in living cells, the elliptic operator in coupled diffusion is generally asymmetric, detailed balance is not satisfied, and the system will reach a nonequilibrium steady state after a long time.

9.4 Limit Cases of Coupled Diffusion Processes

For biochemically interesting applications, the coupled diffusion equations are in general a asymmetric Sturm–Liouville problem. Without the detailed balance, one does not have a routine mathematical tool to find analytical solutions. However, in the limit of fast reaction or fast diffusion, asymptotic solutions can still provide much information about the model.

9.4.1 Limit Case: Fast Jump Process

In the limit case of fast jump between the two states, the jump process reaches steady state immediately and holds the detailed balance thereafter for each and every x. The general form of a spatially continuous coupled diffusion with fast jump and slow diffusion is

$$\frac{\partial}{\partial t}\begin{pmatrix} p_1 \\ p_2 \end{pmatrix} = \varepsilon \begin{pmatrix} \mathcal{L}_1 & 0 \\ 0 & \mathcal{L}_2 \end{pmatrix}\begin{pmatrix} p_1 \\ p_2 \end{pmatrix} + \begin{pmatrix} -\alpha & \beta \\ \alpha & -\beta \end{pmatrix}\begin{pmatrix} p_1 \\ p_2 \end{pmatrix}, \tag{9.23}$$

where

$$\mathcal{L}_i = D_i \frac{\partial^2}{\partial x^2} + \frac{\partial}{\partial x} u_i', \quad i = 1, 2.$$

The probability density functions $p_1(x)$, $p_2(x)$ can be written in perturbation series

$$p_1 = p_1^0 + \varepsilon p_1^1 + \varepsilon^2 p_1^2 + \cdots, \qquad p_2 = p_2^0 + \varepsilon p_2^1 + \varepsilon^2 p_2^2 + \cdots.$$

Plugging into the steady-state equation, we get a sequence of equations by equating the coefficients of powers of ε. The zero-order equation is

$$\varepsilon^0: \quad \alpha p_1^0 = \beta p_2^0.$$

Therefore, in terms of the total probability density of being at position x, $p^0(x) = p_1^0(x) + p_2^0(x)$, we have

$$p_1^0(x) = \frac{\beta(x)}{\alpha(x) + \beta(x)} p^0(x), \qquad p_2^0(x) = \frac{\alpha(x)}{\alpha(x) + \beta(x)} p^0(x).$$

The equation of order ε^1 is

$$\begin{aligned} \varepsilon^1: \quad & \mathcal{L}_1 p_1^0 - \alpha p_1^1 + \beta p_2^1 = 0, \\ & \mathcal{L}_2 p_2^0 + \alpha p_1^1 - \beta p_2^1 = 0. \end{aligned}$$

Adding up the two equations, we get

$$\begin{aligned}&\mathcal{L}_1 p_1^0 + \mathcal{L}_2 p_2^0 = 0\\ \Rightarrow\quad &\mathcal{L}_1\left(\frac{\beta(x)}{\alpha(x)+\beta(x)}p^0(x)\right) + \mathcal{L}_2\left(\frac{\alpha(x)}{\alpha(x)+\beta(x)}p^0(x)\right) = 0.\end{aligned}$$

The equation can be integrated once, and we get a first-order ordinary differential equation in terms of $p^0(x)$, which can be solved directly with corresponding boundary conditions. The asymptotic solutions in the limit of fast reaction for the three examples are given in Sect. 9.8.2.

9.4.2 Limit Case: Fast Diffusion

If diffusion is much faster than the jumps, then the diffusion reaches a steady state very soon. Though the reaction is slow, probability mass is still exchanging. Newly exchanged mass will spread out inside the new state following the diffusion process immediately, and this process reaches a steady state when there is no total net mass exchange between the two states.

For a coupled diffusion with fast diffusion and slow jumps, the general form is

$$\frac{\partial}{\partial t}\begin{pmatrix} p_1 \\ p_2 \end{pmatrix} = \begin{pmatrix} \mathcal{L}_1 & 0 \\ 0 & \mathcal{L}_2 \end{pmatrix}\begin{pmatrix} p_1 \\ p_2 \end{pmatrix} + \varepsilon \begin{pmatrix} -\alpha & \beta \\ \alpha & -\beta \end{pmatrix}\begin{pmatrix} p_1 \\ p_2 \end{pmatrix}. \tag{9.24}$$

Again, we write the probability density functions in perturbation series

$$p_1 = p_1^0 + \varepsilon p_1^1 + \varepsilon^2 p_1^2 + \cdots, \qquad p_2 = p_2^0 + \varepsilon p_2^1 + \varepsilon^2 p_2^2 + \cdots$$

and the zero-order equation

$$\mathcal{L}_1 p_1^0 = 0, \qquad \mathcal{L}_2 p_2^0 = 0$$

provides two solutions for the two steady-state diffusion processes $\mathcal{L}_{1,2}$. Since the solution is not unique, the weight coefficient for each solution is still undetermined.

The first-order equation is

$$\mathcal{L}_1 p_1^1 - \alpha p_1^0 + \beta p_2^0 = 0, \qquad \mathcal{L}_2 p_2^1 + \alpha p_1^0 - \beta p_2^0 = 0.$$

We integrate either equation. Since the differential term disappears by boundary conditions, we get

$$\int \alpha(x) p_1^0(x)\, dx = \int \beta(x) p_2^0(x)\, dx,$$

which, together with the total probability condition, determines the coefficients for dominant terms p_1^0, p_2^0. Complete solutions of the three examples in the limit of fast diffusion are shown in Sect. 9.8.3.

9.4.3 NESS Flux

The most significant feature of a living biological system is the existence of nonzero net flux in NESS, i.e., external energy is consumed to drive the biological process in an organized way continuously. In another word, the NESS flux is a key quantity of life phenomenon. To investigate the NESS flux in the limit cases, it is sufficient to consider the flux between the two states $j(x) = \alpha(x)p_1(x) - \beta(x)p_2(x)$.

In the limit of fast jumps,

$$\begin{aligned} j &= \alpha\big(p_1^0 + \varepsilon p_1^1\big) - \beta\big(p_2^0 + \varepsilon p_2^1\big) + \mathcal{O}\big(\varepsilon^2\big) \\ &= \alpha p_1^0 - \beta p_2^0 + \varepsilon\big(\alpha p_1^1 - \beta p_2^1\big) + \mathcal{O}\big(\varepsilon^2\big) \\ &= \varepsilon\big(\alpha p_1^1 - \beta p_2^1\big) + \mathcal{O}\big(\varepsilon^2\big) \\ &= \varepsilon \mathcal{L}_1 p_1^0 + \mathcal{O}\big(\varepsilon^2\big), \end{aligned}$$

where p_1^0 is the asymptotic solution of $p_1(x)$ in the limit of fast jumps.

In the limit of fast diffusion, the reaction rates are $\varepsilon\alpha(x)$ and $\varepsilon\beta(x)$, and so

$$j = \varepsilon\big(\alpha p_1^0 - \beta p_2^0\big) + \mathcal{O}\big(\varepsilon^2\big),$$

where $p_{1,2}^0$ are asymptotic solutions of $p_{1,2}(x)$ in the limit of fast diffusion.

By comparing these approximations with the flux of numerical solutions, we verified that the limit case fluxes are valid, see Fig. 9.3.

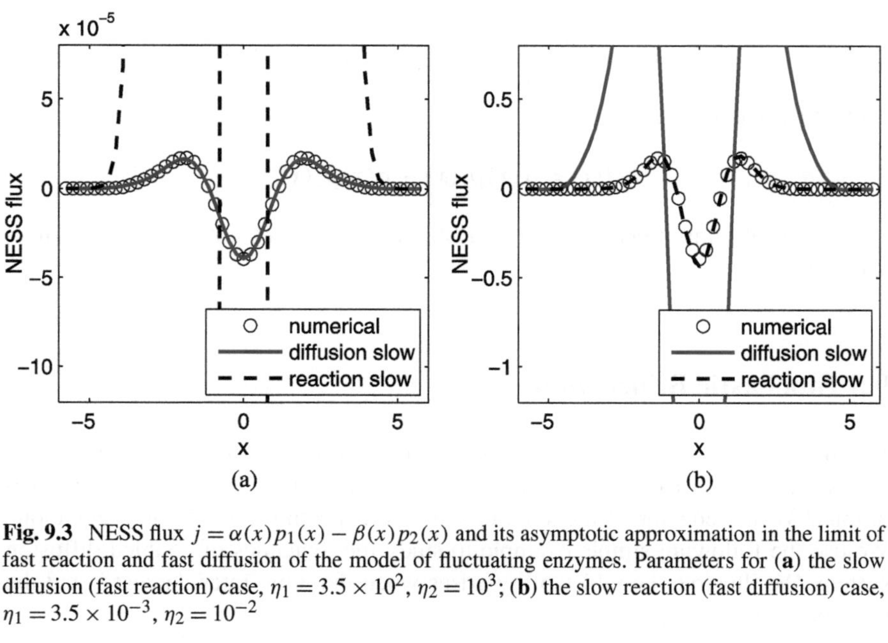

Fig. 9.3 NESS flux $j = \alpha(x)p_1(x) - \beta(x)p_2(x)$ and its asymptotic approximation in the limit of fast reaction and fast diffusion of the model of fluctuating enzymes. Parameters for (**a**) the slow diffusion (fast reaction) case, $\eta_1 = 3.5 \times 10^2$, $\eta_2 = 10^3$; (**b**) the slow reaction (fast diffusion) case, $\eta_1 = 3.5 \times 10^{-3}$, $\eta_2 = 10^{-2}$

9.4.4 Entropy Production

Given the probability distribution of each conformational and enzyme state, the entropy of the system could be defined mathematically. With the general form of coupled diffusion, the entropy is

$$S(t) = -\int p_1(x,t)\log p_1(x,t) + p_2(x,t)\log p_2(x,t)\,dx.$$

Then taking the time derivative and substituting $\partial p_{1,2}/\partial t$ from (9.19), we have

$$\begin{aligned}\frac{dS}{dt} &= \int\left[\frac{J_1^2}{D_1p_1} + \frac{J_2^2}{D_2p_2} + (\alpha p_1 - \beta p_2)\log\frac{\alpha p_1}{\beta p_2}\right]dx\\ &\quad - \int\left[-\frac{u_1'J_1}{D_1} - \frac{u_2'J_2}{D_2} + (\alpha p_1 - \beta p_2)\log\frac{\alpha}{\beta}\right]dx\\ &= epr - hdr.\end{aligned}$$

The entropy production rate and heat dissipation rate are

$$hdr = \int\left[-\frac{u_1'J_1}{D_1} - \frac{u_2'J_2}{D_2} + j\log\frac{\alpha}{\beta}\right]dx, \tag{9.25}$$

$$epr = \int\left[\frac{J_1^2}{D_1p_1} + \frac{J_2^2}{D_2p_2} + j\log\frac{\alpha p_1}{\beta p_2}\right]dx, \tag{9.26}$$

where

$$J_i(x) = -\frac{dp_i(x)}{dt} - u_i'(x)p_i(x), \quad i = 1,2,$$

is the diffusive flux, and

$$j(x) = \alpha(x)p_1(x) - \beta(x)p_2(x)$$

is the vertical jump process flux. In NESS, entropy S does not change with time, and so $hdr = epr$.

9.5 Stochastic Bifurcation

We are able to find asymptotic solutions to the coupled diffusion in both limit cases of fast jump or fast diffusion. But what is the difference between these two solutions? What behavior of the solutions are we expecting for a general equation? We use the following simple example to study the dynamics of coupled diffusion system. We illustrate how bistability arises when the jump rates decrease, and the resulting bifurcation [24].

We consider a simple stochastic dynamics $\mathbf{X}(t)$ with fluctuating $\mu(t)$,

$$d\mathbf{X} = -(\mathbf{X} - \mu)\,dt + \sqrt{2}\,d\mathbf{B}_t, \tag{9.27}$$

where $\mathbf{B}_t$ is the standard Brownian motion, and the fluctuating $\mu(t)$ takes two values $\pm\lambda$ with fluctuation rate q. Then the Fokker–Planck equation for the system is a coupled diffusion on $(-\infty, \infty)$:

$$\begin{aligned}\frac{\partial u_1}{\partial t} &= \frac{\partial^2 u_1}{\partial x^2} + \frac{\partial}{\partial x}(x+\lambda)u_1 - qu_1 + qu_2,\\ \frac{\partial u_2}{\partial t} &= \frac{\partial^2 u_2}{\partial x^2} + \frac{\partial}{\partial x}(x-\lambda)u_2 + qu_1 - qu_2,\end{aligned} \tag{9.28}$$

with noflux boundary conditions at $x = \pm\infty$.

We are interested in the stationary distribution. By a similar asymptotic approach, if $q \gg 1$, then there is a fast equilibration $u_1(x) = u_2(x)$. Therefore, if one sums the two equations in (9.28), one has [21]

$$\frac{\partial u}{\partial t} = \frac{\partial^2 u}{\partial x^2} + \frac{\partial}{\partial x}(xu), \tag{9.29}$$

where $u(x,t) = u_1(x,t) + u_2(x,t)$. The stationary distribution for this equation is a Gaussian centered at $x = 0$ with variance 1. On the other limit, if $q \ll 1$, then the stationary distribution is simply

$$u^{ss}(x) = \frac{1}{2\sqrt{2\pi}}\left(e^{-(x-\lambda)^2/2} + e^{-(x+\lambda)^2/2}\right). \tag{9.30}$$

Hence, if $\lambda > 1$, then $u^{ss}(x)$ has two maxima near $\pm\lambda$.

Therefore, when λ is large, the steady-state distribution has a single peak at $x = 0$ for $q \gg 1$ but two peaks near $x = \pm\lambda$ for $q \ll 1$. There must be a bifurcation from large q to small q. In the case of $\lambda \gg 1$, i.e., the fluctuation in the mean of the Gaussian is much greater than its variance, we can locate this bifurcation by perturbation theory.

We introduce $u(x) = u_1^{ss}(x) + u_2^{ss}(x)$ and $v(x) = u_1^{ss}(x) - u_2^{ss}(x)$. Since we assume that $\lambda \gg 1$, $u(x)$ can be written in perturbation series in terms of λ^{-n} around $x = 0$. Through some calculation in Sect. 9.8.4, we get the leading terms of $u(x)$ as

$$u(x) = c_0\left(1 + \frac{(1-q)}{\lambda^2}x^2\right). \tag{9.31}$$

Therefore, when $q < 1$, $\eta(x)$ is concave at $x = 0$, implying the existence of two peaks on both sides; when $q > 1$, $\eta(x)$ is convex at $x = 0$, i.e., a maximum.

Numerical computation shown in Fig. 9.4 verifies the above analytical result. With increasing q from 0 to 1, the two peaks of $u(x)$ move toward the center $x = 0$ and merge into one at some critical value q_c. The critical point of bifurcation $q_c \approx 1$ for a large λ, as predicted in (9.31). The position of local maximum with parameter

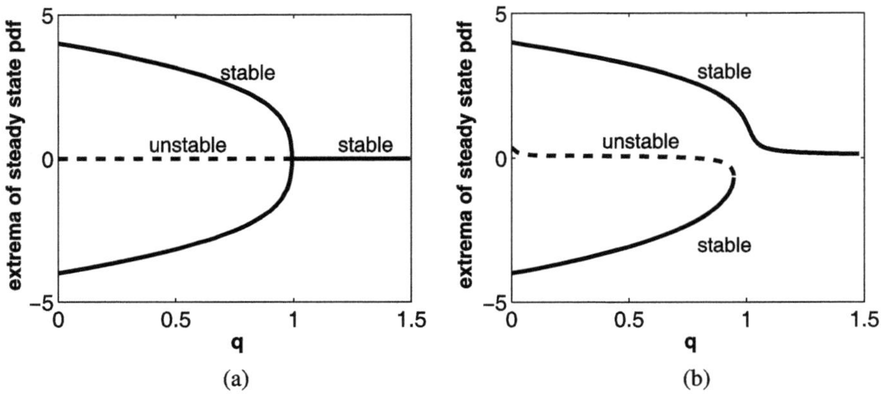

Fig. 9.4 (**a**) Pitchfork bifurcation diagrams of local extrema of $u(x)$, the sum of steady-state probability density functions (pdf) $u_1^{ss}(x)$ and $u_2^{ss}(x)$ according to the toy model in (9.28). q is the rate of fluctuation between the two states. The *solid* and *dashed lines* represent the maxima and the minima of the probability density function, respectively. The parameters $\lambda = 4$ in the calculation. (**b**) Saddle-node bifurcation in the system (9.32) with different diffusion constants $D_1 = 1.2$, $D_2 = 0.8$, and $\lambda = 4$. When the symmetry between the two equations in (9.32) is no longer present, the pitchfork bifurcation is reduced to the saddle-node bifurcation, known as imperfection

q undergoes a pitchfork bifurcation. One should note, however, that pitchfork bifurcation is structurally unstable, and it is due to a high degree of symmetry in (9.28). Suppose that the two Brownian motions have different diffusion constant D_1 and D_2 in the toy model; then the equation becomes

$$\begin{aligned}\frac{\partial u_1}{\partial t} &= D_1\frac{\partial^2 u_1}{\partial x^2} + \frac{\partial}{\partial x}(x+\lambda)u_1 - qu_1 + qu_2,\\ \frac{\partial u_2}{\partial t} &= D_2\frac{\partial^2 u_2}{\partial x^2} + \frac{\partial}{\partial x}(x-\lambda)u_2 + qu_1 - qu_2,\end{aligned} \tag{9.32}$$

and the bifurcation loses symmetry as shown in Fig. 9.4.

9.6 Numerical Methods

Cellular biological processes often happen in a very small time scale—much faster than the variation of the environment. Therefore, the steady-state property usually dominates the behavior of the biochemical system. For the coupled diffusion processes, in such cases, computing the time-independent steady-state solution is sufficient. Solving the resulting coupled diffusion equation is essentially finding the eigenfunction corresponding to the zero eigenvalue of the asymmetric Sturm–Liouville problem.

For the continuous spatial dimension, the most straightforward method is the finite difference method. We discretize the spatial domain by uniform grids $\mathbf{x} = \{x_i\}$,

$i = 0, \ldots, m + 1$, and use the vector

$$\mathbf{u} = \begin{bmatrix} \mathbf{p} \\ \mathbf{q} \end{bmatrix} = [p_0, \ldots, p_{m+1}, q_0, \ldots, q_{m+1}]^T, \tag{9.33}$$

where $p_i = p(x_i), q_i = q(x_i), i = 0, \ldots, m+1$, are the values of p, q at grid points. Replacing the derivatives by midpoint finite difference approximation, we get a linear system

$$\mathbf{Mu} = \mathbf{f}, \quad \mathbf{M} \in \mathbb{R}^{(m+2)^2}, \mathbf{u}, \mathbf{f} \in \mathbb{R}^{m+2}. \tag{9.34}$$

The matrix $\mathbf{M}$ has the shape

$$\mathbf{M} = \begin{bmatrix} T_1 & S_1 \\ S_2 & T_2 \end{bmatrix},$$

where S_1, S_2 are diagonal matrices representing the coupling term between the two equations. With nonflux boundary condition, T_1, T_2 are trigonal matrices; with periodic boundary, the dimension of the linear system is $m + 1$ because only one side boundary value is needed, and T_1, T_2 each have two additional entries at the upper left and lower right corners. The righthand side vector $\mathbf{f}$ is zero.

The numerical problem becomes finding the eigenvector corresponding to the zero eigenvalue of the sparse matrix $\mathbf{M}$. There is a powerful numerical package ARPACK designed for computing several eigenvalues and eigenvectors of large-scale but sparse matrices [25]. Steady-state solution can be computed using this package.

For one-sided boundary problems in discrete space, self-regulating genes as an example, there is a simpler approach. At steady states, the probability transfer in the network can be represented by a sequence of cyclic fluxes. There are many kinds of cyclic flux decomposition of the steady-state network flow; however, we prefer the following one because it leads to a recursive formula for the nth nodes and cyclic flux.

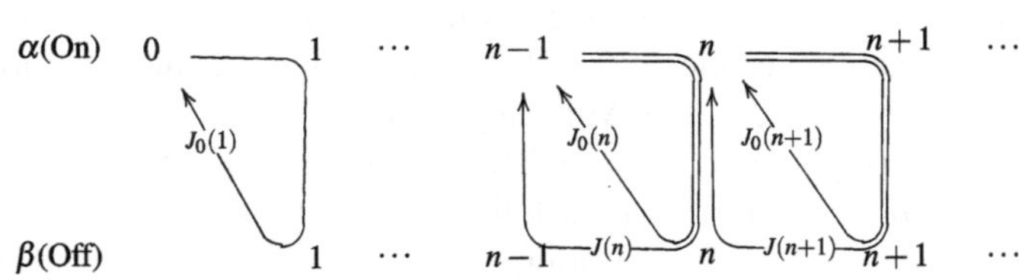

In this representation, we notice that $J_0(n)$ is always a positive flux, and $J_0(n) = p_\beta(n)k$. The probabilities $p_\alpha(n-1)$, $p_\beta(n)$ and flux $J(n)$ satisfy

$$kp_\beta(n+1) + J(n+1) = g_\alpha p_\alpha(n) - k(n+1)p_\alpha(n+1),$$
$$J(n+1) - J(n) = (k+f)p_\beta(n) - hnp_\alpha(n),$$
$$J(n+1) = knp_\beta(n+1) - g_\beta p_\beta(n),$$

$\mathbf{p}(n+1) = A(n)\mathbf{p}(n)$, i.e.,

$$\begin{pmatrix} p_\alpha(n+1) \\ J(n+1) \\ p_\beta(n+1) \end{pmatrix} = \begin{pmatrix} \frac{g_\alpha}{k(n+1)} + \frac{h}{k} & -\frac{1}{kn} & -\frac{(k+f)(n+1)+g_\beta}{kn(n+1)} \\ -hn & 1 & f+k \\ -\frac{h}{k} & \frac{1}{kn} & \frac{f+k+g_\beta}{kn} \end{pmatrix} \begin{pmatrix} p_\alpha(n) \\ J(n) \\ p_\beta(n) \end{pmatrix}. \quad (9.35)$$

The initial condition can be determined by considering the first cycle. We have

$$\begin{aligned} J_0(1) &= kp_\beta(1) = g_\alpha p_\alpha(0) - kp_\alpha(1) = hp_\alpha(1) - fp_\beta(1) - J(2), \\ J(2) &= -2hp_\alpha(1) + (f+k)p_\beta(1) + J(1), \\ J(1) &= 0, \end{aligned}$$

and so

$$\begin{aligned} p_\alpha(1) &= \frac{(2f+2k)g_\alpha}{k(2f+2k+3h)} p_\alpha(0), \qquad J(1) = 0, \\ p_\beta(1) &= \frac{3hg_\alpha}{k(2f+2k+3h)} p_\alpha(0), \end{aligned}$$

and the value of $p_\alpha(0)$ can be determined by the total probability condition

$$p_\alpha(0) + \sum_{n=1}^{\infty} \big[p_\alpha(n) + p_\beta(n)\big] = 1.$$

9.7 Discussion

Irreversible stochastic processes, either the Q-processes without detailed balance or the nonsymmetric diffusion processes, are the appropriate mathematical language for modeling cellular biochemical systems and stochastic biological processes. We have given several examples in this paper for the former; an example for the latter is the theory of population genetics [26–28]. The mathematical studies of irreversible stochastic processes carried out by Min-Ping Qian, Min Qian, and their colleagues at Peking University give rise to several important concepts that include *entropy production* and *probability circulation*. Both are characteristics of any stationary irreversible stochastic process: If the stationarity is not sustained by detailed balance, it has to be sustained by circular balance, the Kirchhoff law. Because of the presence of the circulations, a stationary process is not time-symmetric and thus is nonzero entropy production. Both concepts of entropy production and probability circulation had been in the physics literature in the 1970s [29, 30], but showing that both are mathematical consequences of irreversible stochastic processes and that they are equivalent within given mathematical conditions is the contribution of the Peking University group. The concept of entropy production has also gone through a major development in the 1990s in terms of the *fluctuation theorem* in the West. It is now recognized that it is an integral part of irreversible Markov processes [31, 32]: Entropy production can be defined as a stochastic quantity associated with an irreversible stochastic process. Now with the developed mathematical theory [16], it is clear that the irreversible stochastic processes are applicable to many of the interesting open-system phenomena in chemistry and biochemistry [33, 34], motor protein, fluctuating enzymes in living cells, self-regulating genes, and stochastic resonance not discussed in the present paper [35, 36], just for a few examples.

9.8 Mathematical Methods

9.8.1 Proof of Sturm–Liouville Operator

Consider the Hilbert space $\mathcal{H} = L^2 \times L^2$ with the inner product of $\mathbf{f} = (f_1, f_2)^T$ and $\mathbf{f} = (f_1, f_2)^T$ defined as

$$\langle \mathbf{f}, \mathbf{g} \rangle = \int_{-\infty}^{\infty} \mathbf{f} \cdot \mathbf{g} \, dx = \int_{-\infty}^{\infty} f_1(x) g_1(x) + f_2(x) g_2(x) \, dx.$$

The linear operator on $\mathcal{H}$

$$\mathcal{L} = \partial_x \begin{pmatrix} e^{u_1} \partial_x & 0 \\ 0 & e^{u_2} \partial_x \end{pmatrix} + \begin{pmatrix} e^{u_1(x)}[u_1''(x) - \alpha] & e^{u_1(x)} \beta \\ e^{u_2(x)} \alpha & e^{u_2(x)}[u_2''(x) - \beta] \end{pmatrix}$$

with reflecting or periodic boundary conditions is symmetric if and only if

$$e^{u_1} \beta = e^{u_2} \alpha.$$

Proof Let $f, g \in \mathcal{H}$; then

$$\begin{aligned}
\langle \mathcal{L} f, g \rangle - \langle f, \mathcal{L} g \rangle &= \int_{-\infty}^{\infty} \left[\left(e^{u_1} f_1' \right)' + e^{u_1} (u_1'' - \alpha) f_1 + e^{u_1} \beta f_2 \right] g_1 \, dx \\
&\quad + \int_{-\infty}^{\infty} \left[\left(e^{u_2} f_2' \right)' + e^{u_2} (u_2'' - \beta) f_2 + e^{u_2} \alpha f_1 \right] g_2 \, dx \\
&\quad - \int_{-\infty}^{\infty} \left[\left(e^{u_1} g_1' \right)' + e^{u_1} (u_1'' - \alpha) g_1 + e^{u_1} \beta g_2 \right] f_1 \, dx \\
&\quad - \int_{-\infty}^{\infty} \left[\left(e^{u_2} g_2' \right)' + e^{u_2} (u_2'' - \beta) g_2 + e^{u_2} \alpha g_1 \right] f_2 \, dx \\
&= g_1 \left(e^{u_1} f_1' \right) \big|_{\partial\Omega} + g_2 \left(e^{u_2} f_2' \right) \big|_{\partial\Omega} + \int_{-\infty}^{\infty} e^{u_1} \beta f_2 g_1 \, dx \\
&\quad + \int_{-\infty}^{\infty} e^{u_2} \alpha f_1 g_2 \, dx - f_1 \left(e^{u_1} g_1' \right) \big|_{\partial\Omega} - f_2 \left(e^{u_2} g_2' \right) \big|_{\partial\Omega}
\end{aligned}$$

$$-\int_{-\infty}^{\infty} e^{u_1}\beta g_2 f_1\,dx - \int_{-\infty}^{\infty} e^{u_2}\alpha g_1 f_2\,dx.$$

Applying the boundary condition, if it is reflecting boundary, we have

$$f_1' = -u_1' f_1, \qquad f_2' = -u_2' f_2, \qquad g_1' = -u_1' g_1, \qquad g_2' = -u_2' g_2, \quad x \in \partial\Omega.$$

By substitution, the nonintegral terms cancel; if it is periodic boundary, the nonintegral terms disappear as well. Then we get

$$\langle \mathcal{L}f, g\rangle - \langle f, \mathcal{L}g\rangle = \int_{-\infty}^{\infty} \left(e^{u_1}\beta - e^{u_2}\alpha\right)(f_2 g_1 - f_1 g_2)\,dx.$$

Since the choice of f, g is arbitrary, $\mathcal{L}$ being symmetric requires

$$e^{u_1}\beta = e^{u_2}\alpha. \tag{9.36}$$

□

9.8.2 Asymptotic Solution in the Limit of Fast Jump Process

1. *Fluctuating enzymes*—nonflux boundary.

   By fast reaction, the reaction part in (9.14a)–(9.14d) reaches equilibrium at each conformational position x, and so

   $$p_1 = \frac{k_{21}}{k_{12}+k_{21}} p, \qquad p_2 = \frac{k_{12}}{k_{12}+k_{21}} p.$$

   With nonflux boundary conditions, the steady-state solution of $p(x)$ satisfies

   $$\frac{\partial}{\partial x}\left[a(x)p(x)\right] + b(x)p(x) = 0,$$

   where

   $$a(x) = \frac{k_{21}(x)/\eta_1 + k_{12}(x)/\eta_2}{k_{12}(x)+k_{21}(x)}, \qquad b(x) = \frac{k_{21}(x)u_1'(x)/\eta_1 + k_{12}(x)u_2'(x)/\eta_2}{k_{12}(x)+k_{21}(x)}.$$

   The solution can be expressed as

   $$p(x) = \frac{c}{a(x)} \exp\left(-\int_0^x \frac{b(s)}{a(s)}\,ds\right). \tag{9.37}$$

   The constant c can be addressed by the total probability condition.
2. *Motor protein*—periodic boundary.

   For motor protein (9.16a), (9.16b), the only difference with fluctuating enzyme model is on boundary condition. By setting

   $$p_1 = \frac{g}{f+g} p, \qquad p_2 = \frac{f}{f+g} p,$$

we have the steady-state equation about $p(x)$ as

$$\frac{\partial}{\partial x}\big[a(x)p(x)\big]+b(x)p(x)=c_1,$$

where

$$a(x)=\frac{\beta_2 g+\beta_1 f}{f+g},\qquad b(x)=\frac{\beta_2 u_1' g+\beta_1 u_2' f}{,}k_B T(f+g).$$

The solution will be

$$\begin{aligned}p(x)=&\frac{c_1}{a(x)}\exp\left(-\int_0^x b(s)/a(s)\,ds\right)\int_0^x \exp\left(\int_0^s b(\xi)/a(\xi)d\xi\right)ds\\&+\frac{c_2}{a(x)}\exp\left(-\int_0^x b(s)/a(s)\,ds\right).\end{aligned}\tag{9.38}$$

At boundaries,

$$\begin{aligned}&a(0)p(0)=c_2,\\&a(L)p(L)\exp\left(\int_0^L b(s)/a(s)\,ds\right)\\&\quad=c_1\int_0^L \exp\left(\int_0^s b(\xi)/a(\xi)\,d\xi\right)ds+c_2.\end{aligned}\tag{9.39}$$

The periodic boundary condition provides $a(0)p(0)=a(L)p(L)$, which will give a relation between c_1 and c_2. Together with the total probability condition, the constants can be uniquely determined.

3. *Self-regulating genes*—one-sided reflecting boundary.

   The asymptotic solution of (9.17a), (9.17b) can be computed recursively. Taking self-regulating gene as an example, fast reaction leads to $hnp_\alpha(n)=fp_\beta(n)$, and so

$$p_\alpha(n)=\frac{f}{hn+f}p(n),\qquad p_\beta(n)=\frac{hn}{hn+f}p(n),$$

   where $p(n)=p_\alpha(n)+p_\beta(n)$. Adding the two equations for states α and β, we assume that the detailed balance strictly holds between $p_\alpha(n)$ and $p_\beta(n)$ and obtain a chemical master equation in terms of $p(n)$, which describes the diffusion between neighboring $p(n)$'s. Since the left boundary $n=0$ is reflecting, the net flux between $p(0)$ and $p(1)$ is zero, and then deductively, the net flux between any two adjacent spatial position is zero. In this way, we can express $p(n)$ in terms of $p(0)$. We have

$$\frac{fg_\alpha+h(n-1)g_\beta}{h(n-1)+f}p(n-1)=knp(n).$$

Denote $G(n) = [fg_\alpha + h(n-1)g_\beta]/[h(n-1)+f]$. Then

$$p(n) = c\prod_{\ell=1}^{n} \frac{G(\ell-1)}{k\ell}. \tag{9.40}$$

9.8.3 Asymptotic Solution in the Limit of Fast Diffusion

1. *Fluctuating enzymes*—nonflux boundary.
   Since the diffusion is fast and reaction between E and ES is slow in (9.14a)–(9.14d), we solve the two diffusion equations separately and get

$$p_1(x) = c_1 e^{-u_1(x)}, \qquad p_2(x) = c_2 e^{-u_2(x)}.$$

   Although the detailed balance cannot be satisfied anyway, the total probability of being at both states can be balanced by the chemical reaction, that is,

$$\int k_{12}(x)p_1(x)\,dx = \int k_{21}(x)p_2(x)\,dx.$$

   So

$$c_1\int k_{12}(x)e^{-u_1(x)}\,dx - c_2\int k_{21}(x)e^{-u_2(x)}\,dx = 0, \tag{9.41}$$

$$c_1\int e^{-u_1(x)}\,dx + c_2\int e^{-u_2(x)}\,dx = 1, \tag{9.42}$$

   whence we can compute the coefficients c_1, c_2.
2. *Motor protein*—periodic boundary.
   For the motor protein case in (9.16a), (9.16b), the leading-order equation becomes two separated diffusion equations with periodic boundary conditions

$$p_i'(x) + \frac{u_i'(x)}{k_B T}p_i(x) = c_i, \quad i = 1, 2,$$

   and so

$$p_i \exp\bigl(u_i(x)/k_B T\bigr) = c_i\int_0^x \exp\bigl(u_i(s)/k_B T\bigr)\,ds + d_i, \quad i = 1, 2.$$

   The periodic boundary condition states

$$\begin{aligned} d_i &= p_i(0)\exp\bigl(u_i(0)/k_B T\bigr) = p_i(L)\exp\bigl(u_i(L)/k_B T\bigr) \\ &= c_i\int_0^L \exp\bigl(u_i(s)/k_B T\bigr)ds + d_i, \quad i = 1, 2, \end{aligned} \tag{9.43}$$

   and so $c_i = 0$. As a result, the solution will be the same as the solution of fluctuating enzyme model.

3. *Self-regulating genes*—one-sided reflecting boundary.

In the case of fast birth-death jump and slow interstate transition in (9.17a), (9.17b), the producing and degradation of repressor is fast, while the binding and releasing is slow. Since bound repressor is degradable, we have a nontrivial interstate transition from β state to α state with rate k. Even if the binding and release rate is zero, there still would be a nonzero transition flux from β to α, until $p_\beta(n) \equiv 0$. The steady-state probability distribution is

$$p_\alpha(n) = \frac{g_\alpha^n}{k^n n!} e^{-g_\alpha/k}, \qquad p_\beta(n) = 0. \tag{9.44}$$

9.8.4 Bifurcation of the Toy Model

We introduce $u(x) = u_1^{ss}(x) + u_2^{ss}(x)$ and $v(x) = u_1^{ss}(x) - u_2^{ss}(x)$. Then from (9.28) we have

$$\frac{d^2u(x)}{dx^2} + \frac{d}{dx}(xu + \lambda v) = 0, \tag{9.45}$$

$$\frac{d^2v(x)}{dx^2} + \frac{d}{dx}(xv + \lambda u) - 2qv(x) = 0. \tag{9.46}$$

Equation (9.45) can be integrated, and noting that $[\frac{du}{dx} + xu + \lambda v]_{x=\pm\infty} = 0$, we have

$$\lambda v(x) = -\frac{du(x)}{dx} - xu(x). \tag{9.47}$$

Substituting this into (9.46), we have

$$\frac{d^3u(x)}{dx^3} + 2x\frac{d^2u(x)}{dx^2} + \left(x^2 - \lambda^2 - 2q + 3\right)\frac{du(x)}{dx} + 2(1-q)xu(x) = 0. \tag{9.48}$$

We are interested in the behavior of $u(x)$ around $x = 0$: $u'(0) = 0$ since the solution to (9.48) is an even function. Furthermore, if $u''(x) > 0$, $u(x)$ is concave and bistable since $u(x) \geq 0$ and $u(\pm\infty) = 0$. If $u''(x) < 0$, $u(x)$ is convex with a peak at $x = 0$. Since $\lambda \gg 1$, much larger than other coefficients around $x = 0$, the dominant term of (9.48) is

$$-\lambda^2\frac{du(x)}{dx} = 0,$$

which gives a constant solution $u(x) = c_0$. As a result, in terms of the small parameter λ^{-2}, the solution of (9.48) has the form

$$u(x) = c_0 + \frac{\eta(x)}{\lambda^2}, \qquad c_0 > 0,$$

and then the equation for $\eta(x)$ becomes

$$\frac{1}{\lambda^2}\left(\frac{d^3\eta}{dx^3}+2x\frac{d^2\eta}{dx^2}+\left(x^2-2q+3\right)\frac{d\eta}{dx}+2(1-q)x\eta\right)$$
$$-\frac{d\eta}{dx}+2c_0(1-q)x=0. \tag{9.49}$$

Since $\lambda \gg 1$, we discard the small terms and obtain

$$-\frac{d\eta(x)}{dx}+2c_0(1-q)x=0,$$

and so

$$\eta(x)=c_0(1-q)x^2, \qquad u(x)=c_0\left(1+\frac{(1-q)}{\lambda^2}x^2\right). \tag{9.50}$$

Acknowledgement P.-Z. Shi would like to thank Prof. Min-Ping Qian for her instructions, encouragement, and inspiration during his study and work at Peking University.

References

1. M.-P. Qian and M. Qian. The decomposition into a detailed balance part and a circulation part of an irreversible stationary Markov chain. *Sci Sin A*, **22**:69–79, 1979.
2. M.-P. Qian and M. Qian. Circulation for recurrent Markov chain. *Z Wahrscheinlichkeitstheor Verw Geb*, **59**:203–210, 1982.
3. Y. Zhang, M.-P. Qian, Q. Ouyang, M. Deng, F. Li, and C. Tang. Stochastic model of yeast cell-cycle network. *Physica*, **219**:35–39, 2006.
4. E. Schrödinger. *What is Life?* Cambridge University Press, Cambridge, 1944.
5. P.J. Choi, L. Cai, K. Frieda, and X.S. Xie. A stochastic single-molecule event triggers phenotype switching of a bacterial cell. *Science*, **322**:442–446, 2008.
6. M.W. Deem. Mathematical adventures in biology. *Phys Today*, January: 42–47, 2007. (Feature article).
7. D.A. Beard and H. Qian. *Chemical Biophysics: Quantitative Analysis of Cellular System*. Cambridge University Press, London, 2008.
8. M. Qian and B. Zhang. Multi-dimensional coupled diffusion process. *Acta Math Appl Sin*, **2**:168–179, 1984.
9. Z. Guo, M. Qian, and M.-P. Qian. Minimal coupled diffusion process. *Acta Math Appl Sin*, **3**:58–69, 1987.
10. F. Zhang. Exponential convergence of coupled diffusion processes. *J Math Phys*, **46**:063304, 2005.
11. G.E. Briggs and J.B.S. Haldane. A note on the kinetics of enzyme action. *Biochem J*, **19**:338–339, 1925.
12. I.H. Segel. *Enzyme Kinetics, Behavior and Analysis of Rapid Equilibrium and Steady-State Enzyme Systems*. Wiley-Interscience, New York, 1993.
13. X.S. Xie and H.P. Lu. Single-molecule enzymology. *J Biol Chem*, **274**:15967–15970, 1999.
14. H. Qian. Open-system nonequilibrium steady state: statistical thermodynamics, fluctuations, and chemical oscillations. *J Phys Chem B*, **110**:15063–15074, 2006.
15. H. Qian. Cooperativity and specificity in enzyme kinetics: a single-molecule time-based perspective. *Biophys J*, **95**:10–17, 2008.

16. D.-Q. Jiang, M. Qian, and M.-P. Qian. *Mathematical Theory of Nonequilibrium Steady States: On the Frontier of Probability and Dynamical Systems*. Springer, New York, 2004.
17. K. Kamata, M. Mitsuya, T. Nishimura, J. Eiki, and Y. Nagata. Structural basis for allosteric regulation of the monomeric allosteric enzyme human glucokinase. *Structure*, 429–438, 2004.
18. G.R. Welch. *The Fluctuating Enzyme*. Wiley, New York, 1986.
19. J.A. Hanson, H. Yang, et al. Illuminating the mechanistic roles of enzyme conformational dynamics. *Proc Natl Acad Sci*, **104**:18055–18060, 2007.
20. H. Qian and P.-Z. Shi. Fluctuating enzyme and its biological functions: positive cooperativity without multiple states. *J Phys Chem B*, **113**:2225–2230, 2009.
21. H. Qian. The mathematical theory of molecular motor movement and chemomechanical energy transduction. *J Math Chem*, **27**(3), 2000.
22. F. Jacob and J. Monod. Genetic regulatory mechanisms in the synthesis of proteins. *J Mol Biol*, **3**:318–356, 1961.
23. H. Qian, M. Qian, and X. Tang. Thermodynamics of the general diffusion process: Time-reversibility and entropy production. *J Stat Phys*, **107**:1129–1141, 2002.
24. H. Qian, P.-Z. Shi, and J. Xing. Stochastic bifurcation, slow fluctuations, and bistability as an origin of biochemical complexity. *Phys Chem Chem Phys*, **11**:4861–4870, 2009.
25. R.B. Lehoucq, D.C. Sorensen, and C. Yang. *ARPACK User's Guide. Solution of Large-Scale Eigenvalue Problems with Implicitly Restarted Arnoldi Method*. SIAM, Philadelphia, 1998.
26. E.V. Koonin. Darwinian evolution in the light of genomics. *Nucleic Acids Res*, **37**:1011–1034, 2009.
27. W.J. Ewens. *Mathematical Population Genetics*, 2nd edition. Springer, Berlin, 2004.
28. J.H. Gillespie. *The Causes of Molecular Evolution*. Oxford University Press, London, 1991.
29. J. Schnakenberg. Network theory of microscopic and macroscopic behaviour of master equation systems. *Rev Mod Phys*, **48**:571–585, 1976.
30. K. Tomita and H. Tomita. Irreversible circulation of fluctuation. *Prog Theor Phys*, **51**:1731–1749, 1974.
31. R.K.P. Zia and B. Schmittmann. Probability currents as principal characteristics in the statistical mechanics of non-equilibrium steady states. *J Stat Mech Theor Exp*, 07012, 2007.
32. D. Andrieux and P. Gaspard. Fluctuation theorem for currents and Schnakenberg network theory. *J Stat Phys*, **127**:107–131, 2007.
33. G. Nicolis and I. Prigogine. *Self-Organization in Nonequilibrium Systems*. Wiley-Interscience, New York, 1977.
34. T.L. Hill. *Free Energy Transduction in Biology: The Steady-state Kinetic and Thermodynamic Formalism*. Academic Press, New York, 1977.
35. M. Qian, G.X. Wang, and X.J. Zhang. Stochastic resonance on a circle without excitation: Physical investigation and peak frequency formula. *Phys Rev E*, **62**:6469, 2000.
36. H. Qian and M. Qian. Pumped biochemical reactions, nonequilibrium circulation, and stochastic resonance. *Phys Rev Lett*, **84**:2271–2274, 2000.

Chapter 10
Probability Modeling and Statistical Inference in Periodic Cancer Screening

Dongfeng Wu and Gary L. Rosner

10.1 Background

Early detection and efficient treatments are the most effective ways to increase the cure rate or prolong survival of cancer patients. The primary technique for implementing early detection is screening exams, by which the disease may be found before symptoms are present. There are many different kinds of cancer screenings in the world, such as mammogram and clinical breast exam (CBE) for breast cancer, chest X-ray or computed tomography (CT) scanning for lung cancer, fecal occult blood test (FOBT) for colon cancer, prostate-specific antigen (PSA) for prostate cancer, etc.

Probability modeling in cancer screening was dated back to 1969, right after the Health Insurance Plan of Greater New York (HIP) study was close to an end. The HIP study, which began at the end of 1963, was the first randomized clinical trial to examine regular screening exams that include mammography as a diagnostic screening test for breast cancer. The study randomized initially asymptomatic women aged 40 to 64 years without a history of breast cancer to two groups: the study group and the control group. Each group consisted of about 31,000 women. The screening program for the study group called for up to 4 annual breast cancer screening exams, with each screening exam including both a mammogram and a clinical breast exam. The control group received usual care [16].

Zelen and Feinleib [32] in their first paper on screening proposed the disease progressive model: the disease develops by progressing through 3 states, denoted by $S_0 \rightarrow S_p \rightarrow S_c$ (Fig. 10.1). S_0 refers to the disease-free state or the state in

D. Wu (✉)
Department of Bioinformatics and Biostatistics, University of Louisville, Louisville, KY 40292, USA
e-mail: dongfeng.wu@louisville.edu

G.L. Rosner
Department of Biostatistics, MD Anderson Cancer Center, University of Texas, Houston, TX 77030, USA

J. Feng et al. (eds.), *Frontiers in Computational and Systems Biology*, Computational Biology 15,
DOI 10.1007/978-1-84996-196-7_10,

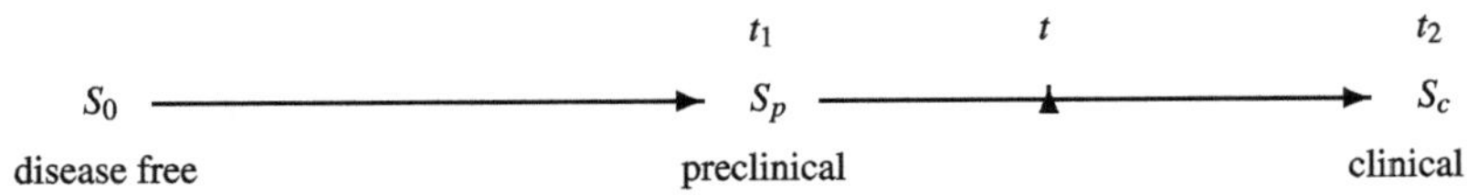

Fig. 10.1 Illustration of disease development and lead time

which the disease cannot be detected; S_p refers to the preclinical disease state, in which an asymptomatic individual unknowingly has disease that a screening exam can detect; and S_c refers to the disease state at which the disease manifests itself in clinical symptoms. The progressive disease model describes the natural history of lesions detected by screening for cancer. The goal of screening programs is to detect the cancer in the preclinical state (S_p). The paper discussed probability modeling in cancer screening where an individual is examined only once. It clearly defined some concepts or key parameters in cancer screening programmes that are widely used until today, such as sensitivity, transition probability from the disease-free state to the preclinical state, and sojourn time distribution in the preclinical state.

Sensitivity is the probability that the screening exam is positive, conditional on the individual being in the preclinical stage S_p. The sensitivity cannot be estimated readily from data collection in screening. For a disease with low incidence, confirmation of the status of disease in a seemingly healthy population is not cost effective or ethical. Also, a screened negative individual who has been followed and found to be positive later may represent either a false negative on the previous screening exams or a newly developed case. **Sojourn time** is the time from when the disease first develops to the manifestation of clinical symptoms. The nature of data collection in a screening program precludes exact observation of the onset of either S_p or S_c. Therefore, estimation of the sojourn time distribution is difficult. However, this information can be obtained under model assumptions, and we believe that the preclinical phase of breast cancer may last from 1 to 5 years [4, 18, 22–24], and it may last longer for colorectal cancer [27]. Hence there is a good chance that cancer can be detected in its preclinical stage, which is the goal of implementing a screening program. **Transition probability** from the disease free state (S_0) to the preclinical state (S_p) is continuously changing with one's age [22–24] and is difficult to estimate without proper modeling. Another vital characteristic is **Lead time**, which is the length of time the diagnosis is advanced by screening. The effectiveness of the screening program directly depends on the lead time. If one enters the preclinical state (S_p) at age t_1 and becomes clinically incident (S_c) later at age t_2, then $(t_2 - t_1)$ is the *sojourn time*. If she is offered a screening exam at time t within the time interval (t_1, t_2), and cancer is diagnosed, then the length of the time $(t_2 - t)$ is the *lead time*. This is illustrated in Fig. 10.1.

An individual with a longer lead time usually has a better prognosis than one with a shorter lead time. For a particular case detected by the screening, the lead time is unobservable.

There are many papers to estimate the sensitivity, the sojourn time distribution, and the transition probability from the disease-free to the preclinical state in a screening program. Walter and Day [20] estimated the incidence, the sensitivity, and the sojourn time from the HIP data. They modeled the sojourn times in the preclinical state as exponentially distributed. They found a high marginal correlation between the sensitivity and the parameter for the sojourn time distribution. This high marginal correlation may reflect a common dependence on age. However, they did not include a dependence on age in their model.

Shen and Zelen [17] presented two models they called stable and nonstable. The stable model assumed constant transition probabilities across all ages (i.e., $w(t) \equiv w$). Their nonstable model considered $w(t)$ to be a step-function of age, with $w(t)$ to be a constant within each 5-year age group (40–44, 45–49, etc.). They assumed constant sensitivity and mean sojourn time across all ages in their models. They analyzed the HIP data and estimated sensitivity under the stable model to be 0.70 (standard error 0.20). The estimated sensitivity with the nonstable model was 0.72 (standard error 0.17). The nonstable disease model requires more assumptions, however. For example, a person's age at the initial and the last screening exams should fall into the same age group. Their stable estimate of the mean sojourn time was 2.5 years (standard error 1.2 years). With the nonstable model, the estimate was 2.2 years, with a standard error of 0.89 years. The innovative part in this paper was that they used a likelihood function to estimate these parameters.

Chen et al. [3] provided a method for estimating the mean sojourn time without using data from interval cases. They analyzed screening data from Taiwan. Their estimated mean sojourn time was 1.90 years, with a 95% C.I. (1.18, 4.86). They assumed that the transition probabilities were exponentially distributed and 100% sensitivity, which they deemed unrealistic.

There have been questions concerning the efficient design of periodic cancer screening programs: at what age should screening programs be initiated [1], and at what frequency [5]. If screening programs were specifically designed for people with various risk factors, then more cases could be diagnosed at scheduled exams. However, optimal design of screening exams is hindered because the theory of screening has not been well developed to date. Very often the recommended frequency in a periodic screening program is arbitrary.

In this chapter, we will briefly review the current status of the probability model and the statistical methods in cancer screening and their limitations, and we will also point out some challenging problems and future developments in this area.

10.2 Current Methods in Periodic Cancer Screening

We will use the HIP data for breast cancer as an example in this section. Consider a cohort of initially asymptomatic individuals who enroll in a breast cancer screening program. Assume there are K ordered screenings that, for a specific individual, occur at ages $t_0 < t_1 < \cdots < t_{K-1}$. We define $\beta_i = \beta(t_i)$ as the sensitivity at age t_i; $w(t)\,dt$ is the transition probability from S_0 to S_p during $(t, t+dt)$; $q(x)$ is the PDF of the sojourn time in S_p, and $Q(z) = \int_z^\infty q(x)\,dx$ is the survivor function. We define the ith generation of women as those who enter S_p during the ith screening interval (t_{i-1}, t_i), $i = 1, 2, \ldots, K-1$. The 0th generation includes all who enter S_p

before t_0. We let $t_{-1} = 0$ and $t_K = T$, where T represents the span of the human life, a fixed value in this section.

10.2.1 MLE and Bayesian Inference of Age-Dependent Sensitivity and Transition Probability in Periodic Screening

We will briefly review the statistical inference procedures that Wu, Rosner, and Broemeling [23] developed under the progressive disease model. The purpose was to provide statistical inference for the sojourn time, the age-dependent sensitivity and the age-dependent transition probability from the disease-free state S_0 to the preclinical state S_p. We used age as a covariate in the estimation of the sensitivity and the transition probability simultaneously, both in a frequentist point of view and in a Bayesian framework.

Consider a cohort of initially asymptomatic individuals who are all aged t_0 at study entry, and there are K ordered screening exams that occur at ages $t_0 < t_1 < \cdots < t_{K-1}$. Let n_{i,t_0} be the total number of individuals examined at the ith screening; s_{i,t_0} is the number of cases detected and confirmed at the ith screening; and r_{i,t_0} is the number of cases diagnosed in the clinical state S_c within (t_{i-1}, t_i), the interval cases. The likelihood function for this age group is proportional to

$$L(\cdot|t_0) = \prod_{k=1}^{K} D_{k,t_0}^{s_{k,t_0}} I_{k,t_0}^{r_{k,t_0}} (1 - D_{k,t_0} - I_{k,t_0})^{n_{k,t_0} - s_{k,t_0} - r_{k,t_0}}, \tag{10.1}$$

where D_{k,t_0} is the probability that an individual will be diagnosed at the kth scheduled exam given that she is in the state S_p; and I_{k,t_0} is the probability of being incident in the kth screening interval. The probability that an individual in S_p is detected at the first scheduled exam (i.e., $k = 1$) at age t_0 is

$$D_{1,t_0} = \beta_0 \int_0^{t_0} w(x) Q(t_0 - x)\, dx. \tag{10.2}$$

The integral in the equation arises because she must have entered the preclinical state S_p before t_0 and remained in that state at least until t_0.

To get D_{k,t_0}, we consider an ith generation individual who was detected at the kth screening exam ($1 \le i < k$). These are the possibilities: either she passed her previous $(k - i - 1)$ exams undetected and had a sojourn time of at least $(t_{k-1} - x)$, where $x \in (t_{i-1}, t_i)$ is her age at onset of S_p; or she entered S_p in the $(k - 1)$th screening interval (t_{k-2}, t_{k-1}). Hence the probability is

$$D_{k,t_0} = \beta_{k-1} \left\{ \sum_{i=0}^{k-2} \left\{ [1 - \beta_i] \cdots [1 - \beta_{k-2}] \int_{t_{i-1}}^{t_i} w(x) Q(t_{k-1} - x)\, dx \right\} \right.$$
$$\left. + \int_{t_{k-2}}^{t_{k-1}} w(x) Q(t_{k-1} - x)\, dx \right\} \quad \text{for } k = 2, \ldots, K. \tag{10.3}$$

To calculate I_{k,t_0}, we consider an ith generation women $i < k$. She must have gone undetected in her $(k-i)$ previous screening exams and have a sojourn time longer than $(t_{k-1} - x)$ but shorter than $(t_k - x)$, where x is her age at onset of S_p. Alternatively, she may have entered S_p after the kth exam and developed clinical symptoms before $(t_k - x)$. Hence,

$$I_{k,t_0} = \sum_{i=0}^{k-1} \big[1-\beta(t_i)\big]\cdots\big[1-\beta(t_{k-1})\big] \int_{t_{i-1}}^{t_i} w(x)\big[Q(t_{k-1}-x) - Q(t_k - x)\big]\,dx$$
$$+ \int_{t_{k-1}}^{t_k} w(x)\big[1 - Q(t_k - x)\big]\,dx \quad \text{for } k = 1,\ldots,K. \tag{10.4}$$

The likelihood for the whole HIP study group is proportional to

$$L = \prod_{t_0=40}^{64} L(\cdot|t_0) \tag{10.5}$$

according to the people's initial age t_0, ranging from 40 to 64 in the HIP study. We carefully chose parametric models for $\beta(t)$, $w(t)$, and $q(x)$ as follows:

$$\beta(t) = \frac{1}{1+\exp\{-b_0 - b_1(t-\bar{t})\}}, \tag{10.6}$$

$$w(t) = \frac{0.2}{\sqrt{2\pi}\sigma t} \exp\big\{-(\log t - \mu)^2/\big(2\sigma^2\big)\big\}, \tag{10.7}$$

$$q(x) = \frac{\kappa x^{\kappa-1}\rho^{\kappa}}{(1+(x\rho)^{\kappa})^2}, \tag{10.8}$$

where $\bar{t}$ is the average age at entry in the study group. The unknown parameters $\theta = (b_0, b_1, \mu, \sigma^2, \kappa, \rho)$ were to be estimated from the likelihood function. Simulations were performed to evaluate the reliability of the proposed likelihood [23, 24]. We applied our model to the HIP data using both Markov Chain Monte Carlo (MCMC) and the Maximum Likelihood Estimate (MLE). The Bayesian posterior and the MLE were very close to each other. Our results show that the sensitivity increases with age; it is 0.6 in age 40 and 0.87 at age 64. The transition probability is not a monotone function of age but has a single maximum at about age 60. The posterior mean sojourn time is 1.88 years. Lee and Zelen [8] used SEER information [19] of breast cancer incidence to estimate $w(t)$, assuming that the sojourn time is exponentially distributed with a mean of 4 years. Our estimated transition probabilities are larger than theirs, as might be expected, since we estimated lower sensitivities among younger women. This is the first time that the HIP data have been used directly to obtain estimates for the transition probability into S_p.

We applied our model to the MCCCS data for colon cancer. The sensitivity appears to increase with age for both genders. The age-dependent transition probability has a single maximum at age 72 for males and age 75 for females. The age-dependency seems more dramatic for females than for males. The posterior mean sojourn time is 4.08 years for males and 2.41 years for females [27].

We applied this model to the Mayo Lung Project for male heavy smokers [26]. Since there is no evidence of age effect for sensitivity of chest X-ray in clinical studies, we slightly modified our model to fit this fact. The posterior mean sensitivity is 0.734; the 95% highest posterior density (HPD) interval is (0.647, 0.813). The posterior mean sojourn time is 9.1 years for males heavy smokers; the 95% HPD interval for the sojourn time is (4.9, 23.8) years. The age-dependent transition probability has a single maximum at age 63. The physicians at the lung cancer clinic at the Brown Cancer Center, University of Louisville, are very interested in our findings, since they are compatible with their clinical observations for the sojourn time.

10.2.2 Bayesian Inference for the Lead Time in Periodic Cancer Screening

Many researchers have proposed methods to infer lead time among participants in a screening program, whether within a randomized study or not. Prorok [14] made a major contribution by deriving the conditional probability distribution of the lead time, given detection at the ith screening exam. He then applied his model in simulations to study the properties of the lead time, assuming different sojourn time distributions. He noted that when one increases the number of exams, keeping the between-exam interval fixed, the local lead time properties appeared to stop changing after 4 or 5 screening exams in his examples. The stabilization of the local lead time properties suggested a stopping rule for comparative studies, in that further screening exams will not yield more information about the benefit of screening versus no screening. His work, however, has limited applicability. He considered only screen-detected cases, ignoring interval cases for whom the lead time is zero. His results apply to cases who are screen-detected at the ith screening exam. He did not estimate the whole proportion of cases who were not detected by the periodic screening.

We will briefly review the statistical inference for the lead time that Wu, Rosner, and Broemeling [25] developed under the progressive disease model. The aim was to provide statistical inference of the lead time for the whole cohort, including both the screen-detected and the interval incident case. The lead time is distributed as a mixture of a point mass at zero and a piecewise continuous distribution. Simulations were carried out, using the HIP data for breast cancer, to make inference under different screening time intervals, the proportion of breast cancer patients who might benefit from the periodic screening exams (i.e., those whose lead time is bigger than zero) and the proportion that do not. The model provides policy makers with important information regarding the screening frequency and the possible benefit to women who take part in a periodic screening program.

We let $D = 1$, indicating the development of clinical disease, and $D = 0$, indicating the absence of the clinical disease before death. We use L to denote the lead time. The lead time distribution is a mixture of the conditional probability $P(L=0|D=1)$ and the conditional PDF $f_L(z|D=1)$:

$$P(L=0|D=1) = \frac{P(L=0, D=1)}{P(D=1)}, \tag{10.9}$$

$$f_L(z|D=1) = \frac{f_L(z, D=1)}{P(D=1)}. \tag{10.10}$$

We need $P(D=1)$, the probability of developing breast cancer during one's lifetime after age t_0, the joint probability $P(L=0, D=1)$, and the joint probability density function $f_L(z, D=1)$, to compute the distribution of the lead time explicitly. We assume that the individual is asymptomatic at age t_0.

The probability of developing breast cancer after the initial screening exam (at age t_0) is the following. Suppose a woman is incident with clinical disease at age $t \in (t_0, T)$. Then, one must move from S_0 to S_p before age t, say at age x. The sojourn time in the preclinical state is $(t-x)$. Hence,

$$P(D=1) = \int_{t_0}^{T} \int_{0}^{t} w(x) q(t-x)\, dx\, dt. \tag{10.11}$$

The lead time is zero if and only if an individual is an interval case. Let $I_{K,i}$ denote the probability of being an interval case in the ith interval (t_{i-1}, t_i) in a sequence of K screening exams. Then

$$P(L=0, D=1) = I_{K,1} + I_{K,2} + \cdots + I_{K,K},$$

where

$$I_{K,j} = \sum_{i=0}^{j-1} (1-\beta_i)\cdots(1-\beta_{j-1}) \int_{t_{i-1}}^{t_i} w(x)\big\{Q(t_{j-1}-x) - Q(t_j - x)\big\}\, dx$$
$$+ \int_{t_{j-1}}^{t_j} w(x)\big\{1 - Q(t_j - x)\big\}\, dx \quad \text{for all } j = 1, \ldots, K. \tag{10.12}$$

It can be proved by mathematical induction that, if the sensitivity is less than 1, then for any fixed sequence $t_0 < t_1 < \cdots < t_{K-1} < T$,

$$I_{1,1} \geq (I_{2,1} + I_{2,2}) \geq \cdots \geq (I_{K,1} + \cdots + I_{K,K}).$$

In other words, more screening reduces the probability that the lead time equals zero among women who would go on to develop cancer.

For the cases whose lead times are greater than zero, we calculate the joint PDF $f_L(z, D=1)$, where $z \in (0, T-t_0]$. When $T - t_1 < z \leq T - t_0$, detection must have been occurred at t_0. In general, when $T - t_j < z \leq T - t_{j-1}$, $j = 2, 3, \ldots, K$, one was screen-detected at t_i, $i = 0, \ldots, j-1$ (i.e., $t_c = t_i + z < T$, $i = 0, \ldots, j-1$),

$$f_L(z, D=1) = \sum_{i=1}^{j-1} \beta_i \left\{ \sum_{r=0}^{i-1} (1-\beta_r)\cdots(1-\beta_{i-1}) \int_{t_{r-1}}^{t_r} w(x) q(t_i + z - x)\, dx \right.$$
$$\left. + \int_{t_{i-1}}^{t_i} w(x) q(t_i + z - x)\, dx \right\} + \beta_0 \int_0^{t_0} w(x) q(t_0 + z - x)\, dx,$$
$$\text{for } z \in (T - t_j, T - t_{j-1}],\ j = 2, 3, \ldots, K, \tag{10.13}$$

$$f_L(z, D=1) = \beta_0 \int_0^{t_0} w(x) q(t_0 + z - x)\, dx, \quad \text{for } z \in (T - t_1, T - t_0]. \tag{10.14}$$

The validity of this distribution can be verified by

$$P(L=0|D=1) + \int_0^{T-t_0} f_L(z|D=1)\, dz = 1.$$

We obtained the same result as in Prorok [14] when conditioning on detection by the ith screenings. However, our deriving procedure is greatly simplified. See Sect. 10.2.2 in Wu et al. [25].

It is clear that the lead time distribution depends on the sensitivity $\beta(t)$, the transition probability $w(t)$, and the sojourn time distribution $q(x)$ in the preclinical state. We know that these three key parameters were modeled by $\theta = (b_0, b_1, \mu, \sigma^2, \kappa, \rho)$ in (10.6) to (10.8) in Sect. 10.2.1, and θ can be estimated by Bayesian posterior samples through Markov Chain Monte Carlo (MCMC) simulation [23].

Let H represent the HIP study group data. The likelihood function $L(\theta|H)$ was defined in (10.5) in Sect. 10.2.1. Let $\pi(\theta)$ be the prior distribution of θ. The posterior distribution of θ is $f(\theta|H) \propto L(\theta|H)\pi(\theta)$. The posterior predictive distribution of the lead-time z can be estimated as follows:

$$f(z|H) = \int f(z, \theta|H)\, d\theta = \int f(z|\theta, H) f(\theta|H)\, d\theta \approx \frac{1}{n} \sum_i f(z|\theta_i^*). \tag{10.15}$$

Where $f(z|\theta_i^*)$ is the mixture distribution of the lead time in (10.9) and (10.10), and θ_i^* is the posterior samples.

Simulation studies using the HIP study data provide predictive inference under different screening frequencies. The time interval between screens was 6, 9, 12, 18, and 24 months from age 50 (t_0) to 80 years (T). From the results we see that, if a woman begins annual screening when she is 50 years old and continues until she reaches 80, then there is a 23.37% chance that she will not benefit (i.e., interval incident case) from early detection by the screening program if she develops breast cancer during those thirty years; however, if she will benefit from the program, her most possible lead time (mode) will be 3.6 months. Her chance of no benefit from the screenings drops to 8.95% if the exams are 6 months apart, and her most possible lead time is 6 months. It seems necessary for a woman to take the exam every six months to guarantee a 90% chance of earlier detection based on our simulation result.

We applied this method to MCCCS for colon cancer [28]. The results show that if a man begins annual screening when he is 50 years old and continues until he reaches 80, then there is a 18.87% chance that he will not benefit from early detection by the screening program if he develops colorectal cancer during those thirty years. His chance of no benefit from the screening program decreases to 6.45% if the exams are 6 months apart. While for the females, the chance of no early detection is 9.48% for the annual test and 2.39% for the 6-month test to guarantee a 90% chance of early detection, it maybe necessary for the males to take the FOBT every 9 months, while the females can take it annually.

We applied the lead-time method to Mayo Lung Project data for male heavy smokers [26]. We found that if a male heavy smoker begins screening when he is 50 years old with a 3-year screening interval and continues until he reaches 80, then there is a 14.78% chance that he will not benefit from early detection by the screening program if he develops lung cancer during those thirty years. His chance of no benefit from the screening program decreases to 7.46% if the exams are two years apart. It is not necessary to take the screening exam every 4 months as the Mayo Lung Project was carried out, but taking the exam every 2 years for male heavy smokers is probably enough to guarantee a 90% of early detection under the then-not-so-high assumption of sensitivity. The reason is that lung cancer tends to have a much longer sojourn time than that of other kinds of cancer.

10.2.3 Testing the Dependence of Two Screening Modalities

We developed a hypothesis testing procedure in [18] under a stable disease model. The parameters to be estimated and tested are the sensitivity of mammogram β_1, the sensitivity of physical exam β_2, and their correlation coefficient ρ when the transition probability $w(t) = w$ was a constant (the stable disease model).

Three cases can be identified by screening exams: cases detected by mammogram only; cases detected by physical exam one; and cases identified by both. Define α_1, α_2, and α_3 as the probabilities of these three mutually exclusive events. Then we can express α_i as a function of β_1, β_2, and ρ, for example, $\alpha_3 = \beta_1\beta_2 + \rho\sqrt{\beta_1\beta_2(1-\beta_1)(1-\beta_2)}$, and the overall sensitivity $\beta = \beta_1 + \beta_2 - \alpha_3$.

It is clear that a positive correlation reduces the overall sensitivity and the opposite is true for a negative correlation. A test for independence can be made by considering $H_0 : \alpha_3 = \beta_1\beta_2$, which is equivalent to $H_0 : \rho = 0$.

Let $t_0 < t_1 < \cdots < t_{K-1}$ represent K ordered screening exam times. Define the ith screening interval (t_{i-1}, t_i) for $i = 1, 2, \ldots, k-1$, with $\Delta_i = t_i - t_{i-1}$. Adopt the following notation: n_i is the total number of individuals examined at t_{i-1}; s_i is the number of cases detected at the exam given at t_{i-1}; and r_i is the number of cases diagnosed within the interval (t_{i-1}, t_i).

Let $D_i(\beta)$ be the probability of an individual diagnosed at the ith scheduled exam given at t_{i-1}, and let $I_i(\beta)$ be the probability of an interval case occurring in the ith interval. The full likelihood is

$$L(\beta_1, \beta_2, \rho) = \prod_{i=1}^{k} D_i(\beta)^{s_i} I_i(\beta)^{r_i} \left[1 - D_i(\beta) - I_i(\beta)\right]^{n_i - s_i - r_i} \prod_{j=1}^{3} (\alpha_j/\beta)^{s_{ij}},$$

where s_{i1}, s_{i2}, s_{i3} denote the number of cases detected by modality 1 only, by modality 2 only, and by both modalities, respectively.

Under the stable disease model, the transition probability $w(t)$ is assumed to be some unknown constant w. The values of $D_i(\beta)$ and $I_i(\beta)$ under the stable disease model are greatly simplified:

$$D_i(\beta) = \begin{cases} \beta P\left\{1 - \beta \sum_{j=1}^{i-1} (1-\beta)^{i-j-1} Q(t_{i-1} - t_{j-1})\right\} & i > 1, \\ \beta P & i = 1, \end{cases}$$

$$I_i(\beta) = P\left[\frac{\Delta_i}{\mu} - \beta \sum_{j=0}^{i-1} (1-\beta)^{i-j-1} \left\{Q(t_{i-1} - t_j) - Q(t_i - t_j)\right\}\right],$$

where P is the prevalence of breast cancer; $Q(\cdot)$ is the survival function of the sojourn time in the pre-clinical state, and the exponential distribution with parameter μ is adopted for the sojourn time.

A simplified conditional likelihood is

$$L_c(\beta_1, \beta_2, \rho) = \prod_{i=1}^{k} \frac{D_i(\beta)^{s_i} I_i(\beta)^{r_i}}{(D_i(\beta) + I_i(\beta))^{s_i + r_i}} \prod_{j=1}^{3} (\alpha_j/\beta)^{s_{ij}},$$

where the prevalence parameter P and n_i are eliminated by virtue of $\frac{D_i(\beta)}{D_i(\beta)+I_i(\beta)}$.

A likelihood ratio test was developed to test the independence of the two screening modalities. Under the null hypothesis H_0 and under some regular conditions, the log-likelihood ratio test is

$$-2\log(LR) = 2\left\{l\left(\hat{\beta}_1, \hat{\beta}_2, \hat{\rho}, \hat{\mu}\right) - l\left(\tilde{\beta}_1, \tilde{\beta}_2, \tilde{\mu}\right)\right\} \sim \chi_1^2 \quad \text{approximately},$$

where the estimators are the maximum likelihood estimators (MLEs) over the intervals on which the parameters are defined. The conditional likelihood ratio test statistic can be derived in a similar fashion.

We applied the methodologies to the HIP and CNBS trials (Table 5, [18]). Our main finding was that the correlation coefficient is zero, or even slightly negative. This means that mammogram and physical exams appear to contribute independently to the detection of breast cancer on screening. The two procedures are complementary to each other. Our analysis shows that it is important to emphasize the contribution of physical exams in addition to mammography in breast cancer screening practice, especially for women under 50 years old.

10.3 Future Developments in Cancer Screening

We will provide some unsolved problems, possible solutions, or our future endeavors in periodic cancer screening.

10.3.1 Evaluate Long Term Benefits of Periodic Cancer Screening

Cancer screening programs have existed for a long time, and NIH recommends annual mammograms for women over 50. However, a specific knowledge gap is: how to evaluate the long-term benefit due to screening? For example, if a woman in her fifties wants to know the benefit of taking annual screening exams in her life time, can policy makers or physicians offer her a quick answer, such as what is the probability of early detection if she would develop breast cancer in the future? And if it is an early diagnosis, how early could it be? Or what could be her risk of over-diagnosis? What is the possibility of no early diagnosis? What is the chance that she might have just wasted her money and her time, and finally die of causes other than breast cancer? These questions are related to the long-term benefit due to periodic cancer screening, and there is almost no literature to fully address these issues so far, hence there is a knowledge gap here.

Wu, Rosner, and Broemeling [25] provided a very limited solution to some of the questions above. The lead-time model can provide answers to some questions, such as what is the probability that a woman's cancer will be detected early if she has cancer later in her lifetime? How does changing the screening time interval affect the lead-time distribution? For example, we found that if a woman begins annual breast cancer screening when she is 50 years old and continues until she reaches 80, then there is a 23.4% chance that she will not benefit from early detection by the program if she develops breast cancer during those 30 years. Her chance of no benefit decreases to 9% if the exams are 6 months apart [25]; For male heavy smokers, it is necessary to take the screening every two years to guarantee a 92.5% chance of early detection [26]. However, there is a big limitation in Wu et al. [25], where the human life time was treated as a fixed value, which means, for anyone who takes the screening exam, it was assumed that cancer symptoms will always appear before death. In other words, it ignores the possibility of over-diagnosis. In reality, people can die of causes other than the targeted cancer.

We want to address these problems by classifying all initially asymptomatic participants in a periodic cancer screening program into four mutually exclusive categories: Over-Diagnosis (or False-Benefit), True-Benefit, No-Benefit, and Unnecessary. Using breast cancer as an example, we will assume that an individual is asymptomatic and without a history of breast cancer before she takes any screening exam. Based on the diagnosis status and whether she would have breast cancer before death or not, we will categorize people who take part in periodic screening into 4 mutually exclusive groups:

- *Case 1* (Unnecessary): A woman who took part in screening exams, no breast cancer was diagnosed, and finally she died of other causes.
- *Case 2* (No-Benefit): A woman who took part in screening exams and who was a clinical incidence case between two scheduled exams.
- *Case 3* (True-Benefit): A woman whose breast cancer was diagnosed at a scheduled screening exam, and her clinical symptom would have appeared before her death.
- *Case 4* (Over-Diagnosis): A woman who was diagnosed with breast cancer at a scheduled screening exam; however, her clinical symptom would NOT have appeared before her death.

For each of the four situations, we will derive the probability and then evaluate the long-term benefit due to screening by simulations. Therefore this is a prospective study based on the existed data for evaluation purposes.

10.3.2 Sensitivity as a Function of Age, Time Spent in S_p and Sojourn Time

Many studies have suggested that screening sensitivity increases with age at diagnosis, especially for breast cancer [5, 9, 10, 16, 20, 22, 23]. Walter and Day [20] also found that screening sensitivity is negatively correlated with sojourn time. More intuitively, when the tumor cell is just formed, the sensitivity is fairly small; while at the late stage, that is, when the preclinical stage S_p comes to an end and the clinical stage S_c will soon start, the sensitivity might be very close to one. Hence the sensitivity not only depends on age at diagnosis and the sojourn time but also depends on the time that the individual spent in the preclinical state.

How to combine this information into the sensitivity is very challenging. We propose a new model to allow the sensitivity to vary with people's age, sojourn time, and time spent in the preclinical state. More specifically, let sensitivity

$$\beta(t, s|T) = \frac{1}{1 + \exp(-b_0 - b_1 * (t - m) - \eta * g(\frac{s}{T}))}, \tag{10.16}$$

where T = the sojourn time, a random variable; t = age at diagnosis, s = time spent in S_p, $s \in [0, T]$, m = the average age at entry, and $g(\frac{s}{T})$ is a generalized linear model to associate the sensitivity β with s and T. The model is chosen such that our previous model on β is nested inside. (b_0, b_1, η) are the parameters need to be estimated. We will use the same parametric model for $w(t)$ and $q(t)$ as in Sect. 10.2.1.

If we keep in mind that $0 \le s \le T$, $g(\frac{s}{T}) = 1$ when $s = T$, and $g(\frac{s}{T}) = 0$ when $s = 0$, then a few candidates for the function $g(\cdot)$ could be:

$$g\left(\frac{s}{T}\right) = \frac{s}{T}, \qquad g\left(\frac{s}{T}\right) = \left(\frac{s}{T}\right)^2,$$

$$g\left(\frac{s}{T}\right) = \sqrt{\frac{s}{T}}, \qquad g\left(\frac{s}{T}\right) = \left(\frac{s}{T}\right)^2 \times \left(3 - \frac{2s}{T}\right).$$

Under this new model, the probabilities D_{k,t_0} and I_{k,t_0} will be changed to

$$\begin{aligned} D_{k,t_0} = & \sum_{i=0}^{k-2} \int_{t_{i-1}}^{t_i} w(x) \int_{t_{k-1}-x}^{\infty} q(t) \left\{ \prod_{j=i}^{k-2} [1 - \beta(t_j, t_j - x|t)] \right\} \\ & \times \beta(t_{k-1}, t_{k-1} - x|t)\, dt\, dx \\ & + \int_{t_{k-2}}^{t_{k-1}} w(x) \int_{t_{k-1}-x}^{\infty} q(t)\beta(t_{k-1}, t_{k-1} - x|t)\, dt\, dx \quad \text{for } k = 2, \ldots, K, \end{aligned} \tag{10.17}$$

$$D_{1,t_0} = \int_0^{t_0} w(x) \int_{t_0-x}^{\infty} q(t)\beta(t_0, t_0 - x|t)\, dt\, dx, \tag{10.18}$$

$$\begin{aligned} I_{k,t_0} = & \sum_{i=0}^{k-1} \int_{t_{i-1}}^{t_i} w(x) \int_{t_{k-1}-x}^{t_k-x} q(t) \left\{ \prod_{j=i}^{k-1} [1 - \beta(t_j, t_j - x|t)] \right\} dt\, dx \\ & + \int_{t_{k-1}}^{t_k} w(x)\big[1 - Q(t_k - x)\big]\, dx, \quad \text{for } k = 1, \ldots, K. \end{aligned} \tag{10.19}$$

The double integral arises because the sensitivity is changing with the sojourn time and the time spent in the preclinical stage, as well as with age. A big challenge might be in the area of computing or MCMC simulation.

10.3.3 Optimal Scheduling for the Next Exam

So far, an important and unaddressed problem in cancer screening is: for an individual who has taken some screening exams in the past, and who is asymptomatic right now, when to schedule his/her next screening exam? Using breast cancer screening as an example: Should she take it after 3 months, 6 months, 9 months, or 12 months? What would her best choice be? Physicians face this question almost every day: how to provide informed and satisfying advice to a woman in such a situation?

Some research work has been done in optimal scheduling for screenings before. Zelen [31] made a major contribution. He developed a utility function to find the optimal scheduling for $(n + 1)$ exams. "This is equivalent to a fixed budget which allows only $(n + 1)$ examinations." (quote from Zelen [31]). The utility function needs to assign different weights to cases diagnosed by the first exam, cases diagnosed at subsequent exams, and the interval cases. The optimal spacing of the exams is to find a sequence of time $(t_0, t_1, \ldots, t_n)$ that maximize the utility function. Zelen found that for the optimal intervals to be equal, the sensitivity must be 1, which cannot be achieved in reality. The other issue is the choice of the weights, which is mostly subjective.

Lee and Zelen [8] developed the threshold method and the schedule sensitivity. Their threshold method calculates the probability of being in S_p, and exams are scheduled whenever this probability reaches the same value as that at age 50 (which is 0.0018 in their simulation). They found that the screening interval gets smaller as people get older. The schedule sensitivity is the ratio of the expected number of cases diagnosed on scheduled exams to the expected number of the total cancer cases. Hence schedule sensitivity will increase if more screenings would be scheduled in a fixed time interval. Again, costs or weights were involved in their schedule sensitivity. There are many other papers (Parmigiani [11, 12], Parmigiani et al. [12]) on optimal scheduling. They all use some kind of utility function that involves cost or weight. Their major contributions are on the optimal scheduling for $(n+1)$ exams as a group but not focusing on the next coming exam.

We propose to use a totally different approach to handle the scheduling of screening. We will not use weight or cost, or utility function. We will not focus on $(n + 1)$ exams but focus only on the next coming exam instead. More specifically, we propose to derive the conditional probability of incidence before the next exam, given one's screening history. Then the next screening interval shall be chosen, such that this probability will be limited by some preselected small value, say 10%, 5%, or less. Hence, with 90% or more chance, a woman will not become a clinically incident case before her next scheduled exam. We will also derive the conditional lead-time distribution, conditional on that one would be diagnosed with cancer at the next screening exam. This could provide individuals (based on her screening history) some predictive information regarding how early the diagnosis could be if she would develop cancer and follow this schedule. The research may provide a theoretical and practical basis to guide individuals or physicians to make informed decision in screening exam. Specifically, the research may solve the problems of when cancer screening should be performed for different individuals with different risk factors in the near future.

10.3.4 Survival Benefit due to Periodic Screening

The effectiveness of a screening program directly depends on the availability of effective therapy and improved outcome if one receives treatment for the disease in its earliest stages of development. When evaluating the effectiveness of a screening program, one should account for the fact that the age at diagnosis is earlier if the disease is detected by screening rather than by the onset of clinical symptoms.

The difference between the age of diagnosis with screening and the future onset of clinical disease without screening is called the lead time. Even in the absence of effective therapy, screening will appear to lengthen the time from diagnosis until death. If one does not account for the lead time when analyzing the benefit of screening, then one's inference is subject to lead-time bias (Prorok [15]).

Much research has been done in this area [2, 6, 7, 15, 21]. Xu and Prorok [29] and Xu et al. [30] made a major contribution in statistical modeling in this area. For a screen-detected case, they used the *total observed survival time* data in HIP, to estimate the distribution of *the time survived post-lead-time* by a deconvolution method. They developed two models, one assumed that the post-lead-time survival is independent of the lead time, and the other assumed that they are positively correlated.

However, their work has big limitations too. They assumed that the lead time was distributed as the exponential (λ) random variable, which is not realistic. And their method is difficult to be generalized if the exponential distribution was modified to other distributions. In Wu et al. [25], the exact lead-time distribution was derived. We hope to combine our new findings on the lead-time distribution and develop a likelihood method to solve this problem. It is hoped that this will provide more accurate measurement for the survival benefit due to screenings.

References

1. S.G. Baker. Evaluating the age to begin periodic breast cancer screening using data from a few regularly scheduled screenings. *Biometrics*, **54**:1569–1578, 1998.
2. S.G. Baker and K.C. Chu. Evaluating screening for the early detection and treatment of cancer without using a randomized control group. *J Am Stat Assoc*, **85**:321–327, 1990.
3. T.H.H. Chen, H.S. Kuo, M.F. Yen, M.S. Lai, L. Tabar, and S.W. Duffy. Estimation of sojourn time in chronic disease screening without data on interval cases. *Biometrics*, **56**:167–172, 2000.
4. N.E. Day and S. Walter. Simplified models of screening for chronic disease: Estimation procedures from mass screening programmes. *Biometrics*, **40**:1–13, 1984.
5. D.M. Eddy. *Screening for Cancer: Theory, Analysis and Design*. Prentice Hall, Englewood Cliffs, 1980.
6. K. Kafadar and P.C. Prorok. Computer simulation of randomized cancer screening trials to compare methods of estimating lead time and benefit time. *Comput Stat Data Anal*, **23**:263–291, 1996.
7. K. Kafadar and P.C. Prorok. Alternative definitions of comparable case groups and estimates of lead time and benefit time in randomized cancer screening trials. *Stat Med*, **22**:83–111, 2003.
8. S.J. Lee and M. Zelen. Scheduling periodic examinations for the early detection of disease: Applications to breast cancer. *J Am Stat Assoc*, **93**:1271–1281, 1998.
9. A.B. Miller, C.J. Baines, T. To, and C. Wall. Canadian National Breast Screening Study: 2. Breast cancer detection and death rates among women aged 50 to 59 years. *Can Med Assoc J*, **147**(10):1477–1488, 1992.
10. A.B. Miller, C.J. Baines, T. To, and C. Wall. Canadian National Breast Screening Study: 1. Breast cancer detection and death rates among women aged 40 to 49 years. *Can Med Assoc J*, **147**(10):1459–1476, 1992.
11. G. Parmigiani. On optimal screening ages. *J Am Stat Assoc*, **88**:622–628, 1993.

12. G. Parmigiani. Timing medical examinations via intensity functions. *Biometrika*, **84**:803–816, 1997.
13. G. Parmigiani, S. Skates, and M.B. Zelen. Modeling and optimization in early detection programs with a single exam. *Biometrics*, **58**:30–36, 2002.
14. P.C. Prorok. Bounded recurrence times and lead time in the design of a repetitive screening program. *J Appl Prob*, **19**:10–19, 1982.
15. P.C. Prorok. Evaluation of screening programs for the early detection of cancer. In R.G. Cornell, editor, *Statistical Methods for Cancer Studies*, pages 267–328. Dekker, New York, 1984. Chap. 7.
16. S. Shapiro, W. Venet, P. Strax, and L. Venet. *Periodic Screening for Breast Cancer. The Health Insurance Plan Project and its Sequelae, 1963–1986*. Johns Hopkins Press, Baltimore, 1988.
17. Y. Shen and M. Zelen. Parametric estimation procedures for screening programmes: Stable and nonstable disease models for multimodality case finding. *Biometrika*, **86**:503–515, 1999.
18. Y. Shen, D. Wu, and M. Zelen. Testing the independence of two diagnostic tests. *Biometrics*, **57**:1009–1017, 2001.
19. National Cancer Institute. Surveillance, Epidemiology, and End Results Program. Age-specific rates for breast cancer. Bethesda, MD, 1994.
20. S.D. Walter and N.E. Day. Estimation of the duration of a preclinical disease state using screening data. *Am J Epidemiol*, **118**:856–886, 1983.
21. S.D. Walter and L.W. Stitt. Evaluating the survival of cancer cases detected by screening. *Stat Med*, **6**:885–900, 1987.
22. D. Wu, G.L. Rosner, and L. Broemeling. Bayesian inference of age-dependent sensitivity, sojourn time and transition rate in screening. Technical Report, UTMDABTR-015-01, Dept. of Biostatistics, Univ. Texas, M.D. Anderson Cancer Center, 2001.
23. D. Wu, G.L. Rosner, and L. Broemeling. MLE and Bayesian inference of age-dependent sensitivity and transition probability in periodic screening. *Biometrics*, **61**:1056–1063, 2005.
24. D. Wu, X. Wu, I. Banicescu, and R. Carino. Simulation procedure in periodic cancer screening trials. *J Mod Appl Stat Methods*, **4**(2):522–527, 2005.
25. D. Wu, G.L. Rosner, and L. Broemeling. Bayesian inference for the lead time in periodic cancer screening. *Biometrics*, **63**:873–880, 2007.
26. D. Wu, D. Erwin, and G.L. Rosner. Sojourn time and lead time in lung cancer screening. Technical Report, 2009.
27. D. Wu, D. Erwin, and G.L. Rosner. Estimating key parameters in FOBT screening for colorectal cancer. *Cancer Causes Control*, **20**:41–46, 2009. doi:10.1007/s10552-008-9215-9.
28. D. Wu, D. Erwin, and G.L. Rosner. A projection of benefits due to fecal occult blood test for colorectal cancer. *Cancer Epidemiology*, 2009. doi:10.1016/j.canep.2009.08.001.
29. J.L. Xu and P.C. Prorok. Non-parametric estimation of the post-lead-time survival distribution of screen-detected cancer cases. *Stat Med*, **14**:2715–2725, 1995.
30. J.L. Xu, R.M. Fagerstrom, and P.C. Prorok. Estimation of post-lead-time survival under dependence between lead-time and post-lead-time survival. *Stat Med*, **18**:155–162, 1999.
31. M. Zelen. Optimal scheduling of examinations for the early detection of disease. *Biometrika*, **80**:279–293, 1993.
32. M. Zelen and M. Feinleib. On the theory of screening for chronic diseases. *Biometrika*, **56**:601–614, 1969.

Chapter 11
On Construction of the Smallest One-sided Confidence Intervals and Its Application in Identifying the Minimum Effective Dose

Weizhen Wang

11.1 Introduction

Suppose that a random vector $\underline{X}$ is observed from a distribution with a known cumulative distribution function $F(\underline{x};\underline{\theta})$ with an unknown $k \times 1$ parameter vector $\underline{\theta} = (\theta, \underline{\eta})$, where θ is the parameter of interest, and $\underline{\eta}$ is the nuisance parameter vector. The parameter space is

$$\Theta = \left\{\underline{\theta} = (\theta, \underline{\eta}) : \underline{\eta} \in D(\theta) \text{ for each } \theta \in [A, B]\right\},$$

where $[A, B]$ is a given interval in R^1 and is open if the corresponding ending is infinity, and $D(\theta)$ is a subset of R^{k-1} depending on θ. In this paper, we describe how to search for the optimal one-sided $1-\alpha$ confidence interval of the form $[L(\underline{X}), B]$ for θ.

Example 1 (One-sided confidence interval for two independent proportions) In clinical trial, it is often of interest to compare a treatment with a control. Let p_1 and p_0 be the proportions of showing improvement for the treatment and control groups, respectively. Let X be the number of subjects in the treatment sample of size n showing improvement, and define Y similarly for the control sample of size m. Then X follows a binomial distribution, denoted by $Bin(n, p_1)$, and $Y \sim Bin(m, p_0)$, and X and Y are independent. In this case the observed random vector is $\underline{\theta} = (X, Y)$. The researcher wishes to see p_1 larger than p_0 by a certain amount, say δ, a predetermined nonnegative number, and has no (or a little) interest in the upper bound of the difference. Therefore, a one-sided confidence interval for the parameter of interest $\Delta = p_1 - p_0$ of the form $[L(\underline{X}), 1]$ is needed, and for example, p_0 is a nuisance parameter in this case. Following the general setting in the previous paragraph, the

W. Wang (✉)
Department of Mathematics and Statistics, Wright State University, Dayton, OH 45435, USA
e-mail: weizhen.wang@wright.edu

J. Feng et al. (eds.), *Frontiers in Computational and Systems Biology*, Computational Biology 15,
DOI 10.1007/978-1-84996-196-7_11,

parameter space $\Theta = \{(p_1, p_0) : 0 \le p_1, p_0 \le 1\}$ can be rewritten as

$$\Theta = \big\{(\Delta, p_0) : p_0 \in D(\Delta) \text{ for each } \Delta \in [-1, 1]\big\} \tag{11.1}$$

with $[A, B] = [-1, 1]$ and

$$D(\Delta) = \begin{cases} [0, 1 - \Delta] & \text{if } \Delta \in [0, 1], \\ [-\Delta, 1] & \text{if } \Delta \in [-1, 0). \end{cases} \tag{11.2}$$

We will discuss the construction of a smallest interval $[L_S(\underline{X}), 1]$ in Sect. 11.3.

Example 2 (Identifying the minimum effective dose with binary data) Suppose that we have a sequence of independent binomial random variables $X_i \sim Bin(n_i, p_i)$ for $i = 1, \dots, k$ and $Y \sim Bin(m, p_0)$. The goal here is to identify the smallest positive integer i_0 such that $p_i > p_0 + \delta$ for any $i \in [i_0, k]$. Each p_i is the proportion of patients who show improvement using a drug at dose level i. A large i associates with a large dose level, and p_0 is the proportion for the control group. Then i_0 is called the minimum effective dose (*MED*). Finding the *MED* is important since high doses often turn out to have undesirable side effects. Typically, the *MED* is to be found when the observation follows a normal distribution with the comparison in proportions replaced by that in means. Thus, the assumption of normality is an issue to be addressed. See, for example, Tamhane, Hochberg, and Dunnett [10], Bretz, Pinheiro, and Branson [4], and Wang and Peng [16] for results under this setting. Now we search for the *MED* with a binary response without such concerns on the distribution, see Tamhane and Dunnett [9]. A sequence of hypotheses can be formulated to detect the *MED* as follows:

$$H_{0i} : \min_{j \ge i}\{p_j - p_0\} \le \delta \quad \text{vs.} \quad H_{Ai} : \min_{j \ge i}\{p_j - p_0\} > \delta \quad \text{for } i = 1, \dots, k, \tag{11.3}$$

which is similar to the one in Hsu and Berger [6, p. 471]. It is easy to see that the *MED* equals the smallest i for which H_{0i} is not true. It is also well known that there is a one-to-one relationship between test and confidence interval. Therefore, we will use the smallest interval derived in Sect. 11.4 for testing each H_{0i} and propose a modification to conduct simultaneous tests for (11.3) with the experimentwise error rate controlled at level α in Sect. 11.4.

Typically, there are two requirements for a confidence interval: the accuracy and precision. For the accuracy, we employ $1 - \alpha$ confidence interval $C(\underline{X})$, i.e., the coverage probability of interval $C(\underline{X})$ is no less than $1 - \alpha$ for all parameter configurations as given below,

$$Cover_C(\theta, \underline{\eta}) = P_{\underline{\theta}}(\theta \in C(\underline{X})) \ge 1 - \alpha \quad \forall \underline{\theta} \in \Theta. \tag{11.4}$$

For the precision, we employ the set inclusion criterion by Wang [11], i.e., for two $1 - \alpha$ confidence intervals $C_1(\underline{X})$ and $C_2(\underline{X})$, $C_1(\underline{X})$ is said to be no worse than $C_2(\underline{X})$ if

$$C_1(\underline{x}) \text{ is a subset of } C_2(\underline{x}) \text{ for any } \underline{x} \in R_{\underline{X}}, \tag{11.5}$$

where $R_{\underline{X}}$ is the range of $\underline{X}$. This is easy to use because no expectation computation is needed to check the superiority of C_1 over C_2 as in two commonly used criteria: the minimum expected length and the minimum false coverage probability. Also, it is clear in interpretation since C_2 is discarded because it is always larger than C_1.

In consequence, we should search for the *smallest* $1-\alpha$ confidence interval which is a subset of any other $1-\alpha$ confidence interval on all values $\underline{x}$ of $\underline{X}$, perhaps in a certain class of intervals. In other words, if the intersection of all $1-\alpha$ confidence intervals in a class still belongs to that class, then it is the smallest interval, and it is the best in the strongest sense.

The interval class from which we search for the best interval cannot be too small or too large. If it is too small, the "best" interval in that class is useless; but if it is too large, then the "best" interval simply does not exist. To characterize the appropriate class, suppose that an ordering, denoted by $\preceq$, is defined on $R_{\underline{X}}$, i.e.,

(a) For any two vectors $\underline{x}_1$ and $\underline{x}_2$ in $R_{\underline{X}}$, one and only one of the following relationships is true:

$$\underline{x}_1 \prec \underline{x}_2 \quad \text{or} \quad \underline{x}_1 \equiv \underline{x}_2 \quad \text{or} \quad \underline{x}_2 \prec \underline{x}_1.$$

(b) If three vectors $\underline{x}_1$, $\underline{x}_2$, and $\underline{x}_3$ in $R_{\underline{X}}$ satisfy

$$\underline{x}_1 \preceq (\text{i.e., } \prec \text{ or } \equiv) \underline{x}_2 \quad \text{and} \quad \underline{x}_2 \preceq \underline{x}_3,$$

then

$$\underline{x}_1 \preceq \underline{x}_3.$$

(c) For any vector $\underline{x}$ in $R_{\underline{X}}$, the sets $U_{\underline{x}} = \{\underline{y} \in R_{\underline{X}} : \underline{y} \preceq \underline{x}\}$ and $V_{\underline{x}} = \{\underline{y} \in R_{\underline{X}} : \underline{x} \equiv \underline{y}\}$ are σ-measurable.

Example 3 An ordering on $R_{\underline{X}}$ can be easily generated by a given σ-measurable function $J(\underline{X})$, including continuous functions as follows. For two vectors $\underline{x}_1$ and $\underline{x}_2$, define

$$\underline{x}_1 \prec (\equiv, \succ) \underline{x}_2 \quad \text{if and only if} \quad J(\underline{x}_1) < (=, >) J(\underline{x}_2), \tag{11.6}$$

respectively, and denote this ordering by $\preceq_J$.

We will search for the smallest interval in an interval class defined below.

Definition 11.1 For a given ordering $\preceq$ on $R_{\underline{X}}$, define a class of one-sided $1-\alpha$ confidence intervals for θ:

$$\mathcal{B}_l = \left\{[L(\underline{X}), B] : L(\underline{x}_1) = L(\underline{x}_2), \text{ if } \underline{x}_1 \equiv \underline{x}_2;\ L(\underline{x}_1) \le L(\underline{x}_2), \text{ if } \underline{x}_1 \preceq \underline{x}_2\right\}.$$

Definition 11.2 A confidence interval $[L_S(\underline{X}), B]$ in $\mathcal{B}_l$ is the smallest if for any $[L(\underline{X}), B]$ in $\mathcal{B}_l$, $L(\underline{x}) \le L_S(\underline{x})$ for all $\underline{x} \in R_{\underline{X}}$.

We call the interval above the smallest interval in $\mathcal{B}_l$ or the smallest interval under ordering $\preceq$.

Remark 1 The idea for the above two definitions is very simple. To illustrate it using Example 1, suppose $n = 10 = m$. For two possible observation values $(X, Y) = (10, 0)$ and $(X, Y) = (0, 10)$, the parameter value of $\Delta = p_1 - p_0$ should be larger if the former is observed. One reason for this is that the maximal likelihood estimator for $\Delta = p_1 - p_0$ at (10, 0) equals 1, larger than that at (0, 10), which is equal to 0. Therefore, the confidence limit $L(X, Y)$ at (10, 0) should be larger. Once the ordering on all values of (X, Y) is available, to determine the confidence limit L_S of the smallest interval at a given point (x, y), one holds all other $L_S(x', y')$ still and simply raises $L_S(x, y)$ until the minimum coverage probability of $[L_S(X, Y), 1]$ decreases to $1 - \alpha$. This justifies the existence of the smallest confidence interval under the specified ordering.

There have been several efforts to construct the smallest one-sided $1 - \alpha$ confidence intervals. When the distribution of the one-dimensional observed random variable X only involves a single parameter θ, Bol'shev [2] first constructed the so-called most accurate confidence interval, which is equivalent to the smallest interval, for θ under the assumption that the distribution of X is stochastically nondecreasing in θ. He employed the natural ordering by the identity function $J(x) = x$, which is intuitively correct for a nondecreasing distribution family in θ. Wang [12] generalized Bol'shev's result under any given ordering, and his result did not need the assumption of nondecreasing distribution family in θ. When there exist nuisance parameters, Bol'shev and Loginov [3] offered an unsuccessful solution because they did not realize the importance of the ordering on the confidence limits. Wang [13] proposed the smallest interval for the difference of two independent proportions in the setting of Example 1, and Wang [14] obtained the general construction of the smallest interval in the presence of nuisance parameters.

The rest of the paper is organized as follows. In Sect. 11.2, we introduce a general construction of the smallest interval in $\mathcal{B}_l$ proposed in Wang [14]. In Sect. 11.3, one reasonable ordering on the sample space in Example 1, $R_{\underline{X}} = \{(x, y) : 0 \leq x \leq n, 0 \leq y \leq m\}$, is introduced following Wang [13], and the smallest interval is derived under this ordering. In Sect. 11.4, we describe how to identify the MED in Example 2 as given in Wang [13].

11.2 The Smallest One-sided Confidence Interval

The following theorem provides the existence and the construction of the smallest interval in $\mathcal{B}_l$.

Theorem 11.1 (Wang [14, Theorem 1]) *Suppose that a vector $\underline{X}$ is observed from a distribution with known cdf $F(\underline{x}; \underline{\theta})$ and unknown parameter vector $\underline{\theta}$. For $\alpha \in [0, 1]$, an ordering $\preceq$ on $R_{\underline{X}}$, and any $\underline{x} \in R_{\underline{X}}$, let*

$$f_{\underline{x}}(\theta) = \inf_{\underline{\eta} \in D(\theta)} \left[1 - P_{(\theta,\underline{\eta})}(\underline{y} \in R_{\underline{X}} : \underline{x} \preceq \underline{y})\right], \tag{11.7}$$

and let

$$G_{\underline{x}} = \{\theta \in [A, B] : f_{\underline{x}}(\theta') \geq 1 - \alpha \ \forall \theta' < \theta\}. \quad (11.8)$$

Define

$$L_S(\underline{x}) = \begin{cases} \sup G_{\underline{x}} & \text{if } G_{\underline{x}} \neq \emptyset, \\ A & \text{otherwise.} \end{cases} \quad (11.9)$$

Then (1) $[L_S(\underline{X}), B] \in \mathcal{B}_l$; (2) $[L_S(\underline{X}), B]$ *is the smallest in* $\mathcal{B}_l$.

Equations (11.7), (11.8), and (11.9) carry out the idea described in the second part of Remark 1.

Remark 2 The assumption on the underlying distribution for Theorem 11.1 being valid is mild. One only needs the set $\{\underline{y} \in R_{\underline{X}} : \underline{x} \preceq \underline{y}\}$ to be σ-measurable so that $f_{\underline{x}}(\theta)$ in (11.7) is well defined. This is implied by condition (c) of the ordering. However, a reasonable ordering may not be easy to obtain if one does not know the underlying distribution and/or the parameter of interest well. When $R_{\underline{X}}$ is finite, the ordering seems relatively easy to obtain. But when $R_{\underline{X}}$ is infinite, it is challenging to derive a reasonable ordering.

Remark 3 Theorem 11.1 can be used as a polishing tool to improve any given interval as follows. For any $1-\alpha$ confidence interval for $[L(\underline{X}), B]$, let $\preceq_L$ be the ordering on $R_{\underline{X}}$ by a function L as in (11.6). Then the smallest $1-\alpha$ confidence interval $[L_S(\underline{X}), B]$ exists under this ordering by Theorem 11.1. Therefore, $L(\underline{x}) \leq L_S(\underline{x})$ for any $\underline{x}$. Also, a complete class of one-sided $1-\alpha$ intervals, under the set inclusion criterion, can be obtained by collecting all $[L_S(\underline{X}), B]$ for any $1-\alpha$ interval $[L(\underline{X}), B]$.

The smallest interval is constructed for any given ordering $\preceq$ on $R_{\underline{X}}$ as in Theorem 11.1. Can this interval be further improved? Yes, if there exists an ordering $\preceq^*$ finer than $\preceq$.

Definition 11.3 For two orderings $\preceq$ and $\preceq^*$ on $R_{\underline{X}}$, $\preceq^*$ is finer than $\preceq$ if $\{\underline{y} \in R_{\underline{X}} : \underline{y} \prec \underline{x}\}$ is a subset of $\{\underline{y} \in R_{\underline{X}} : \underline{y} \prec^* \underline{x}\}$ for any $\underline{x}$ in $R_{\underline{X}}$. An ordering $\preceq$ is finest if $V_{\underline{x}} = \{\underline{y} \in R_{\underline{X}} : \underline{y} \equiv \underline{x}\}$ contains only one element for any $\underline{x} \in R_{\underline{X}}$.

Theorem 11.2 *Let* $[L_S(\underline{X}), B]$ *and* $[L_S^*(\underline{X}), B]$ *be the smallest* $1-\alpha$ *intervals for* θ *with respect to orderings* $\preceq$ *and* $\preceq^*$, *respectively. If* $\preceq^*$ *is finer than* $\preceq$, *then*

$$L_S(\underline{x}) \leq L_S^*(\underline{x}) \quad (11.10)$$

for any $\underline{x} \in R_{\underline{X}}$.

Example 4 Consider two orderings (both unreasonable) on $R_X = \{0, 1, 2, 3\}$, where $X \sim Bin(3, p)$:

$$\preceq_1 : 0 \equiv 1 \equiv 3 \prec 2$$

Table 11.1 Three smallest intervals under three orderings: $\preceq_1$, $\preceq_2$ and $\preceq$

x	0	1	2	3
$L_{S,1}(x)$	0	0	0.13914	0
$L_{S,2}(x)$	0	0.01695	0.13914	0.13535
$L_{S,\leq}(x)$	0	0.01695	0.13535	0.36840

and

$$\preceq_2 : 0 \prec 1 \prec 3 \prec 2$$

and the natural ordering of $\leq$. Following Theorem 11.1, the 95% smallest intervals under these three orderings are obtained and listed in Table 11.1.

The third one is the famous one-sided 95% Clopper and Pearson interval. Since $\preceq_2$ is finer than $\preceq_1$, as implied in Theorem 11.2, $[L_{S,1}(X), 1]$ is uniformly smaller than $[L_{S,2}(X), 1]$. However, the second and third orderings are both finest, so neither one of the two corresponding smallest intervals uniformly dominate the other. A more interesting fact is that the same conclusion holds for the two intervals if the expected length of these intervals are computed for comparison on the precision.

11.3 A Smallest Interval for the Difference of Two Independent Proportions

In this section, we will derive a smallest interval for $\Delta = p_1 - p_0$ in the setting of Example 1. Let $p_B(x; n, p)$ denote the pmf of $Bin(n, p)$. Recall $R_{\underline{X}} = \{\underline{x} = (x, y) : x \in [0, n], y \in [0, m]\}$. Which ordering on $R_{\underline{X}}$ provides an interval that cannot be uniformly improved? Roughly speaking, first, by Theorem 11.1, we prefer an ordering on $R_{\underline{X}}$ that yields a large smallest solution of

$$f_{\underline{x}}(\Delta) = 1 - \alpha \quad \text{for all } \underline{x}\text{'s}, \tag{11.11}$$

where

$$f_{\underline{x}}(\Delta) = \min_{p_0 \in D(\Delta)} \sum_{(x',y') \prec \underline{x}} p_B(x'; n, p_0 + \Delta) p_B(y'; m, p_0); \tag{11.12}$$

secondly, due to Theorem 11.2, each set $V_{\underline{x}}$ would contain only one point; lastly, because of the specialty of binomial distributions, the ordering $\preceq$ should satisfy: (1) $(x_1, y) \preceq (x_2, y)$ (so $L(x_1, y) \leq L(x_2, y)$) for $x_1 \leq x_2$; and (2) $(x, y_2) \preceq (x, y_1)$ (so $L(x, y_2) \leq L(x, y_1)$) for $y_1 \leq y_2$. Let B_B denote the class of all one-sided $1 - \alpha$ intervals for Δ satisfying (1) and (2). We will search for optimal intervals, perhaps admissible ones, from B_B in this section. Let $\preceq_D$ be the desirable ordering on $R_{\underline{X}}$. It is clear that $(n, 0)$ must be the largest among all (x, y)'s under $\preceq_D$. The second largest (x, y) should be either $(n-1, 0)$ or $(n, 1)$, or both (if $n = m$). We, by induction, construct an ordering $\preceq_D$ that satisfies (1) and (2) and starts at point $(n, 0)$

as follows. Step 1: Let $C_1 = \{(n, 0)\}$ be the subset of $R_{\underline{X}}$ that contains the largest value in $R_{\underline{X}}$. Step 2: Let $m_0 = 0$ and $m_1 = 1$. Suppose, by induction, that $\{C_j\}_{j=1}^{k}$ are available for some positive integer k, where

$$C_j = \{(x_i, y_i)\}_{i=m_{j-1}+1}^{m_j}$$

for some nonnegative integers $m_0, m_1, \ldots, m_k$ satisfying: (I) $C_j = \{(x.y) \in R_{\underline{X}} : (x, y) \equiv_D (x_{m_j}, y_{m_j})\}$, and (II) $(x_{m_j}, y_{m_j}) \prec_D (x_{m_{j-1}}, y_{m_{j-1}})$ for each $j \le k$. So C_j contains the jth largest value in $R_{\underline{X}}$ under $\preceq_D$, $L(x, y)$ is a constant on C_j, and $L(x_{m_j}, y_{m_j})$ is nonincreasing in j. Now we determine C_{k+1}, i.e., the set of the $(k+1)$th largest value in $R_{\underline{X}}$ under $\preceq_D$. Let $S_k = \bigcup_{j=1}^{k} C_j$, and let N_k be the "neighbor" set of S_k, i.e.,

$$N_k = \{(x, y) \in S : (x, y) \notin S_k;\ (x+1, y) \in S_k \text{ or } (x, y-1) \in S_k\}.$$

Due to (1) and (2), some points in N_k are disqualified to be in C_{k+1}. To exclude these points, let NC_k be the "candidate" set within N_k satisfying

$$NC_k = \{(x, y) \in N_k : (x+1, y) \notin N_k \text{ and } (x, y-1) \notin N_k\}. \tag{11.13}$$

Therefore, C_{k+1} must be a subset of NC_k, and a point selected from NC_k automatically guarantees (1) and (2). For each point $\underline{x}_0 = (x_0, y_0)$ in NC_k, consider

$$f_{\underline{x}_0}(\Delta) = \min_{p_0 \in D(\Delta)} \sum_{z \in (\{z_0\} \cup S_k)^c} p_B(x; n, p_0 + \Delta) p_B(y; m, p_0).$$

Let

$$E_{\underline{x}_0} = \{\Delta \in [-1, 1] : f_{\underline{x}_0}(\Delta') \ge 1 - \alpha\ \forall \Delta' < \Delta\} \tag{11.14}$$

and

$$L_o(\underline{x}_0) = \begin{cases} \sup E_{\underline{x}_0} & \text{if } E_{\underline{x}_0} \ne \emptyset, \\ -1 & \text{otherwise.} \end{cases} \tag{11.15}$$

Define

$$C_{k+1} = \Big\{\underline{x} \in NC_k : L_o(\underline{x}) = \max_{\underline{x}_0 \in NC_k} L_o(\underline{x}_0)\Big\} \quad \text{and} \tag{11.16}$$

$$m_{k+1} = m_k + \text{ the number of elements in } C_{k+1}. \tag{11.17}$$

Note that C_{k+1} may contain more than one point especially when $n = m$. By induction, an ordering $\preceq_D$ characterized by $\{C_j = \{(x_i, y_i)\}_{i=m_{j-1}+1}^{m_j}\}_{j=1}^{k_0}$ with some positive integer k_0 is constructed. Therefore, the smallest one-sided $1-\alpha$ confidence interval under this ordering, denoted by $[L_S^D(\underline{X}), 1]$, is constructed for estimating Δ following Theorem 11.1.

Remark 4 Although the description of the construction on the ordering $\preceq_D$ is fairly long, the idea behind the construction is very simple. Since $R_{\underline{X}}$ is finite, we easily identify the largest value in $R_{\underline{X}}$, $(n,0)$, by a common sense, and this is the starting point of the entire construction. Given that the largest k values on $R_{\underline{X}}$ are identified, which is the next largest value (the $(k+1)$th largest) on $R_{\underline{X}}$? There must be some points next to the already constructed ones, also yielding the largest possible confidence limit. These two are done by (11.13) and (11.16). Furthermore, If C_j always contains a single point for any j, then, for $\underline{x} \in C_j$, $L_S^D(\underline{x})$ equals the largest of $L_o(\underline{x}_0)$'s in the previous step, and we have the following result.

Proposition 11.1 (*Wang* [13, *Proposition* 3]) *For ordering* $\preceq_D$ *and interval* $[L_S^D(\underline{X}), 1]$ *constructed in Steps* 1 *and* 2, *if each* C_j *contains only one sample point (i.e.,* $m_j = m_{j-1} + 1$ *for all* j*'s), then* $[L_S^D(\underline{X}), 1]$ *is admissible in* B_B. *i.e., for an interval* $[L(\underline{X}), 1] \in B_B$, *if* $L_S(\underline{x}) \le L(\underline{x})$ *for any* $\underline{x} \in R_{\underline{X}}$, *then* $L_S(\underline{x}) = L(\underline{x})$ *for any* $\underline{x} \in R_{\underline{X}}$.

Remark 5 Conditions (1) and (2) were first proposed in Barnard [1] and called the "C" condition. He constructed an optimal rejection region for the hypothesis testing problem

$$H_0(0): \Delta \le 0 \quad \text{vs.} \quad H_A(0): \Delta > 0,$$

using a special ordering on $R_{\underline{X}}$. His ordering satisfies conditions (1) and (2) and is generated also by induction starting at $C_1 = (n,0)$, and is similar to ours, except that he focused on $\Delta = 0$, but we deal with all $\Delta \in [-1, 1]$. Pointed out by Martin Andres and Silva Mato [8], Barnard's test is the (overall) most powerful existing test for comparing two independent proportions. The drawback of this test is the complexity on the determination on the ordering. Ours seems worse. When n and m are large, the determination of ordering $\preceq_D$ requires extensive numerical computation in (11.16).

Example 5 Suppose $n = 4$ and $m = 1$. Now construct the smallest 95% confidence interval with ordering specified by $\{C_j\}_{j=1}^{k_0}$. First, $C_1 = \{(4,0)\}$ following Step 1, and $L_S(4,0) = -0.095$ by solving

$$f_{(4,0)}(\Delta) = \min_{\{p_0 \in D(\Delta)\}} \big(1 - p_X(4; 4, \Delta + p_0) p_Y(0; 1, p_0)\big) = 0.95$$

because $f_{(4,0)}$ now is nonincreasing in Δ. In Step 2, N_1, the neighbor set of $S_1 (= C_1)$, is equal to $\{(3,0), (4,1)\}$, and $NC_1 = N_1$. Following (11.15),

$$L_o(3,0) = -0.345, \qquad L_o(4,1) = -0.527.$$

Thus, $C_2 = \{(3,0)\}$ by (11.16). In Step 3, three sets S_2, N_2 and NC_2 are needed, and they are given in Table 11.2. Note that here $NC_2 \ne N_2$.

Again, for each point in NC_2, we have $L_o(2,0) = -0.561$, $L_o(4,1) = -0.527$, following (11.15). Then $C_3 = \{(4,1)\}$ by (11.16). The rest of the interval construction is given in Table 11.3. Following Remark 4, since each C_j contains a single point, $L_S^D(x,y)$ on C_j is equal to the largest $L_o(\underline{x}_0)$ in the previous step, and is

Table 11.2 Three sets S_2, N_2, NC_2 needed in Step 3 of the interval construction

y	x				
	0	1	2	3	4
1	–	–	–	N_2	N_2, NC_2
0	–	–	N_2, NC_2	S_2	S_2

Table 11.3 The details of the construction of partition $\{C_j\}_{j=1}^{k_0}$ when $n = 4$ and $m = 1$

j	C_j	N_j	NC_j $L_o(\underline{x}_0)$	$L_S^D(x, y)$
1	(4,0)	(3,0), (4,1)	(3,0), (4,1) **−0.345**, −0.527	−0.095
2	(3,0)	(2,0), (3,1), (4,1)	(2,0), (4,1) −0.561, **−0.527**	**−0.345**
3	(4,1)	(2,0), (3,1)	(2,0), (3,1) **−0.578**, −0.752	**−0.527**
4	(2,0)	(1,0), (2,1), (3,1)	(1,0), (3,1) −0.757, **−0.752**	**−0.578**
5	(3,1)	(1,0), (2,1)	(1,0), (2,1) **−0.770**, −0.902	**−0.752**
6	(1,0)	(0,0), (1,1), (2,1)	(0,0), (2,1) −0.950, **−0.902**	**−0.770**
7	(2,1)	(0,0), (1,1)	(0,0), (1,1) **−0.950**, −0.987	**−0.902**
8	(0,0)	(0,1), (1,1)	(1,1) **−0.987**	**−0.950**
9	(1,1)	(0,1)	(0,1) **−1**	**−0.987**
10	(0,1)			**−1**

reported in the last column of Table 11.3, and the construction is complete at the 10th ($= k_0$) step. This interval is admissible in B_B due to Proposition 11.1.

11.4 Identifying the Minimum Effective Dose

In this section, we discuss how to identify the MED in the setting of Example 2. Recall that we observe a sequence of independent binomial random variables $X_i \sim Bin(n_i, p_i)$ for $i = 1, \ldots, k$ and $Y \sim Bin(m, p_0)$, and we are interested in a sequence of hypotheses given in (11.3). The goal is to identify the smallest positive integer i_0 such that H_{0i_0} is not true. This can be done by conducting simultaneous tests for (11.3) with the experimentwise error rate controlled at α.

Note that there is a special structure on $\mathcal{C} = \{H_{0i} : i = 1, \ldots, k\}$: H_{0i} is decreasing in i. Thus $\mathcal{C}$ is closed under the operation of intersection, i.e., the intersection of any two hypotheses in $\mathcal{C}$ still belongs to $\mathcal{C}$. Suppose that a level α nondecreasing (in i) rejection region R_i for H_{0i} is constructed. Then, for a multiple test problem for testing all null hypotheses in $\mathcal{C}$, we define a multiple test procedure: assert H_{Ai} if R_i occurs. This procedure automatically controls the experimentwise error rates at level α following the closed test procedure by Marcus, Peritz, and Gabriel [7].

Now we can apply the interval constructed in Sect. 11.3 to obtain a level α test for H_{0i} with a nondecreasing rejection region in i. Let $L^D_{S,i}(X_i, Y)$ be the smallest one-sided $1-\alpha$ confidence interval for $p_i - p_0$ obtained in Sect. 11.3 before Remark 4. Define a rejection region for H_{0i}

$$R_i = \left\{(x_1, \ldots, x_k, y) : \min_{i \le j \le k} \left\{L^D_{S,j}(x_j, y)\right\} > \delta\right\}. \tag{11.18}$$

It is clear that R_i is nondecreasing in i.

Theorem 11.3 (Wang [13, Theorem 2]) *Rejection region R_i is nondecreasing in i and is of level α for H_{0i}. Therefore, the multiple test procedure, which asserts not H_{0i} (i.e., asserts H_{Ai}) if R_i occurs for any $H_{0i} \in \mathcal{C}$, controls the experimentwise error rate at level α.*

Remark 6 The multiple test procedure with rejection regions $\{R_i\}_{i=1}^k$ in (11.18) is equivalent to the following step-down test procedure.

Step 1. If R_k does not occur, conclude that the *MED* does not exist and stop; otherwise go to the next step.
Step 2. If R_{k-1} does not occur, conclude that $MED = k$ and stop; otherwise go to the next step.

⋮

Step k. If R_1 does not occur, conclude that $MED = 2$ and stop; otherwise conclude that $MED = 1$ and stop.

11.5 Discussion

In this paper, we discuss how to derive the smallest confidence interval under any ordering on the sample points for a parameter of interest in the presence of nuisance parameter(s). The interval construction is based on a direct analysis on the coverage probability and only needs mild assumptions on the underlying distribution besides an ordering on the sample points. A generalization to the two-sided confidence interval construction can also be found in Wang [15]. This indicates that the interval construction based on the coverage probability has potentials to be a general method besides the five methods discussed in Casella and Berger [5, Chap. 9] for constructing confidence intervals. For one-sided intervals, the proposed construction is an automatic algorithm, provided that an ordering is given.

The set inclusion criterion is employed for searching for good intervals because it has a clear interpretation. Under this criterion, the smallest interval is the best in the strongest sense, provided its existence. It is well known that the existence of the best interval depends on the class of intervals from which the best is searched. We successfully characterize such classes by (a) considering one-sided $1-\alpha$ confidence intervals and (b) requiring an ordering on the random confidence limits. Bol'shev and Loginov [3] did not construct the interval under (b). Another general application of the proposed method is to improve any given one-sided $1-\alpha$ confidence interval as explained in Remark 3. As a special application, we discuss how to construct optimal one-sided confidence interval for the difference of two dependent proportions. An ordering similar to Barnard's [1] is derived, however, a computation issue remains. One can also use the proposed method to construct confidence interval for the difference of two dependent proportions. See more details in Wang [14]. The future research question is how to derive an optimal ordering for a given parameter of interest, especially when $R_{\underline{X}}$ is continuous.

References

1. G.A. Barnard. Significance tests for 2×2 tables. *Biometrika*, **34**:123–138, 1947.
2. L.N. Bol'shev. On the construction of confidence limits. *Theory Probab Appl*, **10**:173–177, 1965. (English translation)
3. L.N. Bol'shev and E.A. Loginov. Interval estimates in the presence of nuisance parameters. *Theory Probab Appl*, **11**:82–94, 1966. (English translation)
4. F. Bretz, J.C. Pinheiro, and M. Branson. Combining multiple comparisons and modeling techniques in dose-response studies. *Biometrics*, **61**:738–748, 2005.
5. G. Casella and R.L. Berger. *Statistical Inference*. Duxbury, Belmont, 1990.
6. J.C. Hsu and R.L. Berger. Stepwise confidence intervals without multiplicity adjustment for dose-response and toxicity studies. *J Am Stat Assoc*, **94**:468–482, 1999.
7. R. Marcus, E. Peritz, and K.R. Gabriel. On closed testing procedures with special reference to ordered analysis of variance. *Biometrika*, **63**:655–660, 1976.
8. A.A. Martin and M.A. Silva. Choosing the optimal unconditional test for comparing two independent proportions. *Comput Stat Data Anal*, **17**:555–574, 1994.
9. A.C. Tamhane and C.W. Dunnett. Stepwise multiple test procedures with biometric applications. *J Stat Plan Inference*, **82**:55–68, 1999.
10. A.C. Tamhane, Y. Hochberg, and C.W. Dunnett. Multiple test procedures for dose finding. *Biometrics*, **52**:21–37, 1996.
11. W. Wang. Smallest confidence intervals for one binomial proportion. *J Stat Plan Inference*, **136**:4293–4306, 2006.
12. W. Wang. A Note on one-sided confidence interval for a single-parameter distribution family. Manuscript. Department of Mathematics and Statistics, Wright State University, 2008, submitted.
13. W. Wang. On construction of the smallest one-sided confidence interval for the difference of two proportions. *Ann Stat*, **38**(2):1227–1243, 2010.

14. W. Wang. On construction of the smallest one-sided confidence interval in the presence of nuisance parameter. Manuscript. Department of Mathematics and Statistics, Wright State University, 2009, submitted.
15. W. Wang. On construction of confidence interval for a proportion. Manuscript. Department of Mathematics and Statistics, Wright State University, 2009, submitted.
16. W. Wang and J. Peng. A step-up test procedure to identify the minimum effective dose. Manuscript, 2008.

Chapter 12
Group Variable Selection Methods and Their Applications in Analysis of Genomic Data

Jun Xie and Lingmin Zeng

12.1 Introduction

Regression is a simple but the most useful statistical method in data analysis. The goal of regression analysis is to discover the relationship between a response y and a set of predictors $x_1, x_2, \ldots, x_p$. When fitting a regression model, besides prediction accuracy, parsimony is another important criterion of goodness. Simpler models are preferred by researchers for easier interpretation of the relationship between x and y. Moreover, discarding irrelevant predictors often improves prediction accuracy [10]. Variable selection methods have long been used in regression analysis, for example, forward selection, backward elimination, best subset regression. The number of variables p in the traditional setting is typically 10 or at most a few dozens. Modern scientific technology, led by the microarray, has produced data dramatically above the conventional scale. We have $p = 1{,}000$ to 10,000 in gene expression microarray data, and p up to 500,000 in single nucleotide polymorphism (SNP) data.

To make things more complicated, the large number of variables in the biological data are dependent. For example, it is well known that for genes that share a common biological function or participate in the same metabolic pathway, the pairwise correlations among them can be very high [11]. Traditional variable selection methods that select variables one by one may miss important group effects on pathways. Consequently, when traditional variable selection methods are applied in multiple data sets from a common biological system, the selected variables from the multiple studies may show little overlap. To overcome the challenges, we have developed a series of group variable selection methods, which construct highly correlated genes

J. Xie (✉)
Department of Statistics, Purdue University, 250 N. University Street, West Lafayette, IN 47907, USA
e-mail: junxie@stat.purdue.edu

L. Zeng
MedImmune, 1 MedImmune Way, Gaithersburg, MD 20878, USA

J. Feng et al. (eds.), *Frontiers in Computational and Systems Biology*, Computational Biology 15,
DOI 10.1007/978-1-84996-196-7_12,

into a group and select the whole group once one gene among them is in the model. In this chapter, we introduce the idea of group variable selection and illustrate its utility by applying the methods to genomic data analysis.

12.2 Background

12.2.1 Existing Variable Selection Methods

We consider the linear regression model

$$\mathbf{y} = \mathbf{x}_1\beta_1 + \mathbf{x}_2\beta_2 + \cdots + \mathbf{x}_p\beta_p + \varepsilon,$$

where the response $\mathbf{y}$ is predicted by p predictors $\mathbf{x}_1, \ldots, \mathbf{x}_p$. Without loss of generality, the response and the predictors are all centered so that there is no intercept in the model. Assume that n observations and the error term $\varepsilon = (\varepsilon_1, \ldots, \varepsilon_n)$ are i.i.d. with mean 0 and variance σ^2. The regression model is often expressed in a matrix format

$$\mathbf{y} = \mathbf{X}\beta + \varepsilon,$$

where $\mathbf{y} \in R^n$, $\mathbf{X} \in R^{n \times p}$, and $\beta \in R^p$.

A traditional variable selection method is known as the best subset selection. The procedure first determines a criterion of model goodness, for example, residual sum of squares, adjusted R^2, Mallow's C_p, the Akaike information criterion (AIC), or the Bayesian information criterion (BIC). Then all possible subsets of variables are evaluated by the criterion, and the subset that optimizes the criterion is selected. However, when the number of variables p is large, the best subset selection is computationally intensive. Huo and Ni [3] prove that the best subset selection is an NP-hard (nondeterministic polynomial-time hard) problem. That is, the best subset solution cannot be obtained in computation times as a polynomial of the number of variables. Alternatively, sequential approaches can be used, including forward selection, backward elimination, and stepwise regression. The sequential approaches are computationally less demanding than the best subset selection. However, their heuristic searches of variables cannot guarantee an optimal solution to the regression model.

More recently, penalized least squares methods have been used for variable selection. The most popular one is Lasso (Least absolute shrinkage and selection operator) proposed by Tibshirani [14]. The Lasso estimators are defined by

$$\hat{\beta}_{\text{Lasso}} = \text{argmin}_\beta \|\mathbf{y} - \mathbf{X}\beta\|^2 + \lambda \sum_{j=1}^{p} |\beta_j|,$$

where λ is a nonnegative regularization parameter. The second term of the sum of the absolute regression coefficients is usually called L_1 penalty. Equivalently, Lasso is a constrained ordinary least squares that minimizes

$$\|\mathbf{y} - \mathbf{X}\beta\|^2 \quad \text{subject to} \sum_{j=1}^{p} |\beta_j| \le s,$$

where s is a corresponding regularization parameter. Due to the nature of the L_1 penalty, Lasso shrinks the regression coefficients toward 0 and produces some coefficients that are exactly 0 and hence implements variable selection. Many researchers have studied properties of Lasso [5, 6, 20, 21]. Under certain conditions, Lasso is shown to select the right set of variables with a probability going to 1. However, Lasso's conditions are violated for a group of highly correlated variables. In this situation, Lasso tends to select only one variable from the group and does not care which one is selected.

Efron et al. [1] propose a new variable selection algorithm, Least Angle Regression (LARS), which is a less greedy version of traditional forward selection methods. A special feature of LARS is that a simple modification of the LARS algorithm calculates all possible Lasso estimators but uses computer times an order of magnitude less than Lasso. The efficiency of the LARS algorithm makes it an attractive variable selection method.

Besides Lasso, other penalized least squares approaches have been proposed, using penalty functions more general than the L_1 penalty. Fan and Li [2] define a special penalty function that is singular at the origin to produce sparse coefficient estimators, satisfies certain conditions to produce continuous models, and is bounded by a constant to produce nearly unbiased estimators for large coefficients. Their penalty function is called SCAD. Fan and Li [2] show that, with a proper choice of the regularization parameter, SCAD possesses oracle properties, which are referred to that the probability of selecting the right set of variables (with nonzero coefficients) converges to 1 and that the estimators of the nonzero coefficients are asymptotically normal with the same means and covariances as if the zero coefficients were known in advance. Kim et al. [4] also apply SCAD in certain high-dimensional data.

The traditional variable selection methods and the later additions Lasso, LARS, and SCAD do not select variable groups. In fact, they all ignore the correlation between the variables. Elastic net proposed by Zou and Hastie [23] is the first variable selection method that works for groups of predictors. Elastic net is also a member of penalized least squares. The penalty function is a linear combination of L_1 and L_2 penalties. By introducing a L_2 penalty term, elastic net encourages strongly correlated variables to be in or out of the model at the same time. This phenomenon is termed "grouping effect." In theory, a strictly convex penalty function provides a sufficient condition for the grouping effect. The L_2 penalty guarantees strict convexity. On the other hand, elastic net does not reveal the underlying group structure in its solution and does not possess the properties introduced by Fan and Li [2] in SCAD.

When people have prior knowledge on variable groups, group Lasso proposed by Yuan and Lin [18] is designed to select predefined groups of predictors. Suppose that p predictors are divided into J groups with sizes $k_1, \dots, k_J$. The group Lasso estimators are obtained by minimizing

$$\left\| \mathbf{y} - \sum_{j=1}^{J} \mathbf{X}_j \beta_j \right\|^2 + \lambda \sum_{j=1}^{J} \|\beta_j\|_{K_j},$$

where λ is the regularization parameter, and $\|z\|_K = (z' K z)^{1/2}$ with a symmetric $k \times k$ positive definite matrix K. The positive definite matrices $K_1, \dots, K_J$ for the J groups can be chosen as identity matrices of sizes $k_1, \dots, k_J$, respectively. Yuan and Lin [18] also propose group LARS as an extension of LARS. Group Lasso and group LARS have been used in multifactor ANOVA models, in which each factor may have several levels and can be expressed through a group of dummy variables.

In addition to the frequentist methods, Bayesian interpretations of the penalized regression Ridge and Lasso have been proposed. It can be shown that the Bayesian estimators of the coefficients $\beta_1, \dots, \beta_p$ are equivalent to Ridge when we assume normal prior distributions for β's, and are equivalent to Lasso when we assume Laplace prior distributions [8, 17]. Bayesian approach offers an alternative framework of variable selection. Theoretically, Bayesian methods can deal with high-dimensional inter-correlated variables through generalized prior distributions. In practice, Bayesian methods will encounter the same difficulty as its frequentist counterpart.

12.2.2 Large Scale Genomic Data

High-throughput gene expression microarray techniques have now been routinely used in biological applications. An array measures expression levels of thousands of genes simultaneously. Differences between experiment conditions (treatments) are implied by expression variations of a large number of genes. In medical research, microarray is used to detect associations between gene expression profiles and clinical outcomes, for example, cancer types or stages. Consider a clinical outcome as the response variable y and all genes measured in the microarray as the predictor variables $x_1, \dots, x_p$. Then the variables are of high dimension, with complicated dependent structures. Identifying a subset of significant genes that affect the clinical outcome will be a good application of our proposed group selection methods.

In one of our previous projects, we have developed a suite of statistical methods [19] for inferring cis-regulatory modules, which are groups of transcription factors binding in the promoter regions to regulate gene expression. Our approach is an integrative analysis that combines information from multiple types of biological data, including genomic DNA sequences, genome-wide location analysis (ChIP-chip experiments), and gene expression microarray. We first use a hidden Markov model by Wu and Xie [16] to predict a cluster of transcription factor binding sites in DNA sequences. The predictions are refined by regression analysis on gene expression microarray data and/or ChIP-chip binding experiments. We have constructed a regression model that describes a gene of interest as a function of its TFs. The response variable is the gene expression level. The predictor variables are the TF binding levels approximated by the TF gene expression values. We view a combinatorial effect of multiple TFs on the gene through multiple regression analysis. However, the difficulty is to select an appropriate set of TFs which has significant effects on the gene.

Due to complicated dependence among TFs, the problem of selecting TF covariates for a gene posts a challenge to the standard variable selection procedures. Consider a regression example of gene ACE2 versus a set of TFs consisting of Fkh1, Fkh2, Mcm1, Ndd1, Swi4, and Swi6 using Spellman et al.'s [13] yeast cell cycle microarray data. It is known that ACE2 was bound by Fkh1, Fkh2, and the complex Mcm1/Fkh2/Ndd1 [12]. Hence, a good variable selection method is to select the corresponding four TFs as much as possible. The expression levels of FKH1 and FKH2 are highly correlated with a correlation coefficient of 0.63. Using forward selection, Lasso, and elastic net, Fkh2 always enters the model first. However, the standardized regression coefficient of Fkh2 is 0.677 and that of Fkh1 is −0.045 when both Fkh2 and Fkh1 are in the model. The available methods fail to select both Fkh2 and Fkh1 as a group of covariates. In fact, variable selection in regression analysis tends to keep only one variable in the model, whenever there is a group of highly correlated covariates. The group variable selection methods attempt to solve this problem.

Another application of the proposed group variable selection methods is SNP data analysis. SNPs are the most common genetic variations in the human genome and occur once in several hundred base pairs. An SNP is a position at which two alternative bases occur at appreciable frequency ($>1\%$) in the human population. The NCBI dbSNP database currently stores 5 million human SNPs identified by comparing the DNA of different individuals, making it possible to use them for genome-wide SNP genotyping. SNPs can serve as genetic markers for identifying disease genes by linkage studies in families, linkage disequilibrium in isolated populations, association analysis of patients and controls, and loss-of-heterozygosity studies in tumors [15]. Oligonucleotide SNP microarrays have been developed for high-throughput genotyping of human SNPs with marker number ranging from 10,000 (Mapping 10-K array) to 500,000 (Mapping 500-K array set). With the technique advances, genome-wide association studies become popular to detect specific DNA variants that contribute to human phenotypes and particularly human diseases. In SNP data analysis, we assume a phenotype of interest as the response variable y, and a large number of SNPs as the predictor variables. The proposed group variable selection methods will be used to identify genetic variants that associate with variation in the phenotype.

12.3 gLars and gRidge Algorithms

Consider the linear regression model

$$\mathbf{y} = \mathbf{X}\beta + \varepsilon,$$

where $\mathbf{y} = (y_1, y_2, \ldots, y_n)^T$ is the response variable, $\mathbf{X} = (\mathbf{x}_1, \mathbf{x}_2, \ldots, \mathbf{x}_p)$ is the predictor matrix, and ε is a vector of independent and identically distributed random errors with mean 0 and variance σ^2. There are n observations and p predictors. We center the response variable and standardize the column vectors of the predictor matrix. Hence, there is no intercept in our model:

$$\sum_{i=1}^{n} y_i = 0, \qquad \sum_{i=1}^{n} x_{ij} = 0, \qquad \sum_{i=1}^{n} x_{ij}^2 = 1 \quad \text{for } j = 1, 2, \ldots, p.$$

The LARS algorithm proposed by Efron et al. [1] is a less greedy forward model selection procedure. At the beginning of LARS, a predictor enters the model if its absolute correlation with the response is the largest one among all the predictors. The coefficient of this predictor grows in its ordinary least squares direction until another predictor has the same correlation with the current residual (i.e., equal-angle). Next, both coefficients of the two selected predictors begin to move along their ordinary least squares directions until a third predictor has the same correlation with the current residual as the first two. The whole process continues until all predictors enter the model. In each step, one variable adds into the model, and the solution paths, which are the coefficient estimators as functions of the tuning parameter (defined later in Formula (12.1)), are extended in a piecewise linear fashion. After all variables enter the model, the whole LARS solution paths complete.

For data with dependent structures, we propose gLars and gRidge algorithms that construct groups simultaneously along the variable selection process. We first give a grouping definition. Predictors form a group if they satisfy both of the two criterions:

- They are highly correlated with the response variable (or current residual).
- They are highly correlated with each other.

The correlation thresholds for the two criterions will be determined from the data. For instance, the threshold of the first criterion is suggested to be the 75th percentile of all correlations (in absolute value) between the current residual and unselected predictors. The correlation threshold (absolute value) for the second criterion is either the 75th percentile of all pairwise correlations among the predictors or chosen from a set of grids, for example, 0.9, 0.8, 0.7, 0.6. An important difference of the proposed method from the standard forward selection procedures is that our variable selection criterion has two components and hence is defined by a region in the two-dimensional space, (t_1, t_2) in Step 3 in the following algorithm. In addition, the first requirement of selecting a variable highly correlated with the response variable is not affected by collinearity among predictors.

In the gLars algorithm, we start as LARS to select a predictor which has the largest correlation with the response. We call this predictor a "leader element." We then build a group based on this leader element and the current residual according to the two grouping criterions. Note that both criterions have to be satisfied when selecting a variable into a group. Once a group has been constructed, it will be represented by a unique direction in R^n as the linear combination of the ordinary least squares directions of all variables in the group. Next, we choose another leader element, analogous to the equal-angle requirement of the LARS algorithm. A new group is formed again following the grouping definition. We refine the solution paths in a piecewise linear format. The whole process continues until all predictors enter the model. The detailed algorithm is described below.

1. Initialization: Set the step index $k = 1$, $\beta^{[0]} = 0$, residual $r^{[0]} = Y$, active set $A_0 = \emptyset$, inactive set $A_0^C = \{X_1, X_2, \ldots, X_p\}$.
2. Identify the leader predictor x for the first group, where $x = \text{argmax}_{x_i} |x_i' r^{[k-1]}|$, $x_i \in A_{k-1}^C$.
3. Construct the group G_k with the leader predictor x from Step 2 according to the two criterions: $x_j' r^{[k-1]} > t_1$ and $x_j' x > t_2$, $x_j \in A_{k-1}^C$, where $t_1 = 0.75$th percentile of all correlations between x_j and $r^{[k-1]}$, and $t_2 = t \in \{0.9, 0.8, 0.7, 0.6\}$.
   Set $A_k = A_{k-1} \cup G_k$, $A_k^C = A_{k-1}^C \setminus G_k$.
4. Compute the current direction γ with components
$$\gamma_{A_k} = \left(X_{A_k}' X_{A_k}\right)^{-1} X_{A_k}' r^{[k-1]}, \qquad \gamma_{A_k^C} = 0,$$
   where X_{A_k} denotes the matrix comprised of the columns of $\mathbf{X}$ corresponding to A_k.
5. Calculate how far the gLars algorithm progresses in direction γ. It divides into two small steps:
   Find $x_{j'}$ in A_k^C which corresponds to the smallest $\alpha \in (0, 1]$ such that
$$\frac{\|X_{G_j}'(r^{[k-1]} - \alpha X\gamma)\|_{L_1}}{p_j} = \left|x_{j'}'\left(r^{[k-1]} - \alpha X\gamma\right)\right|,$$
   where G_j is a group from A_k, p_j is the number of variables in group G_j, and $\|.\|_{L_1}$ represents the sum of absolute values.
   Justification. As in Step 3, find the group with the leader predictor $x_{j'}$ selected above and denote the group as $X_{G_{j'}}$. Recalculate $\alpha \in (0, 1]$ for this selected new group such that
$$\frac{\|X_{G_j}'(r^{[k-1]} - \alpha X\gamma)\|_{L_1}}{p_j} = \frac{\|X_{G_{j'}}'(r^{[k-1]} - \alpha X\gamma)\|_{L_1}}{p_{j'}}.$$
   Update $\beta^{[k]} = \beta^{[k-1]} + \alpha\gamma$, $r^{[k]} = Y - X\beta^{[k]}$.
6. Update k to $k + 1$, and $A_k = A_{k-1} \cup G_{j'}$, $A_k^C = A_{k-1}^C \setminus G_{j'}$.
7. If $A_k^C \neq \emptyset$, return to Step 4. Otherwise, set γ, β, and r to be the OLS solutions and stop.

Ordinary least squares would perform poorly when the correlations among the predictors are high and/or the noise level is high. Since both LARS and gLars move towards ordinary least squares direction in each step, they face the same shortage. Ridge estimators, on the other hand, perform better in this situation. We propose a gRidge algorithm, which moves towards ridge estimator direction in each step. The relationship between ridge estimator $\hat{\beta}(\lambda)$ and ordinary least squares estimator $\hat{\beta}$ can be shown as

$$\hat{\beta}(\lambda) = (X'X + \lambda I)^{-1}X'Y = \left(X'X\left(I + \lambda(X'X)^{-1}\right)\right)^{-1}X'Y$$
$$= \left(I + \lambda(X'X)^{-1}\right)^{-1}\hat{\beta} = C\hat{\beta},$$

where $C = (I + \lambda(X'X)^{-1})^{-1}$, and λ is the ridge parameter. The gRidge algorithm is thus a simple modification of the gLars algorithm. When a group is constructed, gRidge represents the group by a unique direction from the linear combination of the ridge directions of all variables in the group. The variable coefficients are moving towards the ridge directions.

As we run simulations, we notice that gRidge outperforms other methods in terms of relative prediction errors (RPEs, defined below in the simulations). However, this method is limited by its comparably larger false positives due to an over-grouping effect. We propose to add a hard threshold δ to gRidge estimators so that small (but nonzero) coefficients will be removed, i.e., $\tilde{\beta}_j = \hat{\beta}_j I\ (\hat{\beta}_j > \delta)$. Based on simulations, we define the threshold $\delta = \sqrt{\sigma \log(p)/n}$. Hence smaller error term, smaller number of predictors, or larger sample size give smaller threshold. We name the modified gRidge algorithm gRidge_new, with this hard threshold filtering. Simulation studies show that gRidge_new not only preserves low RPE but also greatly reduces false positives.

Both gLars and gRidge produce the entire piecewise linear solution paths as LARS does. Groups of variables are selected when we stop the paths after a certain number of steps. The number of step k is the tuning parameter. Equivalently, we may use a tuning parameter as the fraction of the L_1 norm of the coefficients

$$s = \Sigma_{j\_\mathrm{selected}} \left\|\hat{\beta}_j\right\|_{L_1} / \Sigma_j \left\|\hat{\beta}_j\right\|_{L_1}. \tag{12.1}$$

For gLars, s (or k) is the only tuning parameter. It is determined by a standard five-fold cross-validation (CV). For gRidge, there are two tuning parameters, the ridge parameter λ in addition to s (or k). Similar to elastic net, we cross-validate on two dimensions. First, we choose a grid for λ, say $\{0.01, 0.1, 1, 10, 100, 1000\}$. Then for each λ, gRidge produces the entire solution path. The parameter s (or k) is selected by five-fold CV. At the end, we choose the λ value which gives the smallest CV error.

12.3.1 Simulation Studies

Simulation studies are used to compare the proposed gLars and gRidge with ordinary least squares, ridge regression, LARS, and elastic net. The simulated data are generated from the true model $\mathbf{y} = \mathbf{X}\beta + \sigma\varepsilon$, $\varepsilon \sim N(0, 1)$. We have studied many examples for different scenarios but only present four here due to the space limit. For each example, we simulate 100 data sets. Each data set consists of a training set and a test set. The tuning parameters are selected on the training set by five-fold cross-validation. The variable selection methods are compared in terms of relative prediction error (RPE) [22] and selection accuracy on the test set. The relative prediction error is defined as $\text{RPE} = (\hat{\beta} - \beta)^T \Sigma (\hat{\beta} - \beta)/\sigma^2$, where Σ is the population covariance matrix of $\mathbf{X}$. The four scenarios are given by:

Example 1 (Adopted from [23]) There are 100 and 200 observations in the training and test sets, respectively. The true parameter $\beta = (3, 1.5, 0, 0, 2, 0, 0, 0)$, and $\sigma = 3$. The pairwise correlation between $\mathbf{x}_i$ and $\mathbf{x}_j$ is set to be $corr(\mathbf{x}_i, \mathbf{x}_j) = 0.5^{|i-j|}$. This example creates a sparse model with a few large effects, and the covariates have first-order autoregressive correlation.

Example 2 (Adopted from Daye and Jeng unpublished) We simulate 100 and 400 observations in the training and test sets, respectively. We set the true parameters as

$$\beta = (\underbrace{3, \dots, 3}_{15}, \underbrace{1.5, \dots, 1.5}_{5}, \underbrace{0, \dots, 0}_{20})$$

and $\sigma = 6$. The predictors are generated as

$$\begin{aligned}
&\mathbf{x}_i = Z + \varepsilon_i^x, \quad Z \sim N(0, 1), \quad i = 1, \dots, 15, \\
&\mathbf{x}_i \sim N(0, 1), \quad i.i.d., \quad i = 16, \dots, 40,
\end{aligned}$$

where ε_i^x are independent identically distributed $N(0, 0.01)$, $i = 1, \dots, 15$. This example creates one group from the first 15 highly correlated covariates. The next five covariates are independent but provide signals on the response variable.

Example 3 (Adopted from [23]) We simulate 100 and 400 observations in the training and test sets, respectively. We set the true parameters as

$$\beta = (\underbrace{3, \dots, 3}_{15}, \underbrace{0, \dots, 0}_{25})$$

and $\sigma = 15$. The predictors are generated as

$$\begin{aligned}
&\mathbf{x}_i = Z_1 + \varepsilon_i^x, \quad Z_1 \sim N(0, 1), \quad i = 1, \dots, 5, \\
&\mathbf{x}_i = Z_2 + \varepsilon_i^x, \quad Z_2 \sim N(0, 1), \quad i = 6, \dots, 10, \\
&\mathbf{x}_i = Z_3 + \varepsilon_i^x, \quad Z_3 \sim N(0, 1), \quad i = 11, \dots, 15, \\
&\mathbf{x}_i \sim N(0, 1), \quad i.i.d., \quad i = 16, \dots, 40,
\end{aligned}$$

where ε_i^x are independent identically distributed $N(0, 0.01)$, $i = 1, \dots, 15$. There are three equally important groups with five members in each. There are also 25 noise variables.

Table 12.1 Median relative prediction errors (RPE) and median number of nonzero coefficients/median number of zero coefficients misspecified as nonzero coefficients for the four examples based on 100 replications. The best results are emphasized in italic fonts

Methods	Example 1		Example 2		Example 3		Example 4	
OLS	0.5843	3/5	0.6364	20/20	0.6390	15/25	0.6458	9/31
Ridge	0.2832	3/5	0.2519	20/20	0.0993	15/25	0.1971	9/31
LARS	0.4640	*3/0*	0.3208	12/1	0.1620	6/2	0.1200	3/6
Elastic net	*0.1714*	3/1	0.2587	15/2	0.0800	*15/1*	0.1110	7/8
gLars	0.2616	3/2	0.4235	*20/3*	0.2220	*15/3*	0.2121	*9/8*
gRidge	0.1806	3/3	*0.1963*	20/12	*0.0700*	15/10	*0.0700*	9/13
gRidge_new	0.1816	3/1	*0.1988*	*19/3*	*0.0700*	*15/2*	*0.0690*	*9/8*

Example 4 We simulate 100 and 200 observations in the training and test sets, respectively. We set the true parameters as

$$\beta = (\underbrace{3,3,3,0,0}_{5},\underbrace{3,3,3,0,0}_{5},\underbrace{3,3,3,0,0}_{5},\underbrace{0,\dots,0}_{25}).$$

The predictors and the error terms are the same as in Example 3. There are also three equally important groups with five members in each of them. However, in each group, there are two noise variables, which have no effect on the response variable but are highly correlated with the other three important variables. There are totally 31 noise variables.

Table 12.1 summarizes the prediction results. The median RPE from 100 simulations is reported. The smallest RPE is emphasized in italic font, which indicates the most accurate method for each example. We also report the median number of nonzero coefficients versus the median number of zero coefficients misspecified as nonzero, which imply the true positive and false positive of a method. The simulation results indicate that LARS tends to produce very sparse models but does not work for collinearity. Elastic net improves LARS when predictors are correlated. But elastic net misses the five true signals with the small coefficients 1.5 in Example 2. The first proposed method gLars improves elastic net in terms of true positives, especially in Examples 2 and 4. gRidge and gRidge_new produce the smallest RPEs in all the examples and therefore are the most accurate models in terms of prediction. We also notice that while preserving the large coefficients close to the true coefficients, gRidge tends to select more variables than elastic net, due to its over grouping effect. After we add a hard threshold to gRidge, the gRidge_new estimators achieves the best performance.

12.4 Unbiased Variable Selection via SCAD_$\ell 2$

SCAD is proposed by Fan and Li [2] as a variable selection method via penalized least squares. The SCAD penalty function is specially defined to satisfy three properties for the coefficient estimators: unbiasedness, sparsity, and continuity. To address the challenges of genomic data analysis, we add another property of grouping effect and propose a new penalty function named SCAD_$\ell 2$. Instead of defining grouping criterions as we have proposed in gLars and gRidge, we achieve the grouping effect in SCAD_$\ell 2$ through a strictly convex penalty function, which is a linear combination of the L_2 norm and the SCAD function. More specifically, we propose a naive SCAD_$\ell 2$ estimator $\hat{\beta}_{\text{naive}}$ as the minimizer of the penalized least squares function

$$Q(\beta) = \frac{1}{2}\|\mathbf{y} - \mathbf{X}\beta\|^2 + \sum_{j=1}^{p} f_{\lambda_1}(\beta_j) + \lambda_2 \|\beta\|^2, \tag{12.2}$$

where $\|\beta\|^2 = \sum_{j=1}^{p} \beta_j^2$, and $f_\lambda(\theta)$ is the SCAD function defined as

$$f_\lambda(\theta) = \begin{cases} \lambda|\theta| & \text{if } 0 \le |\theta| < \lambda, \\ -\dfrac{\theta^2 - 2a\lambda|\theta| + \lambda^2}{2(a-1)} & \text{if } \lambda \le |\theta| < a\lambda, \\ (a+1)\lambda^2/2 & \text{otherwise.} \end{cases}$$

Here a is a real number larger than 2. Under the condition that the columns of $\mathbf{X}$ are orthonormal, we can obtain the explicit expression of the naive SCAD_$\ell 2$ estimator $\hat{\beta}_{\text{naive}}$. Specifically, $\hat{\beta}_{\text{naive}} = \hat{\beta}_{\text{OLS}}/(1+2\lambda_2)$ for large $|\hat{\beta}_{\text{OLS}}|$ and hence is a biased estimator. The true SCAD_$\ell 2$ estimator $\hat{\beta}_{\text{SCAD\_}\ell 2}$ is defined as $\hat{\beta}_{\text{SCAD\_}\ell 2} = (1+2\lambda_2)\hat{\beta}_{\text{naive}}$, to attain unbiasedness.

For a general predictor matrix $\mathbf{X}$ not orthonormal, including situations with correlated predictors, SCAD_$\ell 2$ estimator is defined by the naive SCAD_$\ell 2$ estimator multiplying a matrix depending on λ_2 and the covariance matrix of $\mathbf{X}$. We can show that the SCAD_$\ell 2$ estimator satisfies the following four properties:

1. *Unbiasedness*: $\hat{\beta}_{j,\text{SCAD\_}\ell 2} = \hat{\beta}_{j,\text{OLS}}$ for large components of $|\hat{\beta}_{j,\text{OLS}}|$.
2. *Sparsity*: $\hat{\beta}_{j,\text{SCAD\_}\ell 2} = 0$ when $|\hat{\beta}_{j,\text{OLS}}|$ is small.
3. *Continuity*: $\hat{\beta}_{\text{SCAD\_}\ell 2}$ is a continuous function with respect to $\hat{\beta}_{\text{OLS}}$.
4. *Grouping effect*: Two coefficients $\hat{\beta}_{j,\text{SCAD\_}\ell 2}$ and $\hat{\beta}_{i,\text{SCAD\_}\ell 2}$ tend to be equal if the two respective variables $\mathbf{x}_j$ and $\mathbf{x}_i$ are highly correlated.

Following Fan and Li's [2] discussion, for sparsity, it is sufficient to prove that $\min_{\theta \ne 0}\{(|\theta| + p'_\lambda(|\theta|)\} > 0$; and for continuity, it is sufficient to prove that $\text{argmin}_\theta\{|\theta| + p'_\lambda(|\theta|)\} = 0$, where $p_\lambda(|\theta|)$ is the penalty function of SCAD_$\ell 2$ as defined by the last two terms in Formula (12.2). To prove the grouping effect, we use the fact that the penalty function is strictly convex. In addition, let $\hat{\beta}_{i,\text{naive}}$ and $\hat{\beta}_{j,\text{naive}}$ denote the ith and jth elements of $\hat{\beta}_{\text{naive}}$, respectively. Define $D(i,j) = |\hat{\beta}_{i,\text{naive}} - \hat{\beta}_{j,\text{naive}}|/|\mathbf{y}|$. The following theorem implies that strongly correlated variables will be in or out of model together through SCAD_$\ell 2$.

Theorem 12.1 *Assuming* $\hat{\beta}_{i,\text{naive}} \cdot \hat{\beta}_{j,\text{naive}} > 0$ *and a regularity condition for* λ_2, *we have*

$$D(i, j) \leq C \cdot \sqrt{2(1-\rho)},$$

where C *is a constant that may depend on* λ_2, *and the sample correlation* $\rho = \mathbf{x}_i^T \mathbf{x}_j$.

The quantity $D(i, j)$ measures the difference between the coefficients of two predictors $\mathbf{x}_i$ and $\mathbf{x}_j$. In an extreme case where the absolute value of the correlation between the two predictors is close to 1, Theorem 12.1 guarantees that the coefficients of the two predictors will be almost identical except the sign difference. In other words, naive SCAD_$\ell 2$ has the group effect. The true SCAD_$\ell 2$ estimator equals a scalar multiplying the naive SCAD_$\ell 2$ estimator. Therefore, SCAD_$\ell 2$ has the grouping effect as well.

We establish asymptotic theories for SCAD_$\ell 2$, when the number of variables p is fixed and the sample size n goes to infinity. Note that the larger n becomes, the heavier the least squares part in Formula (12.2) weighs. As an adjustment, we consider the following penalized least squares function:

$$Q(\beta) = \frac{1}{2}(\mathbf{y} - \mathbf{X}\beta)^T(\mathbf{y} - \mathbf{X}\beta) + n\sum_{j=1}^{p} f_{\lambda_1}(\beta_j) + n\lambda_2\|\beta\|^2.$$

Let $\beta^* = (\beta_1^*, \ldots, \beta_p^*)^T$ denote the true value of β in the linear regression problem. Without loss of generality, we assume the first p_1 elements $\beta_1^*, \ldots, \beta_{p_1}^*$ are nonzeros, and the remaining $p - p_1$ elements are zeros. Denote $\beta_N^* = (\beta_1^*, \ldots, \beta_{p_1}^*)^T$ and $\beta_Z^* = (\beta_{p_1+1}^*, \ldots, \beta_p^*)^T$. We use $\hat{\beta}(n) = (\hat{\beta}_1(n), \ldots, \hat{\beta}_p(n))^T$ to denote the minimizer of $Q(\beta)$ and denote $\hat{\beta}_N(n) = (\hat{\beta}_1(n), \ldots, \hat{\beta}_{p_1}(n))^T$ and $\hat{\beta}_Z = (\hat{\beta}_{p_1+1}(n), \ldots, \hat{\beta}_p(n))^T$ as the estimators of the nonzero and zero coefficients, respectively. We rewrite λ_1 and λ_2 as $\lambda_1(n)$ and $\lambda_2(n)$ to emphasize that they vary as n changes. The following asymptotic theorems hold.

Theorem 12.2 (Estimation consistency) *If* $\lambda_1(n) \to 0$ *and* $\sqrt{n}\lambda_2(n) \to 0$ *as* $n \to \infty$, *then there exists a local minimizer* $\hat{\beta}(n)$ *of* $Q(\beta)$ *such that* $\|\hat{\beta}(n) - \beta^*\| = \mathcal{O}_p(n^{-1/2})$.

Theorem 12.3 (Selection consistency) *If* $\lambda_1(n) \to 0$, $\sqrt{n}\lambda_1(n) \to +\infty$, *and* $\sqrt{n}\lambda_2(n) \to 0$ *as* $n \to +\infty$, *then* $\lim_{n\to\infty} Prob\{\hat{\beta}_Z(n) = 0\} = 1$.

Theorem 12.4 (Oracle property) *If* $\lambda_1(n) \to 0$, $\sqrt{n}\lambda_1(n) \to +\infty$, *and* $\sqrt{n}\lambda_2(n) \to 0$ *as* $n \to +\infty$, *then the root-n consistent local minimizer* $\hat{\beta}(n) = \binom{\hat{\beta}_N(n)}{\hat{\beta}_Z(n)}$ *satisfies the following with probability tending to* 1:

1. *Sparsity*: $\hat{\beta}_Z(n) = \mathbf{0}$.
2. *Asymptotic normality*: $\sqrt{n}(\hat{\beta}_N(n) - \beta_N^*) \xrightarrow{D} \mathcal{N}(\mathbf{0}, \sigma^2 \Sigma_N^{-1})$, *where* Σ_N *is the covariance matrix of the first* p_1 *predictors.*

The basic ideas in the proofs of these asymptotic theorems include applying Taylor expansion of the penalized least squares function $Q(\beta)$ and the law of large number or the central limit theorem. These results build strong theoretical backgrounds for the proposed group variable selection method. To implement SCAD_$\ell 2$, we use local quadratic approximation similar to the algorithm of SCAD.

12.5 Applications in Genomic Data Analysis

12.5.1 SNP Data Analysis

The proposed group variable selection methods are particularly useful in high-dimensional data with dependent structures, for instance, gene expression microarray data, genetic variation SNP data, and transcription factor binding ChIP-chip data. Statisticians have been playing important roles in gene expression microarray data analyses in the past decade. With the advance of SNP techniques and the stride in SNP detections and the international HapMap project, SNP data analysis becomes another interesting field for statisticians to explore.

As an initial example, we study genetic variation (SNPs) for human gene expression. Natural variation in the baseline expression of many genes can be considered as heritable traits. Morley et al. [7] have collected microarray and SNP data to localize the DNA variants that contribute to the expression phenotypes. The data consists of 14 families with 56 unrelated individuals (the grandparents). There are $\sim$8,500 genes on the array and 2,756 SNP markers genotyped for each individual. The expression level of a given gene is the response variable and the 2,756 SNP markers are the predictors in our model. We apply the proposed group selection methods to search for optimal set of SNPs for gene ICAP-1A, which is the top gene in Morley et al.'s [7, Table 1] with the strongest linkage evidence.

We code each SNP as 0, 1, 2 for wild-type homozygous, heterozygous, and mutation (rare) homozygous genotypes, respectively, according to the genotype frequency. We first screen data to exclude SNPs that have a call rate less than 95% or minor allele frequency less than 2.5%. The number of SNPs is reduced to 1,739 after screening. Then we select 500 most "variable" SNPs as the potential predictors. The variability of an SNP is measured by its sample variance.

We split data into the training set with 42 observations and the test set with 14 observations. Model fitting and choices of the tuning parameters are based on the training set. The first grouping criterion is set up to be the 75th percentile of all correlations between x and y. The second grouping criterion requires the correlation with the leader element to be greater than 0.6. The prediction error (residual sum of squares) is evaluated on the test data. Table 12.2 shows that gLars and gRidge have lower prediction errors than LARS and elastic net with about 30 SNPs selected for gene ICAP-1A. We notice that R^2 of the LARS fitted model with 24 SNP covariates is 0.779, which supports the hypothesis of Morley et al.'s [7] study that gene expression phenotypes are controlled by genetic variants. On the other hand,

Table 12.2 Test prediction errors of Lasso, elastic net, gLars, and gRidge for the SNP data

Methods	Test prediction error	Number of genes	Tuning parameter s
LARS	1.569	24	0.4343
Elastic net ($\lambda_2 = 0.01$)	1.872	23	0.2323
gLars	1.360	33	0.3838
gRidge ($\lambda_2 = 0.01$)	1.531	28	0.4141

Table 12.3 The first 12 steps of predictors selected by Lasso, elastic net, gLars, and gRidge for the SNP data

Methods	LASSO	Elastic net	gLars	gRidge
Step 1	458	458	458	458
Step 2	481	481	481	481
Step 3	321	321	321,131	321,131
Step 4	287	287	287	287
Step 5	240	240	240	240
Step 6	76	76	406	406
Step 7	406	131	76	76
Step 8	131	406	102	102
Step 9	345	345	345	345
Step 10	102	102	498	498
Step 11	498	498	30,27	30,27
Step 12	30	30	167	167

the coefficient estimators of all SNP covariates are very small, in the scale of 0.01, suggesting small additive effects of multiple SNPs.

Table 12.3 lists the first 12 steps that the predictors are selected in each algorithm. The numbers in the table are the indices of the variables. For instance, 458 in Step 1 means that variable x_{458} enters the model at the first step. At Step 3, gLars and gRidge depart from LARS and elastic net due to the grouping effect. The first group consists of two SNPs, x_{321} SNP rs1004620, and x_{131} SNP rs1868237. The correlation of these two variables is 0.65. The two SNPs are in chromosome 3 with 14 K base pairs apart. They are in an intergenic region. The two closest genes have no functional annotation. Another group consisting of two SNPs x_{30} rs1882600 and x_{27} rs1001396 are selected by gLars and gRidge at Step 11. These two SNPs are in chromosome 7 with over 2.5 million base pairs apart. SNP x_{30} rs1882600 is in an intergenic region, whereas x_{27} rs1001396 resides in the gene FOXK1. According to Swiss-Prot functional annotation, FOXK1 is a transcriptional regulator that binds to the upstream of myoglobin gene.

Our results suggest more SNP associations with a gene expression phenotype than the simple linkage analysis. For example, variables x_{240} SNP rs1446297 and x_{76} SNP rs933602 are jointly selected as important covariates for the expression

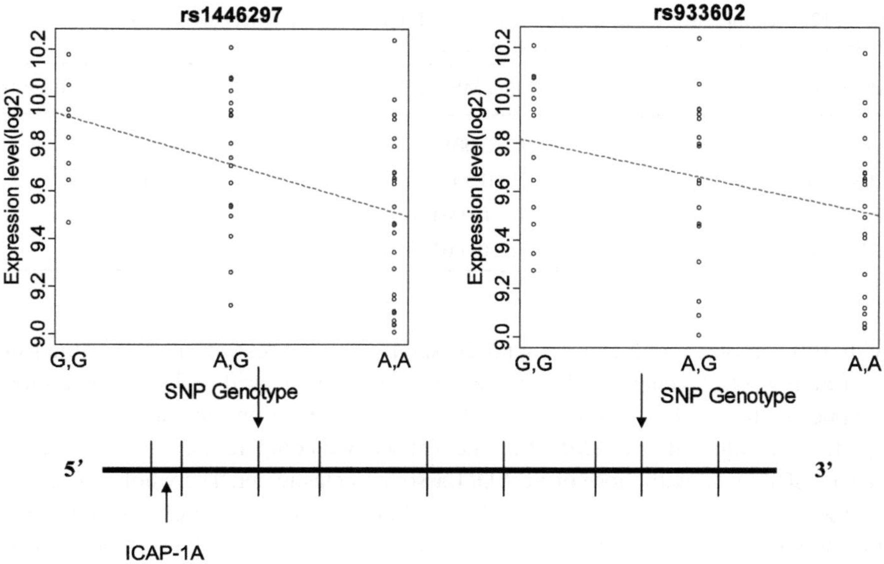

Fig. 12.1 Regression of the expression phenotype of ICAP-1A on the two nearby SNPs

of ICAP-1A according to all four variable selection methods. However, they are not significant in simple regression analysis with a p-value cutoff 0.01. These two SNPs locate in the same chromosome as ICAP-1A (chromosome 2) with nearly 27 million base pairs and 220 million base pairs, respectively, away from ICAP-1A. Figure 12.1 shows their locations and regression plots. SNP rs1446297 is in the promotor region (about 200 base pairs upstream) of gene FAM82A1. SNP rs933602 is in the promotor region (about 300 base pairs upstream) of gene DNER. The significant effects of these two SNPs on gene ICAP-1A may suggest associations among the corresponding genes.

12.5.2 Gene Expression Data Analysis

We apply SCAD_$\ell 2$ to a genomic data set in a study of rat eye disease by Scheetz et al. [9]. The data set consists of 120 rats generated from two highly inbred parental rat strains. Among 31,000 genes that express in eyes, we are interested in finding relevant genes which are correlated with gene TRIM32 known to cause the eye disease Bardet–Biedl syndrome.

We first exclude genes that lack sufficient variation to result in 3,000 most variable genes. Then we order the 3,000 genes based on their absolute correlations with gene TRIM32 from the largest to the smallest. The top 90 genes are selected as the potential predictors for the response variable gene TRIM32.

Next, we apply SCAD_$\ell 2$ to select groups of genes that may influence the expression of TRIM32. The 120 rat samples are randomly split into a training set

Table 12.4 Comparison of SCAD_$\ell 2$ with SCAD, Lasso, and elastic net based on 100 simulations in the analysis of gene expression data of rat eye disease

Methods	Median MSE (SE)	Median nonzero
Lasso	0.017 (0.0012)	13
Elastic net	0.016 (0.0015)	10
SCAD	0.0439 (0.0043)	39
SCAD_$\ell 2$	0.0383 (0.0043)	25

with 100 samples and a test set with 20 samples. A regression model is fitted in the training set. A generalized cross validation method is used to decide the tuning parameters, $(a, \lambda_1, \lambda_2)$, on the training data. Model prediction accuracy is measured by the mean squared error (MSE) on the test set. We compare the prediction accuracy of SCAD_$\ell 2$ with those of SCAD, Lasso, and elastic net. The whole processes are repeated 100 times. The median MSE and the median number of selected genes are shown in Table 12.4. Elastic net and Lasso produce more sparse models with fewer numbers of predictors than those of SCAD and SCAD_$\ell 2$. Elastic net and Lasso also provide similar results, without an obvious group effect in elastic net. Although elastic net and Lasso give small MSEs, their sparse set of variables may miss important signals, due to the fact that the methods may only select one variable from a group. On the other hand, SCAD_$\ell 2$ performs better than its nongroup effect counterpart SCAD. Specifically, SCAD_$\ell 2$ outperforms SCAD by offering a moderate size model (with 25 predictors) and 13% reduction of MSE.

12.6 Discussion

Although large-scale genomic data have been routinely created in biomedical research, extracting useful information from the data remains a challenge. Available statistical and computational tools encounter major difficulties of high dimensionality and complicated dependence in the data. This chapter discusses variable selection approaches for high dimensions and, more importantly, new ideas of group variable selection. The group information naturally embedded in biological systems or pathways helps to enhance signals in analysis of genomic data.

Traditional forward selection is a heuristic approach, not guaranteeing an optimal solution. LARS, a less greedy version of traditional forward selection method, however, is shown by Efron et al. [1] to be closely related to Lasso, which possesses optimal properties under appropriate conditions [5, 20, 21]. Our proposed gLars and gRidge take advantage of the LARS procedure while aiming at group selections for dependent data. The methods do not require prior information on the underlying group structures but construct groups along the selection procedure. Our grouping criterions consider the joint information of x and y and therefore better fit the context of variable selection than standard clustering on x alone. On the other hand, any prior information on the model or groups can be easily incorporated into the algorithms of gLars and gRidge by manually selecting certain variables at specific steps. The current methods may be improved by exploring different thresholds (t_1, t_2) in the grouping definition.

SCAD_$\ell 2$ is a combination of the unbiased approach SCAD and the ridge regression. It is not computationally efficient as the forward procedure but possess good properties in terms of coefficient estimation. One of our future works is to extend the proposed group selection methods to general regression models, where y may depend on x through any nonlinear function. The proposed methods are more appropriate than other variable selection algorithms for data with complicated dependent structures.

Acknowledgement This work is supported by National Science Foundation Grant DMS-0604776.

References

1. B. Efron, I. Johnstone, T. Hastie, and R. Tibshirani. Least angle regression. *Ann Stat*, **32**:407–499, 2004.
2. J. Fan and R. Li. Variable selection via nonconcave penalized likelihood and its oracle properties. *J Am Stat Assoc*, **96**:1348–1360, 2001.
3. X. Huo and X. Ni. When do stepwise algorithms meet subset selection criteria? *Ann Stat*, **35**(2):870–887, 2006.
4. Y. Kim, H. Choi, and H.-S. Oh. Smoothly clipped absolute deviation on high dimensions. *J Am Stat Assoc*, **103**(484):1665–1673, 2008.
5. K. Knight and W. Fu. Asymptotics for lasso-type estimators. *Technometrics*, **12**(1):69–82, 2000.
6. C. Leng, Y. Lin, and G. Wahba. A note on the lasso and related procedures in model selection. *Stat Sin*, **16**:1273–1284, 2006.
7. M. Morley, C.M. Molony, T.M. Weber, J.L. Devlin, K.G. Ewens, R.S. Spielman, and V.G. Cheung. Genetic analysis of genome-wide variation in human gene expression. *Nature*, **430**:743–747, 2004.
8. T. Park and G. Casella. The Bayesian Lasso. Technical report, University of Florida, Gainesville, FL, 2008.
9. T.E. Scheetz, K.Y. Kim, R.E. Swiderski, A.R. Philp, T.A. Braun, K.L. Knudtson, A.M. Dorrance, G.F. DiBona, J. Huang, T.L. Casavant, V.C. Sheffield, and E.M. Stone. Regulation of gene expression in the mammalian eye and its relevance to eye disease. *Proc Natl Acad Sci*, **103**:14429–14434, 2006.
10. G.A. Seber and J. Alan. *Linear Regression Analysis*, 2nd edition. Wiley-Interscience, New York, 2003.
11. M. Segal, K. Dahlquist, and B. Conklin. Regression approach for microarray data analysis. *J Comput Biol*, **10**:961–980, 2003.
12. I. Simon, J. Barnett, N. Hannett, C.T. Harbison, N.J. Rinaldi, T.L. Volkert, J.J. Wyrick, J. Zeitlinger, D.K. Gifford, T.S. Jaakkola, and R.A. Young. Serial regulation of transcriptional regulators in the yeast cell cycle. *Cell*, **106**:697–708, 2001.
13. P.T. Spellman, G. Sherlock, M.Q. Zhang, V.R. Iyer, K. Anders, M.B. Eisen, P.O. Brown, D. Botstein, and B. Futcher. Comprehensive identification of cell cycle-regulated genes of the yeast Saccharomyces cerevisiae by microarray hybridization. *Mol Biol Cell*, **9**(12):3273–3297, 1998.

14. R. Tibshirani. Regression shrinkage and selection via the lasso. *J R Stat Soc B*, **58**:267–288, 1996.
15. D.G. Wang, J. Fan, C. Siao, A. Berno, P. Young, R. Sapolsky, G. Ghandour, N. Perkins, E. Winchester, J. Spencer, L. Kruglyak, L. Stein, L. Hsie, T. Topaloglou, E. Hubbell, E. Robinson, M.S. Morris, N. Shen, D. Kilburn, J. Rioux, C. Nusbaum, S. Rozen, T.J. Hudson, R. Lipshutz, M. Chee, and E.S. Lander. Large-scale identification, mapping, and genotyping of single-nucleotide polymorphisms in the human genome. *Science*, **280**:1077–1082, 1998.
16. J. Wu and J. Xie. Computation-based discovery of cis-regulatory modules by hidden Markov models. *J Comput Biol*, **15**(3):279–290, 2008.
17. N. Yi and S. Xu. Bayesian Lasso for quantitative trait loci mapping. *Genetics*, **179**:1045–1055, 2008.
18. M. Yuan and Y. Lin. Model selection and estimation in regression with grouped variables. *J R Stat Soc B*, **68**(1):49–67, 2006.
19. L. Zeng, J. Wu, and J. Xie. Statistical methods in integrative analysis for gene regulatory modules. *Stat Appl Genet Mol Biol* **7**(1):28, 2008. http://www.bepress.com/sagmb/vol7/iss1/art28.
20. C. Zhang and J. Huang. The sparsity and bias of the lasso selection in high-dimensional linear regression. *Ann Stat*, **36**(4):1567–1594, 2008.
21. P. Zhao and B. Yu. On model selection consistency of lasso. *J Mach Learning Res*, **7**:2541–2563, 2006.
22. H. Zou. The adaptive lasso and its oracle properties. *J Am Stat Assoc*, **101**(476):1418–1429, 2006.
23. H. Zou and T. Hastie. Regularization and variable selection via the elastic net. *J R Stat Soc B*, **67**:301–320, 2005.

Chapter 13
Modeling Protein-Signaling Networks with Granger Causality Test

Wenqiang Yang and Qiang Luo

13.1 Introduction

The development of computational techniques to identify the gene networks, such as regulatory networks and protein–protein interaction networks, underlying observed gene expression patterns, and protein image data is a major challenge in the analysis of high-throughput data. Gene interaction networks can be critical in the analysis and treatment of complex diseases. Significant progresses have been made in the last few years in characterizing regulatory interactions at the genomic level [2, 6, 9], including methods for identifying gene and protein interactions, regulatory modules occurring with a high frequency in the genome, and the identification of transcription motifs [3, 17, 19, 24]. Methods for gene network reconstruction have been proposed based upon statistical methods such as Bayesian networks [16, 20, 21, 26], Boolean models [18], graphical Gaussian models [8, 23], etc.

The key of reconstructing gene networks is to identify causal relations among simultaneously acquired signals. Karen Sachs et al. [22] reconstruct the causal protein signaling network with the Bayesian network structure inference algorithm. However, Bayesian networks have several limitations. First, the computational cost of Bayesian network inferences is usually very high, and the obtained results are not always accurate, in comparison with other reverse engineering approaches (see, for example, [4]). Second, Bayesian network can only be applied to signaling pathways that they are acyclic, whereas signaling pathways are known to be rich in feedback loops.

One major approach to analyze the causality between two signals is to examine if the prediction of one signal could be improved by incorporating information of the

W. Yang (✉)
Department of Mathematics, Hunan Normal University, Changsha, Hunan 410875, China
e-mail: yangwq@pku.org.cn

W. Yang · Q. Luo
Department of Mathematics, National University of Defense Technology, Changsha, Hunan 410073, China

J. Feng et al. (eds.), *Frontiers in Computational and Systems Biology*, Computational Biology 15,
DOI 10.1007/978-1-84996-196-7_13,

other, as proposed by Granger [5, 10]. In particular, if the prediction error of the first signal is reduced by including measurements from the second signal in the linear regression model, then the second signal is said to have a causal influence on the first signal. By exchanging the roles of the two signals, one can address the question of the causal influence in the opposite direction. In this article, we extend the Granger causality definition to nonlinear problems to infer causal protein-signaling networks from single-cell data. Our approach can construct cyclic relations among proteins. Furthermore, the traditional Granger causality is only applicable to timer series data. Our setup is applicable to static data, and therefore it opens a full and new spectrum of applications.

In the next section we review the original approach by Granger while describing our point of view about its nonlinear extension. In Sect. 13.3 we show application of the proposed method to reconstruct a causal protein network. Some conclusions are drawn in Sect. 13.4.

13.2 Granger Causality and Approach

We briefly recall the Granger causality [11]. Let X and Y be two time series. Denote F_t^X be the entire information up to and including time t, with $F_t^X = \sigma\{X_s, s \leq t\}$, and let F_t be the information set available at time t. $F_t \setminus F_t^X$ indicates the information set excluding F_t^X. The Granger causality is defined as follows:

1. X does not cause Y in mean with respect to F_{t-1} if

$$E(Y_t|F_{t-1}) = E\big(Y_t|F_{t-1} \setminus F_t^X\big). \tag{13.1}$$

2. X is a prima facie cause in mean of Y with respect to F_{t-1} if

$$E(Y_t|F_{t-1}) \neq E\big(Y_t|F_{t-1} \setminus F_t^X\big). \tag{13.2}$$

A conventional approach of testing Granger causality is to consider the conditional mean $E(Y_t|F_{t-1})$ to be a parametric linear model and to test the null hypothesis that the coefficients on lagged values of X are all zero. The conventional linear tests are powerful in uncovering linear causal relations. However, there is a disadvantage. The tests require modeling assumptions such as the linearity of the regression structure. To infer a nonlinear causal relationship is somewhat limited. In order to circumvent the nonlinearity issue, some approaches have been proposed [1, 15].

In order to extend the Granger causal testing to reconstruct the protein causal networks, we consider a random variable set $\mathbf{P} = \{P_1, P_2, \ldots, P_N\}$, with P_i being the measurement of the expression level for the ith protein, $i = 1, 2, \ldots, N$. To simplify the statement, we only consider the causal relation between P_1 and P_2. Let $Y = P_2$ be the variable of the target protein, $X = P_1$ be the variable of the Granger cause candidate protein, and let $\mathbf{Z} = P \setminus \{X, Y\} = (P_3, \ldots, P_4)$ be the variable of the rest proteins; then we can test whether Y is the Granger cause of X or not.

Define two regression models for Y:

$$Y = \mathrm{E}(Y|\mathbf{Z}) + \varepsilon, \tag{13.3}$$

$$Y = \mathrm{E}(Y|X,\mathbf{Z}) + \varepsilon', \tag{13.4}$$

where ε and ε' are random noises. Denoting the expectation function for model (13.3) by $f(\mathbf{Z})$, and for model (13.4), by $g(X,\mathbf{Z})$, the two models can be reformulated as follows:

$$Y = f(\mathbf{Z}) + \varepsilon, \tag{13.5}$$

$$Y = g(X,\mathbf{Z}) + \varepsilon'. \tag{13.6}$$

By the definition of the Granger causality (13.2), X is a Granger cause of Y if

$$\mathrm{E}(Y|\mathbf{Z}) \neq \mathrm{E}(Y|X,\mathbf{Z}). \tag{13.7}$$

Therefore, we introduce a log-likelihood ratio measure as follows:

$$R(X \to Y|\mathbf{Z}) = \log\left\{\frac{\mathrm{E}(Y - \mathrm{E}(Y|\mathbf{Z}))^2}{\mathrm{E}(Y - \mathrm{E}(Y|X,\mathbf{Z}))^2}\right\}. \tag{13.8}$$

Apparently, X is a Granger cause of Y if $R(X \to Y|\mathbf{Z}) > 0$. In this case, a directed arc from X to Y may be interpreted as a causal influence from the first protein to the second one.

Next, we need to compute the log-likelihood ratio $R(X \to Y|\mathbf{Z})$. The expectations in the numerator and the denominator of (13.8) are the error variances of the regression models in (13.3) and (13.4), respectively. If we denote the residuals of the two regression models by $e_{k,\mathbf{Z}}$ and $e_{k,X\mathbf{Z}}$, respectively, where K is the number of the data points, the estimates of the error variances can be given by the averages of the squared residuals:

$$S_{\mathbf{Z}}^2 = \frac{1}{K-2}\sum_{k=1}^{K} e_{k,\mathbf{Z}}^2, \tag{13.9}$$

$$S_{X,\mathbf{Z}}^2 = \frac{1}{K-2}\sum_{k=1}^{K} e_{k,X\mathbf{Z}}^2, \tag{13.10}$$

where

$$e_{k,\mathbf{Z}} = y_k - f(\mathbf{z}_k), \tag{13.11}$$

$$e_{k,X\mathbf{Z}} = y_k - g(x_k,\mathbf{z}_k). \tag{13.12}$$

Then the log-likelihood ratio can be well estimated by

$$\hat{R}(X \to Y|\mathbf{Z}) = \log\left(\frac{S_{\mathbf{Z}}^2}{S_{X,\mathbf{Z}}^2}\right). \tag{13.13}$$

Although for the target protein Y, the Granger cause candidate protein X, and the rest proteins $\mathbf{Z}$, the observation sample $(y_k, x_k, \mathbf{z}_k)$, $k = 1, \ldots, K$, is available, we know nothing about the regression functions f and g, and they must be very sophisticated for the complex nature of the life phenomena, so neither the common linear nor nonlinear regression method is suited in this case. Fortunately, in the Granger causality test the explicit formulations of the regression functions are not necessary. So the Artificial Neural Networks (ANN) as black-boxes can be employed to fit the input and output data of the regression functions. The best estimates $\hat{f}$ and $\hat{g}$ will minimize the averages of the squared error at each data point for the regression models. Mathematically, let

$$\hat{S}^2_{\mathbf{Z}} = \frac{1}{K-2}\sum_{k=1}^{K}\left(y_k - \hat{f}(\mathbf{z}_k)\right)^2, \tag{13.14}$$

$$\hat{S}^2_{X\mathbf{Z}} = \frac{1}{K-2}\sum_{k=1}^{K}\left(y_k - \hat{g}(x_k, \mathbf{z}_k)\right)^2. \tag{13.15}$$

If the ANNs are feedforward ANNs with one hidden layer, we can train the ANNs with the observation sample by some training algorithm to give the best estimates $\hat{f}$ and $\hat{g}$ for the regression function f and g, respectively. Obviously, the input layer has as many pure linear neurons as the different proteins in the input vector, and only one output pure linear neuron for the target protein Y. The number of the tangent sigmoid neurons in the hidden layer is a model parameter, which will be specified empirically.

For example, the data points $(T_k, I_k)_{k=1}^{K}$, where $T_k = y_k$ and $I_k = \mathbf{z}_k$, are used to train the ANN for the regression model defined by (13.3) with the target vector T_k and the input vector I_k. Here, the high-throughput proteomic data are available for the modeling of the protein-signal networks.

Together with the definitions given by (13.9), (13.10), (13.11), (13.12), (13.14), and (13.15), the averages of the squared residuals $S^2_{\mathbf{Z}}$ and $S^2_{X\mathbf{Z}}$ will be good estimate by the averages of the squared error $\hat{S}^2_{\mathbf{Z}}$ and $\hat{S}^2_{X\mathbf{Z}}$ after training, respectively, and hence (13.13) can be computed by

$$\hat{R}(X \to Y|\mathbf{Z}) = \log\left(\frac{\hat{S}^2_{\mathbf{Z}}}{\hat{S}^2_{X,\mathbf{Z}}}\right). \tag{13.16}$$

13.3 Data and Results

We use our method to reconstruct the protein causality network discussed in [22] (experimental data are available on Science Online, which include 14 data files. We only use the first data file). Data are expression levels of 11 phosphorylated molecules which are simultaneously measured from single cells by flow cytometry.

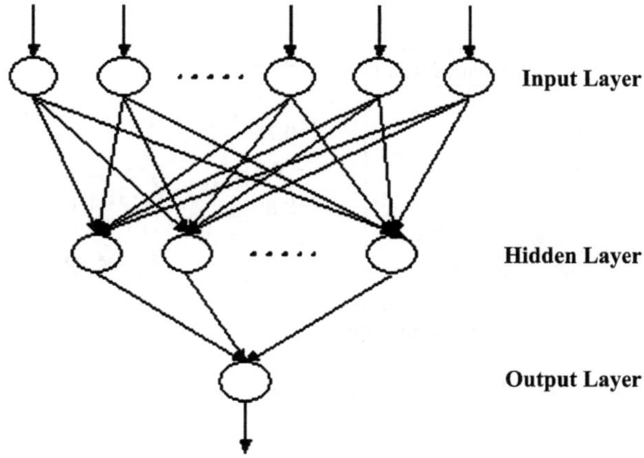

Fig. 13.1 The structure of the three-layer feedforward neural networks with 6 hidden neurons

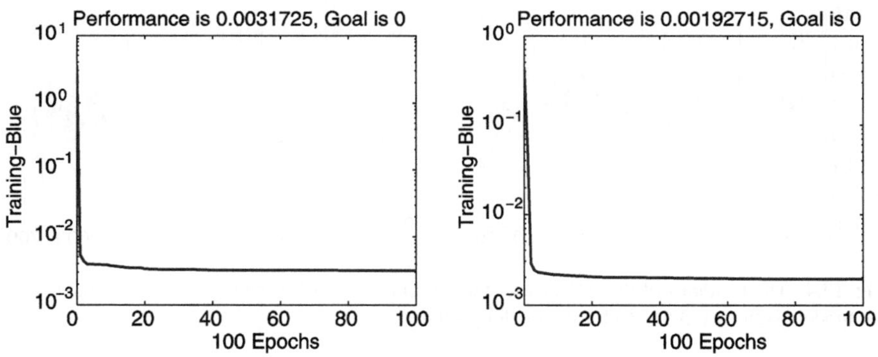

Fig. 13.2 The training processes of the ANNs for the regression models: (**a**) for the model given by (13.5); (**b**) for the model given by (13.6)

Here, we set the random variables $\mathbf{P} = \{P_1, \ldots, P_{11}\}$ to denote the expression levels for the 11 proteins. The three-layer feedforward neural networks with 6 hidden neurons (as shown in Fig. 13.1) are employed in our algorithm to estimate the log-likelihood ratio measure $R(P_i \to P_j | \mathbf{P} \setminus \{P_i, P_j\})$ $(i, j = 1, 2, \ldots, 11)$.

After some pretreatment, the regularization and outlier-elimination exactly, on the expression data, the ANN can be trained by the Levenberg–Marquardt algorithm [14].

Figure 13.2 shows the training processes of the ANNs for regression models given by (13.5) and (13.6). The outputs of the ANNs after training are presented on Fig. 13.3, where X is the protein Mek, and Y, Z are the protein Raf and the others, respectively. Figure 13.4 shows the residuals given by ANNs after training for the regression models. The averages of the squared residuals can be estimated by $S^2_{\mathbf{Z}} = 0.0019$ and $\hat{S}^2_{\mathbf{Z}} = 0.0031$, thereby the log-likelihood ratio for $(X, Y, \mathbf{Z})$ is $\hat{R}(X \to Y|\mathbf{Z}) = \log(0.0031/0.0019) = 0.4895 > 0$. Therefore, in this example,

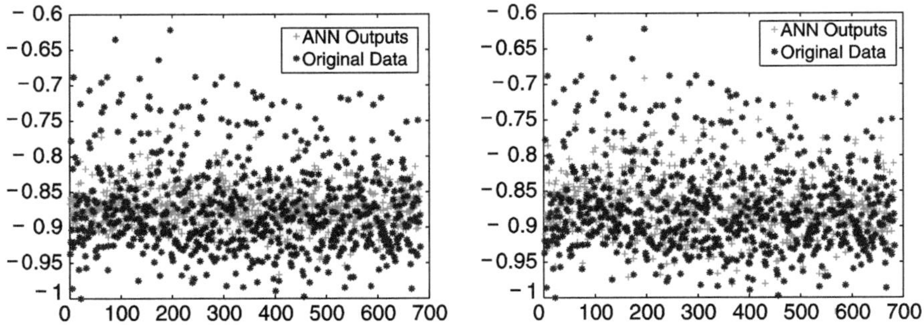

Fig. 13.3 The output of the ANNs after the training process for the regression models: (**a**) for the model given by (13.5); (**b**) for the model given by (13.6)

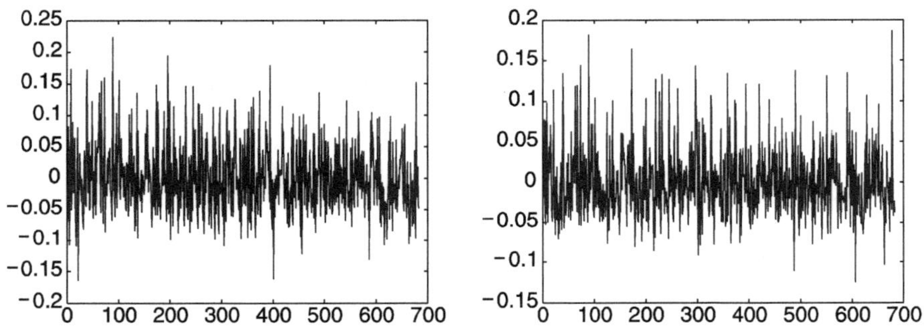

Fig. 13.4 The residuals of the ANNs after the training process for the regression models: (**a**) for the model given by (13.5); (**b**) for the model given by (13.6)

X(Raf) is the Granger cause of Y(Mek) given the background information $\mathbf{Z}$, i.e., given the rest proteins, a causal influence from the protein Raf to the protein Mef is inferred by the Granger causality test model.

The resulting protein causal network was reconstructed by Granger causality tests (Fig. 13.5). To evaluate the validity of the model, we compared the network with those described in the literature. According to [22], we categorized the arcs as the follows: (i) reported, for connections well-established in the literature that have been demonstrated under numerous conditions in multiple model systems; (ii) reported, for connections that are not well known, but for which we were able to find at least one literature citation; (iii) added, for connections that are not demonstrated; (iv) missed, which indicates an expected connection that our Granger causality test failed to find. Of the 29 arcs in our model, 16 were expected, 2 were reported, 11 were additional pathways, and 2 were missed.

Several of the additional connections from our model (Plcγ $\rightarrow$ Raf, Plcγ $\rightarrow$ Jnk, Plcγ $\rightarrow$ PKA, Plcγ $\rightarrow$ Erk, PIP2 $\rightarrow$ Jnk, PIP2 $\rightarrow$ Akt, PIP2 $\rightarrow$ Mek, PIP3 $\rightarrow$ PKC, PIP3 $\rightarrow$ P38) demonstrate that there is causal influence among those proteins. For example, Plcγ $\rightarrow$ Raf means that the protein Plcγ is causal influence of Jnk,

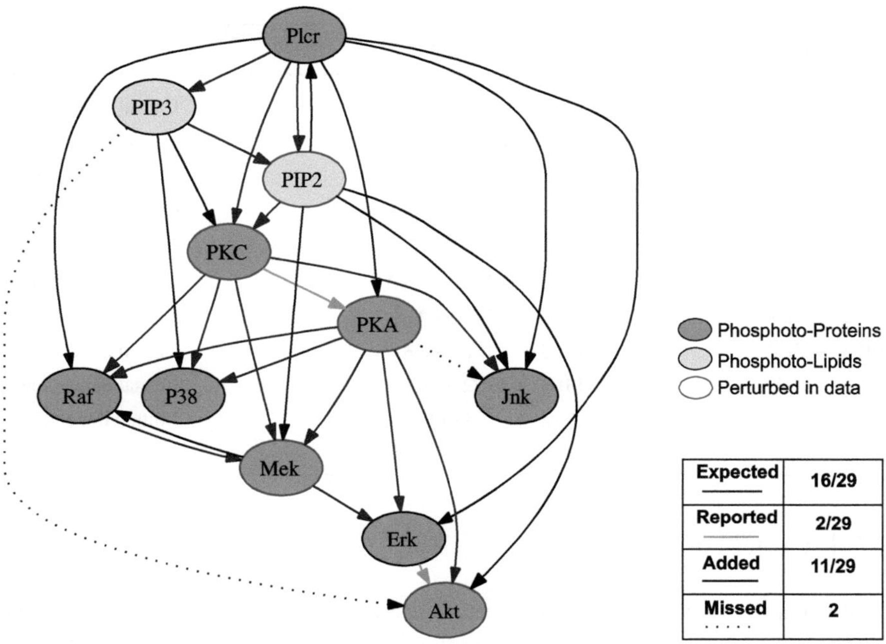

Fig. 13.5 The graph of protein-signaling network reconstructed by the Granger causality tests

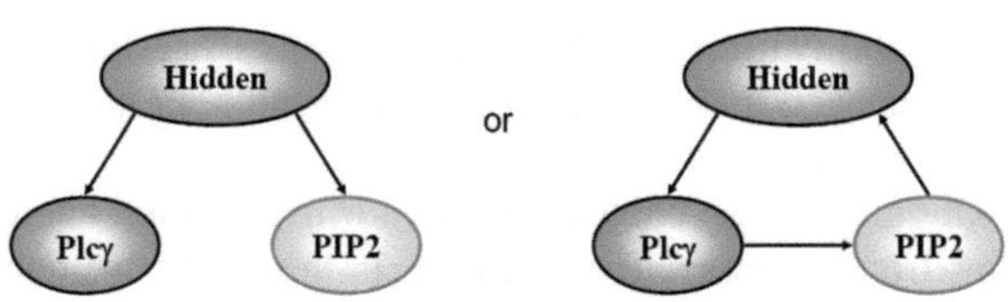

Fig. 13.6 The causal relationship between proteins PIP2 and Plcγ with hidden variable

because the connection from Plcγ to PKC and the connection from PKC to Raf explain the dependence of Raf on Plcγ.

The other additional connections from our model are from PIP2 to Plcγ and from Mek to Raf. There are known direct connections from Plcγ to PIP2 and Raf to Mek. It is possible that there might be mutual causal, i.e., Plcγ $\rightarrow$ PIP2 as well as PIP2 $\rightarrow$ Plcγ and Raf $\rightarrow$ Mek as well as Mek $\rightarrow$ Raf. On the other hand, maybe there is another cause that causes the mutual influences, but those proteins are not measured (we denote those proteins as hidden variables). For instance, PKC influence Raf and Mek. Maybe the influence of protein PIP2 on Plcγ was mediated by intermediate proteins that were not measured in the data set, so there is a mutual influence between PIP2 and Plcγ. They can be describe as Fig. 13.6. Table 13.1 enumerates possible causal relationships corresponding to all added connections in our model.

Table 13.1 Possible causal relationships represented by added connections in our model

Added connection	Causal relationship	Type
Plcγ → Raf	Plcγ → PKC → Raf	Causal dependence
Plcγ → Jnk	Plcγ → PKC → PKA → Raf	Causal dependence
Plcγ → PKA	Plcγ → PKC → PKA	Causal dependence
Plcγ → Erk	Plcγ → PKC → PKA → Erk	Causal dependence
PIP2 → Jnk	PIP2 → PKC → Jnk	Causal dependence
PIP2 → Akt	PIP2 → PKC → PKA → Akt	Causal dependence
PIP2 → Mek	PIP2 → PKC → Mek	Causal dependence
PIP3 → PKC	PIP3 → PIP2 → PKC	Causal dependence
PIP3 → P38	PIP3 → PIP2 → PKC → P38	Causal dependence
PIP2 → Plcγ	PIP2 → ⋯ → Plcγ → PIP2	Mutual dependence
Mek → Raf	Mek ← PKA → Raf	Mutual dependence

13.4 Conclusion

In this paper, we proposed Granger causal test combining artificial neural network to infer protein causal networks. We correctly reconstructed the protein-signaling network from multivariate flow cytometry data. The result showed that the Granger causal test is suitable for modeling gene networks. The Granger causality test has advantages, including the ability to detect indirect and direct connections, efficient computing, and the ability to capture the feedback loops. Therefore, it is a promising method that can scale well for large, genome-scale gene or protein networks.

Modeling gene or protein networks is one of the central topics in systems biology [7, 12, 13, 25]. Our method can uncover the unknown regulation relationship among genes or proteins from express and image data. In this paper, we only demonstrate the model based on the static data. However, the method can also be applied for time series data. We investigate these topics in our future papers.

References

1. N. Ancona, D. Marinazzo, and S. Stramaglia. Radial basis function approach to nonlinear Granger causality of time series. *Phys Rev*, **70**:056221, 2004.
2. M.M. Babu, N.M. Luscombe, L. Aravind, M. Gerstein, and S.A. Teichmann. Structure and evolution of transcriptional regulatory networks. *Curr Opin Struct Biol*, **14**(3):283–291, 2004.
3. Z. Bar-Joseph, G. Gerber, T. Lee, N. Rinaldi, J. Yoo, F. Robert, D. Gordon, E. Fraenkel, T. Jaakkola, R. Young, and D. Gifford. Computational discovery of gene modules and regulatory networks. *Nat Biotechnol*, **21**:1337–1342, 2003.
4. I. Cantone, L. Marucci, F. Iorio, M. Ricci, V. Belcastro, M. Bansal, S. Santini, M. di Bernardo, D. di Bernardo, and M.P. Cosma. A yeast synthetic network for in vivo assessment of reverse-engineering and modeling approaches. *Cell*, **137**:172–181, 2009.
5. Y.H. Chen, G. Rangarajan, J.F. Feng, and M.Z. Ding. Analyzing multiple nonlinear time series with extended Granger causality. *Phys Lett A*, **324**:26–35, 2004.

6. N.O. Eric. Gene regulatory networks in the evolution and development of the heart. *Science*, **313**:1922–1927, 2006.
7. J.F. Feng, D.Y. Yi, R. Krishna, S.X. Guo, and V. Buchanan-Wollaston. Listen to genes: dealing with microarray data in the frequency domain. *PLoS One*, **4**(4):e5098, 2009. doi:10.1371/journal.pone.0005098.
8. A. Fuente, N. Bing, I. Hoeschele, and P. Mendes. Discovery of meaningful associations in genomic data using partial correlation coefficients. *Bioinformatics*, **20**:3565–3574, 2004.
9. T. Gardner, D. Bernardo, D. Lorenz, and J. Collins. Inferring genetic networks and identifying compound mode of action via expression profiling. *Science*, **301**:102–105, 2003.
10. C. Granger. Investigating causal relations by econometric methods and cross-spectral methods. *Econometrica*, **34**:424–438, 1969.
11. C. Granger. Some recent development in a concept of causality. *Econometrica*, **39**:192–211, 1988.
12. S.X. Guo, A. Seth, K. Kendrick, and J.F. Feng. Partial Granger causality: eliminating exogenous inputs and latent variables. *J Neurosci Methods*, **172**:79–83, 2008.
13. S.X. Guo, J.F.H. Wu, M.Z. Ding, and J.F. Feng. Uncovering interactions in the frequency domain. *PLoS Comput Biol*, **4**(5):e1000087, 2008. doi:10.1371/journal.pcbi.1000087.
14. S. Haykin. *Neural Networks: A Comprehensive Foundation*, 2nd edition. Prentice Hall, New York, 1998. Chapter 4.
15. C. Hiemstra and J. Jones. Testing for linear and nonlinear Granger causality in the stock price volume relation. *J Finance*, **49**:1639–1644, 1994.
16. D. Husmeier. Sensitivity and specificity of inferring genetic regulatory interactions from microarray experiments with dynamic Bayesian networks. *Bioinformatics*, **19**:2271–2282, 2003.
17. J. Ihmels, G. Friedlander, S. Bergmann, O. Sarig, Y. Ziv, and N. Barkai. Revealing modular organization in the yeast transcriptional network. *Nat Genet*, **313**:1922–1927, 2002.
18. S. Kauffman, C. Peterson, B. Samuelsson, and C. Troein. Random Boolean network models and the yeast transcriptional network. *Proc Natl Acad Sci*, **100**:14796–14799, 2003.
19. T. Lee, N. Rinaldi, F. Robert, D. Odom, Z. Bar-Joseph, G. Gerber, N. Hannett, C. Harbison, C. Thompson, I. Simon, J. Zeitlinger, E. Jennings, H. Murray, D. Gordon, B. Ren, J. Wyrick, J. Tagne, T. Volkert, E. Fraenkel, D. Gifford, and R. Young. Transcriptional regulatory networks in saccharomyces cerevisiae. *Science*, **298**:799–804, 2002.
20. D. Pe'er, A. Regev, G. Elidan, and N. Friedman. Inferring subnetworks from perturbed expression profiles. *Bioinformatics*, **17**:215–224, 2001.
21. B.E. Perrin, L. Ralaivola, A. Mazurie, S. Bottani, J. Mallet, and F. d'Alche Buc. Gene networks inference using dynamic Bayesian networks. *Bioinformatics*, **19**:ii138–ii148, 2003.
22. K. Sachs, O. Perez, D. Pe'er, D. Lauffenburger, and G. Nolan. Causal protein-signaling networks derived from multiparameter single-cell data. *Science*, **308**:523–529, 2005.
23. J. Schaefer and K. Strimmer. An empirical Bayes approach to inferring large-scale gene association networks. *Bioinformatics*, **21**:754–764, 2005.
24. E. Segal, M. Shapira, A. Regev, D. Pe'er, D. Botstein, D. Koller, and N. Friedman. Module networks: identifying regulatory modules and their condition-specific regulators from gene expression data. *Nat Genet*, **34**:166–176, 2003.
25. J.H. Wu, X.G. Liu, and J.F. Feng. Detecting causality between different frequencies. *J Neurosci Methods*, **167**:367–375, 2007.
26. M. Zou and S.D. Conzen. A new dynamic Bayesian network (DBN) approach for identifying gene regulatory networks from time course microarray data. *Bioinformatics*, **21**:71–79, 2005.

Chapter 14
DNA Copy Number Profiling in Normal and Tumor Genomes

Nancy R. Zhang

14.1 Introduction

For a biological sample, the DNA copy number of a genomic region is defined as the number of copies of the DNA in that region within the genome of the sample, relative to either a single control sample or a pool of population reference samples. Within the last decade, significant advances in microarray technology have enabled the genome-wide fine-scale measurement of DNA copy number in a high-throughput manner [5, 35, 39, 40, 47]. This enables systematic studies which can lead to a better understanding of the role of DNA copy number changes in human disease and in phenotypic variation in the human population. These high-throughput experiments produce large amounts of data that are rich in structure, motivating the development of new statistical methods for their analysis. This chapter reviews the computational and statistical problems that arise in DNA copy number data and surveys recent advances in their treatment.

First, we review some terms and general concepts relating to DNA copy number. A copy number variant (CNV) is defined as a genomic region where the DNA copy number differs between two or more individuals from a population. CNVs that have so far been catalogued are by convention larger than 1 kilobase, although technologies based on high-throughput sequencing [45] and denser arrays [19] can detect shorter CNVs. Within the last five years, many studies [9, 10, 21, 30, 41] have shown that CNVs are a common type of genetic variation in the human population, with the fraction of the genome covered by CNVs estimated to be between 2% [10] and 15% [13]. Like single nucleotide polymorphisms (SNPs), variants in copy number segregate in a Mendelian fashion and contribute to phenotypic variation. Considering that they cover significantly more genomic territory in terms of base pairs and that they are more likely than SNPs to have a deleterious effect, CNVs are now routinely used alongside SNPs in genetic association studies.

N.R. Zhang (✉)
Department of Statistics, Stanford University, 390 Serra Mall, Stanford, CA 94305-4065, USA
e-mail: nzhang@stanford.edu

J. Feng et al. (eds.), *Frontiers in Computational and Systems Biology*, Computational Biology 15,
DOI 10.1007/978-1-84996-196-7_14,

Changes in DNA copy number have also been highly implicated in tumor genomes. Some of these changes are inherited, but many are due to somatic mutations that occur during the clonal development of the tumor. The copy number changes in tumor genomes are often referred to as copy number aberrations (CNAs), to differentiate them from inherited CNVs. CNAs are usually larger in size than CNVs, often involving gains and losses of entire chromosome arms. Their roles in tumor development are varied, and high-fold amplification of genomic regions containing oncogenes and deletion of regions containing tumor suppressor genes have been widely documented. For example, a search using the terms "copy number" and "tumor" brings up 4421 articles in Pubmed. These evidences suggest that at least some CNAs play a role in driving tumor progression.

Given the raw DNA copy number data from a single sample, an immediate challenge lies in estimating the true underlying copy number from the noisy measurements. This problem, often referred to as segmentation of total copy number, has drawn considerable attention and is reviewed in Sect. 14.2. For data from some array platforms, such as the Affymetrix and Illumina genotyping arrays, it is possible to tease apart the underlying copy numbers of the two distinct sets of chromosomes inherited from the two biological parents. This problem, which we refer to as parent- or allele-specific copy number estimation, is motivated and reviewed in Sect. 14.3. In both total copy number and parent-specific copy number estimation, it is important to distinguish between tumor and normal samples in the formulation of the statistical model. This is a theme that will be reiterated in this chapter.

In many studies, multiple technical platforms or different versions of the same platform are being used to interrogate the same biological samples. Pooling information across these multiple sources can give a more accurate consensus molecular profile for each sample. Section 14.4 looks at recent approaches to multiplatform integration. A more complex problem is the joint analysis of multiple copy number profiles, each coming from a different biological sample. There can be many different goals in such cross-sample analyses, which deserve different statistical approaches. Section 14.4 reviews the modeling issues and recent developments in cross-sample models for DNA copy number.

14.2 Total Copy Number Estimation for One Sample

The total DNA copy number data for any given sample comes in the form of a sequence $\{(x_i, y_i) : i = 1, \ldots, n\}$, where n is the number of probes, and x_i and y_i are respectively the genome location and normalized intensity for probe i. "Probe" and "normalized intensity" mean different things for different experimental platforms, and the reader is referred to [5, 35, 39, 40, 47] for more details. The term "total copy number" refers to the sum of the copy numbers for the chromosomes inherited from the two biological parents. If this number varies over the cells in the sample, then the intensity is a reflection of average copy number over all of the cells. Thus, although total copy number for each individual cell is integer valued, when the sample is

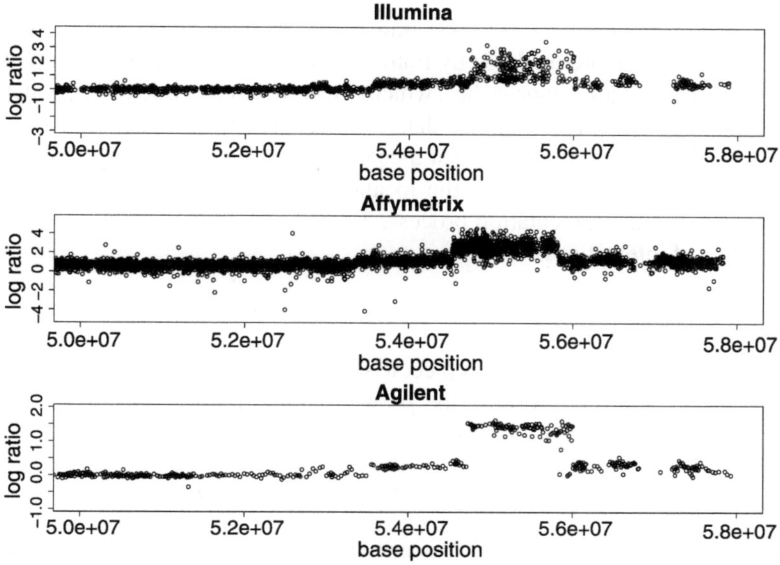

Fig. 14.1 Copy number data for a tumor sample assayed on the Agilent, Illumina, and Affymetrix platforms

genetically heterogeneous, the average copy number can vary over a continuous scale.

The appropriate preprocessing procedure that is necessary to normalize the intensity measurements depends on the technical platform that generated the data, see [1, 35, 48] for some examples of nontrivial preprocessing procedures. The data from most platforms is in the form of a log ratio of the DNA quantity in the target sample versus the DNA quantity in an appropriate control. The "normal" state, where the copy number in the target agrees with that in the control, should have mean 0. A contiguous stretch of measurements that are on average higher (or lower) than 0 suggests a gain (or loss) in copy number. Figure 14.1 shows an example copy number profile for a genomic region from a tumor sample, assayed on three different platforms. Note that different experimental platforms vary in noise variance, responsiveness to signal, and location of probes. Section 14.4 examines these differences between platforms in more detail.

The observed intensities are noisy surrogates of the true copy number at the measured positions. Since chromosomes are gained and lost in segments, adjacent positions in the genome are highly likely to have the same underlying copy number. This is why change-point models [34, 38, 51, 55, 58], smoothing methods [6, 18, 26, 50], Haar-based wavelets [17], spatially restricted clustering [52, 57], and various formulations of hidden Markov models [3, 8, 12, 14, 15, 26] have been proposed for the estimation of DNA copy number. [25] and [56] reviewed and compared the performance of existing approaches in 2005. It is impossible to review in this chapter all of the above approaches. We focus on the change-point formulation for this problem that underlies the Circular Binary Segmentation (CBS) algorithm [34, 51], which is one of the simplest and most transparent methods. CBS was found to be one of the more accurate methods by both [25] and [56]. We then summarize hidden Markov model-based approaches, which, as we will see in Sect. 14.3, generalize naturally to model the more complex data from genotyping arrays.

Since the location of the probes, at a coarse global scale, is approximately uniformly distributed in the genome, the location information $\{x_i : i = 1, \dots, n\}$ is often ignored in the segmentation process. Then, a simple change-point model for the sequence of intensities is

$$y_i = \mu_i + \varepsilon_i, \quad i = 1, \dots, n, \tag{14.1}$$

where $\mu = \{\mu_i : i = 1, \dots, n\}$ is a piecewise constant function of i, and $\{\varepsilon_i : i = 1, \dots, n\}$ are i.i.d. errors. To describe μ, we assume that there exists a series of change-points $0 = \tau_0 < \tau_1 < \cdots < \tau_m < \tau_{m+1} = n$ such that

$$\mu_t = \theta_i, \quad t \in [\tau_i, \tau_{i+1}),\ i = 0, \dots, m. \tag{14.2}$$

For inference, the errors are usually assumed to be Gaussian, although this assumption is not crucial if the distances between successive τ_j's are large. Under this model, the segmentation problem reduces to estimating the change-points and the means within each segment. The number of change-points m is also not known and has been observed to range from below 10 to above 100 in some tumor samples.

If the values of the change-points τ are known, then θ_j can be estimated by the mean of the observations that fall in the jth segment. To estimate τ, the CBS algorithm employs a greedy top-down approach that recursively applies the generalized likelihood ratio statistic for testing a square wave change. In more detail, for any interval $1 \le a < b \le n$, let the null hypothesis be that the observations are i.i.d. Gaussian, and let the alternative be that there is a subinterval with a change in mean and no change in variance. The generalized likelihood ratio statistic is

$$\max_{a<s<t<b} Z_{s,t}, \quad \text{where } Z_{s,t} = \frac{S_t - S_s - \frac{t-s}{b-a}(S_b - S_a)}{\hat{\sigma}\sqrt{(t-s)[1-(t-s)/(b-a)]}}, \tag{14.3}$$

and $S_j = y_1 + \cdots + y_j$. CBS starts by setting $a = 1$ and $b = n$. Let z^{obs} be the observed maximum of $Z_{s,t}$, and (s^*, t^*) be the maximizing interval. If the p-value of the scan, $P(\max_{a<s<t<b} Z_{s,t} > z^{\text{obs}})$, is smaller than some prechosen threshold α, then the maximizing interval is reported, and the intervals $[a, s^*)$, $[s^*, t^*)$, $[t^*, b]$ are recursively scanned using the same procedure. The recursion stops when none of the subregions contain a square wave change that is significant at the level α.

The p-value for the scan statistic in (14.3) can be computed using asymptotic approximations given by [20] and [46], which is quite accurate for tail probabilities. Alternatively, [58] proposed a modified BIC criterion for estimating m and showed that, when used in conjunction with CBS, has more accurate off-the-shelf performance than p-value based thresholds.

In contrast to the change-point formulation, hidden Markov model-based methods assume that the observed intensities are emitted by an underlying Markov chain.

Different published methods assume different dynamics for the underlying Markov chain. The earliest method [14] assumes that the hidden states follow a discrete-state Markov chain and obtains a segmentation using the Viterbi algorithm. The discrete-state model works well for detecting inherited CNVs in normal samples but is not flexible enough for tumor samples, where due to sample heterogeneity, it is hard to predict how many states there should be in the underlying chain. To better accommodate fractional copy number changes, [26] assumes that the underlying mean is a continuous-valued Markov jump process with a baseline state and a changed state, where every time a jump is made to the changed state the Markov chain takes on a new Gaussian value. Exact recursive equations for the posterior expectation of the underlying mean given the entire observed intensity sequence are given in [26], along with a fast linear time approximation. The Bayesian approach allows computation of confidence intervals for the expected copy number at each position and the total number of CNVs. The hidden Markov models in [12, 15] also assume continuous-valued jumps but estimate the underlying states using Markov chain Monte Carlo or pseudo-likelihood based approaches.

The fundamental difference between frequentist approaches such as CBS and hidden Markov models lie in the necessary assumptions about the length and magnitude of jumps. For a hidden Markov model, one must explicitly specify the waiting time distribution between jumps and the distribution of the underlying state sequence. If reliable prior information in this regard is available, then hidden Markov models can more flexibly incorporate them. However, when prior information is not available, they must either be specified arbitrarily by the user or estimated from the data. Frequentist approaches do not require the user to specify these prior distributions, and thus, while being less flexible, may have better off-the shelf performance.

The methods mentioned so far use only total intensity data which measures the sum of the copy numbers of both parental chromosomes. Data from some platforms, such as Illumina and Affymetrix genotyping arrays and Molecular Inversion Probes, can reveal more information. These platforms measure, for targeted bi-allelic single nucleotide polymorphisms, the quantities of both alleles. The total intensity obtained from these platforms is usually the log transform of a sum of the intensities of both alleles, normalized to a group of population control samples. This essentially reduces a two-dimensional data sequence into a one-dimensional sequence of log ratios, resulting in a loss of information. For Illumina data, for example, the B-allele frequency, defined as the normalized ratio of the quantity of the B-allele to the total quantity of both alleles, seems to be more informative for detecting low-amplitude jumps [35]. Methods for detecting inherited CNVs can gain power by incorporating the B-allele frequency, as done in the softwares QuantiSNP [8] and PennCNV [53]. More details of these methods, in the context of parent specific copy number estimation, are given in the next section.

14.3 Parent Specific Copy Number Estimation

The genome of each somatic human cell normally contains two copies of each of the 22 autosomes, one inherited from each biological parent. At any genome loca-

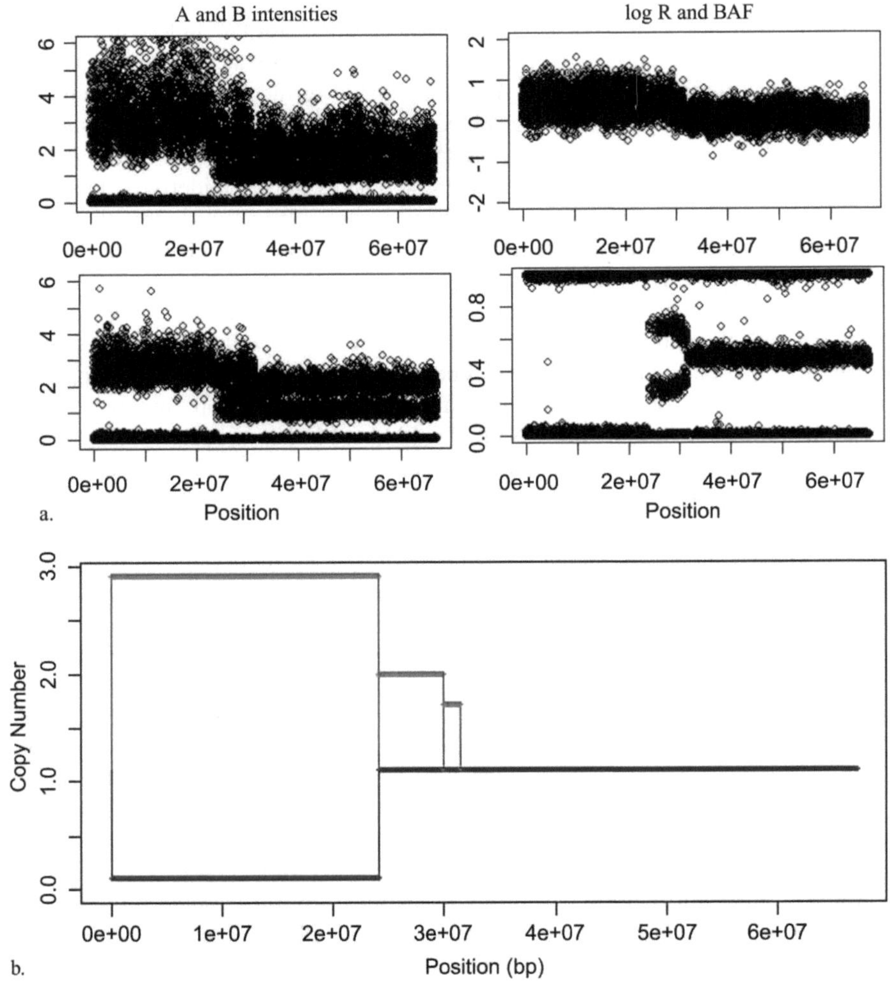

Fig. 14.2 (**a**) An example data sequence taken from a TCGA glioblastoma sample. The *left panel* shows the A- and B-allele intensities. The *right panel* shows the $\log R$ and B-allele frequencies. (**b**) The estimated major and minor copy numbers for the example region from TCGA sample 0258 chromosome 2

tion, one or both of these two chromosomes may gain or lose copies. The methods described in the last section use only the total intensity data $\{y_i\}$, which measure the sum of the copy numbers of the two inherited chromosomes. These methods do not reveal whether both chromosomes have changed copies and, at polymorphic loci, which of the alleles have been affected. The extraction of this information is important, because the allele-specific nature of amplifications and deletions is highly relevant for the biological understanding of cancer and represents a key advantage of SNP array-based assays in comparison to the conventional array-based comparative genomic hybridization experiments that measure only total copy number. Furthermore, copy neutral loss-of-heterozygosity events, defined as the simultaneous gain of one copy and loss of the other copy of the inherited chromosomes, have long been implicated in the actions of oncogenes. These events can only be detected using allelic-specific measurements.

For simplicity, we consider only biallelic SNPs and let the alleles be arbitrarily labeled A and B. Allele "A" may refer to different bases at different SNPs and may also reside on different chromosomes across adjacent SNPs. Genotyping platforms give, at selected SNPs, a bivariate measurement quantifying each of the alleles A and B. An example sequence of normalized A and B intensities for a genomic region of a tumor sample assayed using the Illumina platform is shown in the left panel of Fig. 14.2a. The log ratio and the B-allele frequency are shown in the middle panel of the figure.

As is evident in Fig. 14.2a, the allele-specific measurements at each SNP follow a mixture distribution that depend on the genotype of the sample at that SNP. The genotype is usually unknown and must also be inferred from the data. Without the genotype information, adjacent allele-specific measurements cannot be smoothed. Thus, conventional change-point models cannot be applied directly to this problem. However, the parent-specific copy number, which we define as a bivariate quantity that distinguishes between the chromosomes inherited from the two parents, is smooth across adjacent positions on a chromosome. Without family data, it is impossible to distinguish which chromosome is maternal and which is paternal. Thus the parent specific copy numbers are exchangeable. When there is an imbalance in copy number, the chromosome with the higher copy number is called the "major" chromosome, and the other is called the "minor" chromosome.

La Framboise et al. [24] is one of the earlier methods that make use of allele-specific data. They applied existing segmentation algorithms to the total copy number. Then, the B-allele frequency is used to estimate the allele-specific copy number and loss of heterozygosity status for each segment. Discrete-state hidden Markov models [8, 27, 53] have also been proposed for this problem. In these approaches, the hidden states representing changes in whole copy number or generalized genotypes such as *AA*, *AB*, *BB*, $A-$, $B-$, *AAB*, *ABB*, etc. [8] is one of the earliest methods in this category. Designed for Illumina data, it is based on a hidden Markov model with six underlying states described in Table 14.1. Within each state, the log ratio and B-allele frequency are assumed to be independent. The log ratio is assumed to be a mixture of a uniform distribution and a Gaussian distribution with state-dependent mean and variance. The uniform distribution acts as a noninformative state for capturing outliers in the data. The B-allele frequency follows a mixture distribution that depends on the unknown genotype, with also a uniform component for robustness against outliers. The parameters of this model can be estimated by maximizing the marginal likelihood, thus giving it some desirable frequentist properties while also allowing for flexible Bayesian type inference. The PennCNV software [53] uses a similar model. By utilizing the information in the B-allele frequency, these hidden Markov models can more accurately detect and genotype inherited CNVs, as compared to the procedures in Sect. 14.2.

Methods based on discrete-state hidden Markov models are designed for detecting copy number variants in normal tissue, where the assumption of idealized

Table 14.1 Hidden states, associated copy numbers, genotype states, and biological interpretation in [8]

Hidden state	Copy number	Number of genotypes	Interpretation
1	0	0	Full deletion
2	1	1	Single copy loss
3	2	3	Normal
4	2	2	Normal LOH
5	3	4	Single copy gain
6	4	5	Double copy gain

Table 14.2 Relationship between the inherited allele configuration s_t and the true allele specific copy numbers x_t in [7]

s_t	x_t^A	x_t^B
AA	$\theta_t^1 + \theta_t^2$	0
AB	θ_t^1	θ_t^2
BA	θ_t^2	θ_t^1
BB	0	$\theta_t^1 + \theta_t^2$

unit-copy changes holds because the cells within the samples are usually homogeneous. By requiring a fixed set of predefined discrete states, these methods do not adapt well to data from heterogeneous tumor samples, which produce data with apparently fractional copy number changes. Through simulated titration studies, [48] show that methods relying on idealized genotype states lose sensitivity when tumors are diluted with normal cells.

To treat the heterogeneity in tumor samples, [7] propose a continuous-state hidden Markov model to simultaneously estimate the parent specific DNA copy numbers and the unknown genotypes. To describe their model, let $y = \{y_t = (y_t^A, y_t^B) : t = 1, \ldots, n\}$ be the normalized intensity values for alleles A and B at n SNPs ordered by their location in a reference genome. Let $\theta = \{\theta_t = (\theta_t^1, \theta_t^2) : t = 1, \ldots, n\}$ be the underlying parent specific copy numbers, and $s_t \in S = \{AA, AB, BA, BB\}$ be the configuration at SNP t specifying the alleles carried by the inherited chromosomes. Let x_t be the true copy numbers of alleles A and B at SNP t. The value of x_t is determined by θ_t and s_t, with the mapping shown in Table 14.2. Note that when a somatic event causes a change in copy number in one or both parental chromosomes at SNP t, the allele-specific copy numbers x_t change, but s_t remains fixed. For example, if $s_t = AB$ and if θ_t^1 were amplified two-fold, then the true copy number of allele A would be 2, but s_t would still be AB. The *observed* allele specific intensities y_t are assumed to be equal to the true allele specific quantities plus an independent measurement error,

$$y_t = x_t + \varepsilon_t, \tag{14.4}$$

where $\varepsilon_t \sim N(0, \Sigma_{s_t})$, and Σ_{s_t} are state-specific error covariance matrices. This model is summarized in Fig. 14.3. The inherited allele configurations s_t are as-

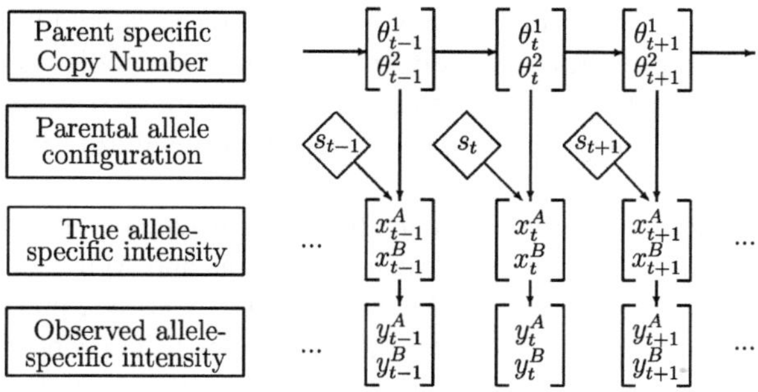

Fig. 14.3 Overview of stochastic segmentation model in [7]

sumed to be i.i.d. multinomial with parameters $(p_t^{AA}, p_t^{BA}, p_t^{AB}, p_t^{BB})$, which can be obtained from the genotyping data of a set of normal control samples. The dynamics of θ is modeled as a Markov jump process that generalizes the one-dimensional model in [26] to two dimensions. Conceptually, each time a jump occurs, θ_t can either go to a baseline state which consists of a point mass at a predefined value or choose a new value in $\Re^2$ according to a bivariate Gaussian distribution. Chen et al. [7] generalizes the estimation procedure in [26] to simultaneously estimate s and θ from y. The parameters of the hidden Markov model can be estimated from the data via expectation maximization. A plot of the major and minor copy numbers estimated from their procedure for the example region given in Fig. 14.2a is given in Fig. 14.2b. From the plot we see that this region contains an unbalanced gain of two copies in one chromosome coupled with an almost complete deletion of the other chromosome. This is immediately followed by gain of one chromosome with the other fixed at normal level. Without analyzing allele-specific data, we would not have known that the gain in total copy number at the left end of this region actually involves an almost complete loss of heterozygosity.

14.4 Integration of Multiple Array Platforms

With the rapid development of new genome-wide profiling platforms, there is now an increasing need of data integration when more than one technical platform, or different replicates on the same platform, are used to assay the same biological samples. For example, the Cancer Genome Atlas (TCGA) project, an NIH-funded initiative to characterize DNA, RNA, and epigenetic abnormalities in tumors, has adopted three independent platforms for studying DNA copy number variants (CNVs) in its pilot phase: Affymetrix SNP 6.0 arrays, Illumina HumanHap 550K SNP arrays, and Agilent CGH 244K arrays. The conventional approach for analyzing these types of data is to apply existing segmentation algorithms to search for copy number changes using the data from each platform separately. The segmentation results from all platforms then need to be combined. However, the platforms often disagree on the calling of a change, either on its significance or on its location and magnitude. In such situations, it is difficult to decide what the consensus should be. Furthermore, by segmenting each platform separately, information is not pooled across platforms for boosting the power of hard-to-detect signals. For these reasons, [2] and [59] proposed methods to integrate the data across platforms during the segmentation process.

To integrate the data during segmentation, the differences between platforms need to be resolved. Some platforms, such as Illumina and Agilent, produce allele-specific data, while others, such as Agilent and cDNA arrays, produce two-color ratio data that measure only total intensity. The probes from each platform map to a different set of locations in the genome, at different densities. For example, the Affymetrix 6.0 array has 1.8 million probes, while the Agilent CGH 244K array has about one-sixth as many probes. The measurements from different platforms also have different signal-to-noise ratios, as can be seen from Fig. 14.1. Furthermore, the different platforms respond with different saturation curves [2]. In regions of high-fold amplification, Illumina and Affymetrix tend to have more pronounced signal saturation than Agilent. In short, each of the three platforms has its advantages and disadvantages, but together they produce a balanced genome-wide survey for each sample and represent a much denser coverage than each platform does alone.

Bengtsson et al. [2] studied the problem of differing degrees of attenuation of the true copy number changes across platforms and proposed a method based on principal curves to correct for between platform differences. Their method brings the unsegmented intensity levels to the same scale across sources. For the same underlying true copy number, each platform should have the same mean value after this normalization procedure. Then, existing segmentation methods can be applied to a combined set of intensity levels over all sources to identify the copy number changes. However, the normalization procedure in [2] does not resolve the issue of differing error variances between sources, and the homogeneous variance model in (14.3) would not be optimal when applied to this combined data set.

Zhang et al. [60] proposed a multiplatform Circular Binary Segmentation (MPCBS) procedure, which has been adopted in the processing of the TCGA glioblastoma samples. MPCBS sums statistical evidence across platforms with proper scaling and do not require a prestandardization of different data sources. The statistics are based on maximizing the likelihood of a simple multiplatform model, which can be formulated as follows. Let the platforms be indexed by $k = 1, \ldots, K$, with K being the total number of platforms.The observed data is $\mathbf{y}_k = y_{k1}, \ldots, y_{kn_k}$ for the n_k snps/clones on the kth platform, which have ordered locations $(t_{k1}, \ldots, t_{kn_k})$ along a chromosome. It is assumed that for each platform, the data has been normalized to be centered at 0 for "normal" copy number and to have Gaussian noise. The fact that all $\{\mathbf{y}_k : k = 1, \ldots, K\}$ are assaying the same biological sample implies that at any genomic location t there is only one true underlying copy number μ_t for all platforms. Let $f_k(\cdot)$ be the response function for platform k, which quantifies the dependence of the mean intensity on the underlying copy number. The observed intensity level for the ith probe of the kth platform is modeled as

$$y_{ki} = f_k(\mu_{t_{k,i}}) + \varepsilon_{k,i}, \tag{14.5}$$

where the noise terms $\varepsilon_{k,i}$ are independently distributed $N(0, \sigma_k^2)$, and σ_k^2 is the platform-specific noise variance. Zhang et al. [60] consider only linear response functions $f_k(\mu) = r_k \mu$, where the parameter r_k, called the response ratio, describes the ratio between the change in signal intensity and the underlying copy number change for platform k. This linearity assumption allows for simple and intuitive test statistics and fast scanning algorithms. The true copy number μ_t is modeled as a piecewise constant function as in (14.2), where the endpoint n is replaced by the length of the chromosomal region in base pairs. The magnitude parameters $\theta = (\theta_0, \ldots, \theta_m)$ and change-points $\tau = (\tau_1, \ldots, \tau_m)$ are all unknown and, like the response ratios, must be estimated from the data.

It is insightful to look at the generalized likelihood ratio statistic that arises from this cross-platform model and compare it to (14.3) for the one sample case and, later, to (14.12) for the multisample case described in the next section. Consider the case where $r = (r_1, \ldots, r_K)$ is known, and the goal is to test whether there is a CNV at a window from s to t. Under the *null* hypothesis that there is no change, the data within this region should have baseline mean $f_k(0) = 0$, i.e.,

$$H_0: \quad y_{ki} \sim N\big(0, \sigma_k^2\big) \quad \text{for } k = 1, \ldots, K \text{ and } i: \; s \le t_{ki} < t. \tag{14.6}$$

If there is a gain (or loss) of magnitude μ, each platform should respond with signal $f_k(\mu) = r_k \mu$. The signal is thus a mean shift in a *common direction* for all platforms, with the observed magnitude of shift being $r_k \mu$ for platform k, i.e.,

$$H_A: \quad y_{ki} \sim N\big(r_k \mu, \sigma_k^2\big) \quad \text{for } k = 1, \ldots, K \text{ and } i: \; s \le t_{ki} < t. \tag{14.7}$$

Let $n_k(s,t) = |\{i : t_{k,i} \in (s,t]\}|$ be the number of probes from the kth platform that falls within $(s,t]$, and $\bar{y}_{k,(s,t]}$ be the mean intensity of these probes. The generalized log-likelihood ratio statistic for testing H_A versus H_0 is

$$Z(s,t) = \frac{[\sum_{k=1}^{K} \delta_{k,s,t} X_{k,s,t}]^2}{\sum_{k=1}^{K} \delta_{k,s,t}^2}, \tag{14.8}$$

where

$$X_{k,s,t} = \frac{\bar{y}_{k,[s,t]} - \bar{y}_{k,[1,n_k]}}{\sigma_k \sqrt{n_k(s,t)^{-1} + n_k^{-1}}} \tag{14.9}$$

and

$$\delta_{k,s,t} = r_k \sqrt{n_k(s,t)} / \sigma_k. \tag{14.10}$$

$X_{k,s,t}$ is equivalent to the statistic in (14.3) computed for the kth platform, and the cross-platform statistic is the projection of $X_{s,t} = (X_{1,s,t}, \ldots, X_{K,s,t})$ on to the vector $\delta = (\delta_{1,s,t}, \ldots, \delta_{K,s,t})$. We thus call (14.8) the projected χ^2 statistic. It can also be viewed as the squared norm of a weighted sum of t-test statistics, where the weight $\delta_{k,s,t}$ for platform k is proportional to the response ratio r_k, the square root of the number of probes from that platform that falls into $[s,t)$, and the inverse of

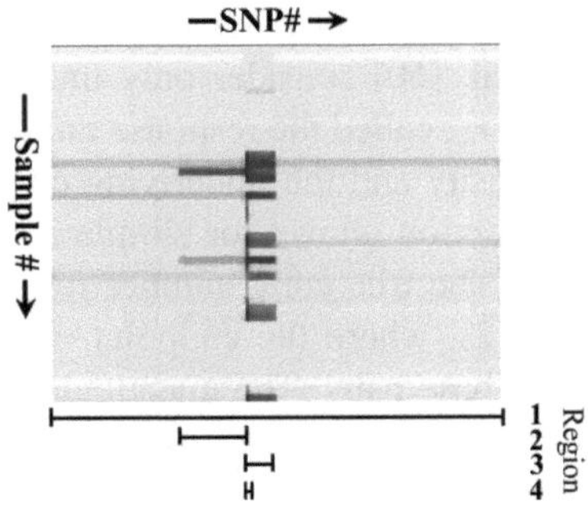

Region	Sample Number		
	1	...	N
1	X_{11}	...	X_{1N}
2	X_{21}	...	X_{2N}
3	X_{31}	...	X_{3N}
4	X_{41}	...	X_{4N}

Fig. 14.4 An example of a joint segmentation of a set of tumor samples. The segmentation outputs a set of common change-points to give the best sparse summary of the set of tumors (for the color version, see Color Plates on p. 393)

the error standard deviation σ_k. When there is only one platform, the statistic (14.8) is equivalent to the chi-square statistic used in the Circular Binary Segmentation algorithm of [34]. As for CBS, σ_k is usually unknown and must be estimated from the data. For more details on the estimation of the platform response ratios, see [59].

14.5 Modeling Recurrence Across Samples

When the same biological sample is assayed using multiple platforms, the underlying signal for the copy number profiles from each platform should be the same. This is the concept that underlies the projected χ^2 statistic (14.8) in the MPCBS algorithm. However, when each copy number profile represents a different biological sample, the underlying signal is no longer shared. Usually, only a fraction of the samples are carriers of any given CNV. In an integrated analysis of copy number data across multiple biological samples, one is often interested in the differences and similarities across samples.

Before introducing the statistical models for cross-sample analysis, we need to examine more carefully the purpose of cross-sample integration. What do we hope to achieve in such an analysis? What types of signals are we aiming to capture across samples? The answer to this question is simple for multiplatform integration, where the goal is simply to combine data across platforms to obtain a better estimate of the shared underlying change-points. In cross sample analyses, how should the concept of a shared signal be defined? When the signals are not shared, how should the variation be characterized?

One goal in copy number studies over a cohort of tumor samples is finding regions of recurrent aberration. Such regions, where a large number of samples of the same type of tumor have gained or lost copies, may contain genes that are key players in the development of the tumor. For example, Fig. 14.4 shows a set of tumor samples, with many samples carrying overlapping deletions covering chromosome 9 [43]. Such commonly deleted regions may carry genes that play a role in cell proliferation or delay apoptosis. Similarly, commonly amplified regions may harbor tumor suppressor genes. In Sect. 14.5.1, we review methods that are geared towards identifying these regions.

The methods reviewed in Sect. 14.5.1 combine information across samples post-segmentation. That is, each sample is segmented on its own, and the cross sample analysis sees only the segmented data. However, in some cases, such as inherited CNVs, the change-points are shared across samples for instances of the same CNV. In such cases, aggregating information across samples can improve the power of detecting shared weak signals. Consideration of power is especially relevant to the detection of inherited CNVs, most of which are very short and may only span a few probe sets or clones, and thus are easily missed in single-sample detection methods. Inherited variants are also hard to detect in the sense that they usually involve single-copy changes, as compared to aberrations in tumors which often consist of high-fold amplifications and homozygous deletions. In Sect. 14.5.2, we discuss the aggregation of data across samples prior to or during segmentation.

In the analysis of both inherited and somatic copy number variants, it is often useful to obtain a sparse cross-sample summary of a complex region for use in downstream analyses. For example, in clinical studies we may have, along with the copy number data, variables such as survival outcome or status of other biomarkers. We may want to find chromosomal regions whose copy number status is correlated with these variables. These types of analyses are often done with gene or protein expression data, but for copy number data, it is unclear what to use as the explanatory variables. If each probe were considered as a variable, then the smoothness of the underlying signal is ignored. Since copy number studies now routinely use platforms containing hundreds of thousands to over a million probes, if each probe were considered as a variable, we would be faced with a very large number of highly correlated variables, which would reduce the sensitivity of downstream analyses. Some studies take the average copy number over each chromosome, chromosome arm, or cytoband as the variables for downstream analysis. This clearly is a coarse method that sacrifices sensitivity. In Sect. 14.5.3, we describe a method for reducing a set of copy number profiles into a set of representative regions, so that the average copy number of each sample in each region gives a good summary of the cohort.

14.5.1 Post-Segmentation Procedures

In post-segmentation procedures [4, 11, 16, 32, 33, 42, 49], each sample is first segmented separately, which reduces them to piecewise constant sequences indicating regions of amplification, deletion, or normal copy number. Then, the samples are aligned, and a statistical model [32, 33] or permutation-based approach [11] is used to identify regions of highly recurrent aberration.

One of the earliest methods is STAC [11], which takes in a binary sequence for each sample that indicates whether the sample contains an aberration at each probe position. Consider first the simple method which considers only the prevalence of

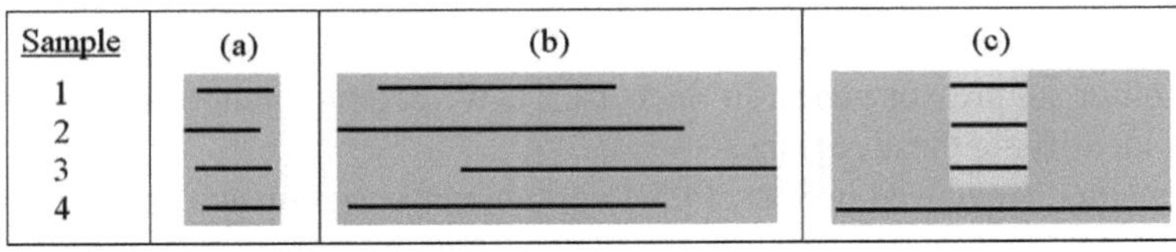

Fig. 14.5 Illustration of the concept of a footprint of in the STAC program. In each of the scenarios (**a–c**), the *gray box* indicates the footprint for the entire stack of four samples. In (**c**), the *light gray box* indicates the footprint for the stack containing only samples 1–3

aberration at each location m, denoted by $F(m)$. The p-value of $F(m)$ can be computed by permutation, where the intervals within each profile are randomly rearranged. Let $F_i(m)$ be the prevalence over samples at location m for permutation i. Then, the p-value of a location m_0 is computed by

$$P_F(m_0) = \frac{|\{i : \max_m F_i(m) > F(m_0)\}}{\text{total number of permutations}}.$$

Locations that are aberrant in a large number of samples have a low p-value. However, the statistic $F(m)$ does not capture the fact that it is more surprising when several samples have tight overlap for a short aberrant interval (Fig. 14.5a), as opposed to when an overlap is a result of long aberrations that do not align tightly (Fig. 14.5b). To differentiate between these two situations, [11] defines a "stack" to be a set of intervals that share at least one common position and the "footprint" of a stack to be the union of the sets of positions covered by the intervals. Each of the set of intervals (a–c) in Fig. 14.5 is a stack, but Fig. 14.5b has a larger footprint than Fig. 14.5a. A smaller footprint provides greater evidence for localization of an important gene in the region. In Fig. 14.5c, the set of all 4 intervals is a stack, but so is any subset, e.g., the intervals from only samples 1–3. The stack for samples 1–3 has a much smaller footprint than the stack containing all four samples and may be more biologically interesting. For each position m, let S_m to be the set of all stacks with m as a common position, i.e., the set of all subsets of samples that are aberrant at position m. For each $s \in S_m$, let $P(s)$ be the p-value for the footprint of the stack s computed by permutation. Then, glossing over details, the score for the most significant footprint is computed as $R(m) = \min_{s \in S_m} P(s)$. A related method is MSA, which builds upon the notions of frequency and footprint but relies on the original intensity data and searches over a set of possible cut-off values in the segmentation procedure. As the permutation-based p-values are computationally intensive, MSA also contains algorithmic improvements which reduce the execution time.

Another example of a method in this category is GISTIC [4]. Unlike STAC and MSA, which consider only the location and length of the aberrant intervals, GISTIC also factors in the amplitude of the aberration in each sample. The rationale given for this is that the prevalence of the aberration among samples and the average amplitude of these events are both positively associated with the likelihood that a region carries driver aberrations. Beroukhim et al. [4] define the G-score as the prevalence of the copy-number change times the average amplitude over the carriers. Permutation tests are used to compute the significance of the observed G-scores, and regions with maximal G-scores are selected.

It is important to note that, since these methods use segmented data as input, the quality of their results depend on the reliability of the underlying segmentation algorithm. All segmentation methods incur errors, and these methods assume that it is more likely for biologically significant aberrations to recur across samples than experimental or statistical errors. However, many errors in segmentation are due to experimental artifacts, such as local trends, which also recur across samples at the same locations. These artifacts, if not carefully removed, may be misconstrued as significant recurrent regions. Another concern is that, while it is likely that the recurrent regions are due to driver mutations, they may also be a result of biases in the DNA mutation or repair machinery. Thus, care must be taken in their interpretation.

14.5.2 Cross-Sample Detection of Inherited Variants

The methods described in the previous section pool information across samples post segmentation. In this section, we consider methods for joint segmentation of a cohort of samples. For detecting inherited CNVs, such joint segmentation methods have been shown to boost power [60]. Also, since different samples have different signal quality and noise characteristics, integrating them during segmentation can account for these differences.

When testing for a change in mean in a single sequence, the two quantities that affect the power of detection are the height of the jump and the width of the changed interval. The generalized likelihood ratio statistic (14.3) is a function of these two quantities. When multiple samples are simultaneously scanned, a third quantity, the number of carriers of the change, should be factored into the scan statistic. When more than one sample show evidence for a change, then the change is more likely to be real. By pooling samples in the segmentation step, we are capitalizing on this fact to boost power.

There are now several models for multisample joint segmentation of copy number data. The HMM-based approach of [54] focuses on the analysis of cancer data and does not assume the change-points to be shared across samples. The authors mention that a shared change-point model would be desirable for the detection of inherited CNVs, and they note the substantial computational task inherent in a satisfactory HMM for this problem. Such an undertaking is reported in [44], where a multilayer hierarchical hidden Markov model is used to segment all samples simultaneously. Shah et al. [44] assumes an underlying "master" Markov chain which decides whether the samples, as a group, should enter a "changed" state. Given that the master has entered the changed state, each sample can choose, with a flip of coin, whether to jump to a shared mean level or to stay at the baseline level. This model assumes that all carriers must change in the same direction with the same magnitude. An MCMC algorithm is proposed to sample from the posterior distribution of the master state.

Hidden Markov models for this problem rest on many assumptions about how the aberrations are shared across samples, e.g., they must be in a common direction or must be present in a majority of the samples. For a less restrictive approach, Zhang et al. [60] proposed a simultaneous change-point model for the detection of inherited changes. To describe the model, let the observed data be a two-dimensional array $\{y_{it} : i = 1, \ldots, N,\ t = 1, \ldots, T\}$, where y_{it} is the data point for the ith sample at probe t, N is the total number of samples, and T is the total number of probes. For each sample i, the random variables $y_i = \{y_{it} : t = 1, \ldots, T\}$ are mutually independent and Gaussian with mean values μ_{it} and variances σ_i^2. The null hypothesis assumes that the means for each profile are identical across locations. The simplest alternative where there is a single changed interval assumes that there exist integer values $1 \leq \tau_1 < \tau_2 \leq T$ and at least one sample i such that

$$\mu_{it} = \mu_i + \delta_i I_{\{\tau_1 < t \leq \tau_2\}}, \tag{14.11}$$

where the δ_i are nonzero constants, and μ_i is the baseline mean level for profile i. For this testing problem, a direct generalization of (14.3) is $\max_{s<t} Z(s,t)$, where

$$Z(s,t) = \sum_{i=1}^{N} U_i^2(s,t), \tag{14.12}$$

and $U_i(s,t)$ is the sequence-specific statistic defined as in (14.3) for the ith sequence. If the variances were known, then (14.12) would be the generalized log-likelihood ratio statistic for testing H_0 versus H_A. For each fixed $s < t$, the null distribution of $Z(s,t)$ is approximately χ^2 with N degrees of freedom. Large values of $\max_{s<t} Z(s,t)$ are evidence against the null hypothesis. If the null hypothesis is rejected, the maximum likelihood estimate of the location of the variant interval is $(s^*, t^*) = \operatorname{argmax}_{s,t} Z(s,t)$. Zhang et al. [60] developed analytic p-value approximations for scans using statistics of the form (14.12).

Similar to the sum of chi-squares statistic is the "interval scores" method of [28], which uses a statistic like $Z(s,t)$ but without the squares. Thus, like [44], the method focuses on common deletions or common amplifications. However, many inherited CNVs have changes in both directions at any given locus. This is because the individual copy numbers are defined relative to the population average, and when two or more copy number levels exist in the population for a given locus, normalizing to the average creates both "gains" and "losses" among the carriers.

To assess the improvement in sensitivity gained from pooling data across samples, [60] used a set of 62 samples assayed using Illumina 550K Beadchips. The experiments were performed on DNA samples extracted from lymphoblastoid cell lines derived from 10 sets of trios consisting of a child and his/her two parents, and 16 pairs of technical replicates for 16 independent DNA samples. Zhang et al. showed that using statistic (14.12) improves the concordance rate between replicates and between the child and parent samples, as compared to single sample scans.

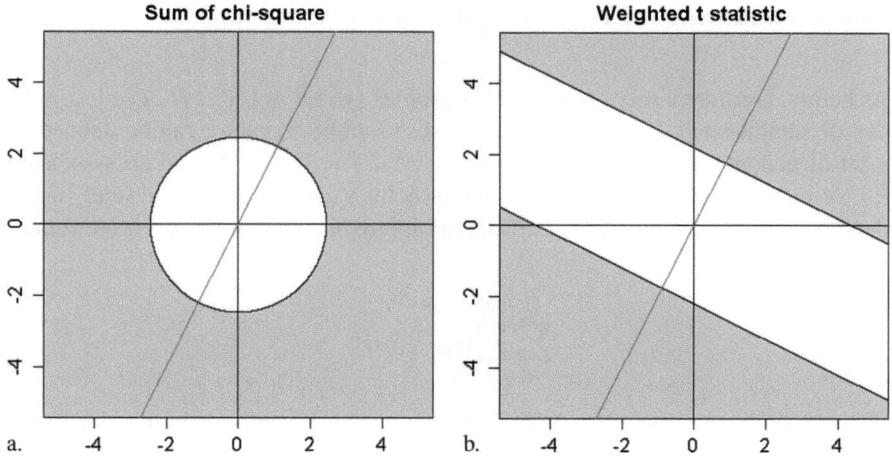

Fig. 14.6 (Color online) Comparison of the null hypothesis rejection regions between the sum of chi-square statistic (14.12) and the weighted t-statistic (14.8) on $K = 2$ platforms. In all figures, the axes are the magnitudes of the X variables (14.9) for platforms 1 and 2. A significance level of 0.05 is used to determine the decision boundaries of both statistics. For (**b**), weights of $\delta_1 = 1$ and $\delta_2 = 2$ are used. The *red line* shows the direction of the weight vector $\delta = (\delta_1, \delta_2)$

Note the differences between the projected chi-squares statistic (14.8) and the sum-of-chi-squares statistic (14.12) for multisample segmentation. When pooling data across independent biological samples, not all samples are expected to carry the same CNV, and often both deletions and amplifications can be observed between the samples at the same genome location. This is why the sum-of chi-squares statistic (14.12) does not "reward" agreement in direction of change between samples. In contrast, the statistic in (14.8) rewards agreement and penalizes disagreement. The difference between the two statistics is shown graphically in Fig. 14.6, where we illustrate the simple case of two samples/platforms with the response ratio of the second platform being twice that of the first platform. Suppressing the dependence on s, t, note that both statistics are functions of $X = (X_1, X_2)$, as defined in (14.9) which, assuming that σ_k is known, is bivariate Gaussian with mean 0 and identity covariance matrix under the null hypothesis. Figures 14.6a–b show in gray the region in the (X_1, X_2) plane where the null hypothesis will be rejected. For example, in Fig. 14.6a, which depicts the situation in (14.12), the rejection boundary is a circle centered at the origin. In Fig. 14.6b, which depicts the situation in (14.8), the rejection boundary is $\{X : \delta' X > t_\alpha\}$, which is perpendicular to the vector δ_2/δ_1. Importantly, Fig. 14.6b awards agreement between the two platforms, while Fig. 14.6a treats all quadrants of the plane equally. The statistic (14.8) (Fig. 14.6b) also allows one platform to dominate the others: In the case where the directions disagree, e.g., in the upper left or lower right quadrants, the consensus can still be made according to the dominant platform.

14.5.3 Obtaining a Cross-Sample Signature

As before, consider a set of copy number profiles $\{y_{it} : i = 1, \ldots, N,\ t = 1, \ldots, T\}$ over n samples and T probes. A sparse cross-sample signature can be defined as a set of genomic regions $R = \{(s_i, t_i) : s_i < t_i,\ i = 1, \ldots, M\}$ and an associated $n \times M$ matrix X such that an approximation $\hat{y}$ for y can be constructed solely using the information in R and X. To evaluate the approximation, one may use the sum of squared errors

$$\sum_{i=1}^{T} \sum_{j=1}^{N} (y_{ij} - \hat{y}_{ij})^2.$$

We seek signatures where $M \ll T$. Then, the matrix X can be used in place of the matrix y in downstream clustering, classification, and regression modeling. This is still largely an open problem, with much ongoing work. Here, we describe a solution given in [60] based on an extension of the CBS algorithm:

Algorithm 1 (Multisample Circular Binary Segmentation) *Fix the global significance level* α, *parameter* p, *and maximum window* $T_0 < T$. *We denote by* $Y_{h:k}$ *the matrix* $\{y_{i,t} : i = 1, \ldots, N,\ t = h, \ldots, k\}$.

1. *Initialize* $T_1 = 1$ *and* $T_2 = T$.
2. *Compute*

$$Z_{\max} = \max_{\substack{T_1 \le s < t \le T_2 \\ 1 \le t - s \le T_0}} \big\{Z(s,t)\big\}.$$

   Let (s^*, t^*) *be the maximizing interval.*
3. *If the* p*-value of* $Z_{\max}$*, as computed using approximations give in* [60], *is less than* α, *then for each* $(u, v) \in \{(T_1, s^* - 1), (s^*, t^*), (t^* + 1, T_2)\}$, *do*:
   a. *Determine the carriers of the variant. For all* $t = u, \ldots, v$, *if a sample carries the variation, let* $\hat{y}_{i,t} = \bar{y}_{i,u:v}$, *and for the other samples, let* $\hat{y}_{i,t} = \bar{y}_{i,T_1:T_2}$. *Let* $Y'_{u:v} = Y_{u:v} - \hat{Y}_{u:v}$, *where* $\hat{Y}_{u:v}$ *is the matrix* $\{\hat{y}_{i,t} : i = 1, \ldots, N,\ t = u, \ldots, v\}$.
   b. *Repeat steps 2–3 for* $T_1 = u$, $T_2 = v$, *and the newly normalized* $Y'_{u:v}$.

This algorithm, like CBS, recursively scans the genome for intervals that maximize the sum of chi-square statistic. If the p-value of the maximum is smaller than a predefined threshold, a cut is made at the maximizing interval. Then, the carriers of the variant, i.e., samples whose mean level actually changes in the interval, are determined. There are many ways of determining the carriers, some simple ad hoc solutions are given in [60]. A box-shaped signal is estimated for the carriers, while a flat line is fitted for the noncarriers. Residuals are taken, and the regions to the left, center, and right of the cut are recursively segmented using the same procedure.

The multisample CBS algorithm was used to analyze the 9p21 deletion in childhood leukemia [43] to give the sparse cross sample signature shown in Fig. 14.4.

14.6 Concluding Remarks

In this chapter, we surveyed some of the recent developments in the analysis of DNA copy number data from microarray platforms. These analyses started with a focus on single-sample total copy number segmentation. Now, new statistical and computational challenges arise in the proper extraction of allele-specific information from DNA genotyping arrays and in the analysis of DNA copy number data from multisample, multiplatform experiments. These applications have inspired new developments in change-point models, especially in the formulation of simultaneous change-point models over multiple sequences. The theory underlying these models may prove useful to other applications, especially other types of genome-wide profiling.

There is much ongoing work on the integrated analysis of DNA copy number data with gene expression, protein expression, and methylation data. Although such studies are well motivated on the biological side, there is a shortage of statistical models in this area. For recent progress on this problem, see [37] and [36].

Since detection of copy number variants is a zero cost by-product of the genotyping arrays used in association studies, there is much motivation for using them in existing association studies. However, further statistical work and biological evidence is need to determine how to utilize the copy number information in studies of genetic inheritance. Advances in this area rest on the understanding of how CNVs segregate in a population [41] and the selective pressures acting on CNVs [10]. For a nice survey on this problem, see [29].

When interpreting the results from microarray-based copy number studies, we must carefully note that these assays are noisy and prone to cross hybridization, especially in repetitive regions or regions with complex rearrangements [10]. Concordance of CNVs detected using microarrays with those detected through sequencing experiments is incredibly low [10, 31, 59]. To date, biological confirmation of CNV detection methods has been limited to small-scale experiments involving, for example, male vs. female copy number on the X chromosome or confirmation of a few known CNVs. New sequencing-based technologies, especially paired end mapping [22, 23], will be invaluable complementary tools in CNV discovery and in characterization of structural variants in the genome.

References

1. H. Bengtsson, R. Irizarry, B. Carvalho, and T.P. Speed. Estimation and assessment of raw copy numbers at the single locus level. *Bioinformatics*, **24**(6):759–767, 2008.
2. H. Bengtsson, A. Ray, P. Spellman, and T.P. Speed. A single-sample method for normalizing and combining full-resolution copy numbers from multiple platforms, labs and analysis methods. *Bioinformatics*, **25**(7):861–867, 2009.
3. R. Beroukhim, M. Lin, Y. Park, K. Hao, X. Zhao, L.A. Garraway, E.A. Fox, E.P. Hochberg, I.K. Mellinghoff, M.D. Hofer, A. Descazeaud, M.A. Rubin, M. Meyerson, W.H. Wong, W.R. Sellers, and C. Li. Inferring loss-of-heterozygosity from unpaired tumors using high-density oligonucleotide snp arrays. *PLoS Comput Biol*, **2**:e41, 2006.

4. R. Beroukhim, G. Getz, L. Nghiemphu, J. Barretina, T. Hsueh, D. Linhart, I. Vivanco, C.L. Jeffrey, J.H. Huang, S. Alexander, J. Du, T. Kau, R.K. Thomas, K. Shah, H. Soto, S. Perner, J. Prensner, R.M. Debiasi, F. Demichelis, C. Hatton, M.A. Rubin, L.A. Garraway, S.F. Nelson, L. Liau, Mischel, T.F. Cloughesy, M. Meyerson, T.A. Golub, E.S. Lander, I.K. Mellinghoff, and W.R. Sellers. Assessing the significance of chromosomal aberrations in cancer: Methodology and application to glioma. *Proc Nat Acad Sci*, 0710052104+, December 2007.
5. G.R. Bignell, J. Huang, J. Greshock, S. Watt, A. Butler, S. West, M. Grigorova, K.W. Jones, W. Wei, M.R. Stratton, P.A. Futreal, B. Weber, M.H. Shapero, and R. Wooster. High-resolution analysis of DNA copy number using oligonucleotide microarrays. *Genome Res*, **14**(2):287–295, 2004.
6. P. Broët and S. Richardson. Detection of gene copy number changes in CGH microarrays using a spatially correlated mixture model. *Bioinformatics*, **22**:911–918, 2006.
7. H. Chen, H. Xing, and N.R. Zhang. Estimation of parent specific DNA copy number in tumors using high-density genotyping arrays. Technical Report, Department of Statistics, Stanford University, 2009.
8. S. Colella, C. Yau, J.M. Taylor, G. Mirza, H. Butler, P. Clouston, A.S. Bassett, A. Seller, C.C. Holmes, and J. Ragoussis. QuantiSNP an objective Bayes hidden-Markov model to detect and accurately map copy number variation using SNP genotyping data. *Nucleic Acids Res*, **35**(6):2013–2025, 2007.
9. D. Conrad, T. Andrews, N. Carter, M. Hurles, and J. Pritchard. A high-resolution survey of deletion polymorphism in the human genome. *Nat Genet*, **38**:75–81, 2006.
10. G.M.M. Cooper, T. Zerr, J.M.M. Kidd, E.E.E. Eichler, and D.A.A. Nickerson. Systematic assessment of copy number variant detection via genome-wide SNP genotyping. *Nat Genet*, **40**:1199–1203, 2008.
11. S.J. Diskin, T. Eck, J. Greshock, Y.P. Mosse, T. Naylor, C.J. Stoeckert Jr., B.L. Weber, J.M. Maris, and G.R. Grant. STAC: A method for testing the significance of DNA copy number aberrations across multiple array-CGH experiments. *Genome Res*, **16**:1149–1158, 2006.
12. D.A. Engler, G. Mohapatra, D.N. Louis, and R.A. Betensky. A pseudolikelihood approach for simultaneous analysis of array comparative genomic hybridizations. *Biostatistics*, **7**:399–421, 2006.
13. X. Estivill and L. Armengol. Copy number variants and common disorders: Filling the gaps and exploring complexity in genome-wide association studies. *PLoS Genet*, **3**(10):e190+, 2007.
14. J. Fridlyand, A. Snijders, D. Pinkel, D.G. Albertson, and A.N. Jain. Application of hidden Markov models to the analysis of the array-CGH data. *J Multivar Anal*, **90**:132–153, 2004.
15. S. Guha, Y. Li, and D. Neuberg. Bayesian hidden Markov modeling of array CGH data. Harvard University Biostatistics Working Paper Series, 2006.
16. M. Guttman, C. Mies, K. Dudycz-Sulicz, S.J. Diskin, D.A. Baldwin, C.J. Stoeckert, and G.R. Grant. Assessing the significance of conserved genomic aberrations using high resolution genomic microarrays. *PLoS Genet*, **3**(8):e143+, 2007.
17. L. Hsu, S.G. Self, D. Grove, T. Randolph, K. Wang, J.J. Delrow, L. Loo, and P. Porter. Denoising array-based comparative genomic hybridization data using wavelets. *Biostatistics*, **6**:211–226, 2005.
18. P. Hupé, N. Stransky, J.P. Thiery, F. Radvanyi, and E. Barillot. Analysis of array CGH data: from signal ratio to gain and loss of DNA regions. *Bioinformatics*, **20**(18):3413–3422, 2004.
19. A.S. Ishkanian, C.A. Malloff, S.K. Watson, R.J. Deleeuw, B. Chi, B.P. Coe, A. Snijders, D.G. Albertson, D. Pinkel, M.A. Marra, V. Ling, C. Macaulay, and W.L. Lam. A tiling resolution DNA microarray with complete coverage of the human genome. *Nat Genet*, **36**(3):299–303, 2004.
20. B. James, K.L. James, and D. Siegmund. Tests for a change-point. *Biometrika*, **74**:71–83, 1987.
21. R. Khaja, J. Zhang, J.R. MacDonald, Y. He, A.M. Joseph-George, J. Wei, M.A. Rafiq, C. Qian, M. Shago, L. Pantano, H. Aburatani, K. Jones, R. Redon, M. Hurles, L. Armengol, X. Estivill, R.J. Mural, C. Lee, S.W. Scherer, and L. Feuk. Genome assembly comparison to identify structural variants in the human genome. *Nat Genet*, **38**:1413–1418, 2007.

22. J.M. Kidd, G.M. Cooper, W.F. Donahue, H.S. Hayden, N. Sampas, T. Graves, N. Hansen, B. Teague, C. Alkan, F. Antonacci, E. Haugen, T. Zerr, A.N. Yamada, P. Tsang, T.L. Newman, E. Tüzün, Z. Cheng, H.M. Ebling, N. Tusneem, R. David, W. Gillett, K.A. Phelps, M. Weaver, D. Saranga, A. Brand, W. Tao, E. Gustafson, K. Mckernan, L. Chen, M. Malig, J.D. Smith, J.M. Korn, S.A. Mccarroll, D.A. Altshuler, D.A. Peiffer, M. Dorschner, J. Stamatoyannopoulos, D. Schwartz, D.A. Nickerson, J.C. Mullikin, R.K. Wilson, L. Bruhn, M.V. Olson, R. Kaul, D.R. Smith, and E.E. Eichler. Mapping and sequencing of structural variation from eight human genomes. *Nature*, **453**(7191):56–64, 2008.
23. J.O. Korbel, A.E. Urban, J.P. Affourtit, B. Godwin, F. Grubert, J.F. Simons, P.M. Kim, D. Palejev, N.J. Carriero, L. Du, B.E. Taillon, Z. Chen, A. Tanzer, A.C. Saunders, J. Chi, F. Yang, N.P. Carter, M.E. Hurles, S.M. Weissman, T.T. Harkins, M.B. Gerstein, M. Egholm, and M. Snyder. Paired-end mapping reveals extensive structural variation in the human genome. *Science*, **318**:420–426, 2007.
24. T. LaFramboise, B.A. Weir, X. Zhao, R. Beroukhim, C. Li, D. Harrington, W.R. Sellers, and M. Meyerson. Allele-specific amplification in cancer revealed by SNP array analysis. *PLoS Comput Biol*, **1**:e65, 2005.
25. W.R. Lai, M.D. Johnson, R. Kucherlapati, and P.J. Park. Comparative analysis of algorithms for identifying amplifications and deletions in array CGH data. *Bioinformatics*, **21**:3763–3770, 2005.
26. T.L. Lai, H. Xing, and N.R. Zhang. Stochastic segmentation models for array-based comparative genomic hybridization data analysis. *Biostatistics*, **9**:290–307, 2007.
27. M. Lin, L.-J. Wei, W.R. Sellers, M. Lieberfarb, W.H. Wong, and C. Li. dChipSNP: significance curve and clustering of SNP-array-based loss-of-heterozygosity data. *Bioinformatics*, **20**(8):1233–1240, 2004.
28. D. Lipson, Y. Aumann, A. Ben-Dor, N. Linial, and Z. Yakhini. Efficient calculation of interval scores for DNA copy number data analysis. *J Comput Biol*, **13**:215–228, 2006.
29. S.A. McCarroll. Copy-number analysis goes more than skin deep. *Nat Genet*, **40**(1):5–6, 2008.
30. S.A. McCarroll, T.N. Hadnott, G.H. Perry, P.C. Sabeti, M.C. Zody, J.C. Barrett, S. Dallaire, S.B. Gabriel, C. Lee, M.J. Daly, and D.M. Altshuler. The International HapMap Consortium. Common deletion polymorphisms in the human genome. *Nat Genet*, **38**:86–92, 2006.
31. S.A.A. McCarroll, F.G.G. Kuruvilla, J.M.M. Korn, S. Cawley, J. Nemesh, A. Wysoker, M.H.H. Shapero, P.I.W.I. de Bakker, J.B.B. Maller, A. Kirby, A.L.L. Elliott, M. Parkin, E. Hubbell, T. Webster, R. Mei, J. Veitch, P.J.J. Collins, R. Handsaker, S. Lincoln, M. Nizzari, J. Blume, K.W.W. Jones, R. Rava, M.J.J. Daly, S.B.B. Gabriel, and D. Altshuler. Integrated detection and population-genetic analysis of SNPs and copy number variation. *Nat Genet*, **40**:1166–1174, 2008.
32. M. Newton and Y. Lee. Inferring the location and effect of tumor suppressor genes by instability-selection modeling of allelic-loss data. *Biometrics*, **56**:1088–1097, 2000.
33. M. Newton, M. Gould, C. Reznikoff, and J. Haag. On the statistical analysis of allelic-loss data. *Stat Med*, **17**:1425–1445, 1998.
34. A.B. Olshen, E.S. Venkatraman, R. Lucito, and M. Wigler. Circular binary segmentation for the analysis of array-based DNA copy number data. *Biostatistics*, **5**:557–572, 2004.
35. D.A. Peiffer, J.M. Le, F.J. Steemers, W. Chang, T. Jenniges, F. Garcia, K. Haden, J. Li, C.A. Shaw, J. Belmont, S.W. Cheung, R.M. Shen, D.L. Barker, and K.L. Gunderson. High-resolution genomic profiling of chromosomal aberrations using infinium whole-genome genotyping. *Genome Res*, **16**(9):1136–1148, 2006.
36. J. Peng, P. Wang, N.F. Zhou, and J. Zhu. Partial correlation estimation by joint sparse regression model. *J Am Stat Assoc*, **104**(486):735–746, 2009.
37. J. Peng, J. Zhu, A. Bergamaschi, W. Han, D.Y. Noh, J.R. Pollack, and P. Wang. Regularized multivariate regression for identifying master predictors with application to integrative genomics study of breast cancer. *Ann Appl Stat*, 2010, in press.
38. F. Picard, S. Robin, M. Lavielle, C. Vaisse, and J. Daudin. A statistical approach for array CGH data analysis. *BMC Bioinform*, **6**:27, 2005.

39. D. Pinkel, R. Segraves, D. Sudar, S. Clark, I. Poole, D. Kowbel, C. Collins, W.L. Kuo, C. Chen, Y. Zhai, S.H. Dairkee, B.M. Ljung, J.W. Gray, and D.G. Albertson. High resolution analysis of DNA copy number variation using comparative genomic hybridization to microarrays. *Nat Genet*, **20**(2):207–11, 1998.
40. J.R. Pollack, C.M. Perou, A.A. Alizadeh, M.B. Eisen, A. Pergamenschikov, C.F. Williams, S.S. Jeffrey, D. Botstein, and P.O. Brown. Genome-wide analysis of DNA copy-number changes using cDNA microarrays. *Nat Genet*, **23**:41–46, 1999.
41. R. Redon, S. Ishikawa, K.R. Fitch, L. Feuk, G.H. Perry, D.T. Andrews, H. Fiegler, M.H. Shapero, A.R. Carson, W. Chen, E.K. Cho, S. Dallaire, J.L. Freeman, J.R. Gonzalez, M. Gratacos, J. Huang, D. Kalaitzopoulos, D. Komura, J.R. Macdonald, C.R. Marshall, R. Mei, L. Montgomery, K. Nishimura, K. Okamura, F. Shen, M.J. Somerville, J. Tchinda, A. Valsesia, C. Woodwark, F. Yang, J. Zhang, T. Zerjal, J. Zhang, L. Armengol, D.F. Conrad, X. Estivill, C. Tyler-Smith, N.P. Carter, H. Aburatani, C. Lee, K.W. Jones, S.W. Scherer, and M.E. Hurles. Global variation in copy number in the human genome. *Nature*, **444**:444–454, 2006.
42. C. Rouveirol, N. Stransky, P. Hupé, P. La Rosa, E. Viara, E. Barillot, and F. Radvanyi. Computation of recurrent minimal genomic alterations from array-CGH data. *Bioinformatics*, **22**:849–856, 2006.
43. J.D. Schiffman, Y. Wang, L.A. Mcpherson, K. Welch, N. Zhang, R. Davis, N.J. Lacayo, G.V. Dahl, M. Faham, and J.M. Ford. Molecular inversion probes reveal patterns of 9p21 deletion and copy number aberrations in childhood leukemia. *Cancer Genet Cytogenet*, **193**(1):9–18, 2009.
44. S.P. Shah, W.L. Lam, R.T. Ng, and K.P. Murphy. Modeling recurrent DNA copy number alterations in array CGH data. *Bioinformatics*, **23**:450–458, 2007.
45. J. Shendure, R.D. Mitra, C. Varma, and G.M. Church. Advanced sequencing technologies: methods and goals. *Nat Rev Genet*, **5**(5):335–344, 2004.
46. D. Siegmund. Tail approximations for maxima of random fields. In L.H.Y. Chen, K.P. Choi, K. Yu, and J.-H. Lou, editors, *Probability Theory: Proceedings of the 1989 Singapore Probability Conference*, pages 147–158. de Gruyter, Berlin, 1992.
47. A.M. Snijders, N. Nowak, R. Segraves, S. Blackwood, N. Brown, J. Conroy, G. Hamilton, A.K. Hindle, B. Huey, K. Kimura, S. Law, K. Myambo, J. Palmer, B. Ylstra, J.P. Yue, J.W. Gray, A.N. Jain, D. Pinkel, and D.G. Albertson. Assembly of microarrays for genome-wide measurement of DNA copy number. *Nat Genet*, **29**:263–264, 2001.
48. J. Staaf, D. Lindgren, J. Vallon-Christersson, A. Isaksson, H. Goransson, G. Juliusson, R. Rosenquist, M. Hoglund, A. Borg, and M. Ringner. Segmentation-based detection of allelic imbalance and loss-of-heterozygosity in cancer cells using whole genome SNP arrays. *Gen Biol*, **9**:R136+, 2008.
49. B.S. Taylor, J. Barretina, N.D. Socci, P. Decarolis, M. Ladanyi, M. Meyerson, S. Singer, and C. Sander. Functional copy-number alterations in cancer. *PLoS ONE*, **3**(9):e3179+, 2008.
50. R. Tibshirani and P. Wang. Spatial smoothing and hot spot detection for CGH data using the fused lasso. *Biostatistics*, **9**:18–29, 2008.
51. E.S. Venkatraman and A.B. Olshen. A faster circular binary segmentation algorithm for the analysis of array CGH data. *Bioinformatics*, **23**:657–663, 2007.
52. P. Wang, Y. Kim, J. Pollack, B. Narasimhan, and R. Tibshirani. A method for calling gains and losses in array-CGH data. *Biostatistics*, **6**:45–58, 2005.
53. K. Wang, M. Li, D. Hadley, R. Liu, J. Glessner, S.F.A. Grant, H. Hakonarson, and M. Bucan. Penncnv: An integrated hidden Markov model designed for high-resolution copy number variation detection in whole-genome SNP genotyping data. *Genome Res*, **17**(11):1665–1674, 2007.
54. H. Wang, J.H. Veldink, R.A. Ophoff, and C. Sabatti. Markov models for inferring copy number variations from genotype data on Illumina platforms. Technical Report, Dept. of Statistics, University of California at Los Angeles, 2008.
55. C. Wen, Y. Wu, Y. Huang, W. Chen, S. Liu, S. Jiang, J. Juang, C. Lin, W. Fang, C.A. Hsiung, and I. Chang. A Bayes regression approach to array-CGH data. *Stat Appl Mol Biol* **5**(1), 2006.
56. H. Willenbrock and J. Fridlyand. A comparison study: applying segmentation to array CGH data for downstream analyses. *Bioinformatics*, **21**:4084–4091, 2005.

57. B. Xing, C.M.T.M. Greenwood, and S.B.B. Bull. A hierarchical clustering method for estimating copy number variation. *Biostatistics*, **8**:632–653, 2007.
58. N.R. Zhang and D.O. Siegmund. A modified Bayes information criterion with applications to the analysis of comparative genomic hybridization data. *Biometrics*, **63**:22–32, 2007.
59. N.R. Zhang, Y. Senbabaoglu, and J.Z. Li. Joint estimation of DNA copy number from multiple platforms. *Bioinformatics*, **26**(2):153–160, 2010.
60. N.R. Zhang, D.O. Siegmund, H. Ji, and J.Z. Li. Detecting simultaneous change-points in multiple sequences. *Biometrika*, 2010, in press.

Chapter 15
Spatial Disease Surveillance: Methods and Applications

Tonglin Zhang

15.1 Introduction

The availability of geographical indexed health and population data and statistical methodologies have enabled the realistic investigation of spatial variation in disease risks, particular at the small unit level. Recently, incidence or mortality counts at county level have been announced on the website of the United States Centers for Disease Control and Prevention (CDC). Due to the reason of privacy, CDC does not announce the exact locations of disease events. Instead, it only announces the disease counts aggregated over geographical units, such as at the county or zipcode level. As this is also common in most countries, spatial disease surveillance therefore focuses on methodology development for aggregated data.

Overall, statistical approaches for disease incidence or mortality fall into two classes of approaches: *cluster detection approaches* and *disease mapping approaches*. Cluster detection approaches, which are also called *hypothesis testing approaches* or *frequentist approaches*, are expected to test and locate spatial clusters in the study area. These approaches can also be classified into two categories: general testing approaches or focused testing approaches [5]. General testing approaches, also called general tests, consist of a single global test statistic associated with the null hypothesis of no spatial clusters [3, 8, 16, 37, 40]. They usually collect evidence of the existence of clusters throughout the whole study area without evaluating the statistical significance of a particular cluster. If the null hypothesis is rejected, a global statistic cannot give the location of the spatial clusters. Focused testing approaches, also called focused tests, are designed to locate the spatial cluster around a prespecified point in the study area [2, 16, 21, 23, 43].

Disease mapping is defined as the estimation and presentation of unit summary measurements of disease counts [26]. The aims of disease mapping include sim-

T. Zhang (✉)
Department of Statistics, Purdue University, 250 North University Street, West Lafayette, IN 47907-2066, USA
e-mail: tlzhang@purdue.edu

J. Feng et al. (eds.), *Frontiers in Computational and Systems Biology*, Computational Biology 15,
DOI 10.1007/978-1-84996-196-7_15,

ple description, hypothesis generation, allocation of health care resources, assessment of inequalities, and estimation of background variability in underlying disease risk [39]. A disease mapping method, which is typically specified under the Bayesian framework, produces smooth estimates of unit-specific disease rates suitable for mapping [6]. Those approaches are usually carried out with Poisson models under the Bayesian framework with random effect having either spatial correlated or uncorrelated extra variation [26]. In general, disease mapping approaches are able to capture gradual and global variations of disease rates and are less useful in detecting abrupt, localized variations [14]. Typical Bayesian approaches address the disease clustering problem including Lawson [25, 26], Green and Richardson [18], and Diggle [10]. Since a disease mapping approach usually is not tractable analytically, the intractability of the posterior distribution has led to the use of Markov Chain Monte Carlo (MCMC) simulation methods to generate samples from joint distributions.

Some Background Suppose that a study area is partitioned into m disjoint (subarea) units. Let Y_i be the disease count, y_i be the observed count, and n_i be the at risk population size in unit i for $i = 1, \dots, m$. Assume that Y_i follows (or approximately follows) a Poisson distribution; it has been known that this is appropriate in most rare disease studies in spatial epidemiology [3, 24, 26, 36, 37, 45]. Then, the statistical model can be specified as

$$Y_i \sim Poisson(\theta_i E_i) \tag{15.1}$$

conditionally independently for $i = 1, 2, \dots, m$, where θ_i is an unknown parameter, and E_i is an expected count proportional to the known population size. In this model, θ_i is often called the relative risk, and the observed value of θ_i, denoted by y_i/n_i, is often called the standard morbidity rate (SMR) [26].

The spatial variation of θ_i is modeled via the generalized linear mixed effect models [28, 35, 44], which is usually modeled differently in cluster detection approaches or disease mapping approaches. In a cluster detection approach, Y_i are usually assumed marginally independently distributed with different expected risks for units within the cluster or outside of the cluster, respectively [4, 18, 21]. In the simplest case, suppose that C is the only cluster in the study area. A statistical model in a hypothesis testing approach assumes that $\theta_i = \theta_c$ if $i \in C$ and $\theta_i = \theta_0$ if $i \notin C$. If $\theta_c > \theta_0$, C is a hot spot. If $\theta_c < \theta_0$, C is a cool spot. Multiple cluster cases can also be modeled similarly. Even though people often focus on the hot spot detection problem, the cool spot detection problem is also important [46].

In a disease mapping approach, θ_i is no longer treated as an unknown parameter. Instead, it is treated as a random variable. Commonly the prior for θ_i in model (15.1) is specified as [39]

$$\log(\theta_i) = \mu_i + U_i, \tag{15.2}$$

where μ_i is a linear combination of fixed effects, and U_i is a spatial dependent random effect (mostly assuming normally distributed). Since Poisson distribution with a lognormal prior has no an analytic solution, numerical methods are applied, such as using quasi-likelihood and MCMC methods (see Lawson [26, p. 42]).

The organization of the sections of this chapter is as follows. In Sect. 15.2, we will review several well-known hypothesis testing approaches in details. In Sect. 15.3, we will review a few disease mapping approaches briefly. In Sect. 15.4, we will compare well-known hypothesis testing approaches by simulation and case studies. In the end, we will provide concluding remarks.

15.2 Review of Cluster Detection Approaches

Even though both hypothesis testing approaches and disease mapping approaches can be used to analyze spatial patterns for disease, hypothesis testing approaches are more commonly used than disease mapping approaches. In most hypothesis testing approaches (e.g., see [37, 39]), the null hypothesis is

$$H_0 : \theta_1 = \theta_2 = \cdots = \theta_m.$$

It means that, under the null hypothesis, the expected count is proportional to the at risk population. The alternative hypothesis is defined differently according to interests of problems. For example, in the hot spot only study, the alternative hypothesis is defined as

$$H_1 : E(Y_i) = \theta n_i (1 + \delta_i)$$

with $\delta_i > 0$ if unit i is within the cluster and $\delta_i = 0$ if unit i is outside of the cluster [18, 21]. Here, θ is the overall average of the disease rate, and θ is an unknown parameter. In the hot and cool spot study, the alternative hypothesis allows δ_i taking negative values, so that $\delta_i > 0$ if unit i is within in a hot spot cluster and $\delta_i < 0$ if unit i is within in a cool spot cluster.

Several traditional hypothesis testing methods can assess the null hypothesis via the examination of the set Y_i and n_i, such as methods of tests by using the Pearson χ^2 goodness-of-fit statistic and the deviance (or likelihood ratio) goodness-of-fit statistic. Assume that $\hat{y}_i$ is the predicted count of the ith unit under the null hypothesis; then the Pearson χ^2 statistic is defined (see [1, p. 22]) as

$$X^2 = \sum_{i=1}^{m} \frac{(y_i - \hat{y}_i)^2}{\hat{y}_i},$$

and the deviance statistic (see [1, p. 142], for detail) is defined as

$$G^2 = \sum_{i=1}^{m} 2y_i \log(y_i/\hat{y}_i).$$

Under H_0, both X^2 and G^2 approximately follow a chi-squared distribution with $m - p$ degrees of freedom, where p is the number of parameters contained in the model under the null hypothesis. The null hypothesis is rejected if X^2 or G^2 is large.

Since goodness-of-fit statistics are not specified for test of spatial variations, they cannot be used in cluster detection and localization problems. When the null hypothesis is rejected, goodness-of-fit statistics only indicate heterogeneous disease risks which could have two different interpretations: the first one is the heterogeneous risks without spatial dependence, and the second one is the heterogeneous risk with spatial dependence. In addition, the null hypothesis of no spatial clusters may provide a mechanism for inducing various forms of overdispersion. In particular, if θ_i are an iid sample from a probability density function, X^2 or G^2 may also be significant, but in this case the truth is still the null hypothesis [20]. In the following, we review several well-known hypothesis testing methods that have been extensively used in detection and localization of disease incidence or mortality clusters.

15.2.1 Scan Statistics

Kulldorff developed the spatial scan statistic by combing time series scan statistic [32] and spatial analysis machine (GAM) methods [33]. The spatial scan test uses a moving circle of varying size to find a set of regions or points that maximizes the likelihood ratio test for the null hypothesis of a purely random Poisson or Bernoulli process. Time can be included for a space-time scan test. Categorical covariates, such as age group and sex, can be included as control variables. Recent developments have led to various modifications of scan statistics for cluster shape detections [4].

In Kulldorff's scan method, the local cluster C is unknown and is the key issue to be determined by the method. In the most general case, any connected subset of units can be a candidate of the cluster. Since the number of units is finite, the number of candidates is also finite. However, even if m is not quite large, the number of candidates of clusters can also be extremely large. For example, when $m = 30$, it could be as large as $O(2^m) = O(2^{30})$. Therefore, scan method and its modifications consider only a small portion of the possible candidates. For example, Kulldorff considers circular or rectangle subsets as the candidates of clusters, and his circular and rectangle spatial scan statistic has been used extensively along with his software SaTScan [22]. Tango and Takahashi's modified flex scan considers a small portion of subsets of connected units [38].

Suppose $\mathcal{C}$ is the (small portion of) collect of candidates of clusters. For a $C \in \mathcal{C}$, the scan statistic detects any units within a cluster C in which counts are significantly higher (or lower) than expected. The test compares the total number of disease counts, $y_c = \sum_{i \in C} y_i$, within C with the total number of counts, $y_{\bar{c}} = \sum_{i \notin C} y_i$, outside of C, given the corresponding at-risk populations inside and outside of C, denoted by $n_c = \sum_{i \in C} n_i$ and $n_{\bar{c}} = \sum_{i \notin C} n_i$, respectively. Let $y = y_c + y_{\bar{c}} = \sum_{i=1}^{m} y_i$ and $n = n_c + n_{\bar{c}} = \sum_{i=1}^{m} n_i$. Assume that $\theta_i = \theta_c$ for $i \in C$ and $\theta_i = \theta_0$ for $i \notin C$. To test the null hypothesis of $H_0 : \theta_c = \theta_0$ against the alternative hypothesis of

$H_1 : \theta_c > \theta_0$, Kulldorff developed the likelihood ratio as

$$\Lambda_C = \frac{\max_{\theta_c > \theta_0} L(C, \theta_c, \theta_0)}{\max_{\theta_c = \theta_0} L(C, \theta_c, \theta_0)} = \left(\frac{y_c/n_c}{y/n}\right)^{y_c} \left(\frac{y_{\bar{c}}/n_{\bar{c}}}{y/n}\right)^{y_{\bar{c}}}$$

when $y_c/n_c \geq y_{\bar{c}}/n_{\bar{c}}$ and $\Lambda_C = 1$ otherwise, where $L(C, \theta_c, \theta_0)$ is the likelihood function. Since C is unknown, it can be treated as a parameter, and the maximum likelihood ratio test statistic for unspecified spatial cluster C is given by

$$\Lambda = \max_{C \in \mathcal{C}} \Lambda_C. \tag{15.3}$$

Like most test statistics, Kulldorff's scan statistic also has some limitations. For instance, it cannot directly include ecological covariates and overdispersion. In addition, the spatial scan statistic does not account for overdispersion. We display a modification of Kulldorff's scan statistic below, so that spatial analysts are not only able to include ecological covariates but also able to account for overdispersion [45].

Modifications Recently, Zhang, and Lin [45] have proposed a modification of Kulldorff's scan statistic to account for ecological covariates and overdispersion. Suppose that we only allow the existence of hot spot. The model-based spatial scan statistics use the concept of relative risk to introduce a spatial loglinear model:

$$\log(\theta_i) = \mathbf{x}_i^t \beta + \alpha_i, \tag{15.4}$$

where β is the vector of coefficients of parameters, α_i is the cluster indicator defined as $\alpha_i = \alpha_c$ if $i \in C$ and $\alpha_i = \alpha_0$ if $i \notin C$. By the constraint, α_0 is always set as $\alpha_0 = 0$. Under the null hypothesis of $H_0 : \alpha_c = 0$, model (15.4) becomes

$$\log(\theta_i) = \mathbf{x}_i^t \beta. \tag{15.5}$$

Under the alternative hypothesis of $H_1 : \alpha_c > 0$ for some C, model (15.4) becomes

$$\log(\theta_i) = \mathbf{x}_i^t \beta + \alpha_c I_{i \in C}. \tag{15.6}$$

When C is given, the MLE (maximum likelihood estimate) of β under the null hypothesis and the MLE of α_c and β under the alternative hypothesis can easily be derived (see [1, 13]). Let $\hat{\beta}^0$ be the MLE under the null hypothesis, and $\hat{\alpha}_c$ and $\hat{\beta}^1$ be the MLE under the alternative hypothesis. The predicted counts based on H_0 and H_1 can be derived by $\hat{y}_i^0 = n_i e^{\mathbf{x}_i^t \hat{\beta}_0}$ and $\hat{y}_i^1 = n_i e^{\mathbf{x}_i^t \beta + \hat{\alpha}_c I_{i \in C}}$, respectively. Let G_0^2 and X_0^2 be the deviance and Pearson goodness-of-fit statistics under model (15.5), and G_1^2 and X_1^2 be those under model (15.6). Then, the deviance statistic, the Pearson statistic, and the Wald statistic for the significance test of the cluster C are

$$G_c^2 = G_0^2 - G_1^2; \qquad X_c^2 = X_0^2 - X_1^2; \qquad Z_c = \frac{\hat{\alpha}_c}{\hat{\sigma}_{\hat{\alpha}_c}},$$

where $\hat{\sigma}_{\hat{\alpha}_c}$ is the estimate of the standard error of $\hat{\alpha}_c$ under (15.6). Following Kulldorff, the model-based deviance scan statistic, Pearson scan statistic, and Wald scan statistic can be readily defined respectively as

$$G_s^2 = \max_{C \in \mathcal{C}} G_C^2, \qquad X_s^2 = \max_{C \in \mathcal{C}} X_C^2, \qquad Z_s = \max_{C \in \mathcal{C}} Z_C. \tag{15.7}$$

For G_s^2, X_s^2, or Z_s, we reject H_0 and claim the existence of spatial clusters if the value is large. If we also allow the existence of cool spot, we can take the sign of Z_c into consideration by modifying the definitions in (15.7).

Modification for Overdispersion Overdispersion can occur in the spatial Poisson process either because of spatially correlated data or because of spatially varied cases and populations from a large number of spatial units. To gauge and to reduce the potential effect of overdispersion, a dispersion parameter denoted by ϕ can be introduced via the quasi-Poisson model [30]. When $\phi = 1$, it is a regular Poisson regression without overdispersion. When $\phi > 1$, it is the Poisson regression with overdispersion. An estimate of the dispersion parameter can be used:

$$\hat{\phi}_C = \max\left(\frac{X_1^2}{df_{C,\text{res}}}, 1\right),$$

where $df_{c,\text{res}}$ is the residual degree of freedom of model (15.6). When the dispersion parameter is considered, the G_c^2, X_c^2, and Z_c^2 should be modified accordingly as $G_{c,o}^2 = G_c^2/\hat{\phi}_c$, $X_{c,o}^2 = X_c^2/\hat{\phi}_c$, and $Z_{c,o} = Z_c/\sqrt{\hat{\phi}_c}$, which implies that $G_{s,o}^2$, $X_{s,o}^2$, and $Z_{s,o}$ can also be modified accordingly via similar equations as those in (15.7), respectively.

Similar to Kulldorff's spatial scan statistic Λ, the null distributions of test statistics are not analytically tractable. Therefore, we follow Kulldorff's idea to compute their p-values by the bootstrap method. The null hypothesis of no clustering is rejected for large G_s^2, X_s^2, or Z_s values without adjusting overdispersion, and for large $G_{s,o}^2$, $X_{s,o}^2$, and $Z_{s,o}$ values with adjusting overdispersion. The corresponding bootstrap p-value is the rank of the real observed values in the corresponding bootstrap distribution. Following Kulldorff [21], the p-values are calculated by a bootstrap method conditionally on the total number of counts by the following algorithm.

Bootstrap for the p-Values of G_s^2, X_s^2, Z_s, $G_{s,o}^2$, $X_{s,o}^2$, or $Z_{s,o}$

(i) Let T be the test statistic. Calculate the observed value (denoted by T^*) of T based on the real data.
(ii) Derive the predicted counts $\hat{y}_i$ under the null model (15.5).
(iii) Generate K independent multinomial random variables with total counts equal to y and proportional vector equal to $(\hat{y}_1/y, \ldots, \hat{y}_m/y)$. Calculate the simulated values $T_{k,\text{sim}}^*$ of T.
(iv) Report the p-values of T by $\#\{T_{k,\text{sim}}^* \geq T^* : k = 1, 2, \ldots, K\}/K$, where $\#(A)$ represents the number of elements in a set A.

15.2.2 Permutation Testing Methods

A number of permutation testing methods have been proposed in detecting clusters. Almost all of them need a well-defined measure of the closeness (or weight) between two units. Let w_{ij} be the measure of the closeness between units i and j [7]. In the simplest case, we can take $w_{ij} = 1$ if units i and j are adjacent and 0 otherwise. Another common choice is to use a decreasing function of the distance between the centroids of units i and j. We now describe three permutation testing methods: they are Moran's I [31], Geary's c [15], and Getis–Ord's G [16]. Even though there are quite a few other methods, we choose these three methods because they are the most popular methods in cluster detection.

Let z_i be the variable of interest at unit i. Moran's I is defined as

$$I = \frac{1}{S_0 b_2} \sum_{i=1}^{m} \sum_{j=1, j\neq i}^{m} w_{ij}(z_i - \bar{z})(z_j - \bar{z}), \tag{15.8}$$

where $S_0 = \sum_{i=1}^{m} \sum_{j=1,j\neq i}^{m} w_{ij}$, $\bar{z} = \sum_{i=1}^{m} z_i/m$, and $b_k = \sum_{i=1}^{m}(z_i - \bar{z})^k/m$. Moran's I statistic usually ranges between -1 and 1 even though its absolute value could be over 1 in extreme cases. With a coefficient close to -1, Moran's I indicates neighborhood dissimilarity; with a coefficient close to 1, Moran's I indicates neighborhood similarity. When the coefficient of Moran's I is close to 0, it indicates spatial randomness or independence [7].

Geary's c statistic is defined as

$$c = \left(\frac{m-1}{2mb_2 S_0}\right) \sum_{i=1}^{m} \sum_{j=1, j\neq i}^{m} w_{ij}(z_i - z_j)^2. \tag{15.9}$$

Geary's c is always positive, with a small value indicating neighborhood similarity, a large value indicating neighborhood dissimilarity, and a value close to 1 indicating spatial randomness or independence [41]. Getis–Ord's G statistic is given by

$$G = \frac{\sum_{i=1}^{m} \sum_{j=1,j\neq i}^{m} w_{ij} z_i z_j}{\sum_{i=1}^{m} \sum_{j=1,j\neq i}^{m} z_i z_j}, \tag{15.10}$$

with a small value indicating neighbor dissimilarity and a large value indicating neighbor similarity.

The p-values of Moran's I, Geary's c, and Getis's G are computed under random permutation test schemes. A random permutation test generally calculates the moments of the test statistic under every possible arrangement of the data. These arrangements would be used to generate the distribution of the test statistic under the null hypothesis. A related approach uses many Monte Carlo rearrangements of the data rather than enumeration of all of the possible arrangements [12]. If the number of Monte Carlo rearrangements is large and each arrangement has equal probability in each Monte Carlo replicate, then the exact test and Monte Carlo permutation test will have similar results, and the Monte Carlo permutation test is asymptotically equivalent to the exact permutation test [17, pp. 185–187]. Note that for a general m, there are $m!$ possible permutation arrangements in the exact permutation test. It is generally impossible to obtain the exact moments of a test statistic under the permutation test scheme. However, since the numerators of Moran's I, Geary's c, and Getis–Ord's G are in quadratic form and their denominators are permutation invariant, the exact expressions of their moments under random permutation test scheme are available and have been included in the many textbooks (e.g., see Cliff and Ord [7]). In the following, we denote by $E_R(\cdot)$ and $V_R(\cdot)$ the expected value and variance of a statistic under the exact permutation test scheme. Formulae of the moments then are given below accordingly:

$$E_R(I) = -\frac{1}{m-1}, \qquad E_R(c) = 1, \qquad E_R(G) = \frac{S_0}{m(m-1)},$$

and

$$\begin{aligned}
E_R(I^2) &= \frac{S_1(mb_2^2 - b_4)}{S_0^2 b_2^2 (m-1)} + \frac{(S_2 - 2S_1)(2b_4 - mb_2^2)}{S_0^2 b_2^2 (m-1)(m-2)} \\
&\quad + \frac{(S_0^2 - S_2 + S_1)(3mb_2^2 - 6b_4)}{S_0^2 b_2^2 (m-1)(m-2)(m-3)}, \\
E_R(c^2) &= \frac{S_1(m-1)(b_4 + 3b_2^2)}{2mS_0^2 b_2^2} + \frac{(S_2 - 2S_1)(m-1)(3b_2^2 + b_4)}{4S_0^2 b_2^2} \\
&\quad + \frac{(S_0^2 - S_2 + S_1)(m-1)[(m^2 - 3m + 3)b_2^2 - (m-1)b_4]}{(m-2)(m-3)S_0^2 b_2^2}, \\
E_R(G^2) &= \frac{S_1(mB_2^2 - B_4)}{(m-1)(m^2B_1^2 - mB_2)^2} \\
&\quad + \frac{(S_2 - 2S_1)(m^2B_1^2B_2 - mB_2^2 - 2mB_1B_3 + 2B_4)}{(m-1)(m-2)(m^2B_1^2 - mB_2)^2} \\
&\quad + \frac{(S_0^2 - S_2 + S_1)(m^3B_1^4 - 6m^2B_1^2B_2 + 8mB_1B_3 + 3mB_2^2 - 6B_4)}{(m-1)(m-2)(m-3)(m^2B_1^2 - mB_2)^2},
\end{aligned}$$

where $B_k = \sum_{i=1}^m z_i^k/m$, $S_1 = \sum_{i=1}^m \sum_{j=1, j\neq i}^m (w_{ij} + w_{ji})^2/2$, and $S_2 = \sum_{i=1}^m [\sum_{j=1, j\neq i}^m (w_{ij} + w_{ji})]^2$. Their variances are $V_R(I) = E_R(I^2) - E_R^2(I)$, $V_R(c) = E_R(c^2) - E_R^2(c)$, and $V_R(G) = E_R(G^2) - E_R^2(G)$. Assuming that I, c, and G are asymptotically normal, their p-values are calculated by a two-sided z-test as $2[1 - \Phi(|\frac{I - E_R(I)}{\sqrt{V_R(I)}}|)]$ for I statistic, $2[1 - \Phi(|\frac{c - E_R(c)}{\sqrt{V_R(c)}}|)]$ for c statistic, and $2[1 - \Phi(|\frac{G - E_R(G)}{\sqrt{V_R(G)}}|)]$ for G statistic. If the p-values are less than the significance level, then the null hypothesis of spatial independence is rejected, and spatial dependence is concluded.

The common choice of z_i can be the observed rate, the Pearson residual, or the deviance residual in model (15.5). The statistics given by (15.8), (15.9), and (15.10) then become the rate based as denoted by I_r, c_r, or G_r, the Pearson residual based as denoted by I_{PR}, c_{PR}, or G_{PR} or deviance residual based as denoted by I_{DR}, c_{DR}, and G_{DR}. It has been shown that the Pearson residual-based and the deviance residual-based test statistics are more reliable than the rate-based test statistics [46].

Simply looking at the values of Moran's I, Geary's c, and Getis–Ord G is not able to locate the detected cluster. Therefore, local versions of these statistics have been proposed. The most common used local versions (called focused tests) are local Moran's I_i. Together with local Geary's c_i, it is called LISA in the spatial statistical literature [2]. Another commonly used focused test is Getis–Ord's local G_i [16]. Local I_i, c_i, and G_i are complementary of the global test statistics.

Anselin's LISA [2] has two components of the statistics: the local I_i is defined as

$$I_i = \frac{(z_i - \bar{z})}{b_2} \sum_{j=1, j\neq i}^{m} w_{ij}(z_j - \bar{z});$$

the local c_i is defined as

$$c_i = \frac{1}{b_2} \sum_{j=1, j\neq i}^{m} w_{ij}(z_i - z_j)^2$$

for $i = 1, \ldots, m$; and Getis–Ord's G_i [16] is defined as

$$G_i = \frac{\sum_{j=1, j\neq i}^{m} w_{ij} z_j}{\sum_{j=1, j\neq i}^{m} z_j}.$$

Similarly to the method used in computation of the moments of the global statistics, the moments of I_i, c_i, and G_i are also computed under the permutation test scheme. The results are

$$E_R(I_i) = -\frac{w_{i\cdot}}{m-1}, \qquad E_R(c_i) = \frac{2mw_{i\cdot}}{m-1}, \qquad E_R(G_i) = \frac{w_{i\cdot}}{m-1},$$

and

$$E_R(I_i^2) = \frac{w_{i(2)}(mb_2^2 - b_4)}{b_2^2(m-1)} + \frac{(w_{i\cdot}^2 - w_{i(2)})(2b_4 - mb_2^2)}{b_2^2(m-1)(m-2)},$$

$$E_R(c_i^2) = \frac{w_{i(2)}(2mb_4 + 6mb_2^2)}{b_2^2(m-1)} + \frac{(w_{i\cdot}^2 - w_{i(2)})m^2(3b_2^2 + b_4)}{b_2^2(m-1)},$$

$$E_R(G_i^2) = \frac{1}{(mB_1 - z_i)^2}\left\{\frac{w_{i(2)}(mB_2 - z_i^2)}{(m-1)} + \frac{(w_{i\cdot}^2 - w_{i(2)})}{(m-1)(m-2)}\left[(mB_1 - z_i)^2 - \left(B_2^2 - z_i^2\right)\right]\right\},$$

where $w_{i\cdot} = \sum_{j=1,j\neq i}^{m} w_{ij}$ and $w_{i(2)} = \sum_{j=1,j\neq i}^{m} \sum_{k=1,k\neq i}^{m} w_{ij} w_{ik}$.

As those for the global statistics, the p-values of the local statistics are also suggested to be calculated under the asymptotic normality assumption. Based on the z-test scheme for I_i and G_i, local similarity or dissimilarity is obtained if their values are greater than or less than its permutation expected value with a significance p-value, respectively. For local c_i, the conclusion is opposite: local similarity is obtained if c_i is less than its permutation expected value with a significance p-value, and local dissimilarity is obtained if c_i is greater than its permutation expected value with a significance p-value. Since all of them encounter the multiple testing problems, I_i, c_i, and G_i are only recommended to use when their global statistics are significant [2, 16].

15.2.3 Other Methods

Even though Kulldorff's scan, Moran's I, Geary's c, and Getis–Ord's G are the most commonly used statistics, a few other testing methods are also proposed in spatial statistical literature. Examples include Tango's C_G [37], Whittemore's T statistic [42], Rogerson's R statistic [36], and Besag–Newell's R statistic [5]. However, we are not able to describe any of them here. The reason for us to introduce in Sects. 15.2.1 and 15.2.2 Kulldorff's scan and permutation tests is that they have already been incorporated in many statistical software packages, which make them more popular than other statistics.

15.3 Review of Disease Mapping Approaches

In disease mapping, the rare disease counts Y_i are assumed conditionally independently following Poisson distributions, but their conditional expected values are spatially dependent. Disease mapping approaches are often formulated under the framework of Bayesian hierarchical models. As the unit-specified MLEs of the relative risks given by $\hat{\theta}_i = y_i/n_i$ is highly unstable, more robust estimation is provided by specifying a joint model for $\theta = (\theta_1, \theta_2, \ldots, \theta_m)$, which allows the estimate of each θ_i to borrow strength from its neighboring units. This can be achieved by using a multivariate prior distribution for θ.

A common model in disease mapping methods is usually modified from model (15.2) as

$$\log(\theta_i) = \alpha + X_i^t \beta + U_i, \tag{15.11}$$

where U_i is a spatial random effect term to be specified by a prior distribution. This method has been previously studied by Diggle et al. [11] and Wakefield et al. [39].

In model (15.11), spatial dependence is described by random effects U_i. Two common methods are proposed to model spatial dependence: the first one models spatial dependence by a spatial autoregressive (SAR) model [34], and the second one models spatial dependence by a geostatistical model [11].

Let $U = (U_1, U_2, \dots, U_m)^t$. The SAR model is formulated as

$$U = \rho W U + \varepsilon. \tag{15.12}$$

In (15.12), U is an m-dimensional vector of dependent variables, the scalar ρ is called the spatial autoregressive parameter, the error term ε is assumed to be iid $N(0, \sigma^2)$ distributed, and $W = (w_{ij})_{m\times m}$ is the spatial weight matrix with $w_{ii} = 0$ for the description of the closeness between units i and j as before. By convention, we usually set $w_{ii} = 0$, $w_{ij} \geq 0$, and $\sum_{j=1}^{m} w_{ij} = 1$.

An SAR model is usually not stationary. It describes the spatial autocorrelation between neighboring units. Spatial autocorrelation is a measure of spatial dependence between values of random variables over geographic units. The sign of the spatial autoregressive parameter ρ in model (15.12) indicates two types of spatial autocorrelation, positive autocorrelation and negative autocorrelation, in which a positive autocorrelation captures the existence of high-value clustering or low-value clustering (or hot spots and cool spots), while a negative autocorrelation captures the juxtaposition of high values next to low values [19, 27, 44].

A geostatistical model assumes

$$U \sim N\big(0, \sigma^2 R\big),$$

where σ^2 is the common marginal variance. The correlation matrix R is modeled by an isotopic correlation function. Let h be the distance between two points over the space. We say the correlation function $C(h)$ is isotopic if it only depends on the L_2 norm value $\|h\|$ of h. When $C(\cdot)$ is chosen, the correction matrix R is defined by $R = (c(d_{ij}))_{m\times m}$, where d_{ij} is the distance between units i and j. A common way to define $C(h)$ is to use the parametric way. For example, $C(h)$ can be the famous Matérn correlation function

$$C(h) = c_\nu(h) = \frac{\nu_1}{2^{\alpha-1}\Gamma(\alpha)}\left(\frac{h}{\nu_2}\right)^{\alpha} K_\alpha\left(\frac{h}{\nu_2}\right), \quad \|h\| > 0,$$

for $0 \leq \nu_1 < 1$, $\nu_2 > 0$, and $K_\alpha(\cdot)$ is the modified Bessel function.

In disease mapping, the prior distribution for U is implemented in the Bayesian computation of the posterior risks. This leads to the use of MCMC simulation methods to generate samples from the joint posterior distribution. Details of the computational algorithms using MCMC methods are provided in many textbooks (e.g., see [26]).

15.4 Simulation and Case Study

Some results of this section have been published in a journal article joint with Ge Lin [45].

In both simulation and case studies, we used real-world data based on 110 counties in Guangxi Zhuang ethnic Province. Guangxi is one of five autonomous ethnic minority regions in China. We obtained the county-level infant birth and death data from the 2000 Census in China. Infant mortality rates vary substantially among these counties, ranging from 248 to 7260 per 100,000 (Fig. 15.1). Guangxi for the most part is mountainous and is considered a less developed region in southwestern China. The county-level elevations range from 20 to 1,140 meters above sea level, lower in the southeast and higher toward the west. In the absence of other variables, elevation is a good measure of access to care, education, and other socioeconomic resources. For this reason, we included elevation in both simulation and the case study.

15.4.1 Simulation

We evaluated the type I error probabilities and the power functions of well-known hypothesis testing methods reviewed in Sect. 15.2. For a pure spatial situation, the model-based likelihood ratio scan statistic is equivalent to the original spatial scan statistic, and its statistical powers have been thoroughly evaluated [24]. For this reason, we opted to evaluate the performance of these model-based scan statistics by including an ecological covariate. For simplicity, we define the size of a local cluster or spatial association by a county and its adjacent counties [29], which can be indexed by a column vector (w_i) in the spatial weight matrix. We used 110 counties to generate the spatial weight matrix based on spatial adjacency and used infant births (n_i) and deaths (Y_i) and elevations x to fit a baseline model. Assuming that $E(Y_i) = \theta_i n_i$, we fitted a baseline model with the quadratic function of elevation as ecological covariates. The fitted model is

$$\log(\theta_i) = -4.160 + 2.07 \times 10^{-3}x - 1.27 \times 10^{-6}x^2. \tag{15.13}$$

The coefficient of the quadratic term in the fitted model was negative. The maximum predicted mortality rate was $(2.07 \times 10^{-3})/(2 \times 1.27 \times 10^{-6}) = 815$. Since the elevation of most counties (103 out of 110) is lower than 815 meters, model (15.13) suggests that the infant mortality rate increased as elevation increased.

Model (15.13) serves as a baseline model for generating a new response and a local cluster term. We inserted a seven-county cluster C in the baseline model with its center at Pingnan, a nonborder county, and the new response variable Y_i' was then generated from

$$\log(\theta_i') = \eta_i + U_i = -4.160 + 2.07 \times 10^{-3}x - 1.27 \times 10^{-6}x^2 + \delta I_{i \in C} + U_i,$$

where $\theta_i' = E(Y_i')/n_i$, $I_{i \in C}$ is the indicator function, which is 1 if $i \in C$ and 0 otherwise. The random effect U_i is generated independently from $N(-\sigma_i^2/2, \sigma_i^2)$ with $\sigma_i^2 = \log[1 + (\phi - 1)e^{-\eta_i}]$, so that marginally we have the quasi-Poisson model as

$$\frac{V(Y_i)}{E(y_i)} = \phi, \quad \phi \geq 1.$$

Table 15.1 Rejection rate based on 1000 simulated data sets when $\phi = 1$

δ	G_s^2	Z_s	$G_{s,o}^2$	$Z_{s,o}$	I_{DR}
0.0	6.0	5.8	4.5	4.3	5.8
0.025	7.1	6.8	7.1	6.8	6.4
0.05	14.9	14.9	12.6	12.6	8.3
0.075	37.4	37.9	39.4	39.4	23.1
0.1	71.6	72.3	72.4	73.5	48.7
0.125	93.9	94.3	94.2	95.0	75.0
0.15	99.1	99.2	99.3	99.3	92.9
0.175	100.0	100.0	100.	100.0	99.0
0.20	100.0	100.0	100.0	100.0	100.0

We used δ to measure the strength of the cluster from 0 to 0.20 or 20% more than the logarithm of the relative risk outside of the cluster. We used ϕ to measure the overdispersion. Since the model was exactly Poisson when $\phi = 1$, there was no overdispersion for the simulated data. However, when $\phi > 1$, there was overdispersion. As ϕ is larger, the overdispersion becomes stronger.

In the simulation, we compared the type I error rates G_s^2, Z_s, $G_{s,o}^2$, $Z_{s,o}$, and I_{PR} when $\delta = 0$ and their power functions based on 1000 runs for each selected δ (Table 15.1). Since deviance residual-based test statistics and Pearson residual-based test statistics are defined from goodness-of-fit statistics in nature, we did not include X_s^2, $X_{s,o}^2$, and I_{PR} in Table 15.1. We obtained p-values of I_{DR} from permutation testing methods, and p-values of scan statistics from bootstrap method with 1000 replications. Since a spatial cluster can be treated as a spatial association term or an explanatory variable in the loglinear model, we scanned all counties and searched for the spatial association term that yielded the largest likelihood ratio or Wald z-value. We fixed the significance level at 0.05. We evaluated the power of the inserted cluster by the percentage of p-values that were less than or equal to the significance level.

When no overdispersion was not present ($\phi = 1$), the type I error and power functions of the four model-based spatial scan statistics were consistent. All the model-based scan statistics had acceptable type I error probabilities, with 6.0%, 5.8%, 4.5%, and 4.3% respectively for G_s^2, Z_s, $G_{s,o}^2$, and $Z_{s,o}$. As δ or cluster strength increased, the power functions of detectability increased rapidly while accounting for ecological covariates. They were all able to detect the existence of the cluster with moderate cluster strength. When the relative risk inside the cluster was 0.175 more than outside the cluster, the powers were all 100%. The Wald-based spatial scan statistics also performed well in comparison to the likelihood ratio-based scan statistic. As shown in Table 15.1, I_{DR} was not as powerful as scan statistics.

We also did a simulation study when overdispersion was present by taking $\phi = 2$. In this case when $\delta = 0$, the rejection rate of G_s^2 or Z_s^2 was over 50%, the rejection rate of $G_{s,o}^2$ or $Z_{s,o}$ was around 10%, but the rejection rate of I_{PR} was still around 5%. This indicates that the type I error probabilities of scan statistics are inflated, but the type I error probabilities of permutation testing method are not inflated when overdispersion is present, and the proposed scan statistics with the adjustment of overdispersion significantly reduced type I error probabilities.

To briefly summarize, type I error probabilities and power functions for G_s^2, Z_s, $Z_{s,o}^2$, and $Z_{s,o}$ were all comparable for a given cluster strength when overdispersion was not present. In the presence of a strong cluster, all three were able to correctly identified the cluster center 99% of the time. Even though we did not evaluate the pure spatial process without any ecological covariates, we expected the same results based on the formulations for G_s^2 and Z_s; for example, see details on pages 37–40 in [9]. When overdispersion is present, $G_{s,o}^2$ and $Z_{s,o}$ have lower type I error probabilities but are still higher than the nominal value. Moran's I always keeps type I error probabilities as low as the nominal value. In general, scan statistics are more powerful than permutation testing methods, but they are sensitive to the presence of overdispersion.

15.4.2 Case Study

Higher infant mortality in the Guangxi is a sensitive political issue for the central government, which tries to improve living standards for all ethnic groups and nationalities. It is already known that access to adequate primary care is an issue, and primary care is more available in lowland counties than in highland and mountainous counties. An interesting question, in addition to access to care, is what other factors might play a role for high infant mortality in the province. Since local socioeconomic and health care data were not available, we used average county elevation to approximate the lack of access. One of the coauthors was from the region, and our task was to first identify local pockets of counties by controlling for the elevation effect and then turn our findings over to provincial health officials for further health disparity analyses.

We first fitted infant mortality rates with a loglinear model that includes a quadratic elevation term as

$$\log(\theta_i) = \beta_0 + \beta_1 x + \beta_2 x^2,$$

where x represented the county level elevation in meters. The estimated values were $\hat{\beta}_0 = -4.16$, $\hat{\beta}_1 = 2.07 \times 10^{-3}$, and $\hat{\beta}_2 = -1.27 \times 10^{-6}$, with standard errors 2.22×10^{-2}, 1.21×10^{-4}, and 1.23×10^{-7}. Since both the linear term and quadratic term of elevation were highly significant, we considered the following model for cluster detection:

$$\log(\theta_i) = \beta_0 + \beta_1 x + \beta_2 x^2 + \alpha I_{i \in C},$$

where $C \in \mathcal{C}$, and $\mathcal{C}$ is the collection of all candidates for a potential circular cluster.

In the analysis, we also considered potential overdispersion by using $G_{s,o}^2$, $Z_{s,o}$, and I_{DR} [29]. Since I_{DR} does not suffer overdispersion, we use I_{DR} to check for the consistency of $G_{s,o}^2$ and $Z_{s,o}$ for adjustment of overdispersion. We used a stepwise

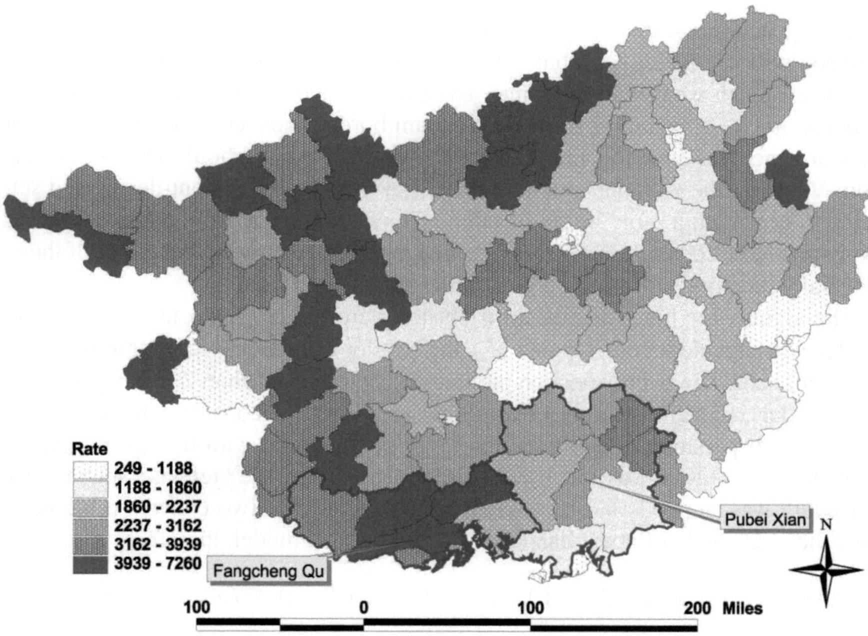

Fig. 15.1 County level infant mortality rate per 100,000 in Guangxi, China in 2000 (for the color version, see Color Plates on p. 393)

Table 15.2 Results from the stepwise scan for infant mortality in Guangxi based on 1000 simulated data sets for the significance of G_s^2, Z_s, and $Z_{s,0}$, where p-value is given in the brackets

$\hat{\phi}$	C	G_s^2	Z_s	$G_{s,o}^2$	$Z_{s,o}$	I_{DR}
29.1	w_{45}	638 (0.001)	27.9 (0.001)	29.9 (0.001)	6.03 (0.001)	0.294 (0.001)
21.3	w_{51}	196 (0.001)	14.3 (0.001)	9.74 (0.052)	3.17 (0.53)	0.124 (0.036)
20.2	w_{81}	81.4 (0.001)	9.39 (0.001)	4.17 (0.690)	2.13 (0.638)	0.047 (0.381)

scan method searching for local clusters by treating a local cluster as an explanatory variable indexed by w_i. This method is identical to the spatial scan method for detecting the first cluster. However, if multiple clusters exist, our method to scan for the secondary cluster might be slightly different from the spatial scan statistic. In SatScan, the secondary cluster is searched by considering the first cluster. In our method, when the first cluster is identified, its effect is taken out by the first cluster spatial association term, and an additional cluster term is used to scan and test for the existence of an additional cluster. The stepwise scan stops if the test statistic is no longer significant.

In the scan process, the first cluster was signaled with a p-value less than 0.001 for G_s^2, Z_s, $Z_{s,o}$, and I_{DR} (Table 15.2). This cluster, centered at Fangcheng Qu (w_{45}), is close to the border with Vietnam, and the area had dilapidated infrastructure for years due to the 1979 Sino–Vietnam Border War. The second cluster was signaled with a p-value less than 0.001 for G_s^2 and Z_s. The p-values for I_{DR} and $Z_{s,o}$ were much weaker in the range. Even though the second cluster, centered at Pubei Xian (w_{51}), is not far from the Vietnam border, most counties were not near the war zone. However, due to the war and prewar refugees who flooded into China from Vietnam, the counties around Pubei Xian were a government-designated settlement area, with a large influx of ethnic Chinese refugees from Vietnam. It was likely that after 20 years, some refugees had resettled elsewhere, but many of them, especially the poor, still lived on government-sponsored farms.

The signals for the potential third cluster were mixed. Even though the third cluster was signaled from G_s^2 and Z_s, the results from $G_{s,o}^2$, $Z_{s,o}$, and I_{DR} were not significant. The overdispersion parameters were greater than 20 among the first three models, signaling a strong effect. Since $G_{s,o}^2$ and $Z_{s,o}$ considers the adjustment of the overdispersion problem, and the result is consistent with that from I_{DR}, the third possible cluster was much less trustworthy. Thus, the stepwise spatial scan procedure was stopped, and it concluded the existence of two clusters centered at Fangcheng Qu and at Pubei Xian (Fig. 15.1). The final model, therefore, is

$$\log(\theta_i) = -4.52 + 3.24 \times 10^{-3}x - 2.13 \times 10^{-6} + 0.927w_{45} + 0.365w_{51}.$$

Based on the final model with two spatial association terms w_{45} and w_{51}, we calculated parameter estimates for elevation and its quadratic terms, 3.24×10^{-3} and -2.13×10^{-6}, respectively (with corresponding z-values 24.31 and -16.31, respectively). With the parameter estimate of overdispersion $\hat{\phi} = 20.16$, the adjusted z-values were 5.42 and -3.65, which were highly significant (with p-values less than 0.001). Based on these parameter estimates, we further examined mortality variation inside and outside of the clusters. The odds ratios of the first and second clusters were 2.53 and 1.44, respectively. Taking elevation into consideration the mortality rates of the counties within the first and second cluster were 153% or 44% higher, respectively, than the expected value. Since the elevation that corresponds to the maximum predicted mortality rate was about 760 meters, and there were only 12 counties above this level, the two detected clusters were the net effect of a general positive relationship between elevation and infant mortality.

15.5 Concluding Remarks

We have outlined a number of cluster detection approaches and disease mapping approaches in this book chapter. Certain issues are worth emphasized on. Given the nature of disease incidence or mortality, it is important to point out that any current method can be modified to adjust ecological covariates by a generalized linear model. This is particular interesting after a spatial cluster has been identified since the following up study is to link the identified cluster with risk factors. Even though there are many methods for cluster detection, their statistical properties have not been understood. This makes the interpretation of discoveries difficult. Although most studies focus on detection of hot spot clusters, the presence of cool spot cluster significantly influences the results of hot spot analysis. Modeling spatial dependence is a typical difficult problem in disease mapping, because most disease mapping approaches are extremely computationally intensive. In most cases, estimation cannot be found analytically, and thus an MCMC algorithm is applied. This is extremely time consuming if we want to compare methods numerically.

References

1. A. Agresti. *Categorical Data Analysis*. Wiley, New York, 2002.
2. L. Anselin. Local indicators of spatial association-LISA. *Geogr Anal*, **27**:93–115, 1995.
3. R. Assuncao and E. Reis. A new proposal to adjust Moran's I for population density. *Stat Med*, **18**:2147–2162, 1999.
4. R. Assuncao, M. Costa, A. Tavares, and S. Ferreira. Fast detection of arbitrarily shaped disease clusters. *Stat Med*, **25**:723–745, 2006.
5. J. Besag and J. Newell. The detection of clusters in rare diseases. *J R Stat Soc A*, **154**:143–155, 1991.
6. D.G. Clayton and J. Kaldor. Empirical Bayes estimates of age-standardized relative risks for use in disease mapping. *Biometrics*, **43**:671–681, 1987.
7. A.D. Cliff and J.K. Ord. *Spatial Processes: Models And Applications*. Pion, London, 1981.
8. N. Cressie and N. Chen. Spatial modeling of regional variables. *J Am Stat Assoc*, **84**:393–401, 1989.
9. A.C. Davison and D.V. Hinkley. *Bootstrap Methods and Their Application*. Cambridge University Press, Cambridge, 1997.
10. P.J. Diggle. Overview of statistical methods for disease mapping and its relationship to cluster detection. In P. Elliott, J.C. Wakefield, N.G. Best, and D.J. Briggs, editors, *Spatial Epidemiology: Methods and Applications*, pages 87–103. Oxford University Press, London, 2000.
11. P. Diggle, J.A. Tawn, and R.A. Moyeed. Model-based Geostatistics with discussion. *Appl Stat*, **47**:299–350, 1998.
12. M. Dwass. Modified randomization tests for nonparametric hypothesis. *Ann Math Stat*, **28**:181–187, 1957.
13. J. Faraway. *Extending the Linear Model with R*. Chapman & Hall, Boca Raton, 2006.
14. R.E. Gangnon and M.K. Clayton. Cluster modeling for disease rate mapping. In A.B. Lawson and G.T. Denison, editors, *Spatial Cluster Modelling*, pages 147–162. Chapman & Hall/CRC, London, 2002.
15. R.C. Geary. The contiguity ratio and statistical mapping. *Inc Stat*, **5**:115–145, 1954.
16. A. Getis and J. Ord. The analysis of spatial association by use of distance statistics. *Geogr Anal*, **24**:189–206, 1992.
17. P. Good. *Permutation Tests*. Springer, New York, 2000.
18. P. Green and S. Richardson. Hidden Markov models and disease mapping. *J Am Stat Assoc*, **96**:1055–1070, 2002.
19. R. Haining. *Spatial Data Analysis in the Social and Environmental Sciences*. Cambridge University Press, Cambridge, 1990.
20. R. Haining, J. Law, and D. Griffith. Modelling small area counts in the presence of overdispersion and spatial autocorrelation. *Comput Stat Data Anal*, **53**:2923–2947, 2009.
21. M. Kulldorff. A spatial scan statistic. *Commun Stat Theory Methods*, **26**:1481–1496, 1997.
22. M. Kulldorff. *SaTScan v7.0.1: Software for the Spatial and Space-time Scan Statistic*. Information Management Services Inc., 2006, http://www.satscan.org/.
23. M. Kulldorff and N. Nagarwalla. Spatial disease clusters: detection and inference. *Stat Med*, **14**:799–810, 1995.

24. M. Kulldorff, T. Tango, and P. Park. Power comparisons for disease clustering tests. *Comput Stat Data Anal*, **42**:665–684, 2003.
25. A.B. Lawson. *Statistical Methods in Spatial Epidemiology*. Wiley, New York, 2001.
26. A.B. Lawson and A. Clark. Spatial mixture relative risk models applied to disease mapping. *Stat Med*, **21**:359–370, 2002.
27. A.B. Lawson and D.G.T. Denison. *Spatial Clustering Modeling*. CRC Press, New York, 2002.
28. G. Lin and T. Zhang. Loglinear residual tests of Moran's I autocorrelation and their applications to Kentucky Breast Cancer Data. *Geogr Anal*, **38**:209–225, 2006.
29. G. Lin and T. Zhang. Loglinear residual tests of Moran's I autocorrelation and their applications to Kentucky Breast Cancer Data. *Geogr Anal*, **39**:293–310, 2007.
30. P. McCullagh. Quasi-likelihood functions. *Ann Stat*, **11**:59–67, 1983.
31. P.A.P. Moran. The interpretation of statistical maps. *J R Stat Soc Ser B*, **10**:243–251, 1948.
32. J.I. Naus. The distribution of the size of the maximum cluster of points on the line. *J Am Stat Assoc*, **60**:532–538, 1965.
33. S. Openshaw, M. Charlton, C. Wymer, and A.W. Craft. A mark I geographical analysis machine for the automated analysis of point data sets. *Int J GIS*, **1**:335–358, 1987.
34. K. Ord. Estimation methods for models of spatial interaction. *J Am Stat Assoc*, **70**:120–126, 1975.
35. S.C. Richardson and C. Monfort. Ecological correlation studies. In P. Elliott, J.C. Wakefield, N.G. Best, and D.J. Briggs, editors, *Spatial Epidemiology*, pages 205–220. Oxford University Press, London, 2000.
36. P.A. Rogerson. The detection of clusters using a spatial version of the chi-square goodness-of-fit statistics. *Geogr Anal*, **31**:130–147, 1999.
37. T. Tango. A class of tests for detecting general and focused clustering of rare diseases. *Stat Med*, **14**:2323–2334, 1995.
38. T. Tango and K. Takahashi. A flexibly shaped spatial scan statistic for detecting clusters. *Int J Health Geogr*, 2005. doi:10.1186/1476-072X-4-11.
39. J.C. Wakefield, J.E. Kelsall, and S.E. Morris. Clustering cluster detection, and spatial variation in risk. In P. Elliott, J.C. Wakefield, N.G. Best, and D.J. Briggs, editors, *Spatial Epidemiology: Methods and Applications*, pages 205–220. Oxford University Press, London, 2000.
40. T. Waldhor. The spatial autocorrelation coefficient Moran's I under heteroscedasticity. *Stat Med*, **15**:887–892, 1996.
41. S.D. Walter. The analysis of regional patterns in health data. *Am J Epidemiol*, **136**:730–741, 1992.
42. A. Whittemore, N. Friend, B. Brown, and E. Holly. A test to detect clusters of disease. *Biometrika*, **74**:631–635, 1987.
43. T. Zhang and G. Lin. A supplemental indicator of high-value or low-value spatial clustering. *Geogr Anal*, **38**:209–225, 2006.
44. T. Zhang and G. Lin. A Decomposition of Moran's I for clustering detection. *Comput Stat Data Anal*, **51**:6123–6137, 2007.
45. T. Zhang and G. Lin. Scan statistics in loglinear models. *Comput Stat Data Anal*, **53**:2851–2858, 2009.
46. T. Zhang and G. Lin. Cluster detection based on spatial associations and iterated residuals in generalized linear mixed models. *Biometrics*, **65**:353–360, 2009.

Chapter 16
From QTL Mapping to eQTL Analysis

Wei Zhang and Jun S. Liu

16.1 Introduction

Genetic loci that affect mRNA expression levels of other genes are referred to as expression quantitative trait loci (eQTL). Discovering eQTLs by combining gene expression data with genetic marker data is an important means to understand gene regulation and to study disease mechanisms. EQTL mapping has been studied in many species, e.g., yeast [1, 2], eucalyptus [3], mice [4–6], rats [7], and human [8, 9]. Results from eQTL studies have been used for identifying hot spots [1, 4–7, 9, 54], constructing causal networks [6, 10–14], prioritizing lists of candidate genes for clinical traits [5, 7, 13], and elucidating subclasses of clinical phenotypes [4, 5].

Most eQTL studies are based on linear regression models [15] in which each trait variable is regressed against each marker variable. The p-value of the regression slope is reported as a measure of significance for the association. In the context of multiple traits and markers, procedures such as false discovery rate (FDR) controls [16, 17] can be used to control family-wise error rates. Despite the success of this type of regression approach, a number of challenging problems remain. First, these methods cannot easily assess the joint effect of multiple markers, i.e., epistatic effects, beyond additive effects. Storey et al. [85] developed a step-wise regression method to find eQTL pairs. This procedure, however, tends to miss eQTL pairs with small marginal effects but a strong interaction effect. Second, there are often strong correlations among expression levels for certain groups of genes, partially reflecting coregulation of genes in biological pathways that may respond to common genetic loci and environmental perturbations [2, 4, 14, 18, 19]. Previous findings of eQTL "hot spots," i.e., loci affecting a larger number of expression traits than expected by chance, and their biological implications further enhance this notion and highlight the biological importance of finding such gene "modules." Mapping genetic loci for multiple traits simultaneously is more powerful than mapping single traits

W. Zhang · J.S. Liu (✉)
Department of Statistics, Harvard University, Cambridge, MA, USA
e-mail: jliu@stat.harvard.edu

J. Feng et al. (eds.), *Frontiers in Computational and Systems Biology*, Computational Biology 15,
DOI 10.1007/978-1-84996-196-7_16,

at a time [20]. Although for a known small set of correlated traits, one can conduct QTL mapping for the principal components [21], this type of method becomes ineffective when the set size is moderately large, or one has to enumerate all possible subsets. An alternative approach is to identify subsets of genes by a clustering method and then fit mixture models to clusters of genes [22] or linear regression by treating genes as multivariate responses [23]. The eQTL mapping then depends on whether the clustering method can find the right number of clusters and the right gene partitions.

In contrast, aforementioned issues can be partially addressed by the Bayesian partition (BP) model [24]. In this framework, we introduce three sets of latent indicator variables for genes, markers, and individuals and then systematically infer the association between groups of genes and sets of markers. A Markov chain Monte Carlo (MCMC) algorithm is designed to traverse the space of all possible partitions. Simulation studies show that the proposed method achieves significantly improved power in detecting eQTLs compared to traditional regression-based methods. A particular strength of the BP model is its ability to detect epistasis when the marginal effects are weak, addressing a key weakness of all other eQTL mapping methods.

This chapter is organized as follows. We first give a brief description of the biological background of eQTL mapping in Sect. 16.2. Then in Sect. 16.3, we provide a brief review of both QTL mapping methods and eQTL analysis methods. We explain the BP model in Sect. 16.4 and show some simulation results in Sect. 16.5. We conclude the chapter with a short discussion in Sect. 16.6.

16.2 Biological Background

16.2.1 Genetic Experiments for eQTL Studies

Natural variation in gene expression is extensive in species from yeast to human. The goal of the eQTL mapping is to correlate variations in the gene expression with DNA variations. In such cases we say that the gene is linked to or mapped to the corresponding DNA region. One justification for studying genetics of gene expression is that transcript abundance may act as an intermediate phenotype between genomic sequence variation and more complex whole-body phenotypes.

The idea of carrying out genome-wide eQTL mapping was introduced by Jansen and Nap [25] and Brem et al. [1]. The principal procedure for studying mice is outlined in Fig. 16.1. First two distinct inbred strains of mice are crossed to produce the F_1 generation, which has heterozygous alleles across the whole genome. In backcross design (Fig. 16.1(a)), the F_1 generation is crossed with one of the parents to produce the B_1 generation. In intercross design (Fig. 16.1(b)), the F_1 generation is crossed among themselves to produce the F_2 generation. mRNA abundances of the offspring are measured from microarray experiment, and the whole genome is scanned generating thousands of SNPs. Due to recombination, the chromosome in the derived offspring is a mosaic of the two grandparental chromosomes. Most

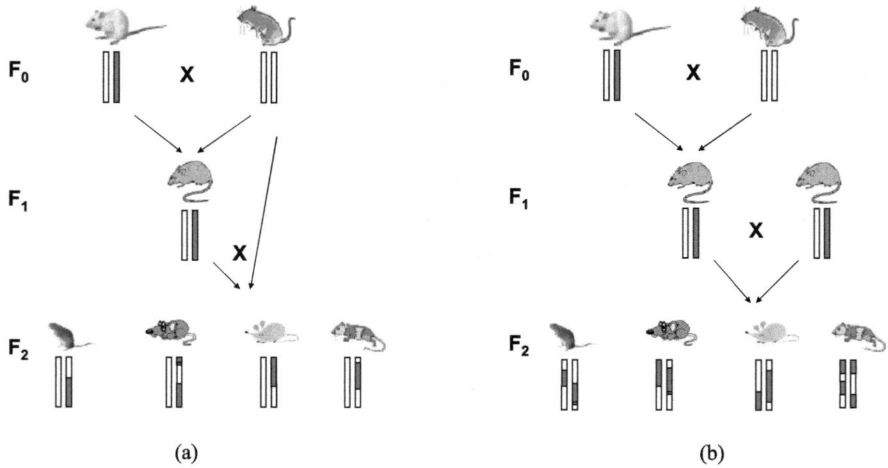

Fig. 16.1 Backcross and intercross. All individuals within an inbred strain (F_0) are genetically identical and are homozygous at all loci. The two parental strains are crossed to produce the F_1 generation. The F_1 individuals are also genetically identical and are heterozygous at all loci. In (**a**), the F_1 generation is then crossed back to one of the parental strains to produce the backcross. In (**b**), the F_1 generation is crossed with itself to produce the intercross

eQTL mapping methods for backcross designs can be easily extended to intercross designs. Intercross designs, in addition to being able to estimate dominant effect, are usually more powerful than backcross designs [26].

The genetic markers are chosen to cover the whole genome. Two distance measures are used to specify the linear order of the marker loci along each chromosome: the physical distance (b), which is the number of nucleotides between two loci, and the genetic distance, measured in centimorgans (cM), which represents the average number of crossovers between the two loci in 100 meioses. The probability of recombination r (also called the recombination rate or recombination fraction) can be calculated from genetic distance d using Haldane's map function $r = (1 - e^{-2d})/2$.

16.2.2 EQTL Hot Spots

A common feature of eQTL studies is the detection of eQTL hot spots, i.e., genomic regions that affect the expression of a much large number of genes than expected by chance. For example, Brem et al. [1] detected eight eQTL hot spots in yeast that affect the expression of a group of 7–94 genes of related functions. Additional five hot spots were predicted using a larger sample size [2]. The existence of eQTL hot spots is also prominent in arabidopsis [27, 28], mice [4, 6, 54], rats [7], and human [9, 19]. The typical procedure in claiming a hot spot is dividing the genome of the species under study into multiple fixed length windows and counting the number of transcripts mapped to each window. A Fisher's exact test is then used to test whether a window contains significantly more eQTLs than expected.

The existence of the hot spots is evidence for master regulators of the gene expression. Many genes that map to the same hot spots are enriched for functional gene sets derived from Gene Ontology, and causal regulators are predicted near the hot spot [2, 14], suggesting that the eQTL hot spots are biologically coherent. However, complicated experiments are prone to misinterpretation. As Darvasi [29] pointed out, the hot spots can be explained by clustering of genes with highly correlated expression, or the phenomenon is falsely inflated by the high false discovery rate due to multiple testing. This is illustrated by a recent simulation study using real expression data from human pedigrees with an independently simulated SNP map [30]. Their analyses showed strong clustering of eQTLs, demonstrating that "ghost" hot spots may result simply from a high correlation among mRNA levels.

16.2.3 eQTL and cQTL

EQTL mapping is a promising technique for the identification of genes relevant to a complex disease. Major loci controlling complex phenotypes, such as obesity, may affect genes of related function in the pathway. Thus, the mRNA expressions of these downstream genes in the relevant pathways will be linked to the major causative loci. By finding colocalized eQTL and traditional clinical QTL (cQTL), it is possible to identify a list of candidate genes for the follow-up study of the disease. For example, in a rat eQTL mapping study [7], 255 cis-acting genes were mapped to regions of physiological QTL affecting blood pressure related traits. Among these genes, 73 have human orthologs that reside within known human blood pressure loci, serving as candidates for follow-up studies of human hypertension.

In addition, there may be heterogeneity among the causative loci for a given disease in a population of interest. When present, this heterogeneity impacts the ability to detect linkages to the causative loci, as the significance of any one locus is diminished when the population is considered as a whole. EQTL data serves as an alternative source to define a trait more accurately, generating genetically more homogeneous groups of individuals. In a study of obesity, Schadt et al. [4] mapped eQTLs and cQTLs for fat pad mass (FPM) in an F_2 mice population. The mice were raised in two different conditions with high versus low FPM. By clustering a collection of genes differentially expressed between the two groups, they identified two distinct subgroups within the high FPM group. Separate genetic analyses were performed on two sets of individuals: (1) those classified as high FPM group 1 or low FPM; (2) those classified as high FPM group 2 or low FPM. QTLs of the FPM trait fell in two nonoverlapping genomic regions, suggesting heterozygous control of the FPM trait. When using the whole population to map FPM, the second peak was missed, and the primary peak had reduced significance level. Despite its success in identifying differential FPM QTLs, how to automatically detect genetic heterogeneity using eQTL data remains an interesting open question.

16.3 Methods for QTL and eQTL Mappings

Quantitative trait loci (QTL) mapping generally refers to identifying the genetic loci that are responsible for variation in a quantitative traits (such as the yield from a crop, the body fat mass of a mouse, etc.). Pioneering genetic mapping studies can be traced back to over 80 years ago, when Sax [31] showed that the association between seed weight and seed coat color in beans is due to the linkage between genes controlling weight and the genes controlling color. However, systematic and accurate mapping of QTLs has not been possible because of the difficulty in arranging crosses with genetic markers densely spaced throughout the whole genome. Recently, advances in genetics have made it possible to genotype markers on the genome scale [32]. Large amount of QTL mapping studies follow the advent of statistical methods [15, 33] for experimental crosses, in which confounding effects are fully controlled so that phenotypic variations are attributed mainly to genetic factors.

Detection and estimation of the effects of genetic factors contributing to a certain trait help one understand the biochemical basis of the trait and may aid in the design of selection experiments to improve the trait. For example, agricultural traits, such as resistance to diseases and pests, tolerance to heat, drought and cold, could be mapped and introgressed into domestic strains from exotic relatives [34]. Aspects of mammalian physiology, such as hypertension, diabetes, predispositions to cancer, and drug sensitivities, could be investigated in animal strains differing widely for these traits [35, 36].

We already introduced the experimental design in Sect. 16.1. In the following text, we consider only the backcross design. When just looking at the QTL effect on the mean of the trait of interest, only one parameter needs to be estimated in the backcross design. We use 0/1 to denote the two possible genotypes at each marker.

16.3.1 Single QTL Model

Consider a backcross of n individuals measured with M markers $\{X_1, \dots, X_M\}$ and a univariate phenotype Y. When studying the association between the trait Y and marker X_j, one can compare the phenotypic means for two classes of progeny: those with $X_j = 0$ and those with $X_j = 1$. The difference between the means provides an estimate of the QTL effect at marker j. The significance of the association can be obtained using the likelihood ratio test, or its equivalence such as the chi-square or F-tests. Specifically, for a test at the jth marker, the single QTL model is

$$Y = \mu + \beta X_j + \varepsilon, \quad \varepsilon \overset{iid}{\sim} N(0, \sigma^2).$$

A LOD score, which is the log-10 based likelihood ratio, is calculated for each marker to test $H_0 : \beta = 0$ vs. $H_1 : \beta \neq 0$. The LOD score is plotted as a function of genome position and compared to a genome-wide threshold to declare any significance. This approach is conceptually simple and easy to implement.

Table 16.1 Probabilities of the genotypes at the QTL conditional on the genotypes at the two flanking markers

Flanking marker genotype (x_L, x_R)	QTL genotype z^a	
	0	1
(0,0)	$\frac{(1-r_L)(1-r_R)}{1-r}$	$\frac{r_L r_R}{1-r}$
(0,1)	$\frac{(1-r_L)r_R}{r}$	$\frac{r_L(1-r_R)}{r}$
(1,0)	$\frac{r_L(1-r_R)}{r}$	$\frac{(1-r_L)r_R}{r}$
(1,1)	$\frac{r_L r_R}{1-r}$	$\frac{(1-r_L)(1-r_R)}{1-r}$

[a]Here r_L, r_R, and r denote the recombination frequencies between the left marker and the QTL, the QTL and the right marker, and between the two flanking markers. The expected mean of the trait given the genotypes at the two flanking markers is $\mu + \beta c_z$, where the coefficient is c_z the conditional probability that the genotype at the QTL is 1 (the third column in the table)

When linear regression and hypothesis testing were first carried out in QTL analysis [15], large-scale genotyping technology was not available, and genetic markers were usually distantly spaced. A lot of efforts were made to infer the QTLs not located at the genotyped position but somewhere between two adjacent markers. This problem can be viewed as a missing data problem since the genotypes at the QTL are not observed. Lander and Botstein [15] proposed an interval mapping method to identify potential QTLs that might reside between genotyped markers. At any given genetic location where the genotype is not measured, the distribution of the trait is a mixture of two normal distributions with means corresponding to genotype being 0 or 1, and the mixture proportions equal to the probabilities of the genotype being 0 or 1, which can be estimated using observed data from two flanking markers and the genetic map, i.e.,

$$P(y|x_L, x_R) = P(z=0|x_L, x_R)P(y|z=0) + P(z=1|x_L, x_R)P(y|z=1),$$

where z denotes the missing genotype of the QTL, and x_L, x_R denote the observed genotypes at the two flanking markers. The EM algorithm [37] is commonly used by treating the genotype at the QTL as missing data to estimate the maximum likelihood under H_1.

Alternatively one can derive the expected mean trait value in terms of the putative QTL, as given in Table 16.1 The QTL effect β can be directly estimated via multiple linear regression:

$$Y = \mu + \beta c_z + \varepsilon,$$

where $c_z = P(z = 1|x_L, x_R)$, and the log ratio of the sum of square of residuals can be used to assess the significance of the effect [38]. Theoretically, this method should suffer from the inappropriate assumption of normality within marker genotype class due to segregation of QTL, but in practice it seems to give similar results as the EM method.

With the advances of high-throughput technology in the last 10 years, genotyped markers are distributed very densely, usually around 1 cM apart. The advantage of interval mapping in giving more precise QTL locations tends to weigh less than the computational cost involved.

16.3.2 Multiple QTL Model

Single QTL model provides a simple tool to detect the association of the quantitative trait with a given QTL, but it does not take into consideration of the interfering effect of multiple QTLs. As a consequence, the power of detection may be compromised, and the estimates of locations and effects of QTLs may be biased [15]. Even nonexistent "ghost" QTL may appear [38, 39]. Although multiple regression procedures [38, 40] are straightforward, it is computational infeasible to explore every possible model and potential QTLs within genotyped marker regions when the number of QTLs gets large.

Jansen [41, 42] and Zeng [43] independently proposed a hybrid mapping method that fits single-interval mapping QTL model at each putative QTL by using selected markers as covariates to eliminate the effect of other QTLs. The task of finding multiple QTLs thus reduces to one-dimensional search. The idea is to first select a subset of markers, S, to control for background genetic variation and then perform a genome scan at each locus, conditional of the genetic effects of markers in the set S. At each locus in the genome, given the genotype z, the trait Y is distributed as

$$Y = \mu + \beta Z + \sum_{j \in S} \beta_j X_j + \varepsilon,$$

where X_j is the genotype of a marker in the set S. The likelihood ratio test is then performed to test the null hypothesis that there is no QTL among the tested markers versus the alternative hypothesis that there is a QTL. A LOD score is calculated and compared with the genome-wide threshold. When the QTL is located between two adjacent markers so that the genotype is not directly observed, one can use ECM algorithm [44] to get the maximum likelihood estimation to perform likelihood ratio test.

The key problem is the choice of the set of markers to use as regressors. Too many markers will give low power for detection and too few markers will cause low accuracy. Including linked markers as regressors will reduce the chance of interference of possible multiple linked QTLs, but with a possible increase of sampling variance. A general guideline in practice is to use variable selection technique, such as forward selection backward elimination with AIC [46], to select markers into the subset S and then drop those markers that are within 10 cM of the test position [41, 47].

Kao et al. [48] proposed a multiple interval mapping (MIM) method to use multiple intervals simultaneously to search QTLs with possible interactions. The generic model is

$$Y = \mu + \sum_{j=1}^{M} \alpha_j Z_j + \sum_{j<k} \delta_{jk} w_{jk} Z_j Z_k + \varepsilon,$$

where $\{Z_j\}_{j=1}^{M}$ are the genotypes at the M putative QTLs, α_j and w_{jk} are marginal and interaction effects, and δ_{jk} is an indicator for interaction. Given the locations of the M putative QTLs, the maximum likelihood estimates of the QTL effects can be obtained via ECM [44]. Starting with an empty model (no QTL), they adopted stepwise selection with a likelihood ratio test as the selection criterion.

Sen and Churchill [45] described a computationally efficient Monte Carlo algorithm using importance sampling. The goal is to make inference about the QTL effect parameters (μ) and the QTL location parameters (γ, including epistasis) based on the observed marker genotypes (m) and phenotype (y). The joint distribution can be factorized as

$$P(y, m, g, \mu, \gamma) = \underline{P(y|g, \mu)P(\mu)} \times \underline{\underline{P(g|m, \gamma)P(m)P(\gamma)}},$$

where g is the unobserved genotypes of the QTL. This factorization implies that the uncertainty of the unobserved genotype g comes from two sources: the phenotypic effect on y and the linkage with m. Based on this, they proposed a two-step procedure to estimate multiple QTLs. In the first step, they imputed q versions of complete genotype information $\{G_1, \ldots, G_q\}$ on an equally spaced grid of locations spanning the genome conditional on the observed genotyped markers, i.e., $G \sim P(G|m)$. In the second step, they computed weight to the selected QTLs under model γ as

$$w_k(\gamma) = P(\gamma)P(y|G_k, \gamma), \quad k = 1, \ldots, q.$$

Using Bayesian rule, the posterior probability of QTL locations can be estimated from

$$P(\gamma|y, m) \propto \int P(\gamma)P(y|G, \gamma)P(G|m)\,\mathrm{d}G \approx \sum_{k=1}^{q} w_k(\gamma).$$

They used a model scanning followed by model selection using Bayes Factor [50] as criterion to update the QTL models. The advantage of this two-step approach is that by separating the genotype imputation and weights calculation into two parts, the imputed genotype map needs not be recomputed when comparing different candidate models.

Due to recent developments of Markov chain Monte Carlo (MCMC), Bayesian model selection methods have become increasingly popular in multiple QTL mapping. The typical procedure starts by setting up a likelihood function for the observed data and prior distribution on the unobserved quantities and then uses MCMC to sample the parameters of interest from the joint posterior distribution. These methods fall in two categories. One is treating the number of QTLs as a random variable and using Reversible Jump Markov chain Monte Carlo (RJ-MCMC) [87] to explore posterior distributions in different dimensions [49, 86]. The ability to "jump" between models of different dimension requires a careful construction of proposal distribution. An alternative Bayesian variable selection method was developed [51] based upon a composite space [52] representation to avoid dimension change. The dimension of the model space is fixed by placing an upper bound of the number of QTLs. Each genetic effect is modeled by a so-called spike and slab mixture distribution, which has a nonzero probability mass at value zero to promote variable selection [51, 53, 55]:

$$Y = \mu + \sum_{j=1}^{M} \beta_j X_j + e,$$

$$\beta_j \sim p\delta(0) + (1 - p)N\big(0, \sigma^2\big),$$

where $\delta(0)$ represents the degenerate distribution with probability 1 at zero (point mass). This prior is a "spikier" function than the spike-and-slab function used in George and McCulloch [56]. The Bayesian MCMC approaches provide a robust inference of genetic architecture that incorporates model uncertainty by averaging over all possible models [57], but they are computationally intensive compared to traditional regression methods.

Broman and Speed [58] compared various model selection methods using different searching algorithms, including deterministic search (forward selection, backward elimination, stepwise search with both forward selection and backward elimination) and stochastic search via MCMC, with the single QTL mapping method and composite mapping method [59], through intensive simulation studies. They pointed out that: (1) Single QTL mapping performs very poorly in detecting multiple QTLs; (2) Composite mapping method generally has a low false positive rate. But the performance highly depends on the choice of the number of markers used as regressors. A considerable attenuation of power is accompanied by a choice of too many or too few markers to serve as regressors; (3) Forward selection method selects a high proportion of extraneous markers even when the sample size is large. This is because the markers are highly correlated, and once an extraneous marker is selected into the model, it remains in the model; (4) MCMC and forward selection followed by backward elimination perform the best with moderate sample size.

16.3.3 Thresholding

A common issue in all likelihood ratio test method in QTL mapping problems is the difficulty of determining appropriate significance thresholds for the purpose of detecting QTL. The source of this difficulty is twofold. First, there is the problem of determining the distribution of the test statistic under an appropriate null hypothesis. The regularity conditions that ensure an asymptotic chi-square distribution for the likelihood ratio statistic are not satisfied. Many factors, including the sample size, the genetic map, and the underlying true magnitude of the QTL effect, can influence the distribution of the test statistic. The second difficulty is to control multiple hypotheses testing error.

When markers are dense and the sample size is large, Lander and Botstein [15] showed that the LOD score for the single QTL model in a backcross experiment varies according to the square of an Ornstein–Uhlenbeck processes. The approximate threshold for LOD score at the type I error rate α is $2\log(10)t_\alpha$, where t_α solves for

$$\alpha = (C + 2Gt_\alpha)\chi^2(t_\alpha).$$

C is the number of chromosomes; G is the length of the genetic map, measure in Morgans; $\chi^2(t_\alpha)$ is the probability that a χ_1^2-distributed random variable is less than t_α. Similar threshold is studied in other QTL models [26].

Churchill and Doerge [60] described a permutation-based method to estimate a threshold value. The quantitative trait data are permuted with respect to the marker data a large number of times to effectively sample from the distribution of the test statistic under a null hypothesis of no QTL. The approach is statistically sound, robust to departures from standard assumptions and is tailored to the experiment under study.

In Doerge and Churchill [61], they generalized the permutation test to the problem of detecting multiple QTL effects in a sequential way. Conditional Empirical Threshold (CET) is obtained to permute the traits after stratification according to the already detected QTLs. Residual Empirical Threshold (RET) is obtained to permute the residuals from a parametric model among whole population. CET provides a completely nonparametric test and allows for general nonadditive interactions among QTLs. However, markers linked with the first QTL will continue to show association with the trait after the stratification. Thus the application is restricted to regions of the genome that are unlinked to the major QTL. RET-based test may be more powerful than CET-based tests when the structural model is approximately true. In a sequential search procedure for multiple QTLs, the type I error rate may not be controlled.

As the number of markers grows, the number of markers showing significant association with the phenotype by chance is also expected to grow if the type I error rate is controlled. To handle this multiple test issue, false discovery rate (FDR) [16] was introduced to control the expected proportion of false discoveries, which essentially allows multiple false positive declarations. Using the notation in Table 16.2, the false discovery rate is defined as the proportion of false positives among all significant hypotheses, i.e., $E(\frac{V}{R}, R > 0)$. The FDR offers less stringent control over Type I errors than the family-wise error rate $P(V > 0)$ and is therefore usually more powerful. Such a relaxation is driven by the nature of the problem under study: "*It is now often up to the statistician to find as many interesting features in a data set as possible rather than test a very specific hypothesis on one item*" [17].

Table 16.2 Possible outcomes from M hypothesis tests

	Accepted null	Rejected null	Total
Null true	U	V	M_0
Alternative true	T	S	M_1
Total	W	R	M

For M independent tests, Benjamini and Hochberg [16] provided a procedure to control FDR at the desired level α as follows:

1. Sort M p-values from smallest to largest as $P_{(1)} \le \cdots \le P_{(M)}$.
2. Starting from $P_{(M)}$, compare $P_{(i)}$ with $\alpha \frac{i}{M}$.
3. Let k be the first time $P_{(i)} \le \alpha \frac{i}{M}$, reject all $P_{(1)}$ through $P_{(k)}$.

Simulation studies confirmed that the BH procedure works well in single QTL mapping and multiple QTL mapping [62].

Storey [63] introduced a "positive false discovery rate" (pFDR) defined as $E(\frac{V}{R}|R > 0)$ and gave a Bayesian interpretation of pFDR. For M independent hypotheses $\{H_1, \ldots, H_M\}$ with p-values $\{P_1, \ldots, P_M\}$, denote $H_i = 0$ if the ith null hypothesis is true and 1 if it is false. Suppose the rejection region is $(p : p < \gamma)$. We further assume $P(H_i = 0) = \pi_0$. Then using Bayes rule,

$$\text{pFDR}(\gamma) = P(H_i = 0|P_i < \gamma) = \frac{P(H_i = 0)P(P_i < \gamma|H_i = 0)}{P(P_i < \gamma)} = \frac{\pi_0 \gamma}{R/M}.$$

The distribution of the p-values is a mixture from the null and the alternative. For p-values close to 1, the mixture component from the alternative becomes very small. This suggests that we can use

$$\begin{aligned} P(P_i > \lambda) &= \pi_0 P(P_i > \lambda|H_i = 0) + (1 - \pi_0)P(P_i > \lambda|H_i = 1) \\ &\approx \pi_0 P(P_i > \lambda|H_i = 0) = \pi_0(1 - \lambda) \end{aligned}$$

to get a conservative estimate of $\hat{\pi}_0 \approx \frac{\#\{P_i : P_i > \lambda\}}{M(1-\lambda)}$, where λ is close to 1. Thus,

$$\text{pFDR}(\gamma) \approx \frac{\#\{P_i : P_i > \lambda\}\gamma}{(1 - \gamma)R}.$$

To remove arbitrariness in choosing λ, Storey and Tibshirani [17] suggested using a cubic spline to estimate π_0. Finally we can associate each p-value with a q-value, which is the minimum pFDR that can be attained when calling that feature significant, i.e.,

$$q(P_{(i)}) = \min_{t \ge P_{(i)}} q(t) = \min\{q(P_{(i+1)}), \text{pFER}(P_{(i)})\}.$$

If we call all features significant with q-values no greater than α, then for large M, the FDR will be no greater than α. The independence assumption can also be relaxed to weak dependence, such as that genes within a small group are independent of all the other genes [17].

16.3.4 Multiple Trait Mapping

Many data for mapping QTL contain measurements on multiple traits or one trait in multiple environments. Methods for single trait mapping do not take advantage

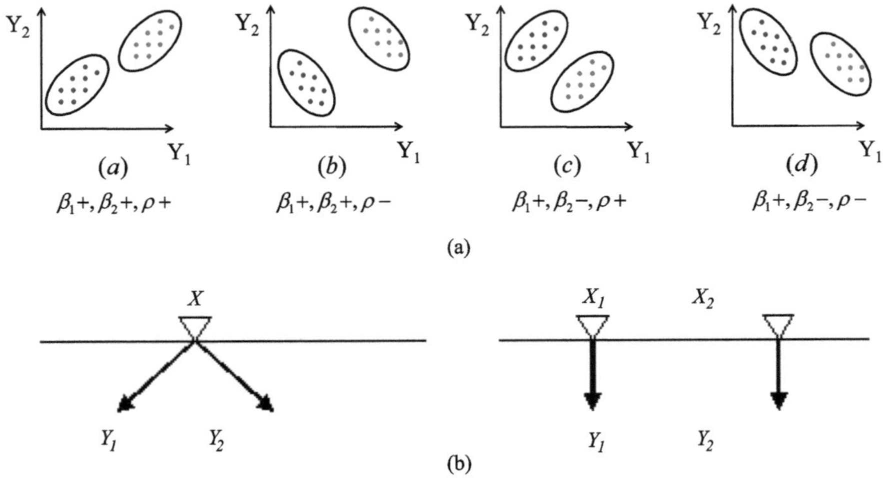

Fig. 16.2 Pleiotropic QTL and linked QTL models. (**a**) Scatter plots of the two quantitative traits evaluated at a common QTL. Individuals are labeled according to the genotypes at the QTL. (**b**) Pleiotropic QTL model (left) vs. linked QTL model [88]

of the correlation structure of the data and are therefore not powerful enough for detecting true QTLs and estimating the location accurately [20, 64–66]. Consider two quantitative traits Y_1 and Y_2 tested at one binary marker X with QTL effects β_1 and β_2, respectively. Denote by ρ the residual correlation between Y_1 and Y_2. Thus the joint distribution of Y_1 and Y_2 can be written as

$$\begin{pmatrix} Y_1 \\ Y_2 \end{pmatrix} = \begin{pmatrix} \mu_1 \\ \mu_2 \end{pmatrix} + \begin{pmatrix} \beta_1 \\ \beta_2 \end{pmatrix} X + \begin{pmatrix} e_1 \\ e_2 \end{pmatrix},$$
$$\begin{pmatrix} e_1 \\ e_2 \end{pmatrix} \sim N\left(0, \begin{pmatrix} \sigma_1^2 & \rho\sigma_1\sigma_2 \\ \rho\sigma_1\sigma_2 & \sigma_2^2 \end{pmatrix}\right). \tag{16.1}$$

Without loss of generality, we let $\beta_1 > 0$. The relationship among Y_1, Y_2 and the QTL can be described in Fig. 16.2(a), corresponding to four combinations of the signs of β_2 and ρ. It has been shown [20] that when $\beta_1\beta_2\rho < 0$, the power of the joint analysis is always higher than that of separate analysis (Fig. 16.2(a2, a3)). When $\beta_1\beta_2\rho > 0$, the power of the joint test may be lower than the higher one of the separate tests due to the fitting of additional parameters in the model. However, empirical studies have suggested that joint mapping is generally more informative than separate mapping for traits moderately or highly correlated. When there are more than two traits to consider simultaneously, joint mapping is even more beneficial.

Another justification for mapping multiple traits simultaneously is that a joint analysis helps one understand the nature of genetic correlations. Generally speaking, two traits are correlated genetically due to pleiotropy or linkage, as illustrated in Fig. 16.2(b). Under a pleiotropic model (16.1), correlation between Y_1 and Y_2 can be driven by a pleiotropic QTL. Under model (16.2),

$$\begin{pmatrix} Y_1 \\ Y_2 \end{pmatrix} = \begin{pmatrix} \mu_1 \\ \mu_2 \end{pmatrix} + \begin{pmatrix} \beta_1 & 0 \\ 0 & \beta_2 \end{pmatrix} \begin{pmatrix} X_1 \\ X_2 \end{pmatrix} + \begin{pmatrix} e_1 \\ e_2 \end{pmatrix},$$
$$\begin{pmatrix} e_1 \\ e_2 \end{pmatrix} \sim N\left(0, \begin{pmatrix} \sigma_1^2 & \rho\sigma_1\sigma_2 \\ \rho\sigma_1\sigma_2 & \sigma_2^2 \end{pmatrix}\right); \tag{16.2}$$

two QTLs in close linkage each influence a trait independently. Separation of these two hypotheses has important implications to our understanding of the nature of genetic correlations between the traits involved.

Likelihood ratio test provides a uniform tool for testing pleiotropic effect and comparing pleiotropic vs. close linkage. When QTL genotype is missing, the EM algorithm can be used to carry out the maximum likelihood estimation [20]. Knott and Haley [66] used multivariate least square estimates as an approximation. When the genotype of the QTL is unknown, the design matrix X is simply a function of the genotype probabilities for each individual. Nonparametric bootstrapping was proposed [70] to construct a confidence interval on the estimated distance between two QTLs to test pleiotropy versus close linkage.

As the number of traits gets larger, dimension reduction techniques have been widely used. For example, Mahler et al. [67] summarized information from numerous histologic phenotypes by principal component analysis in mapping the colitis susceptibility trait in mice. Weller et al. [68] and Mangin et al. [69] applied canonical transformation to obtain uncorrelated canonical traits followed by QTL mapping for the canonical traits.

So far we have reviewed methods for mapping QTL. When it comes to eQTL mapping with thousands of markers and expression of thousands of genes generated from microarray experiments, more sophisticated methods are needed. Ideally, a statistical method for eQTL identification would properly account for multiplicities across the genome, multiplicities across transcripts, epistatic effect, and correlations among transcripts.

16.3.5 Regression Based Methods for eQTL Mapping

Using single trait mapping to large amount of gene expression has been known to suffer from low power in detection, partly due to the multiple testing problem and partly due to its inability to utilize the correlation structure among the gene expression traits. A common practice for handling thousands of transcripts is to select a small number of target genes and map QTLs for these prescreened transcripts. For example, Lan et al. [54] and Yvert et al. [2] applied hierarchical clustering to the gene expression and then used principal component analysis for each gene cluster to reduce the dimension to a few "supergenes" that capture the majority of variations in expression data within each cluster. Biswas et al. [71] used Singular Value Decomposition (SVD) and Independent Component Analysis to reduce the dimension of thousands of expression traits to a few hundred meta-traits. Mapping QTLs for the expression of these "supergenes" or meta-traits can enhance the signal of the genetic association. However, caution should be taken because information is lost during the process of dimension reduction and the transformation of the linkage back to the original gene traits is not always possible.

Partial least square (PLS) regression, introduced by Wold [72], has been used as an alternative approach to the ordinary least square regression in cases where the design matrix X is singular, e.g., X has multi-collinearity, or X has more variables than observations, so that the OLS solution does not exist. In many biological data sets with a large number of covariates and a limited number of samples, commonly referred to as "large p, small n" problems, this is usually the case. The main idea of PLS is to find a set of latent components that performs a simultaneous decomposition of X and Y with the constraint that these components explain as much as possible the covariance between X and Y. The underlying model is as follows:

$$X = TP' + e_X,$$
$$Y = TQ' + e_Y,$$

where $T = XW$ is called the score matrix, and P and Q are the loading matrices. Several algorithms exist for estimating the score matrix and the loading matrices using the successive optimization procedure (NIPALS, SIMPLS), with slightly different constraints. Although dimension reduction via PLS is an appealing way of dealing with ill-posed regression problems, it does not lead to the selection of relevant variables. In eQTL analysis where thousands to millions of SNP markers are under consideration simultaneously, very few are actually linked with the expression traits. The existence of large number of irrelevant variables makes the PLS estimator inconsistent.

To accommodate this problem, a sparse partial least square (SPLS) regression has been proposed by Chun and Keles [23], which imposes a penalty on the L_1 norm of the weight matrix W. The tuning parameter and the number of important latent components are determined via cross-validation to minimize the mean square prediction error. The procedure starts by clustering gene expression into groups of similar expression and then fits SPLS to each cluster by treating the expression of multiple genes as multivariate responses. The final stage is constructing bootstrap confidence interval for the transcript selection using only the selected markers from the original fit. Only those marker/transcript pairs with confidence intervals excluding 0 are claimed as having significant linkage. Simulation studies show that the multivariate SPLS regression leads to increase in power for detecting weak linkage since the inherent correlation among genes are taken into account. An obvious advantage of the SPLS over traditional single-trait-single-marker analysis is that the issue of multiple transcripts and multiple markers is bypassed and therefore it avoids potential multiple-testing errors. However, the performance depends on how well the preclustered groups of gene traits are similar to each other. If the group of genes actually consists of subgroups that are linked to different markers, SPLS may contaminate the structures and generates ambiguous linkages.

16.3.6 Bayesian Methods for Studying eQTLs

Recent efforts in eQTL mapping focus on combined analysis of all transcripts and markers. Mixture-Over-Marker (MOM) model of Kendziorski et al. [22] was the first attempt to allow information sharing across transcripts and to analyze multiple markers jointly by a mixture model over the markers. The MOM model assumes that a transcript t maps to nowhere with probability p_0 and maps to marker m with probability p_m, so that $\sum_{m=0}^{M} p_m = 1$. The marginal distribution of transcript t, $\mathbf{y_t} = (y_{t1}, \ldots, y_{tn})'$, for n observations is given by

$$p_0 f_0(\mathbf{y_t}) + \sum_{m=1}^{M} p_m f_m(\mathbf{y_t}),$$

where $f_0(\mathbf{y_t}) = \int g(\mathbf{y_t}|\mu)\pi(\mu)\,\mathrm{d}\mu$. The underlying mean μ is treated as random effect and integrated out. $f_m(\mathbf{y_t})$ is the distribution given that transcript t is associated with marker m. The genotypes at marker m naturally separate the observations into subgroups, say $\mathbf{y_t} = \{\mathbf{y}_t^0, \mathbf{y}_t^1\}$. Then, $f_m(\mathbf{y_t}) = f_0(\mathbf{y_t^0}) f_0(\mathbf{y_t^1})$. Parameters, $\{p_0, p_1, \ldots, p_M\}$ and those specifying $g(\cdot)$ and $\pi(\cdot)$, are estimated via the EM algorithm. With multiple transcripts present in the data, they proposed to first partition the transcripts into subgroups using k-means clustering. Then for each cluster, the parameters are shared across multiple transcripts. Despite its ability to model associations with multiple markers across multiple transcripts simultaneously, the assumption that each transcript is either associated with one of the markers or not associated with any marker at all is indeed very strong in real applications.

A Bayesian joint analysis of transcripts and markers (BAYES) was proposed by Jia and Xu [73]. To avoid variable selection, they adopted a Bayesian shrinkage analysis so that markers with small effects are forced to shrink their effects to zero. The expression level of the transcript t, Y_t, follows a linear regression model

$$Y_t = X\gamma_t + e_t,$$

where $X = (X_1, \ldots, X_M)$ are the genotypes of M markers, $\gamma_t = (\gamma_{t1}, \ldots, \gamma_{tM})'$ are the regression coefficients of these markers for transcript t, $e_t \sim N(0, R\sigma^2)$, and R is a known positive definite matrix. The coefficient of marker m for transcript t, i.e., γ_{tm}, follows a two-component mixture Gaussian distribution. The strength of the effect of a particular marker is shared across all transcripts. The full model is described as follows:

$$Y_t \sim N\big(X'\gamma_t, R\sigma^2\big);$$

$$\gamma_{tm} \sim (1 - \eta_{tm})N(0, \delta) + \eta_{tm}N\big(0, \sigma_m^2\big), \quad \text{where } \delta = 1e - 4;$$

$$\sigma^2 \sim Inv - \chi^2(0, 0);$$

$$\eta_{tm} \sim Bernoulli(\rho_m);$$

$$\rho_m \sim Dirichlet(1, 1);$$
$$\sigma_m^2 \sim Inv - \chi^2(5, 50).$$

Markov chain Monte Carlo (MCMC) is utilized to sample the parameters from their joint posterior distribution, and certain threshold values are used to select the eQTLs. The posterior mean of the proportion of transcripts associated with each marker (ρ_m) can be used to detect hot spot regions where many transcripts are mapped than expected.

BAYES allows a transcript to be simultaneously associated with multiple markers and a marker to simultaneously alter the expression of multiple transcripts through hierarchical modeling. Simulation studies comparing the method of MOM and BAYES revealed that MOM works well if a transcript is linked to only one marker. However, when a transcript is controlled by multiple markers with different effects, the linkage is detected only at the major eQTL, and the remaining eQTLs will be missed by using MOM. In the full Bayesian approach of BAYES, the multicollinearity problem is not explicitly addressed, and priors for the regression coefficients are assumed to be independent. This assumption is contradictory to the fact that adjacent markers are highly correlated in real data analysis. Therefore, the highly correlated nature of the marker data may hamper the performance of variable selection.

16.3.7 Bayesian Networks

Bayesian network is a graph-based model of joint multivariate probability distributions that captures properties of conditional independence between variables. The network is a directed and acyclic graph so that the joint probability can be decomposed into product of the conditional probabilities of each node given its parents (it is possible for a node to have an empty parent set). Statistical foundations for learning Bayesian networks from observations and computational algorithms to do so are well understood and have been used successfully in many applications [74–76].

Bayesian networks have been used to study causal interaction networks of biological systems based on gene expression data from time series and gene knockout experiments, protein–protein interaction data derived from predicted genomics features, and on other direct experimental interaction data [77, 78]. Recently, Zhu et al. [10] used large-scale liver microarray and genotypic data from the segregating mouse population [4] to construct the gene regulation network in the mouse liver system. The rationale to use eQTL data is that any gene expression trait pair controlled by a common QTL is either (1) independently driven by the same QTL or (2) causally associated in that one is driven by the QTL (upstream gene), while the other responds to the trait driven by the QTL (downstream gene). They employed two assumptions to incorporate the eQTL information into the network construction in order to reduce the computational load. First, only a limited set of genes are allowed to directly interact with any given gene. The candidate genes are selected based on (1) correlation of the LOD score and (2) mutual information of the expression level, with the given gene. Second, eQTL data is used again to provide causal anchors between any gene expression trait pair. For example, cis-acting genes are not allowed to be controlled by the other genes that are mapped to the same eQTL. Genes that have multiple eQTLs are more likely to be in the downstream of the network than genes with fewer but stronger eQTLs. They demonstrated the utility of the resulting network in this system by examining the gene expression behavior of HSD11B1. The predictive capabilities of the network were assessed by comparing the set of genes predicted by the network to respond to perturbations in the expression of HSD11B1, with the set of genes observed to change in response to HSD11B1 inhibition. They showed that involving expression and QTL data in a segregating population leads to optimal networks that possess greater predictive power of causal relationship than similar networks derived from the expression data alone.

A local network construction via eQTL analysis was developed by Li [12] in a mice population study. They first identified a list of 175 transcripts that were mapped to 209 trans-acting QTL regions. The 364 genes that are located in these QTL regions and have SNPs that differ between the two progenitor strains were considered as candidate modulators. By connecting an edge from the modulator gene in the QTL region to the target gene that mapped to the QTL, they constructed a list of 445 QTL-SNP-derived relations. For genes with more than one candidate modulators in a given QTL region, they used the Bayesian network calculation to search for the best modulator and removed the remaining modulators. Among the final list of 145 modulatory relations, they identified two transcription factor binding sites in the two target genes' sequences that were predicted to be regulated by the corresponding transcription factors, confirming the validity of these predictions. However, they made a very strong assumption that the expression of the modulator genes located in the QTL regions "controls" the target genes' expression, which only represents a small portion of eQTL regulation.

16.3.8 Integrative Analysis

Recently, new high-throughput technologies for DNA sequencing and Genomics produce large-scale data sets from diverse sources. Significant progress has been made by integrating multiple sources of data to reconstruct networks that predict complex system behavior. Module network, introduced by Segal et al. [79], combined the known regulator information with the gene expression data to identify regulatory modules and study their condition-specific regulatory program. A *regulatory module* is a set of coregulated genes, associated with a *regulatory program* that explains the expression of the module genes in terms of a set of regulatory contexts. Lee et al. [80] extended the module network approach to incorporate eQTL data into the regulatory network construction. Their algorithm, "Geronemo," takes as input, a list of putative regulators (transcription factors, signal transduction proteins, chromatin modification factors, and mRNA processing factors) for yeast, gene expression profile data, and genetic data measured on a yeast inbred population [81], to build module networks, in which each regulatory program is specified by a combination of both *expression regulators* (from the expression of the putative regulators) and *genotype regulators* (from the genetic data). The algorithm iterates between learning a regulatory program using decision trees for each module and reassigning each gene to the module whose regulation program provides the best prediction for the gene's expression profile.

Zhu et al. [14] combined multiple types of large-scale molecular data, including genotypic, gene expression, TFBS, and PPI data that were previously generated from a number of yeast experiments, to reconstruct causal, probabilistic gene networks for a yeast inbred population [81]. They compared the performances of three Bayesian networks: (1) Bayesian network based on the expression data alone (BN_{raw}); (2) Bayesian network based on expression and eQTL data (BN_{qtl}); and (3) Bayesian network based on expression, eQTL, TFBS, and PPI data (BN_{full}). The networks were constructed using a weighted coexpression network algorithm [82]. As in Zhu et al. [10], the information from eQTL, TFBS, and PPI data was used as prior evidence that two genes were causally related. The obtained networks were divided into sub networks to form gene modules that were comprised of highly interconnected expression traits.

BN_{qtl} and BN_{full} predicted the TF target genes and gene knockout signatures much better than BN_{raw}, suggesting that the latter two represent better causal relationship among the genes. They further used the constructed networks to infer causal regulators for the previously described yeast eQTL hot spots [2]. They first selected putative cis-acting genes for each hot spot regions as candidate causal regulators. They then compared the set of genes directly linked to each candidate regulator in the Bayesian network to the set of genes mapped to the corresponding hot spot region. Again, BN_{full} was demonstrated to be the most predictive network, which inferred a large number of causal regulators consistent with previously proposed results [2], followed by BN_{qtl}. Five previously unknown predictions made by BN_{full} had been experimentally validated.

16.4 A Bayesian Partition Model for eQTL Mapping

Here we briefly describe a Bayesian partition (BP) method for eQTL mapping developed in the PhD thesis of Wei Zhang in Harvard Statistics Department [24, 83]. We define a ***module*** as a set of gene expression traits and a set of DNA markers (e.g., SNPs) such that the expression variation of the genes is associated with the marker variation.

To formally describe the BP model, consider a sample with N individuals. Each individual i is measured with G gene expression values denoted as $\{y_{ig} : g = 1, \ldots, G\}$ and M marker genotypes denoted as $\{x_{im} : m = 1, \ldots, M\}$. We assume that the observed data can be partitioned into D nontrivial modules plus a null component. The number of nonnull modules, D, is prespecified by the user and

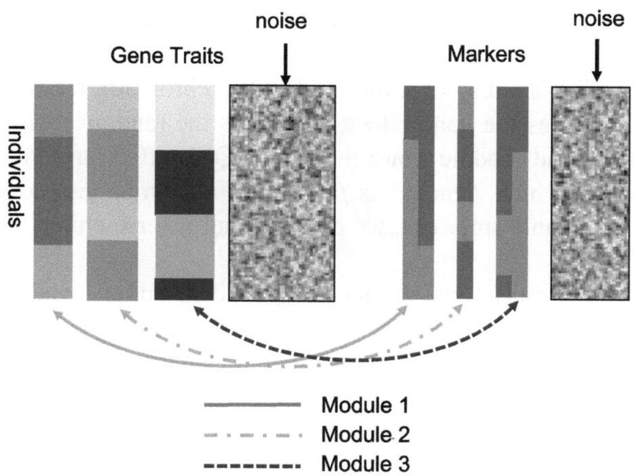

Fig. 16.3 Bayesian Partition Model. Each *row* represents an individual. The *columns* represent gene expression traits (*left*) and markers (*right*). Data is partitioned into three modules plus a null module. Module 1 has two markers associated with a group of genes, represented by a link in *solid line*. In this module individuals are partitioned into three individual types. Genes in module 2 are associated with one marker, with two individual types. Module 3 has two markers linked with a group of genes, with three individual types. Note that different modules have different individual partitions

should reflect the user's prior belief in the higher level structure of the data. Every gene g or marker m belongs to one of the D nontrivial modules or the null module, determined by the gene indicator $I_g \in \{0, 1, \dots, D\}$ and the marker indicator $J_m \in \{0, 1, \dots, D\}$. For each module $d \in \{1, \dots, D\}$, we further partition the N individuals into n_d^T types denoted by the individual indicators $K_{di} \in \{1, \dots, n_d^T\}$ for $i \in \{1, \dots, N\}$. Each module may have a different number of individual types and different ways of partitioning the N individuals. For example, with a single biallelic marker (alleles "A" and "a") in the module, the module may have two individual types corresponding to genotypes aa vs. Aa or AA (dominant model), or three individual types corresponding to genotypes aa, Aa, and AA (additive model). We seek module partitions in which expression patterns are similar for all genes, and gene expression variations across different individuals can be explained by the individual types. The overall partition of genes and markers into modules is determined by gene indicators $\{I_g \in \{0, 1, \dots, D\}, g = 1, \dots, N\}$ and marker indicators $\{J_m \in \{0, 1, \dots, D\}, m = 1, \dots, M\}$, while the module-specific partition for individuals is determined by the individual indicators $\{K_{di} : d = 1, \dots, D, i = 1, \dots, N, K_{di} \in \{1, \dots, n_d^T\}\}$. A cartoon illustration of the partition model is shown in Fig. 16.3.

We model the gene expression traits in module d by an ANOVA model so that each trait value is the sum of the gene effect (α_g), the eQTL effect for individual type k (δ_k), the individual effect (r_i), and an error term:

$$y_{ig} = \delta_k + r_i + \alpha_g + \varepsilon_{ig},$$

where gene g is in module d, and k is the individual type of i; δ_k is the eQTL effect determined by the individual type $k = K_{di}$; r_i is the effect of other regulators, such as transcription factors, signaling molecules, chromatin modification factors, and so on; α_g explains the gene effect; and ε_{ig} is the random measurement error. All genes in the same module share the same eQTL effect and individual effect, the combination of which, denoted as $\beta_{di} = \delta_k + r_i$, can be viewed as the module center. In the Bayesian framework, we put a normal-inverse-chi-square distribution on $\{\delta_k, r_i, \alpha_g, \varepsilon_{ig}\}$.

To account for epistasis, we model the joint distribution of all the associated markers in a module, denoted as $\mathbf{x_i} = \{x_{im} : m \text{ is in module } d, \text{ i.e., } J_m = d\}$, by a multinomial distribution whose frequency parameters are determined by the individual type $k = K_{di}$. We also put a conjugate prior distribution on these parameters:

$$\mathbf{x_i} \overset{iid}{\sim} Multinomial(1; \boldsymbol{\theta}_\mathbf{k}), \quad \boldsymbol{\theta}_\mathbf{k} = \left\{\theta_k^1, \ldots, \theta_k^{L^{n_d^M}}\right\},$$

$$\boldsymbol{\theta}_\mathbf{k} \sim Diri(\boldsymbol{\alpha}_k), \quad \alpha_k^1 = \alpha_k^2 = \cdots = \alpha_k^{L^{n_d^M}} = \frac{\lambda}{L^{n_d^M}},$$

where $\boldsymbol{\theta}_\mathbf{k}$ is the frequency vector of the multinomial distribution for the individual type k in module d; $\boldsymbol{\alpha}_\mathbf{k}$ is the hyper parameters for $\boldsymbol{\theta}_\mathbf{k}$; L is the number of possible genotypes at each marker; $n_d^M = \sum_{m:J_m=d}$ is the total number of linked markers in module d; and λ is the pseudo-count for the Dirichlet prior.

For the null component, we assume that there is no association between genes and markers. Each gene expression trait follows a normal distribution, and each marker follows an independent multinomial distribution. To avoid overfitting, we put an exponential prior on the indicators to penalize the higher complexity partitions:

$$P(\mathbf{I_g}, \mathbf{J_m}, \mathbf{K_{di}}) \infty \exp\left(-c_G \sum_d n_d^g - c_M \sum_d L^{n_d^m} - c_T \sum_d n_d^T\right),$$

where, n_d^g, n_d^m, n_d^T are the numbers of genes, markers, and individual types in module d, and L is the number of genotypes at each marker. Markov chain Monte Carlo algorithms including steps such as parallel tempering and reversible jump MCMC [84] are designed to sample from the above joint distribution (see Zhang [83] for more details).

16.5 Simulation Results

16.5.1 Simulation I

We tested the BP algorithm on a simulated data set in the context of inbred cross of haploid strains. The simulated dataset consists of 120 individuals measured with 1000 genes and 500 markers. Given the haploid nature of the segregants, 500 binary

Table 16.3 Simulation design for simulation I

Module	Model[a]	# Genes[b]	Heritability[c]	Cor.[d]	% of Var.[e]
A	$R = \beta I_{x_1^A = x_2^A} + \beta I_{x_3^A = 1} + e$	60	0.85	0.5	0.236
B	$R = \beta I_{x_1^B = x_2^B} + e$	60	0.7	0.5	0.188
C	$R = \beta I_{x_1^C = 1} + e$	40	0.65	0.5	0.156
D	$R = \beta I_{x_1^D = 1} + \beta I_{x_2^D = 1} + e$	40	0.7	0.5	0.186

[a]Regression models used to simulate the core genes. We denote x_i^d as the ith marker in module d

[b]Number of genes in the module

[c]Heritability of the core gene

[d]Average correlation of the genes in the module with the core gene

[e]Average percentage of variations for genes in the module explained by the true model

markers are equally spaced on 20 chromosomes, each of length 100 cM, using the "*qtl*" package in R. We simulated four modules, A, B, C, and D, each containing 60, 60, 40, and 40 genes, which are associated with 3, 2, 1, and 2 markers, respectively. The associated markers are randomly selected and do not overlap. To mimic the inter-correlation of the genes in real gene expression data, we first generated a core gene R in each module according to the corresponding models depicted in Table 16.3. In each model, $e \sim N(0, \sigma_e^2)$ represents the environmental noise. The regression coefficient β in each model is determined by the corresponding heritability, which is defined as $h^2 = (\sigma_s^2 - \sigma_p^2)/\sigma_s^2$, where σ_s^2 and σ_p^2 are the variances among phenotype values in the segregants and the pooled variance among parental measurements, respectively. We set $\sigma_p^2 = \sigma_e^2 = 1$ and solve other variance parameters based on h^2. After generating the core gene, we simulated the gene expression traits in each module from a Gaussian model where the average correlation to the core gene is set as in Table 16.3 and genes in the same module are independent conditional on the core gene. The percentage of variation explained by the true model averaged over all genes in a module is also listed in Table 16.3. Note that the data simulation model is different from the posited model in the Bayesian analysis.

We ran our algorithm with 15 parallel chains and 100,000 iterations. The trace of the log posterior probability (Fig. 16.4) indicates that the MCMC chain reached the equilibrium after the burn-in period of ~30,000 iterations. To find the genes in each module, we simply counted the number of times a gene appeared in each module from the posterior distribution and assigned genes into modules using the majority vote. From Fig. 16.5 we see that all of the genes in the null component were correctly classified. Most genes in the other four modules were also correctly classified. There were some genes in the nonnull modules that were classified into the null component, most likely due to weak signal among those genes.

To find the linked markers in each module, we not only counted the marginal number of appearances for each marker in each module but also the number of joint

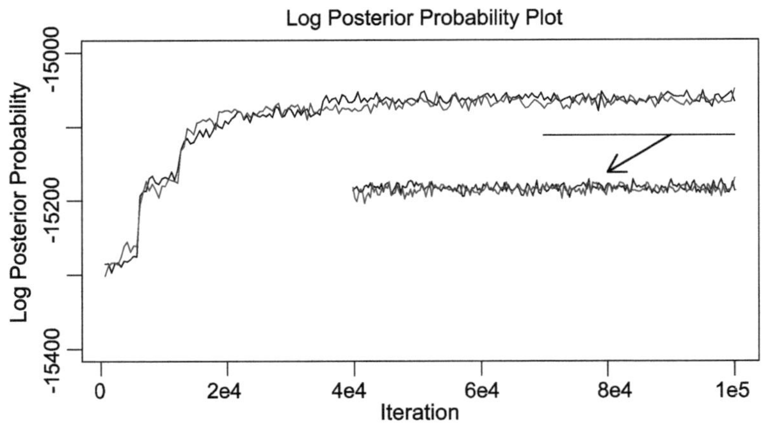

Fig. 16.4 Trace plots and autocorrelation plots of the log posterior probabilities for the simulated data set. The trace plot was generated from two independent runs, each having 100,000 iterations (for the color version, see Color Plates on p. 394)

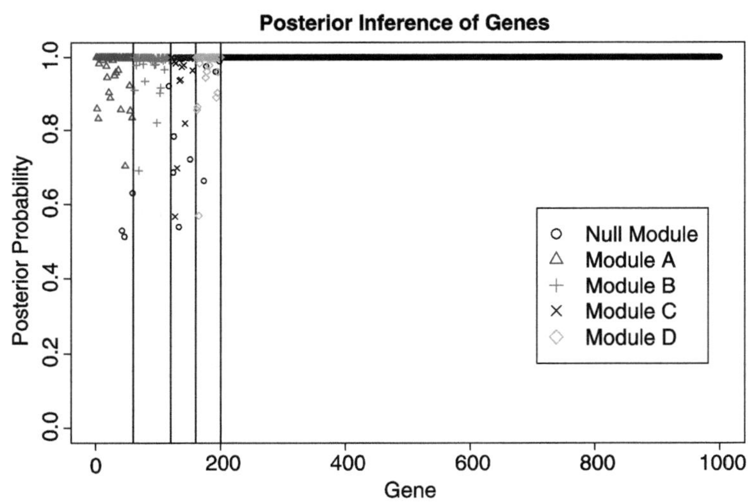

Fig. 16.5 The posterior probability plot for each gene to be included in the corresponding module. The first 200 genes are in one of the four modules, separated by the *red vertical line*. The module membership was determined by the majority vote based on the posterior samples from the last 25,000 iterations (for the color version, see Color Plates on p. 394)

appearances in order to account for the joint effect. The truly linked markers and the posterior inference are summarized in Table 16.4. We see that the truly linked markers were correctly identified for modules A, B, and D. In module B, our method picked the true marker pair (490, 149) and marker pair (491, 149) with probabilities about 0.5 each. This is due to the strong linkage between makers 490 and 491. In

Table 16.4 True markers and inferred markers in each module

Module	True markers	Posterior inference	
		Markers	Posterior Prob.[a]
A	(270, 100, 172)	(270, 100, 172)	0.988
B	(490, 149)	(490, 149)	0.503
		(490, 149)	0.490
C	292	292	0.751
		(292, 61)	0.142
		(292, 62)	0.103
D	(443, 191)	(443, 191)	0.813
		191	0.109

[a]Posterior probabilities are calculated based on joint appearances of the corresponding marker(s) in MCMC iterations

Table 16.5 Simulation design for simulation II

Module	Model[a]	%Var[b]	Locus 1[c]	Locus 2[d]	Epistasis[e]
A	$R = \beta I_{x_1=1 \text{ or } x_2=1} + e$	0.153	0.338	0.339	0.333
B	$R = \beta I_{x_1=x_2} + e$	0.158	0.052	0.052	0.895
C	$R = 2\beta I_{x_1=1 \text{ or } x_2=1} + \beta(x_1 * x_2) + e$	0.160	0.466	0.441	0.088
D	$R = \beta I_{x_1=0,x_2=1} + 2\beta I_{x_1=1,x_2=0} + e$	0.161	0.133	0.128	0.739
E	$R = \beta x_1 + \beta(x_1 * x_2) + e$	0.132	0.748	0.138	0.128
F	$R = 2\beta x_1 + \beta x_2 + e$	0.169	0.736	0.231	0.043
G	$R = 2\beta x_1 + \beta I_{x_1=x_2} + e$	0.168	0.743	0.050	0.211
H	$R = 2\beta I_{01} + 1.5\beta I_{10} + 0.5\beta I_{11} + e$	0.168	0.131	0.048	0.821

[a]Regression models that were used to generate the core gene in each module

[b]Average percentage of variations of genes in the module explained by the true model

[c]Average percentage of genetic variance explained by the first locus

[d]Average percentage of genetic variance explained by the second locus

[e]Average percentage of genetic variance explained by epistasis

all cases, our method correctly identified the truly associated markers with high posterior probabilities.

16.5.2 Simulation II

Here we conducted a comparison study. Similar to the previous simulation, we generated 100 data sets with 120 individuals, 500 binary markers, and 1000 gene expression. Eight different two-eQTL models were used, as summarized in Table 16.5,

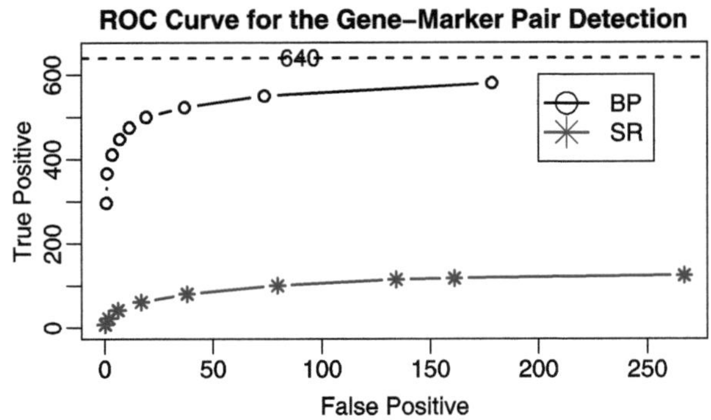

Fig. 16.6 Comparison of the receiver operator characteristic (ROC) curves for the gene-marker pair detection obtained by our Bayesian partition method (BP) and the two-stage regression method (SR). Different points along the ROC curves represent the false positive and true positive counts averaged over 100 simulations at different posterior probability thresholds (for BP) or at different FDR thresholds (for SR). There are 40 genes in each of the eight modules which are linked to two markers, and thus the total number of the true positive gene-marker pairs is 640

each having 40 genes. We fixed h^2 at 0.6 for the core gene and the inter-correlation for genes in the module with the core gene at 0.5 across all eight modules.

We analyzed the simulated data sets using two methods: (1) our Bayesian partition method using parallel tempering with 15 temperature ladders and 100,000 MCMC iterations each, referred to as BP; (2) the two-stage regression method proposed by Storey et al. [85], referred to as SR. As shown from the receiver operating characteristic (ROC) curves in Fig. 16.6, BP achieved a significantly higher power to detect eQTLs compared to SR. There are likely two reasons for this. First, we modeled the coregulated genes as a module so that information from all genes in a given module could be aggregated to improve the signal. Second, we modeled epistatic interactions explicitly so that markers with weak marginal but strong interactive effects could be detected. In contrast, the performance of the SR method strongly depends on the strength of the marginal effect of the major marker.

We compared the total number of the true gene-marker pairs detected in each module at various thresholds (Fig. 16.7). As expected, the SR method had a high failure rate when the marginal effects of both markers are weak, even at a very generous threshold. This can be seen in modules B, D, and H, where no or very weak marginal effect is present, and genetic variations are mainly explained by the epistasis. In modules E, F, and G, where the major marker explains more than 70% of the genetic variation, the SR method detected the major marker in nearly 50% of the simulations at the 0.5 threshold, but not the minor marker. In modules A and C, where the marginal effects of the two markers are almost the same, the SR method detected one of the markers for some genes, but the detection rates were lower than those in modules E, F, and G because neither marker has a very strong marginal effect. In contrast, the BP method performed significantly better than SR in all eight modules.

Since we do not fix the number of individual types in each module, we will encounter the problem of dimension change when we add a new type or remove an

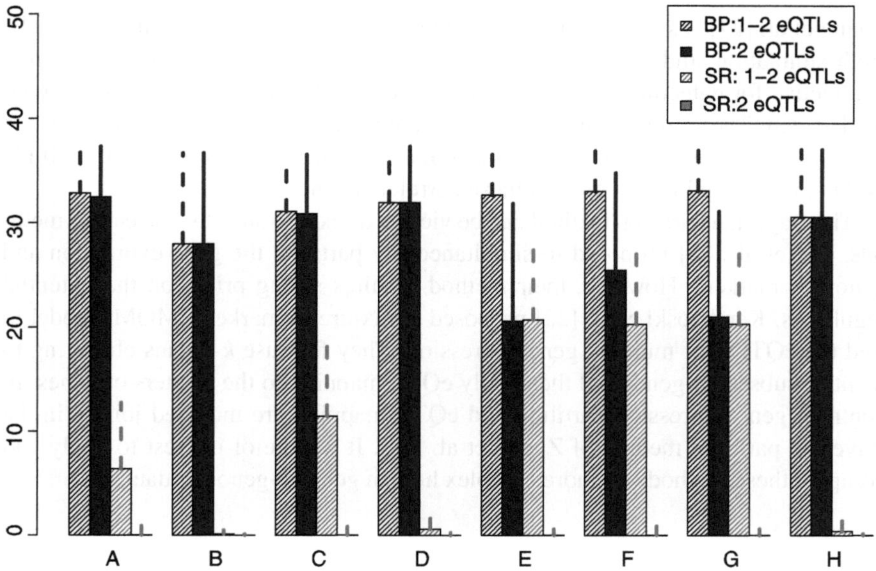

Fig. 16.7 Barplots of the number of true eQTLs detected in each module by the BP method (*blue*) and step-wise regression (SR) method (*green*). The *shaded bar* represents the number of genes detected as mapped to at least one of the true eQTLs, while the *solid bar* represents the number of genes detected as mapped to both eQTLs. The thresholds are 0.5 for both posterior probability (BP) and FDR (SR). From Fig. 16.6 we know that the total number of false positive gene-marker pairs is 11.41 and 38.04 for BP and SR, respectively. When the thresholds are relaxed to 0.1, more eQTLs were detected in each category, as indicated by the *vertical lines* above the *bars*. However, the total number of the false positive gene-marker pairs is still lower using BP (178.37) compared to that using SR (267.07) (for the color version, see Color Plates on p. 395)

old type. The method we provide here is simply adding exponential penalties for the number of individual types. Several parameters need to be specified in the model, including the number of modules D, the penalty parameters, the hyper parameters for the modules, and the hyper parameters for the null component. The module size D is determined based on the prior information about the data set. In simulations, we found that as long as D is as large as or larger than the true number of modules in the data set, the algorithm can always detect module genes and their linked markers. Through simulation studies, we found that the results were not sensitive to the choice of other prior parameters.

16.6 Discussion

In this chapter we introduced the eQTL mapping problem and reviewed a few statistical methods for conducting QTL and eQTL mappings. Whereas conventional linkage analysis has been widely and successfully applied to the study of one or a small number of traits at a time, the new module-based Bayesian partition method of

Zhang et al. [24] is suitable for analyzing thousands of phenotypes simultaneously. Both simulation studies and real data examples demonstrated that the BP method is effective for detecting marker interactions, even when no marginal effects could be detected. These improvements in power are a direct result of accounting for the correlation among gene expression traits and assessing the joint effect of multiple eQTLs, including interactions, on these correlated gene sets.

The Bayesian partition method can be viewed as extensions of some earlier methods. Lee et al. [80] proposed to simultaneously partition the gene expression and genotype markers. However, their method requires strong priors on the potential regulators. Kendzioski et al. [22] proposed a mixture of markers (MOM) model to find the eQTLs for multiple gene expression. They first use k-means clustering to identify subsets of genes and then apply eQTL mapping to the clusters of genes. In contrast, gene expression partition and eQTL mapping are modeled jointly in the Bayesian partition method of Zhang et al. [24]. It will be of interest to apply and compare these methods on more complex human genetic–genomic data.

References

1. R.B. Brem, et al. Genetic dissection of transcriptional regulation in budding yeast. *Science*, **296**(5568):752–755, 2002.
2. G. Yvert, et al. Trans-acting regulatory variation in Saccharomyces cerevisiae and the role of transcription factors. *Nat Genet*, **35**(1):57–64, 2003.
3. M. Kirst, et al. Coordinated genetic regulation of growth and lignin revealed by quantitative trait locus analysis of cDNA microarray data in an interspecific backcross of eucalyptus. *Plant Physiol*, **135**(4):2368–2378, 2004.
4. E.E. Schadt, et al. Genetics of gene expression surveyed in maize, mouse and man. *Nature*, **422**(6929):297–302, 2003.
5. L. Bystrykh, et al. Uncovering regulatory pathways that affect hematopoietic stem cell function using 'genetical genomics'. *Nat Genet*, **37**(3):225–232, 2005.
6. E.J. Chesler, et al. Complex trait analysis of gene expression uncovers polygenic and pleiotropic networks that modulate nervous system function. *Nat Genet*, **37**(3):233–242, 2005.
7. N. Hubner, et al. Integrated transcriptional profiling and linkage analysis for identification of genes underlying disease. *Nat Genet*, **37**(3):243–253, 2005.
8. S.A. Monks, et al. Genetic inheritance of gene expression in human cell lines. *Am J Hum Genet*, **75**(6):1094–1105, 2004.
9. M. Morley, et al. Genetic analysis of genome-wide variation in human gene expression. *Nature*, **430**(7001):743–747, 2004.
10. J. Zhu, et al. An integrative genomics approach to the reconstruction of gene networks in segregating populations. *Cytogenet Genome Res*, **105**(2–4):363–374, 2004.
11. N. Bing and I. Hoeschele. Genetical genomics analysis of a yeast segregant population for transcription network inference. *Genetics*, **170**(2):533–542, 2005.
12. H. Li, et al. Inferring gene transcriptional modulatory relations: a genetical genomics approach. *Hum Mol Genet*, **14**(9):1119–1125, 2005.
13. E.E. Schadt, et al. An integrative genomics approach to infer causal associations between gene expression and disease. *Nat Genet*, **37**(7):710–717, 2005.
14. J. Zhu, et al. Integrating large-scale functional genomic data to dissect the complexity of yeast regulatory networks. *Nat Genet*, **40**(7):854–861, 2008.

15. E.S. Lander and D. Botstein. Mapping Mendelian factors underlying quantitative traits using RFLP linkage maps. *Genetics*, **121**(1):185–199, 1989.
16. Y. Benjamini and Y. Hochberg. Controlling the false discovery rate: a practical and powerful approach to multiple testing. *J R Stat Soc B*, **57**:289–300, 1995.
17. J.D. Storey and R. Tibshirani. Statistical significance for genomewide studies. *Proc Natl Acad Sci USA*, **100**(16):9440–9445, 2003.
18. Y. Chen, et al. Variations in DNA elucidate molecular networks that cause disease. *Nature*, **452**(7186):429–435, 2008.
19. E.E. Schadt, et al. Mapping the genetic architecture of gene expression in human liver. *PLoS Biol*, **6**(5):e107, 2008.
20. C. Jiang and Z.B. Zeng. Multiple trait analysis of genetic mapping for quantitative trait loci. *Genetics*, **140**(3):1111–1127, 1995.
21. D. Mangin. Pleiotropic QTL analysis. *Biometrics*, **54**(1):88–89, 1998.
22. C.M. Kendziorski, et al. Statistical methods for expression quantitative trait loci (eQTL) mapping. *Biometrics*, **62**(1):19–27, 2006.
23. H. Chun and S. Keles. Expression quantitative trait loci mapping with multivariate sparse partial least squares regression. *Genetics*, **182**(1):79–90, 2009.
24. W. Zhang, J. Zhu, E. Schadt, and J.S. Liu. A Bayesian partition model for detecting pleiotropic and epistatic eQTL modules. Technical Report, Harvard University, 2009.
25. R.C. Jansen and J.P. Nap. Genetical genomics: the added value from segregation. *Trends Genet*, **17**(7):388–391, 2001.
26. J. Dupuis and D. Siegmund. Statistical methods for mapping quantitative trait loci from a dense set of markers. *Genetics*, **151**(1):373–386, 1999.
27. R. DeCook, et al. Genetic regulation of gene expression during shoot development in Arabidopsis. *Genetics*, **172**(2):1155–1164, 2006.
28. J.J. Keurentjes, et al. Regulatory network construction in Arabidopsis by using genome-wide gene expression quantitative trait loci. *Proc Natl Acad Sci USA*, **104**(5):1708–1713, 2007.
29. A. Darvasi. Genomics: Gene expression meets genetics. *Nature*, **422**(6929):269–270, 2003.
30. M. Perez-Enciso. In silico study of transcriptome genetic variation in outbred populations. *Genetics*, **166**(1):547–554, 2004.
31. K. Sax. The association of size differences with seed-coat pattern and pigmentation in PHASEOLUS VULGARIS. *Genetics*, **8**(6):552–560, 1923.
32. D. Botstein, et al. Construction of a genetic linkage map in man using restriction fragment length polymorphisms. *Am J Hum Genet*, **32**(3):314–331, 1980.
33. A.H. Paterson, et al. Resolution of quantitative traits into Mendelian factors by using a complete linkage map of restriction fragment length polymorphisms. *Nature*, **335**(6192):721–726, 1988.
34. C.M. Rick. Potential genetic resources in tomato species: clues from observations in native habitats. *Basic Life Sci*, **2**:255–269, 1973.
35. H. Tanase, et al. Genetic analysis of blood pressure in spontaneously hypertensive rats. *Jpn Circ J*, **34**(12):1197–1212, 1970.
36. J. Stewart and R.C. Elston. Biometrical genetics with one or two loci: the inheritance of physiological characters in mice. *Genetics*, **73**(4):675–693, 1973.
37. A.P. Dempster, et al. Maximum likelihood from incomplete data via the EM algorithm. *J R Stat Soc B*, **39**:1–38, 1977.
38. C.S. Haley and S.A. Knott. A simple regression method for mapping quantitative trait loci in line crosses using flanking markers. *Heredity*, **69**(4):315–324, 1992.
39. I. McMillan and A. Robertson. The power of methods for the detection of major genes affecting quantitative characters. *Heredity*, **32**(3):349–356, 1974.
40. S.J. Knapp. Using molecular markers to map multiple quantitative trait loci: models for backcross, recombinant inbred, and doubled haploid progeny. *Theor Appl Genet*, **81**:333–338, 1991.
41. R.C. Jansen. Interval mapping of multiple quantitative trait loci. *Genetics*, **135**(1):205–211, 1993.

42. R.C. Jansen. Controlling the type I and type II errors in mapping quantitative trait loci. *Genetics*, **138**(3):871–881, 1994.
43. Z.B. Zeng. Theoretical basis for separation of multiple linked gene effects in mapping quantitative trait loci. *Proc Natl Acad Sci USA*, **90**(23):10972–10976, 1993.
44. X.L. Meng and D.B. Rubin. Maximum likelihood estimation via the ECM algorithm: A general framework. *Biometrika*, **80**(2):267–278, 1993.
45. S. Sen and G.A. Churchill. A statistical framework for quantitative trait mapping. *Genetics*, **159**(1):371–387, 2001.
46. H. Akaike. A new look at the statistical model identification. *IEEE Trans Autom Control*, **19**(6):716–723, 1974.
47. R.C. Jansen and P. Stam. High resolution of quantitative traits into multiple loci via interval mapping. *Genetics*, **136**(4):1447–1455, 1994.
48. C.H. Kao, et al. Multiple interval mapping for quantitative trait loci. *Genetics*, **152**(3):1203–1216, 1999.
49. J.M. Satagopan, et al. A Bayesian approach to detect quantitative trait loci using Markov chain Monte Carlo. *Genetics*, **144**(2):805–816, 1996.
50. R.E. Kass and A.E. Raftery. Bayes factors. *J Am Stat Assoc*, **90**:773–795, 1995.
51. N. Yi. A unified Markov chain Monte Carlo framework for mapping multiple quantitative trait loci. *Genetics*, **167**(2):967–975, 2004.
52. B.P. Carlin and S. Chib. Bayesian model choice via Markov chain Monte Carlo methods. *J R Stat Soc B*, **57**:473–484, 1995.
53. N. Yi, et al. Bayesian model selection for genome-wide epistatic quantitative trait loci analysis. *Genetics*, **170**(3):1333–1344, 2005.
54. H. Lan, et al. Combined expression trait correlations and expression quantitative trait locus mapping. *PLoS Genet*, **2**(1):e6, 2006.
55. B.S. Yandell, et al. R/qtlbim: QTL with Bayesian interval mapping in experimental crosses. *Bioinformatics*, **23**(5):641–643, 2007.
56. E.I. George and R.E. McCulloch. Variable selection via Gibbs sampling. *J Am Stat Assoc*, **88**(423):881–889, 1993.
57. A.E. Raftery, et al. Bayesian model averaging for regression models. *J Am Stat Assoc*, **92**:179–191, 1997.
58. K.W. Broman and T.P. Speed. A model selection approach for the identification of quantitative trait loci in experimental crosses. *J R Stat Soc B*, **64**(4):641–656, 2002.
59. Z.B. Zeng. Precision mapping of quantitative trait loci. *Genetics*, **136**:1457–1468, 1994.
60. G.A. Churchill and R.W. Doerge. Empirical threshold values for quantitative trait mapping. *Genetics*, **138**(3):963–971, 1994.
61. R.W. Doerge and G.A. Churchill. Permutation tests for multiple loci affecting a quantitative character. *Genetics*, **142**(1):285–294, 1996.
62. C. Sabatti, et al. False discovery rate in linkage and association genome screens for complex disorders. *Genetics*, **164**(2):829–833, 2003.
63. J.D. Storey. The positive false discovery rate: a Bayesian interpretation and the q-value. *Ann Stat*, **31**:1–23, 2003.
64. D.B. Allison, et al. Multiple phenotype modeling in gene-mapping studies of quantitative traits: power advantages. *Am J Hum Genet*, **63**(4):1190–1201, 1998.
65. W.R. Wu, et al. Time-related mapping of quantitative trait loci underlying tiller number in rice. *Genetics*, **151**(1):297–303, 1999.
66. S.A. Knott and C.S. Haley. Multitrait least squares for quantitative trait loci detection. *Genetics*, **156**(2):899–911, 2000.
67. M. Mahler, et al. Genetics of colitis susceptibility in IL-10-deficient mice: backcross versus F2 results contrasted by principal component analysis. *Genomics*, **80**(3):274–282, 2002.
68. J.I. Weller, et al. Application of a canonical transformation to detection of quantitative trait loci with the aid of genetic markers in a multi-trait experiment. *Theor Appl Genet*, **22**:998–1002, 1996.
69. B. Mangin, et al. Pleiotropic QTL analysis. *Biometrics*, **54**(1):88–99, 1998.

70. C.M. Lebreton, et al. A nonparametric bootstrap method for testing close linkage vs. pleiotropy of coincident quantitative trait loci. *Genetics*, **150**(2):931–943, 1998.
71. S. Biswas, et al. Mapping gene expression quantitative trait loci by singular value decomposition and independent component analysis. *BMC Bioinform*, **9**:244, 2008.
72. H. Wold. Estimation of principal components and related models by iterative least squares. In P.R. Krishnaiah, editor, *Multivariate Analysis*, pages 391–420. Academic Press, New York, 1966.
73. Z. Jia and S. Xu. Mapping quantitative trait loci for expression abundance. *Genetics*, **176**(1):611–623, 2007.
74. J. Pearl and T.S. Verma. A theory of inferred causation. In *Principles of Knowledge Representation and Reasoning: Proceedings of the 2nd International Conference*, San Mateo, 1991.
75. D. Heckerman. *A Tutorial on Learning Bayesian Networks. Innovations in Bayesian Networks*, pages 33–82. Springer, Berlin, 1995.
76. N. Friedman, et al. Using Bayesian networks to analyze expression data. *J Comput Biol*, **7**(3–4):601–620, 2000.
77. D. Pe'er, et al. Inferring subnetworks from perturbed expression profiles. *Bioinformatics*, **17**(1):S215–224, 2001.
78. R. Jansen, et al. A Bayesian networks approach for predicting protein–protein interactions from genomic data. *Science*, **302**(5644):449–353, 2003.
79. E. Segal, et al. Module networks: identifying regulatory modules and their condition-specific regulators from gene expression data. *Nat Genet*, **34**(2):166–176, 2003.
80. S.I. Lee, et al. Identifying regulatory mechanisms using individual variation reveals key role for chromatin modification. *Proc Natl Acad Sci USA*, **103**(38):14062–14067, 2006.
81. R.B. Brem and L. Kruglyak. The landscape of genetic complexity across 5,700 gene expression traits in yeast. *Proc Natl Acad Sci USA*, **102**(5):1572–1577, 2005.
82. B. Zhang and S. Horvath. A general framework for weighted gene co-expression network analysis. *Stat Appl Genet Mol Biol*, **4**:17, 2005.
83. W. Zhang. Statistical methods for detecting expression quantitative trait loci (eQTL). PhD. Thesis, Harvard University, 2009.
84. J.S. Liu. *Monte Carlo Strategies in Scientific Computing*. Springer, New York, 2001.
85. J.D. Storey, et al. Multiple locus linkage analysis of genomewide expression in yeast. *PLoS Biol*, **3**(8):e267, 2005.
86. P.J. Gaffney. An efficient reversible jump Markov chain Monte Carlo approach to detect multiple loci and their effects in inbred crosses. Department of Statistics. Madison, WI, University of Wisconsin, 2001.
87. P.J. Green. Reversible jump Markov chain Monte Carlo computation and Bayesian model determination. *Biometrika*, **82**(4):711–732, 1995.
88. S. Wright. Correlation causation. *J Agric Res*, **20**:557–585, 1921.

Chapter 17
An Evaluation of Gene Module Concepts in the Interpretation of Gene Expression Data

Xianghua Zhang and Hongyu Zhao

17.1 Introduction

Essentially all biological functions of a living cell are carried out through the interplay between many genes. Identifying these gene networks and their functions is a main challenge in systems biology, which is a rapidly evolving research area fueled by the recent advances in high-throughput biotechnologies that enable the collection of, large-scale genomics data on gene expressions [1, 2], genome-wide location (or called transcription factor binding sites) [3, 4], protein–protein interactions [5], genetic variations [6, 7], and many other types of data. These data provide valuable system level information on different aspects of the complex biological processes and make it possible to infer the underlying networks. Many computational and statistical methods have been proposed to use these data to dissect transcriptional networks [8, 9]. Despite substantial research on this topic, it remains a great challenge to elucidate the complete network due to the complexity of the transcription processes and the noisy nature of high-throughput data.

Instead of completely characterizing the underlying networks, researchers have found it useful to describe a biological system as consisting of a set of network modules. Each module consists of genes physically or functionally related to each other to perform specific functions. It has been demonstrated that various biological systems, such as transcriptional regulatory networks, metabolic networks, and protein–protein interaction networks, are organized in this modular manner [10–12].

X. Zhang
Biomedical Engineering Institute, Department of Electronic Science and Technology, University of Science and Technology of China, Hefei, Anhui 230027, P.R. China

X. Zhang · H. Zhao (✉)
Department of Epidemiology and Public Health, Yale University, New Haven, CT 06520, USA
e-mail: hongyu.zhao@yale.edu

H. Zhao
Department of Genetics, Yale University, New Haven, CT 06520, USA

J. Feng et al. (eds.), *Frontiers in Computational and Systems Biology*, Computational Biology 15,
DOI 10.1007/978-1-84996-196-7_17,

For example, in transcriptional regulatory networks, modules are commonly used to represent a set of genes with coherent expression patterns or regulated by the same sets of transcription factors. The introduction of the module concept provides useful summaries of gene expression patterns, because it focuses on a group of genes rather than individual genes, thus reducing complexity in a system. It is therefore important to appropriately define and identify modules to facilitate the study of a biological system. We will primarily focus on the transcriptional regulatory network in this chapter.

Early research on module identification in transcriptional regulatory networks mainly relied on gene expression data. Typically, clustering methods, such as hierarchical clustering, k-means clustering, or biclustering methods are used to identify sets of genes with correlated expression patterns from gene expression data using a diverse set of similarity metrics, such as Pearson correlation and mutual information. Each resulting cluster is a set of genes with similar expression patterns and can be treated as a gene module. Based on the assumption that coexpressed genes tend to be coregulated and possibly have similar functions, the clustering results may be helpful to understand the functions of an unknown gene if it falls into a cluster of genes with known functions. Instead of using traditional clustering methods that directly use metrics based on expression profiles, several network-based methods have been proposed, such as weighted gene coexpression network (WGCN) analysis method [13] and ENIGMA method [14]. These methods first map genes to an association network based on a specific correlation metric; then a similarity measure between two genes is defined through their positions in the network. Network modules are inferred from this network-based similarity measure. Note that belonging to the same network module does not guarantee similar functions nor similar regulations for a group of genes.

Instead of using only gene expression data, methods have been developed to integrate diverse data sources for module inference. For example, protein–protein interaction data, genome-wide location data, and sequence data have been used in the literature for this purpose. Benefiting from such additional information, modules can be defined more rigorously, e.g., genes in a module not only need to have coherent expression patterns but also need to be regulated by the same sets of transcription factors. As a result, modules identified by these methods are more related to biological processes because of the use of additional data sources in the process of deriving modules. For example, Segal and colleagues [15] used a probabilistic graph model to identify transcriptional modules from expression data and gene annotation information. More specifically, a precompiled set of candidate regulatory genes, containing both known and putative transcription factors and signal transduction molecules, are considered in this method. Each transcriptional module includes genes whose expression patterns are regulated by the same set of regulatory genes under different conditions. One limitation of their model is that the activity level of a regulator is proportional to its observed expression level, because it is well known that the activity of a molecule may not be correlated with its transcription level due to post-translational modification, protein translocation, and many other reasons.

Because genome-wide location data [3, 4] provide direct physical evidence of regulatory interactions between genes and transcription factors, methods have been proposed recently to integrate gene expression data and genome-wide location data to infer transcriptional modules. Bar-Joseph et al. [16] developed an iterative procedure named GRAM, which combines genome-wide location data and expression data, to discover gene modules of regulatory network. Xu et al. [17] extended Segal et al.'s work to incorporate genome-wide location data as prior knowledge. Liu et al. [18] proposed a similar method using Bayesian hierarchical models.

Other types of data, such as sequence data, have also been combined in identifying transcriptional modules. Tanay et al. [19] proposed a graph bicluster algorithm named SAMBA to simultaneously integrate expression data, genome-wide location data, protein–protein interaction data, and phenotypic data. Lemmens et al. [20] developed an approach named ReMoDiscovery for module discovery based on heterogeneous data, including gene expression data, genome-wide location data, and motif data. Wu et al. [21] developed a method to integrate transcription factor binding site (TFBS), mutant, genome-wide location, and heat shock time series gene expression data to infer transcriptional modules for yeast heat shock response.

Although many different methods have been proposed to identify gene modules from various data sources and many useful results have been obtained, there is no consensus on the definition of modules and a lack of understanding of the biological basis of modules. In this chapter, we present an analysis of gene modules based on WGCN to investigate how they are related to the underlying regulation process. For each expression data set, we first construct a WGCN and then extract gene modules from the constructed network based on some topological measure. To interpret the biological meaning of the extracted modules, we use information from Gene Ontology (GO), Kyoto Encyclopedia of Genes and Genomes (KEGG pathways), and genome-wide location data to study whether each module is enriched for certain categories. Furthermore, we compare the utility between topological overlap and Pearson correlation similarity measures to define modules. Additionally, to study the relationships between modules derived from different expression data sets for the same species, we compare the consistency of gene modules inferred using different expression data sets. Lastly, we perform expression Quantitative Trait Loci (eQTL) analysis to gain a better understanding of the genetic basis of gene modules.

17.2 Methods and Materials

17.2.1 WGCN Construction

In order to identify modules from microarray gene expression data, we used a WGCN analysis method developed by Horvath and colleagues [13] to construct the gene coexpression network. This method has been found to be a useful approach [22, 23]. In a WGCN, each node corresponds to a gene, and two nodes are connected by a weighted edge, which indicates the coexpression similarity of the corresponding genes across the samples.

To construct a WGCN from a given gene expression data set, a similarity measure of how two genes are coexpressed should be defined first. For a pair of genes G_i and G_j, the absolute value of the Pearson correlation coefficient of their expression profiles is usually used as the coexpression similarity measure. Here, we denote it as s_{ij}:

$$s_{ij} = |cor(G_i, G_j)|. \tag{17.1}$$

To derive the coexpression network, an adjacency function is then defined to convert s_{ij} to network connection strength a_{ij}. For an unweighted co-expression network, the following adjacency function is commonly used by hard thresholding:

$$a_{ij} = \begin{cases} 1 & \text{if } s_{ij} \geq \tau, \\ 0 & \text{otherwise,} \end{cases} \tag{17.2}$$

where τ is a hard threshold. Two genes are connected by an edge in the coexpression network if the absolute correlation value between their gene expression profiles is larger than the hard threshold. The use of a hard threshold may lead to information loss. For instance, if τ is selected to be 0.7, there will be no link between two genes with s_{ij} less than but very close to 0.7. Different from this unweighted network construction, WGCN assigns a weight to each edge using a soft threshold instead to reduce information loss. This is accomplished by adopting the following soft adjacency function:

$$a_{ij} = s_{ij}^{\beta}. \tag{17.3}$$

That is, the network adjacency a_{ij} is defined by raising the coexpression similarity to a power with a parameter β. This adjacency measure has a value between 0 and 1. For each node, its network connectivity is calculated as the summation of the connection strengths with the other genes,

$$k_i = \sum_{j \neq i} a_{ij}. \tag{17.4}$$

β is one critical parameter that needs to be determined in the adjacency function because different β values can result in different networks. Most biological networks tend to follow the scale-free topology, and the frequency of connectivity of the nodes in a scale-free network follows an approximate inverse power law distribution. In practice, the value of β is commonly selected to be the smallest integer so that the resulting weighted network has the scale free topology. With this selected β, the soft adjacency can be calculated to derive a WGCN.

17.2.2 Module Identification from WGCN

After constructing a WGCN, the next step is to group highly coexpressed genes into modules. In a WGCN, genes highly coexpressed with each other are subsets of nodes tightly connected to each other, that is, nodes with high topological overlap. To extract sets of nodes with high topological overlap within a WGCN, a topological overlap measure between nodes is defined as

$$w_{ij} = \frac{l_{ij} + a_{ij}}{\min\{k_i, k_j\} + 1 - a_{ij}}, \tag{17.5}$$

where $l_{ij} = \sum_{u \neq i,j} a_{iu} a_{uj}$ represents the number of nodes both connected by nodes i and j. w_{ij} is a measure of similarity in terms of the nodes they are connected to. After the calculation of w_{ij} among all the nodes in a WGCN, traditional clustering methods, such as average linkage hierarchical clustering, can be adopted to cluster the nodes using the topological overlap similarity measure as input to obtain the clustering dendrogram. Gene modules can be defined as discrete branches of the clustering dendrogram. The dynamic cut tree algorithm, a method using the internal structure and lips branches of the dendrogram, is used to automatically identify modules. The algorithm adopted an iterative procedure until the resulting module number becomes stable. Each resulting module denotes a set of genes with coherent expression patterns across samples.

17.2.3 Enrichment Analysis

Coexpressed genes tend to be coregulated and possibly have similar functions. Therefore, genes in the same module are expected to be enriched for some special function categories, pathways or targets genes of specific transcription factors. In order to understand the biological basis of the network modules, we consider each identified gene module for enrichment of annotations from GO [24], KEGG pathways [25], and the physical binding of the transcription factors [4]. In our analysis, GO and KEGG pathway enrichment analysis was performed by the DAVID tool [26]. For transcription factor targets enrichment analysis, the hypergeometric distribution was used to assess the statistical significance. Given a reference set with k genes bound by the same transcription factor, and a testing set with n genes consisting of a module extracted from a WGCN, the probability for observing an overlap with t genes of the two genes sets by chance under the null hypothesis of no enrichment is

$$p = P(x \geq t) = \sum_{t \leq x \leq \min(n,k)} \frac{\binom{n}{x}\binom{N}{k-x}}{\binom{N}{n}}, \tag{17.6}$$

where N is the number of all the genes in the yeast genome. This p-value is then adjusted by the Bonferroni correction to account for multiple comparisons.

17.2.4 eQTL Analysis

To understand the genetic basis of the network modules, we conducted an analysis of gene modules through eQTL analysis. eQTL methods are commonly used to identify genetic variants affecting gene expression variations in different organisms, including yeast, rat, human, and others [6, 27, 28]. Through eQTL analysis, we may infer putative interactions between genes. In this chapter, we used eQTL analysis to study the biological basis of the modules to investigate whether the expression levels of genes in the same module are influenced by the same eQTL or eQTLs located in the same genomic region. In our analysis, eQTL analysis was conducted by performing Student's t-test between each candidate marker and gene expression trait with a significance threshold of p-value 4×10^{-5}. For each gene expression trait, only the marker having the most significant association with the trait across the genome was kept.

17.2.5 Data Sets

We focus on genomics analysis of *S. cerevisiae* in this chapter and analyze two microarray expression data sets. The first data set was generated by Brem and colleagues from a cross between two distinct isogenic strains BY and RM [8]. It contains gene expression measurements for 5,740 genes and genotypes of 2,956 Single Nucleotide Polymorphism (SNP) markers for each of the 112 segregants. The second data set is the Rosetta Compendium generated by Hughes et al. [2], which contains gene expression values for 6,280 genes measured in 300 experiments. These 300 experiments included 276 deletion mutants, 11 tetracycline regulatable alleles of essential genes, and 13 chemical treatments. Hereafter, we denote these two data sets as the Brem data set and the Hughes data set, respectively.

We considered genome-wide location data produced by Harbison and colleagues in the transcription factor enrichment analysis [4]. These data were collected to study the binding specificity of 204 transcription factors in the rich medium condition using ChIP-on-chip. A subset of 84 transcription factors were also profiled in at least one of 12 other experimental conditions, such as heat and amino acid starvation. All genes having binding p-value less than 0.001 with a transcription factor are considered as the target gene set of the transcription factor.

17.3 Results

17.3.1 Identifying Modules from WGCN

For each gene expression data set, we used the 3,500 most differentially expressed genes as input to construct the WGCN and derive modules. This was accomplished by using the R package developed by Horvath and colleagues [29]. Based

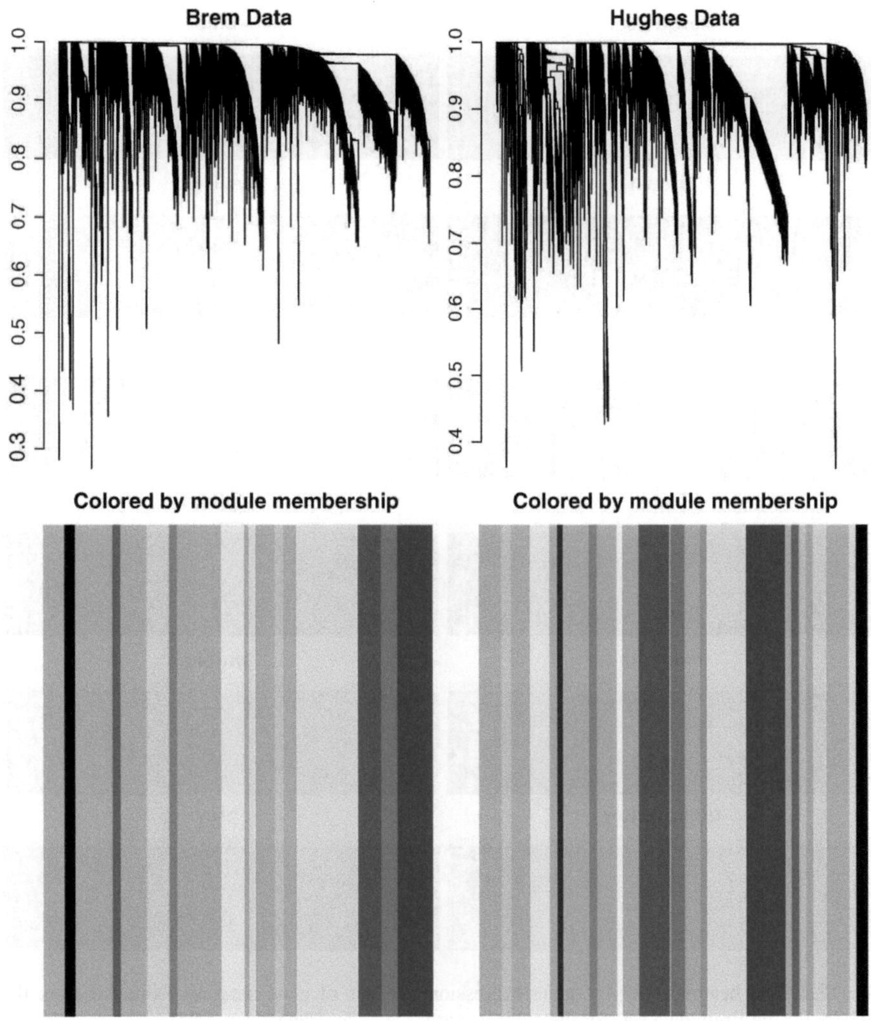

Fig. 17.1 Dendrogram view of hierarchical clustering results for the WGCNs constructed from the Brem data set and the Hughes data set (for the color version, see Color Plates on p. 396)

on the scale-free topology criterion, we set β equal to 6 for both data sets. Figure 17.1 shows the hierarchical clustering dendrograms of genes in the WGCNs based on the topological overlap similarity measurement for the Brem data set and the Hughes data set, where subtrees containing highly connected genes are identified as gene modules. As shown in Fig. 17.1, genes were clustered into distinct modules. We inferred that there were 11 modules for the Brem data set and 17 modules for the Hughes data set. To distinguish different modules, a distinct color was assigned to each module. The number of genes included in each mod-

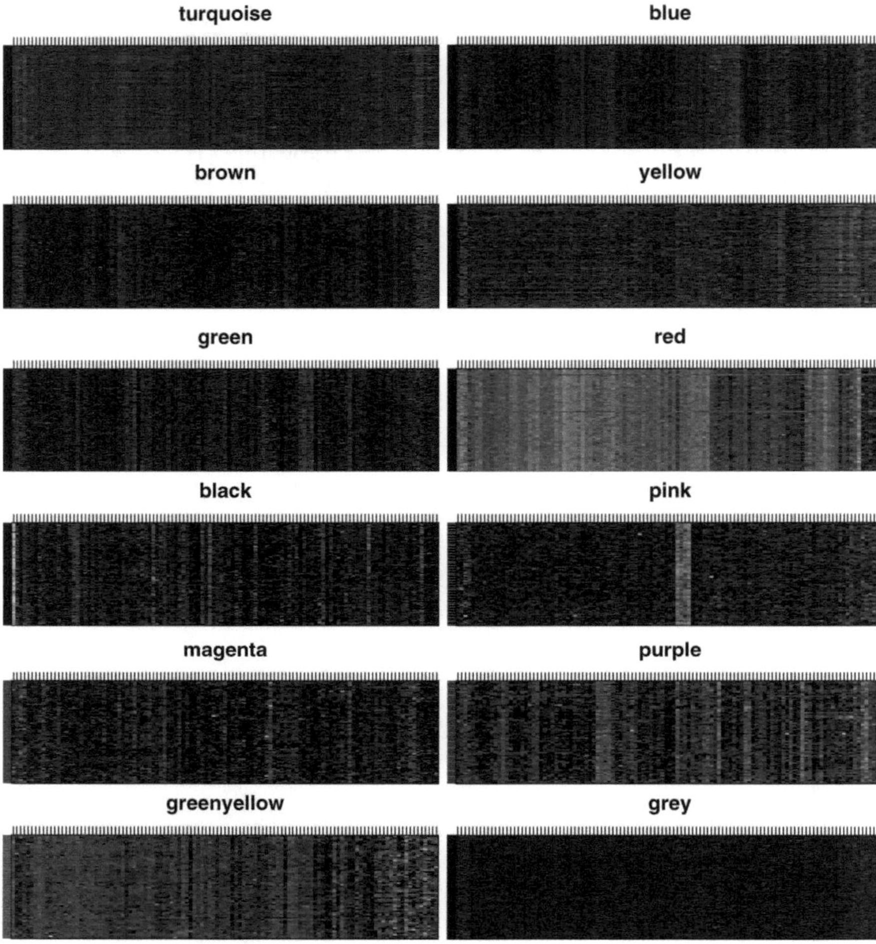

Fig. 17.2 The heatmap view of gene expression patterns of gene modules extracted from the WGCN for the Brem data set (for the color version, see Color Plates on p. 397)

ule varied greatly. For example, the largest turquoise module had 556 genes and the smallest greenyellow one only consisted of 63 genes for the Brem data set, whereas the number ranged from 50 to 358 for the Hughes data set. The heatmaps of expression patterns of the 12 gene modules for the Brem data set are shown in Fig. 17.2 (the grey module, which includes genes not belonging to any of the 11 identified modules, is also presented). Figure 17.2 shows that genes in the same module have similar expression patterns, while those in different module have distinct patterns. For the grey module, there is no apparent expression pattern.

17.3.2 Biological Interpretation of Gene Modules

To assess whether gene modules identified from the WGCN have special biological functions, we performed GO and pathway enrichment analysis. First, enrichment analysis using the yeast GO categories, including molecular functions, cellular components, and biological processes, were examined as described in the method section. For each module, the statistically most significant GO category is presented in Table 17.1 for the Brem data set. More than one GO category was enriched for 10 of 11 modules. For example, 122 of the 556 genes in the turquoise module belong to the cytosolic ribosome cellular component category, and 178 of the 200 genes in the green module belong to the mitochondrion cellular component category. In order to gain further understanding of the functional significance of gene modules, we also conducted a pathway enrichment analysis for each module. Pathways annotated in the KEGG database were used. The results are also shown in Table 17.1. We found that 8 modules were significantly enriched. For example, the turquoise module was enriched for genes within the ribosome pathway with a p-value 8.3×10^{-10}, which is consistent with its GO enrichment results. Similar results were found for the Hughes data set.

Table 17.1 GO and pathway analysis results of gene modules for the Brem data set

Module	Size	Enriched GO category	GO overlap (p-value)	Enriched KEGG pathway	KEGG overlap (p-value)
Turquoise	556	Cytosolic ribosome	122 (9.5×10^{-103})	Ribosome	120 (8.3×10^{-96})
Blue	317	Cytoplasm	23 (1.0×10^{-8})	Starch and sucrose metabolism	15 (1.4×10^{-3})
Brown	209	Nucleolus	33 (5.4×10^{-9})	Purine metabolism	16 (4.0×10^{-5})
Yellow	207	Amino acid metabolic process	46 (5.8×10^{-19})	Valine, leucine and isoleucine biosynthesis	9 (9.2×10^{-5})
Green	200	Mitochondrion	178 (5.5×10^{-96})	Aminoacyl-tRNA biosynthesis	13 (5.0×10^{-8})
Red	144	Nucleolus	87 (6.4×10^{-84})	RNA polymerase	5 (0.02)
Black	96	–	–	–	–
Pink	77	Endoplasmic reticulum	13 (7.7×10^{-3})	–	–
Magenta	72	Mitochondrial membrane part	35 (2.5×10^{-38})	Oxidative phosphorylation	35 (5×10^{-36})
Purple	70	Retrotransposon nucleocapsid	23 (2.1×10^{-29})	–	–
Greenyellow	63	Amino acid biosynthetic process	33 (6.2×10^{-35})	Urea cycle and metabolism of amino groups	7 (8.6×10^{-4})

Table 17.2 Transcription factors are enriched for modules of the Brem data set

Module	Size	Transcription factors with target genes enriched in the module
Turquoise	556	FHL1 HIR1 RAP1 SFP1
Blue	317	CIN5 MSN2 SKN7
Brown	209	–
Yellow	207	BAS1 DAL81 GCN4 LEU3 OPI1 STB1
Green	200	–
Red	144	–
Black	96	KRE33 MBF1 RDR1 YBL054W
Pink	77	HMS1 KRE33 MBF1 RDR1 YBL054W
Magenta	72	HAP1 HAP2 HAP3 HAP4 HAP5 KRE33 MBF1 RDR1 YBL054W
Purple	70	KRE33 KSS1 MBF1 MIG2 MIG3 RDR1 STE12 THI2 YBL054W
Greenyellow	63	ARG80 ARG81 DAL81 GCN4 GLN3 KRE33 MBF1 RDR1 RTG3 YBL054W

We expect that genes regulated by the same transcription factor should have similar expression patterns, and hence the target genes of a transcription factor should more likely appear in the same module. To see whether this is the case in the observed data, we tested if there is any significant enrichment for the target genes of transcription factors in each module. Genome-wide location data generated by Harbison et al. was used to define target genes of different transcription factors [4]. The results are presented in Table 17.2 for the Brem data set. We can see that 8 of the 11 modules from the Brem data set are enriched for more than one transcription factor. Transcription factors enriched for some module are the components of a transcription factor complex. For example, transcription factors enriched for the magenta module, HAP2, HAP3, HAP4, HAP5 are subunits of the heme-activated, glucose-repressed Hap2/3/4/5 CCAAT-binding complex [30]. Transcription factors ARG80 and ARG81, which were enriched for the green–yellow module, are the components of the Mcm1–Arg80–Arg81 protein complex [30]. The enrichment results for GO categories, KEGG pathways, and transcription factor targets suggest that modules identified from the WGCNs include genes coexpressed to perform specific biologic functions, which indicate that these identified gene modules do have functional significance.

17.3.3 Comparison Between Pearson Correlation and Topological Overlap

In WGCN, topological overlap measure is used to define the similarity between genes and to identify modules. We may also use Pearson correlation to define the similarity between genes and use it to find modules instead of topological overlap. Note that topological overlap and Pearson correlation are two distinct ways to measure the similarity of a gene pair. When calculating the Pearson correlation of a

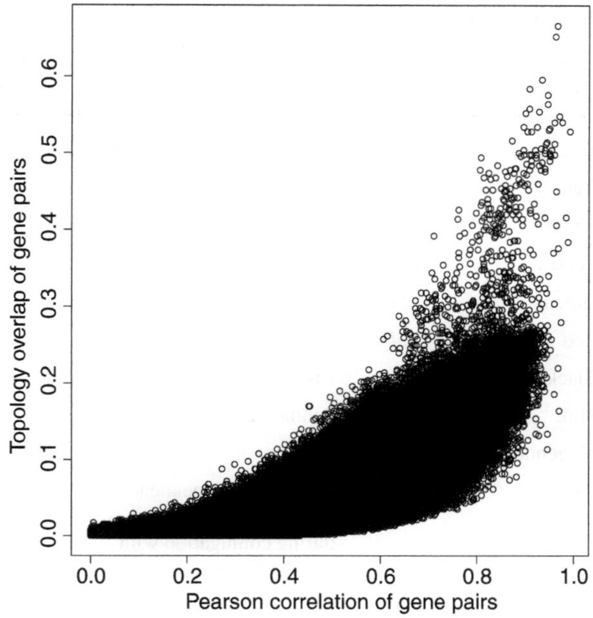

Fig. 17.3 Comparison between Pearson correlation coefficient measure and topological overlap measure for gene pairs of the Brem data set

gene pair, only the information of these two genes is used. While in the calculation of topological overlap, the relationships between this gene pair and all other genes are considered. Genes strongly connected with the same set of genes will have high topological overlap.

Here, to explore the relative utility of topological overlap versus Pearson correlation for module inference, we calculated the similarity of every gene pair for the 3,500 most differentially expressed genes using both kinds of similarity measures. As shown in Fig. 17.3 of the scatter-plot, there is some moderate association between the two similarity measures. For gene pairs with a given Pearson correlation value, their topological overlap had a wide range of values, especially for those having a large Pearson correlation value.

We used Pearson correlation instead of topological overlap to measure the similarity between genes to identify modules for the Brem data set. After the calculation of the Pearson correlations between each pair of genes among the 3,500 most differentially expressed genes, the same procedure was adopted to find modules as before for similarity measures based on topological overlaps. Many more modules, a total of 30 compared to 11 using topological overlap, were detected using Pearson correlation as the similarity measure. The general module size was smaller, with the largest module having 209 genes, and most of the other modules consisted of 50 to 120 genes. As shown in Table 17.3, only 19 modules were enriched for some GO categories, which was relatively low compared to those modules identified through the topological overlap (19/30 versus 10/11). These results suggest that topological overlap may be more informative than Pearson correlation to capture the similarity between genes, because it may borrow information from other genes in addition to the gene pairs considered.

Table 17.3 GO analysis of gene modules using Pearson correlation instead of topological overlap as the similarity measure for the Brem data set

Module	Size	Enriched GO category	Overlap	Enrichment p-value
Turquoise	209	Purine base metabolic process	8	5.6×10^{-3}
Blue	173	Cytoplasm	144	6.0×10^{-10}
Brown	166	Nucleolus	89	2.4×10^{-79}
Yellow	158	–	–	–
Green	129	Cytosolic ribosome	36	8.3×10^{-26}
Red	124	–	–	–
Black	116	Bucleus	65	4.3×10^{-3}
Pink	107	Amino acid metabolic process	34	1.6×10^{-16}
Magenta	105	–	–	–
Purple	104	Pheromone-dependent signal transduction during conjugation with cellular fusion	7	9.9×10^{-4}
Greenyellow	99	–	–	–
Tan	95	Mitochondrial ribosome	51	1.5×10^{-68}
Salmon	93	Tricarbotimesylic acid cycle intermediate metabolic process	6	7.0×10^{-3}
Cyan	92	Nucleosome	8	5.3×10^{-7}
Midnightblue	84	Endoplasmic reticulum	28	3.0×10^{-9}
Lightcyan	78	Amino acid biosynthetic process	46	1.3×10^{-35}
Grey60	77	Mitochondrial membrane part	36	4.2×10^{-38}
Lightgreen	75	Cytosolic ribosome	63	2.3×10^{-95}
Lightyellow	69	–	–	–
Royalblue	67	–	–	–
Darkred	63	–	–	–
Darkgreen	63	–	–	–
Darkturquoise	62	–	–	–
Darkgrey	62	–	–	–
Orange	60	Mating projection tip	7	2.4×10^{-4}
Darkorange	60	Aryl-alcohol dehydrogenase activity	6	3.2×10^{-6}
White	59	Sterol metabolic process	21	8.8×10^{-28}
Skyblue	57	Retrotransposon nucleocapsid	22	2.6×10^{-30}
Saddlebrown	54	–	–	–
Steelblue	53	Translational initiation	7	2.2×10^{-3}

17.3.4 Consistency of Gene Modules

If each module represents a functional unit of some biological process, we expect the modules to be consistent in the coexpression networks built from different expression data sources. To test this hypothesis, we explored the consistency of the modules inferred from the Brem data set and the Hughes data set. For each module detected in the Brem data set, we compared it with all the modules inferred from the Hughes data set to evaluate whether there is any statistically significant overlap between these modules using the hypergeometric test. The module pairs with a p-value less than 0.001 are listed in Table 17.4, and the GO enrichment analysis was carried out to examine if there is any significant functional enrichment of the overlapping genes. Table 17.4 shows that 9 of the 11 modules from the Brem data set have statistically significant overlaps with one or more modules from the Hughes data set. In addition, significant GO enrichment was found for the overlapping genes of 8 module pairs. These results suggest that there is some degree of consistency between modules detected from different data sets.

Table 17.4 Module pairs with significant overlap between the Brem data set and the Hughes data set

Brem Data		Hughes Data		Overlap	Overlap p-value	Enriched GO category
Module	Size	Module	Size			
Turquoise	556	Blue	348	20	1.4×10^{-4}	–
Turquoise	556	Purple	68	56	1.5×10^{-68}	Cytosolic ribosome
Blue	317	Turquoise	358	57	1.6×10^{-11}	Glycogen metabolic process
Blue	317	Tan	60	55	6.6×10^{-144}	Cellular carbohydrate catabolic process
Blue	317	Cyan	55	28	1.3×10^{-4}	Mitochondrion
Brown	209	Green	126	25	1.1×10^{-15}	rRNA processing
Yellow	207	Brown	276	85	1.0×10^{-100}	Amino acid metabolic process
Green	200	Lightcyan	50	43	3.8×10^{-100}	Mitochondrial matrix
Red	144	Green	126	51	6.9×10^{-126}	Nucleolus
Magenta	72	Greenyellow	62	6	4.3×10^{-6}	–
Purple	70	Magenta	72	34	2.6×10^{-212}	Retrotransposon nucleocapsid
Greenyellow	63	Brown	276	57	1.3×10^{-170}	Amino acid biosynthetic process

17.3.5 Genetic Basis of Gene Modules

As detailed above, modules extracted from the WGCNs have certain biological interpretations. For the Brem data set, the genetic marker information was also available. This enables us to investigate whether the eQTLs for the genes in the same module are located in the region near each other in the genome. We conducted eQTL analysis to identify genomic regions affecting gene expression levels for all the genes in each module to 2,896 SNP markers and tested if there is any significant enrichment of eQTL regions for genes in the same module.

A total of 2,567 gene and SNP pairs were identified as significant, which involved 2,567 distinct gene expression traits and 726 markers. We depict the eQTL results using the eQTL viewer (Fig. 17.4) [31]. For each module, the eQTLs of its member genes are represented by its module color. Figure 17.4 shows that the eQTLs of genes in a module are enriched for some genomic region. For example, more than one genomic region contains a relatively large number of eQTLs of genes from the blue module. To further study which genomic regions and to what extent the genomic regions are associated with each module, we divided the genome into 20 kb segments in length, resulting in a total of 602 genomic regions. For each region, we tested the overlap between genes whose eQTL located in that region and genes in a specific module. As shown in Table 17.5, all but one of the 11 modules were significantly enriched for being linked to one genomic region. For example, there were 215 genes linked to the region on chromosome 15 between base pairs 160,000 and 180,000, and 149 of these genes belonged to the blue module. This suggests that genetic variations in a common genomic region can explain the observed expression variations of genes in the same module. It is not unexpected as genes in the same module should have similar expression profiles, therefore more likely to be linked to the same genomic region.

In Fig. 17.4, a dot in the diagonal line represents an eQTL having an SNP marker with genomic location near a gene significantly affected by this marker, which means that the genetic perturbation of the gene leads to a transcription change of itself. These dots represent putative *cis*-regulation, while all others represent putative *trans*-regulation. There are 450 genes having *cis*-eQTLs among the 2756 genes. Most of the eQTLs, totaling 2,306, are *trans*-eQTLs. Intuitively, for the *trans*-eQTL, the expression variation of its trait gene in mutant strains maybe attributed to the genetic variation of genes encoding its transcription factors, and we tested this hypothesis. Genes whose eQTLs were located within the 10-kb upstream and downstream region of genes encoding their transcription factors are taken as affected by the genetic variation of their transcription factor genes. From Table 17.6 we can see that only a relatively small number of eQTLs are located near their transcription factors. Although we have found that some modules are enriched for the target genes of transcription factors, few genetic variations in gene expression of genes in a module are due to polymorphisms in the transcription factors that bind to these genes. This is consistent with previous observation that genes within eQTLs of gene expression traits are not enriched for transcription factors [27]. The expression variation of genes with *trans*-eQTLs in the module may involve genes other than transcription

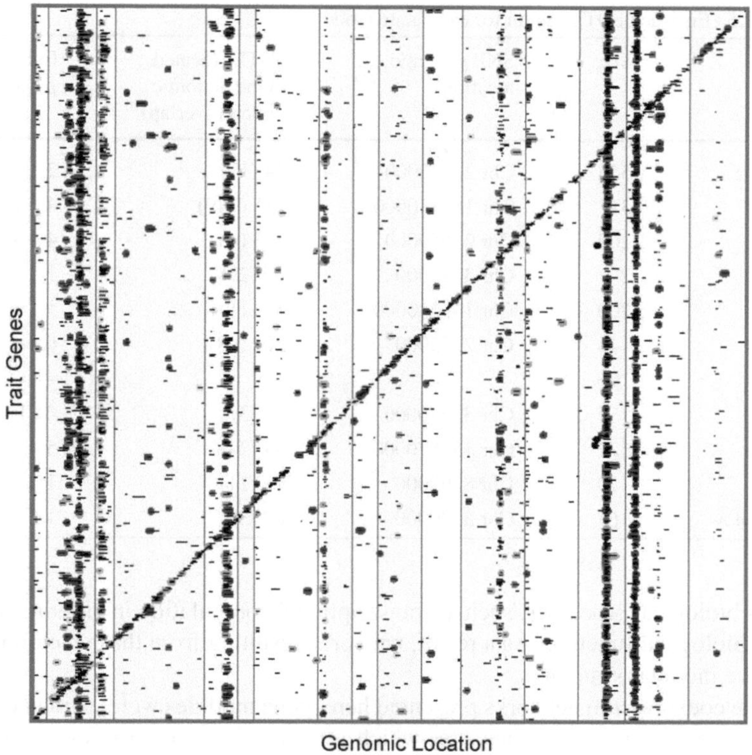

Fig. 17.4 eQTL results for the Brem data set. The *horizontal line* represents the genomic locations of the SNP markers, and the *vertical line* represents the genomic locations of the trait genes. *Color dots* with the same color correspond to the eQTLs of genes in the module with the corresponding color (for the color version, see Color Plates on p. 398)

factors. For example, the products of genes located in the eQTL regions may interact with the activities of some transcription factors that regulate the other genes at the protein level.

17.4 Conclusions

In this chapter, we have analyzed modules extracted from the WGCNs in the context of diverse types of high-throughput data. The goal of this study was to investigate the underlying biological meaning and genetic basis of the inferred modules.

Network methods such as WGCNs have been widely used in systems biology studies, for example, either using gene expression data alone or using gene expression data in combination with other types of data to reconstruct transcriptional regulatory networks. Such methods are promising because they provide a global description of the biological processes. The modularity of biological network is a result of

Table 17.5 Enriched eQTL region for each gene module

Module	Size	SNP genomic location	eQTLs located in the genomic region (overlap)	Enrichment p-value
Turquoise	556	Chr 2, 390000	44 (17)	2.1×10^{-4}
Blue	317	Chr 15, 170000	215 (149)	4.0×10^{-118}
Brown	209	Chr 2, 550000	203 (44)	4.1×10^{-15}
Yellow	207	Chr 3, 90000	87 (29)	1.8×10^{-5}
Green	200	Chr 14, 450000	304 (118)	5.2×10^{-81}
Red	144	Chr 2, 55000	202 (48)	6.4×10^{-26}
Black	96	–	–	–
Pink	77	Chr 3, 570000	22 (3)	0.01
Magenta	72	Chr 15, 570000	34 (24)	5.4×10^{-35}
Purple	70	Chr 8, 90000	61 (11)	1.7×10^{-8}
Greenyellow	63	Chr 3, 70000	83 (38)	1.5×10^{-49}

various biological processes, such as transcriptional coregulation in order to accomplish a biological function. As a result, networks resulting from these methods may also have modular structures.

In the coexpression networks presented here, gene modules were extracted based on the topological overlap measure, which seems to be a more informative metric compared with Pearson correlation to define gene modules. Each gene module represents a set of highly connected genes, that is, a group of highly coexpressed genes. GO and pathway enrichment analyses suggested that gene modules are enriched for genes having similar functions or genes in the same pathways. In addition, gene modules are also enriched for target genes regulated by the same transcription factors. These facts indicate that gene modules extracted from the coexpression network are biologically meaningful. We also found that there were significant overlaps between modules inferred from different data sets and that the overlapping genes were also enriched for GO categories, which suggests that modules may represent an inherent property component of the underlying biological process. On the other hand, we also note that there exist differences between modules inferred from different data sets, which may be induced by different conditions under which the expression data sets were collected.

Integrating genetics and genomics data offers a promising approach to understanding the genetic basis of modules. We showed that some of the modules are directly linked to some genetic regions. This relationship is demonstrated when we conducted eQTL analysis of genes within each module, and the results showed that there existed genomic regions controlling the coexpression network modules. Many of these eQTLs were *trans*-eQTLs, and the genes located near the eQTL region were usually not transcription factors directly regulating the trait genes. Hence, regulators other than transcription factors do contribute to the expression variations. For instance, genes located near the eQTLs may encode some proteins that can in-

Table 17.6 Overlap between genes bound by a transcription factor and genes whose eQTL are located near the transcription factor. All transcription factors with at least one overlap are listed

TF name	Num. of eQTLs located near the TF	Num. of genes bound by the TF	Overlap
HAP1	152	151	44
FKH2	357	169	11
STB5	7	50	3
SWI4	26	161	3
YJL206C	15	35	3
ZAP1	19	22	3
ARG81	46	36	2
CBF1	8	282	2
CIN5	16	223	2
HAL9	162	28	2
HSF1	14	161	2
RFX1	26	25	2
SOK2	15	73	2
ABF1	9	267	1
BAS1	13	52	1
DAL81	3	114	1
DIG1	14	155	1
FKH1	13	142	1
GCR2	5	75	1
MOT3	9	39	1
OAF1	23	61	1
PDR3	11	21	1
RTG3	2	109	1
SMP1	128	91	1
STP2	89	11	1
SWI6	24	158	1
TEC1	23	84	1
YAP1	17	99	1
YAP3	111	18	1

teract with transcription factors at the protein level; then the affected transcription factors control the expression of their target genes. In addition, the eQTL regions likely contain more than one gene. A question of interest is to infer the causal genes for each trait gene or genes in the same module. Integrating other data, such as protein–protein interaction data, we may gain an even better understanding on how such interaction are involved to regulate the expression levels of trait genes [32].

Such knowledge should be informative to understand the transcriptional regulation relationships.

Overall, our results suggest a clear view of gene modules in the WGCN. Gene modules provide us a more global way to interpret the underlying cellular processes. We believe that module-based methods may provide valuable perspectives to understand complex biological processes.

Acknowledgements This study was supported in part by a fellowship award from the China Scholarship Council (X.Z.) and NIH grant GM 59507 (H.Z.).

References

1. P.T. Spellman, G. Sherlock, M.Q. Zhang, V.R. Iyer, K. Anders, M.B. Eisen, P.O. Brown, D. Botstein, and B. Futcher. Comprehensive identification of cell cycle-regulated genes of the yeast saccharomyces cerevisiae by microarray hybridization. *Mol Biol Cell*, **9**(12):3273–3297, 1998.
2. J.D. Hughes, P.W. Estep, S. Tavazoie, and G.M. Church. Computational identification of cis-regulatory elements associated with groups of functionally related genes in saccharomyces cerevisiae. *J Mol Biol*, **296**(5):1205–1214, 2000.
3. T.I. Lee, N.J. Rinaldi, F. Robert, D.T. Odom, Z. Bar-Joseph, G.K. Gerber, N.M. Hannett, C.T. Harbison, C.M. Thompson, I. Simon, J. Zeitlinger, E.G. Jennings, H.L. Murray, D.B. Gordon, B. Ren, J.J. Wyrick, J. Tagne, T.L. Volkert, E. Fraenkel, D.K. Gifford, and R.A. Young. Transcriptional regulatory networks in saccharomyces cerevisiae. *Science*, **298**(5594):799–804, 2002.
4. C.T. Harbison, D.B. Gordon, T.I. Lee, N.J. Rinaldi, K.D. Macisaac, T.W. Danford, N.M. Hannett, J. Tagne, D.B. Reynolds, J. Yoo, E.G. Jennings, J. Zeitlinger, D.K. Pokholok, M. Kellis, P.A. Rolfe, K.T. Takusagawa, E.S. Lander, D.K. Gifford, E. Fraenkel, and R.A. Young. Transcriptional regulatory code of a eukaryotic genome. *Nature*, **431**(7004):99–104, 2004.
5. T. Ito, K. Tashiro, S. Muta, R. Ozawa, T. Chiba, M. Nishizawa, K. Yamamoto, S. Kuhara, and Y. Sakaki. Toward a protein–protein interaction map of the budding yeast: A comprehensive system to examine two-hybrid interactions in all possible combinations between the yeast proteins. *Proc Natl Acad Sci USA*, **97**(3):1143–1147, 2000.
6. R.B. Brem, G. Yvert, R. Clinton, and L. Kruglyak. Genetic dissection of transcriptional regulation in budding yeast. *Science*, **296**(5568):752–755, 2002.
7. R.B. Brem and L. Kruglyak. The landscape of genetic complexity across 5,700 gene expression traits in yeast. *Proc Natl Acad Sci USA*, **102**(5):1572–1577, 2005.
8. N. Bing and I. Hoeschele. Genetical genomics analysis of a yeast segregant population for transcription network inference. *Genetics*, **170**(2):533–542, 2005.
9. J. Zhu, B. Zhang, E.N. Smith, B. Drees, R.B. Brem, L. Kruglyak, R.E. Bumgarner, and E.E. Schadt. Integrating large-scale functional genomic data to dissect the complexity of yeast regulatory networks. *Nat Genet*, **40**(7):854–861, 2008.
10. L.H. Hartwell, J.J. Hopfield, S. Leibler, and A.W. Murray. From molecular to modular cell biology. *Nature*, **402**(6761):C47–C52, 1999.
11. M. Girvan and M.E.J. Newman. Community structure in social and biological networks. *Proc Natl Acad Sci USA*, **99**(12):7821–7826, 2002.
12. J. Ihmels, G. Friedlander, S. Bergmann, O. Sarig, Y. Ziv, and N. Barkai. Revealing modular organization in the yeast transcriptional network. *Nat Genet*, **31**(4):370–377, 2002.
13. B. Zhang and S. Horvath. A general framework for weighted gene co-expression network analysis. *Stat Appl Genet Mol Biol*, **4**:17, 2005.
14. S. Maere, P. Van Dijck, and M. Kuiper. Extracting expression modules from perturbational gene expression compendia. *BMC Syst Biol*, **2**:33, 2008.

15. E. Segal, M. Shapira, A. Regev, D. Pe'er, D. Botstein, D. Koller, and N. Friedman. Module networks: identifying regulatory modules and their condition-specific regulators from gene expression data. *Nat Genet*, **34**(2):166–176, 2003.
16. Z. Bar-Joseph, G.K. Gerber, T.I. Lee, N.J. Rinaldi, J.Y. Yoo, F. Robert, D.B. Gordon, E. Fraenkel, T.S. Jaakkola, R.A. Young, and D.K. Gifford. Computational discovery of gene modules and regulatory networks. *Nat Biotechnol*, **21**(11):1337–1342, 2003.
17. X. Xu, L. Wang, and D. Ding. Learning module networks from genome-wide location and expression data. *FEBS Lett*, **578**(3):297–304, 2004.
18. X. Liu, W.J. Jessen, S. Sivaganesan, B.J. Aronow, and M. Medvedovic. Bayesian hierarchical model for transcriptional module discovery by jointly modeling gene expression and CHIP-chip data. *BMC Bioinform*, **8**:283, 2007.
19. A. Tanay, R. Sharan, M. Kupiec, and R. Shamir. Revealing modularity and organization in the yeast molecular network by integrated analysis of highly heterogeneous genomewide data. *Proc Natl Acad Sci USA*, **101**(9):2981–2986, 2004.
20. K. Lemmens, T. Dhollander, T. De Bie, P. Monsieurs, K. Engelen, B. Smets, J. Winderickx, B. De Moor, and K. Marchal. Inferring transcriptional modules from CHIP-chip, motif and microarray data. *Genome Biol*, **7**(5):R37, 2006.
21. W. Wu and W. Li. Identifying gene regulatory modules of heat shock response in yeast. *BMC Genomics*, **9**:439, 2008.
22. M.C. Oldham, G. Konopka, K. Iwamoto, P. Langfelder, T. Kato, S. Horvath, and D.H. Geschwind. Functional organization of the transcriptome in human brain. *Nat Neurosci*, **11**(11):1271–1282, 2008.
23. A.P. Presson, E.M. Sobel, J.C. Papp, C.J. Suarez, T. Whistler, M.S. Rajeevan, S.D. Vernon, and S. Horvath. Integrated weighted gene co-expression network analysis with an application to chronic fatigue syndrome. *BMC Syst Biol*, **2**:95, 2008.
24. M. Ashburner, C.A. Ball, J.A. Blake, D. Botstein, H. Butler, J.M. Cherry, A.P. Davis, K. Dolinski, S.S. Dwight, J.T. Eppig, M.A. Harris, D.P. Hill, L. Issel-Tarver, A. Kasarskis, S. Lewis, J.C. Matese, J.E. Richardson, M. Ringwald, G.M. Rubin, and G. Sherlock. Gene ontology: tool for the unification of biology, the gene ontology consortium. *Nat Genet*, **25**(1):25–29, 2000.
25. M. Kanehisa, S. Goto, S. Kawashima, Y. Okuno, and M. Hattori. The KEGG resource for deciphering the genome. *Nucleic Acids Res*, **32**:D277–D280, 2004 (Database issue).
26. G. Dennis, B.T. Sherman, D.A. Hosack, J. Yang, W. Gao, H.C. Lane, and R.A. Lempicki. David: Database for annotation, visualization, and integrated discovery. *Genome Biol*, **4**(5):P3, 2003.
27. G. Yvert, R.B. Brem, J. Whittle, J.M. Akey, E. Foss, E.N. Smith, R. Mackelprang, and L. Kruglyak. Trans-acting regulatory variation in saccharomyces cerevisiae and the role of transcription factors. *Nat Genet*, **35**(1):57–64, 2003.
28. E.E. Schadt, S.A. Monks, T.A. Drake, A.J. Lusis, N. Che, V. Colinayo, T.G. Ruff, S.B. Milligan, J.R. Lamb, G. Cavet, P.S. Linsley, M. Mao, R.B. Stoughton, and S.H. Friend. Genetics of gene expression surveyed in maize, mouse and man. *Nature*, **422**(6929):297–302, 2003.
29. P. Langfelder and S. Horvath. WGNCA: an R package for weighted correlation network analysis. *BMC Bioinform*, **9**:559, 2008.
30. K.R. Christie, S. Weng, R. Balakrishnan, M.C. Costanzo, K. Dolinski, S.S. Dwight, S.R. Engel, B. Feierbach, D.G. Fisk, J.E. Hirschman, E.L. Hong, L. Issel-Tarver, R. Nash, A. Sethuraman, B. Starr, C.L. Theesfeld, R. Andrada, G. Binkley, Q. Dong, C. Lane, M. Schroeder, D. Botstein, and J.M. Cherry. Saccharomyces genome database (SGD) provides tools to identify and analyze sequences from saccharomyces cerevisiae and related sequences from other organisms. *Nucleic Acids Res*, **32**:D311–D314, 2004 (Database issue).
31. W. Zou, D.L. Aylor, and Z. Zeng. eQTL Vewer: visualizing how sequence variation affects genome-wide transcription. *BMC Bioinform*, **8**:7, 2007.
32. S. Suthram, A. Beyer, R.M. Karp, Y. Eldar, and T. Ideker. eQED: an efficient method for interpreting eQTL associations using protein networks. *Mol Syst Biol*, **4**:162, 2008.

Chapter 18
Readout of Spike Waves in a Microcolumn

Xuejuan Zhang

18.1 Introduction

Neurons receive and emit spike trains which are typically stochastic in nature, due to the combination of their intrinsic channel fluctuations, their morphologies, and the variability in the input they receive. How to accurately and efficiently read out the input information from spike waves is the central question in (theoretical) neuroscience [12, 18, 22, 25], but the answer has yet been unprovided. For a given neuron or neuronal network, the most commonly used method to read out the input information is undoubtedly the maximum likelihood estimate (MLE), which is optimal under mild conditions. However, in order to perform the MLE, the prerequisite is that we should know the exact expression of the interspike interval (ISI) distribution of efferent spikes of a neuron or a neuronal network. This is, unfortunately, a difficult task in general. Even for the simplest leaky integrate-and-fire (LIF) model, such a distribution, which is equivalent to the distribution of the first passage time of an Ornstein–Uhlenbeck (OU) process, is unknown. Indeed, it has been posed as an open problem for many years in the literature [12, 18, 33].

Besides in neuroscience, the first passage time of an OU process is also of quite interest in many other fields such as physics, engineering and finance, etc., and the topic has been widely addressed in many textbooks [17, 26, 30]. Due to the recent development [1, 19], three expressions of the interspike interval distribution are available. The first two expressions are expressed in terms of infinite series or infinite integrals which are not easy to be calculated numerically. What is suitable for our purpose is the third one, in which the probability density of ISI can be numerically simulated by Monte Carlo method. Based on this, the MLE for the LIF model can be developed. Despite of the fact that the MLE for a LIF model has been discussed intensively by a few authors [8, 10, 15, 24, 28, 32], to the best of our

X. Zhang (✉)
Mathematical Department, Zhejiang Normal University, Jinhua, 321004, Zhejian Province, P.R. China
e-mail: xuejuanzhang@gmail.com

J. Feng et al. (eds.), *Frontiers in Computational and Systems Biology*, Computational Biology 15,
DOI 10.1007/978-1-84996-196-7_18,

knowledge, our approach is based on the exact expression of the ISI distribution of the LIF model, which should open up many interesting issues for further study, as we will partly demonstrate here.

We then employ the MLE to several applications. First, we test the MLE for a single LIF model with Poisson-type inputs. It is found that the input information (rate) can be accurately decoded. To further explore the implications of our approach here, we apply our method to address a long-standing problem: what is the ratio between inhibitory and excitatory inputs in a biological neuron? Although it is found that the number of inhibitory neurons is smaller than the number of excitatory neurons in the cortex [21, 29], it is generally agreed that inhibitory neurons send stronger signals than excitatory neurons [9]. Therefore the exact ratio between inhibitory and excitatory inputs remains elusive. With the MLE developed in the current paper, we are able to reliably estimate the ratio between inhibitory and excitatory inputs to a LIF neuron.

A more interesting problem is to develop the MLE for an array of LIF neurons with dynamical inputs. We first consider the case of an ensemble of identical LIF neurons. With a short time window ($\sim$25 msec), the dynamical input can be read out reliably [20]. This is also an interesting result, and it answers another issue in neuroscience. It was found in [31] that the required time from sensory inputs to motor reactions is around 200 msec [20]. This suggests that only a few spikes can be generated in each layer to (encode) decode the input information and the spikes should be deterministic rather than stochastic. Here we demonstrate that with neuron pools of a reasonable size (100 neurons), the input information can be accurately read out within a very short time window from random spikes if the MLE is employed. Therefore, the stochasticity in spikes does not contradict to the time constraints.

Certainly, neurons in a microcolumn interact with each other. By incorporating lateral inhibition and the time delay of the synaptic inputs, we find that the input information can still be reliably read out from spike waves of an interacting neuronal network, using the MLE strategy mentioned above. The results should open up many new and challenging problems for further research, both in theory and in applications. For example, we would naturally ask how to implement MLE for multilayer interacting spiking neuronal networks.

18.2 Theoretical Results

18.2.1 Distribution of Interspike Interval

We start our discussion from the following single LIF model

$$dV = -\frac{V}{\gamma} dt + \lambda dt + \sqrt{\lambda} dB_t, \quad V \le V_{th}, \tag{18.1}$$

where $V(t)$ is the membrane potential at time t, γ is the decay time, $\lambda \ge 0$ is the input, B_t is the standard Brownian motion, and V_{th} is the threshold. When V exceeds V_{th} from below, V is reset to $V_{re} = 0$, the resting potential.

Our goal is to decode the input λ from the statistical properties of the ISI of the efferent spikes which can be expressed as

$$T = \inf\{t > 0 : V(t) \geq V_{th} | V(0) = 0\}. \tag{18.2}$$

More precisely, we define

$$\begin{aligned} \tau_i &= \inf\{t > \tau_{i-1} : V(t) \geq V_{th} | V(\tau_{i-1}) = 0\}, \quad i \geq 1, \\ \tau_0 &= 0, \end{aligned} \tag{18.3}$$

and $T_i = \tau_i - \tau_{i-1}$, $i \geq 1$. It is ready to see that $\{T_i, i \geq 1\}$ is an i.i.d. sequence and has the identical distribution density as T.

By letting $U = (V - \lambda\gamma)/\sqrt{\lambda}$, the distribution of the ISI of efferent spikes of the LIF model (18.1) is equivalent to the distribution of the first-passage time of the OU process

$$dU = -\frac{U}{\gamma} dt + dB_t \tag{18.4}$$

starting from $U_{re} = -\sqrt{\lambda}\gamma$ to hit $U_{th} = V_{th}\sqrt{\lambda} - \sqrt{\lambda}\gamma$. The interspike interval of efferent spikes can be expressed as

$$T = \inf\{t : U(t) \geq V_{th}/\sqrt{\lambda} - \sqrt{\lambda}\gamma | U(0) = -\sqrt{\lambda}\gamma\}. \tag{18.5}$$

Let $p_\lambda(t)$ be the distribution density of T. As mentioned above, the first-passage time problem, which occurs in many areas, had once been believed to have no general explicit analytical formula (see p. 183 in [18]), except a moment expansion of the first-passage time distribution for constant input. Recently, the knowledge of the sought density is nicely summarized in [1], where three expressions of the distribution of T are presented: the series representation, the integral representation, and the Bessel bridge representation. For numerical approximation, the authors pointed out that the first two approaches are easy to implement but require the knowledge of the Laplace transform of the first hitting time, which can be computed only for some specific continuous Markov processes, while the Bessel bridge approach overcomes the problem of detecting the time at which the approximated process crosses the boundary [1]. For this reason, here we prefer to apply the Bessel bridge method under which the probability density of T has the following form:

$$\begin{aligned} p_\lambda(t) &= \exp\left(\frac{-\lambda^{-1} V_{th}^2 + 2\gamma V_{th} + t}{2\gamma}\right) p^{(0)}(t) \\ &\quad \times E_{0 \to V_{th}} \left\{ \exp\left[-\frac{1}{2\gamma^2 \lambda} \int_0^t (r_s - V_{th} + \lambda\gamma)^2 \, ds \right] \right\}, \end{aligned} \tag{18.6}$$

where

$$p^{(0)}(t) = \frac{V_{th}}{\sqrt{2\pi\lambda t^3}} \exp\left(-\frac{V_{th}^2}{2t\lambda}\right) \tag{18.7}$$

is the distribution density of the nonleaky IF model, $\{r_s\}_{s\leq t}$ is the so-called three-dimensional Bessel bridge starting from point 0 and ending at point V_{th}, which can be constructed from three independent Brownian bridges. Mathematically, it satisfies the stochastic differential equation

$$dr_s = \left(\frac{V_{th} - r_s}{t - s} + \frac{\lambda}{r_s}\right) ds + \sqrt{\lambda}\, dB_s, \quad r_0 = 0,\ s < t. \tag{18.8}$$

In (18.6), $E_{0\to V_{th}}$ represents the expectation with respect to the stochastic process r_s with starting point 0 and ending point V_{th}. So the density function $p_\lambda(t)$ is an ensemble average over many trials of the three-dimensional Bessel bridge.

At a first glance, the expression of (18.6) looks somewhat complicated, as it is an expectation of a singular stochastic process. However, we can resort to Monte Carlo method to numerically evaluate the expectation. To do this, we have to generate a large number, say M, of independent sampling paths of a three-dimensional Bessel bridge. It should be pointed out that in numerical simulations, we do not use (18.8) to directly simulate the process $\{r_s\}_{0\leq s\leq t}$ since it is degenerate at $s = 0$. Instead we consider the process $\{r_s^2\}_{0\leq s\leq t}$ which satisfies

$$d(r_s)^2 = 2r_s\, dr_s + \lambda\, ds. \tag{18.9}$$

Note that the second term on the right-hand side of the equation above is due to the Itô integral. According to this, the iterate procedure to simulate the stochastic process r_s is as follows (we simply denote $r(j)$ as $r(j\Delta t)$):

$$\begin{cases} u(j+1) = u(j) + \Delta t \cdot \left(\dfrac{2r(j)(V_{th} - r(j))}{t - j\Delta t} + 3\lambda\right) + 2r(j) \cdot \sqrt{\lambda} \cdot \Delta B(j), \\ r(j+1) = \sqrt{u(j+1)}, \end{cases}$$

with $r(1) = 0$ and $u(1) = 0$, where Δt is the time step, and $\Delta B(j) \triangleq B(j\Delta t + \Delta t) - B(j\Delta t)$ is the increment of the Brownian motion with distribution $Norm(0, \Delta t)$. Here $Norm(\cdot,\cdot)$ is the normal distribution with corresponding mean and variance. The three-dimensional Bessel bridge r_s has a trajectory as shown in Fig. 18.1A.

Denote by $\{r^i(k\Delta t)\}$ the ith sampling trajectory, and let

$$f_1(t) = \exp\left(\frac{-\lambda^{-1}V_{th}^2 + 2\gamma V_{th} + t}{2\gamma}\right) p^{(0)}(t);$$

then the approximation formula for (18.6) is

$$\bar{p}_\lambda(t) = f_1(t) \cdot \frac{1}{M}\sum_{i=1}^{M} \exp\left[-\frac{1}{2\gamma^2\lambda}\sum_{k=1}^{n}\left(r^i(k\Delta t) - V_{th} + \lambda\gamma\right)^2 \cdot \Delta t\right], \tag{18.10}$$

where $\Delta t = \frac{t}{n}$. The ISI density calculated from $p_\lambda(t)$ is plotted in Fig. 18.1C, which demonstrates that $\bar{p}_\lambda$ matches the histogram obtained from a direct simulation of the LIF model very well.

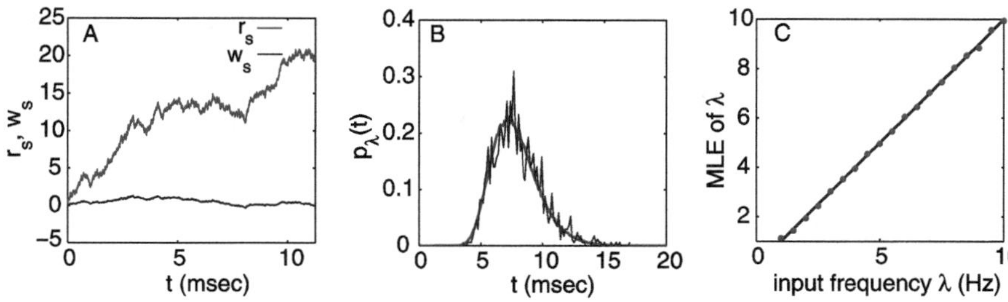

Fig. 18.1 MLE of a single neuron. (**A**) Examples of trajectories of r_s and w_s. (**B**) $p_\lambda(t)$ (*red*) and histogram (*blue*) from a direct simulation of the LIF model. (**C**) The estimated input rate $\hat{\lambda}$ (*dots*) obtained from 600 interspike intervals. Parameters are $V_{th} = 20$ mV and $\gamma = 20$ msec (for the color version, see Color Plates on p. 398)

The Bessel bridge r_s has a trajectory as shown in Fig. 18.1A. The ISI density calculated from $p_\lambda(t)$ is plotted in Fig. 18.1B, which is shown to match with the histogram obtained from a direct simulation of the LIF model.

18.2.2 MLE Decoding Strategy

Having the exact function of the distribution $p_\lambda(t)$, we can perform the MLE decoding procedure. The likelihood function is given by

$$L \triangleq \prod_{i=1}^{N} p_\lambda(T_i), \tag{18.11}$$

where N is the total number of spikes. Then

$$\ln L = \sum_{i=1}^{N} \ln p_\lambda(T_i). \tag{18.12}$$

The optimal estimate of the input information λ corresponds to the root of the equation $d \ln L/d\lambda = 0$. However, calculation of $d \ln L/d\lambda$ will yield a complicated expression, since the density function $p_\lambda(t)$ implies a singular stochastic process r_s whose derivative with respect to λ, $w_s \triangleq dr_s/d\lambda$, is also a singular stochastic process. Actually, w_s satisfies the equation

$$dw_s = \left(\frac{-w_s}{t-s} + \frac{1}{r_s} - \frac{\lambda}{r_s^2} w_s \right) ds + \frac{1}{2\sqrt{\lambda}} dB_s \tag{18.13}$$

with $r_0 = 0$ and $w_0 = 0$. One sampling trajectory of w_s is shown in Fig. 18.1A.

After some calculations, we know that the MLE of λ is the root of the equation

$$\left(N\gamma^{-1}+\sum_{i=1}^{N}T_i^{-1}\right)V_{th}^2-\lambda N+\gamma^{-2}\sum_{i=1}^{N}E_{0\to V_{th}}\big[g_\lambda(T_i)\big]=0, \tag{18.14}$$

where $g_\lambda(t)$ is given by

$$\int_0^t\big[(r_s-V_{th}+\lambda\gamma)^2-2\lambda(r_s-V_{th}+\lambda\gamma)\cdot(w_s+\gamma)\big]ds$$
$$\times\frac{\exp[-\frac{1}{2\gamma^2\lambda}\int_0^t(r_s-V_{th}+\lambda\gamma)^2ds]}{E_{0\to V_{th}}\{\exp[-\frac{1}{2\gamma^2\lambda}\int_0^t(r_s-V_{th}+\lambda\gamma)^2ds]\}}. \tag{18.15}$$

Though it is difficult to find an analytical solution of (18.14), we can numerically find its root, denoted as $\hat{\lambda}$. Figure 18.1C depicts the value $\hat{\lambda}$ vs. its actual value λ for the model defined by (18.1), where each point of $\hat{\lambda}$ is obtained by using 600 interspike intervals. It is clearly shown that the estimated value $\hat{\lambda}$ excellently matches with the true value λ.

Actually, the precision of the estimate can be theoretically decided by one of the nice properties of the MLE, which lies in the fact that we can estimate the confidence intervals according to Fisher information which is defined by

$$I(\lambda)=\int\left(\frac{p'_\lambda(t)}{p_\lambda(t)}\right)^2p_\lambda(t)\,dt. \tag{18.16}$$

From (18.6) we conclude that

$$\frac{p'_\lambda(t)}{p_\lambda(t)}=\frac{1}{2\lambda^2}\big[\big(\gamma^{-1}+t^{-1}\big)V_{th}^2-\lambda+\gamma^{-2}E_{0\to V_{th}}\big[g_\lambda(t)\big]\big].$$

Hence the Fisher information is given by

$$I(\lambda)=\frac{1}{4\lambda^4}\int\big[\big(\gamma^{-1}+t^{-1}\big)V_{th}^2-\lambda+\gamma^{-2}E_{0\to V_{th}}\big[g_\lambda(t)\big]\big]^2p_\lambda(t)\,dt. \tag{18.17}$$

For a given sampling number N, let $\hat{\lambda}_N$ be the estimate of a parameter λ via MLE. We have that

$$\sqrt{N}\big(\lambda-\hat{\lambda}_N\big)\to Norm\big(0,1/I(\lambda)\big)$$

in distribution. Then the confidence intervals of the model parameter λ for a given N can be computed as

$$\left[\lambda-\frac{1}{\sqrt{NI(\lambda)}},\lambda+\frac{1}{\sqrt{NI(\lambda)}}\right]. \tag{18.18}$$

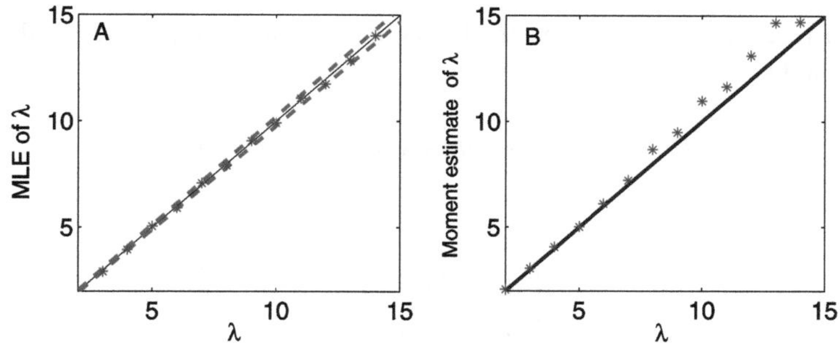

Fig. 18.2 Comparison of two methods. (**A**) Estimate of afferent input λ via the MLE method. *Dashed lines* are confidence intervals calculated from (18.18). (**B**) Estimate of afferent input λ via the "rate coding" method. Here we fix $r = 0$ and $a = 1$ and choose 200 ISI intervals

18.2.3 Comparing with Rate Decoding

One might ask why we do not decode the input information by simply fitting the firing rate and CV since the first and second moments of ISIs are known. Such a "rate decoding" approach has been extensively discussed in the literature. It is known that when the sampling number is large enough, both MLE and rate coding methods give reasonable results. However, from the viewpoint of parameter estimation, the advantage of the MLE method is obvious. By the Cramer–Rao lower bound, we know that the MLE is optimal, but rate decoding is not.

Further numerical comparison of the two approaches is presented in Fig. 18.2. It is known that with the input rate increasing, the variation also increases, and thus both the errors of decoding via the MLE and via the "rate coding" increase in the regime of high λ. However, as can be seen from Fig. 18.2A, the decoding error via the MLE method is bounded by the Cramer–Rao lower bound, while the error via the "rate coding" approach may be out of this range. From this point of view, the MLE is optimal. The advantage of the MLE over the "rate coding" is quite obvious even for a relatively small sampling number (here we take $N = 200$).

18.3 Applications

In this section, we will apply the above developed strategy of MLE to some decoding problems.

18.3.1 Decode Excitatory and Inhibitory Ratio in a Single Neuron with Stationary Input

One can see from Fig. 18.1C that stationary input to a single neuron can be reliably read out by the method of MLE based on the density function of ISI. Generally, a neuron receives excitatory and inhibitory inputs. Hence the corresponding LIF model could be written as

$$dV = -\frac{V}{\gamma}\,dt + \mu\,dt + \sqrt{\beta}\,dB_t, \quad V \le V_{th}, \tag{18.19}$$

where

$$\begin{cases} \mu = a\lambda(1-R), \\ \beta = a^2\lambda(1+R), \end{cases} \tag{18.20}$$

with a being the magnitude of excitatory postsynaptic potentials (EPSP) and R the ratio between inhibitory and excitatory inputs.

Similar to decoding the input rate λ in (18.1), the value of excitatory and inhibitory ratio R can be known from the solution of the following equation:

$$\begin{aligned} &\left(\gamma^{-1} + \frac{1}{N}\sum_{i=1}^{N} T_i^{-1}\right) V_{th}^2 \beta' + 2V_{th}\big(\mu'\beta - \mu\beta'\big) - \beta\beta' \\ &+ \frac{\beta'}{\gamma^2 N}\sum_{i=1}^{N} E_{0\to V_{th}}\big[g_R(T_i)\big] = 0, \end{aligned} \tag{18.21}$$

where the derivative is with respect to R, g_R is a function related to the three-dimensional Bessel bridge and can be calculated in the same way as $g_\lambda(t)$ in (18.15). In Fig. 18.3, the ratio for the model defined by (18.20) is estimated, which is shown to be very accurate.

In the literature, what is the exact value of R has been a long debating issue [9, 29]. On the one hand, it has been argued that R should be around unity [21, 29]. On the other hand, experimentally measured postsynaptic value seems to be much higher [9]. However, since input signals travel through the dendrites and soma (both highly nonlinear media) and finally arrive at the hillock where the spikes generate, the ratio could be very different from the one measured from postsynaptic current.

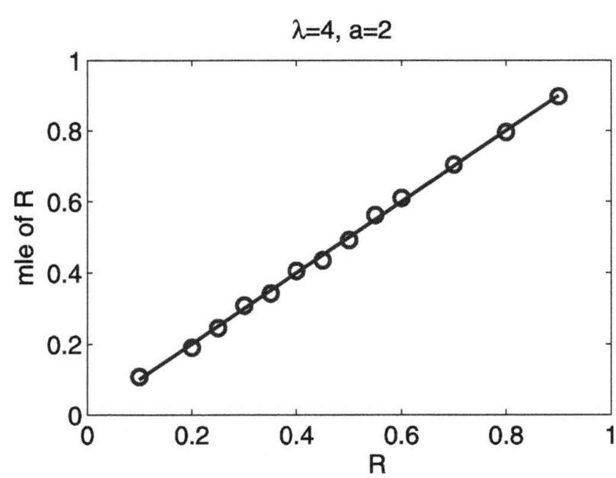

Fig. 18.3 The estimated ratio $\hat{R}$ (*open circles*) vs. the actual value R. Parameters are $V_{th} = 20$ mV, $\gamma = 20$ msec, and $V_{re} = 0$ mV

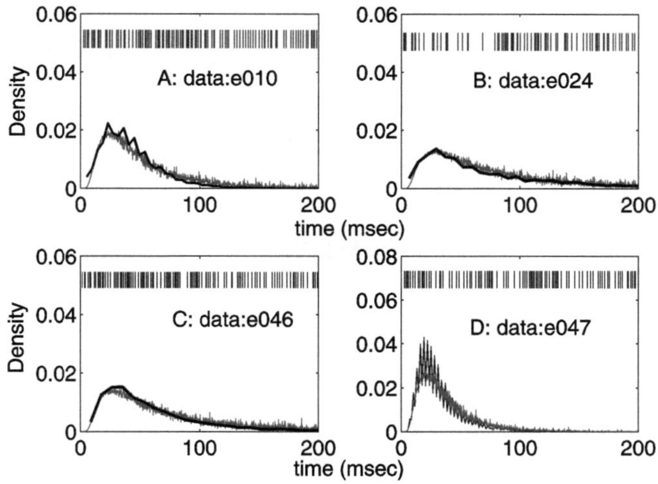

Fig. 18.4 (Color online) MLE of the ratio R. Results for four datasets are shown here. *Black curves* are the histograms from the experimental data, and *red curves* are the density distribution functions

To see whether the debate can be partially answered, let us have a look of some recorded datasets of pyramidal neurons in macaque monkey MT/V5 with random dots stimuli [4] (http://www.neuralsignal.org/index_data.html). In Fig. 18.4, we plot data histograms (black ones) of the interspike intervals from four sets of recorded datasets. In each panel, the inset is a section of the whole recording of the spikes.

We suppose that the input–output relationship of each pyramidal cell can be regarded as a LIF neuron with (18.19) and (18.20). We set $V_{th} = 25$ mV, $\gamma = 20$ msec, $\lambda = 2$, $a = 1.5$ mV, $V_{re} = 0$ mV, all in physiologically reasonable regions. The interspike intervals distribution density can be obtained from the MLE of R, which is the solution to (18.21). The red curves in Fig. 18.4 depict the estimated density distribution functions, which are shown to be very good approximations to the corresponding histograms from the experimental data.

We have analyzed 10 datasets, and the ratios obtained from MLE are 0.682, 0.605, 0.792, 0.78, 0.61, 0.77, 0.51 0.55, 0.905, 0.77, with a mean of $R = 0.6974$, in accordance with the value suggested in the literature [13].

18.3.2 Decode Dynamical Inputs in Networks Without Interactions

In the following, let us further develop the MLE to decode dynamical inputs in pools of neurons, which has been partly solved in [27]. The network as schematically plotted in Fig. 18.5A is composed of 100 neurons. We assume that the input is varying slowly compared with the time scale of the neuronal dynamics, so that in each time window of fixed length T_W, the ISI distribution adiabatically follows the stationary one.

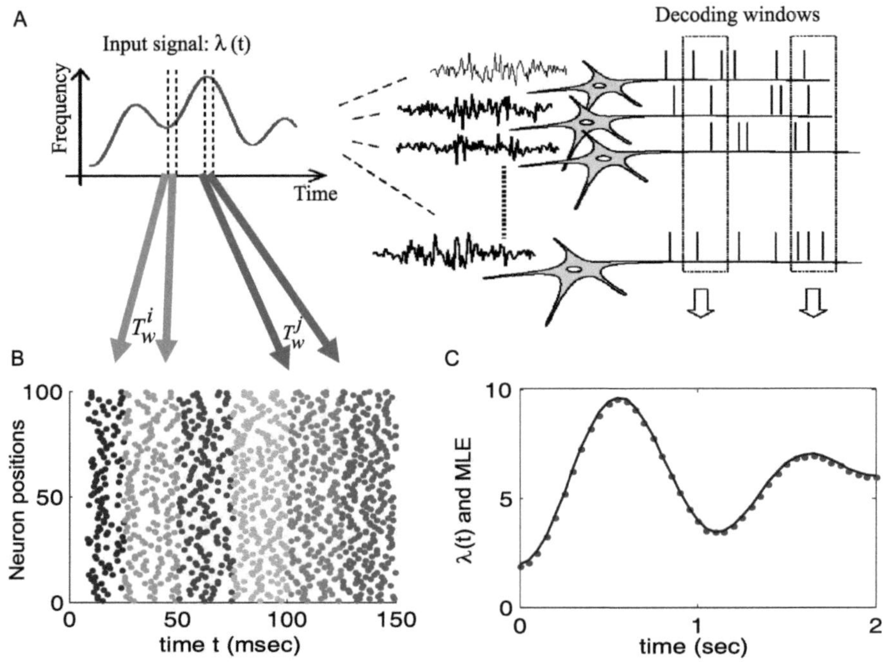

Fig. 18.5 MLE in a network without interactions. (**A**) Schematic plot of reading dynamic inputs from an ensemble of neurons. For a fixed time window (indicated by T_w^i, T_w^j), the spikes are collected, and the input is decoded by means of MLE. (**B**) Raster plots of spikes in five different decoding windows. The information is read out from the spikes in each window. (**C**) An example of reading the input information from ensemble of neurons is shown. The original signal $\lambda(t)$ is plotted in the *continuous line*, while *dots* are estimated values of $\lambda(t)$ (for the color version, see Color Plates on p. 399)

To express the main idea of how to apply the MLE to decode dynamical input information, we assume that each neuron receives a common excitatory Poisson synaptic input, and the procedure below is also valid for other ratios between the inhibitory input and the excitatory input. We suppose that the waveform of input is $\lambda(t) = 2 + 4(\sin^2(2\pi t) + \sin^2(\frac{3}{2}\pi t))$. Here the time scale of the input is measured in the unit of second, and thus it varies slowly compared with the time scale of the neuronal dynamics. Note that in this case any interspike interval longer than T_w will not be included in the procedure, so the estimated input $\hat{\lambda}(t)$ is bound to be biased, a typical situation in survival analysis. To obtain an unbiased estimate, the censored intervals have to be included, and more detailed calculations are required. However, since the numerical results (see Fig. 18.5C) indicate that the bias is very limited, we simply ignore the issue of censored intervals. Figure 18.5C depicts the MLE vs. the input frequency for time windows $T_w = 25$ ms. Although T_w is very short, we can see that the estimate is excellent (except that it is slightly downward biased).

Now we consider a more biologically realistic setup. Assume that an ensemble of neurons, say 1000 leaky LIF neurons, are grouped into $N = 10$ columns, and each

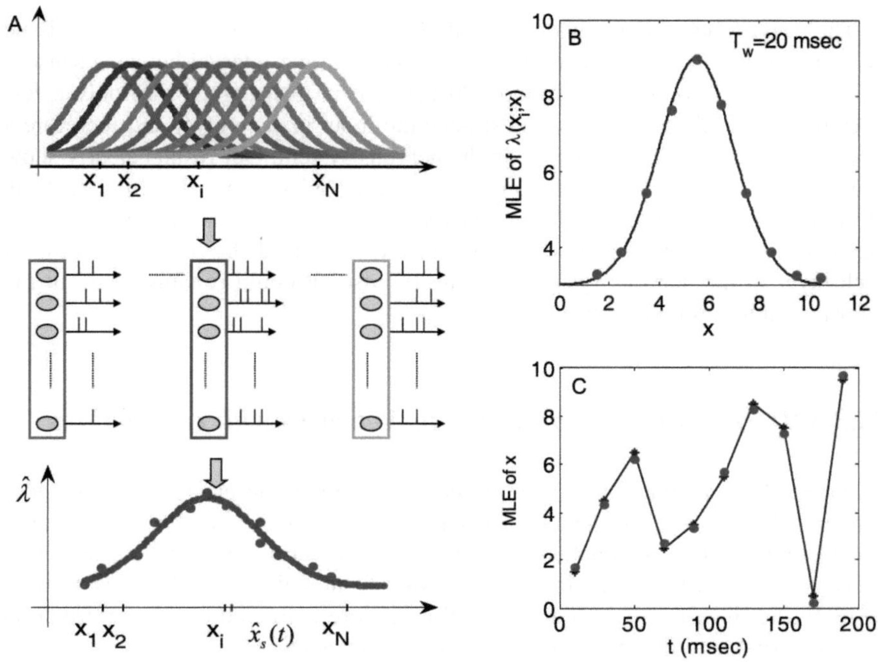

Fig. 18.6 MLE in position tracking. (**A**) The setup of our encoding and decoding. The place information $x(t)$ is feeded to 1000 neurons organized in 10 columns with different preferred positions x_i. For each column, spikes are collected in the window T_w, and the MLE is applied. The position is read out by first solving (18.22), obtaining the corresponding $\hat{x}$ and fitting a Gaussian curve to all 10 data points $\hat{x}$. The maximum point of the fitted Gaussian curve is the decoded position. (**B**) One example of $\hat{\lambda}$ vs. columns. (**C**) Examples of 10 decoded results (for the color version, see Color Plates on p. 400)

100 neurons in a same column share an identical tuning curve defined by

$$\lambda(x_i; x) = \lambda_0 + c \cdot \lambda_0 \exp\left(-\frac{(x - x_i)^2}{2\sigma^2}\right), \tag{18.22}$$

where c is a constant scaling factor, σ is the tuning width, x_i is the column's field center, and x is the position. Hence, for example, the neuron in the ith column receives an input which is the position information $x(t) \in [0, L]$ (we take $L = 11$ here), and the response of the neuron is given by $\lambda(x_i, x(t))$. The task is to read out the position information $x(t)$. The setup mimics the situation of reading out the position of a rat from simultaneous recordings of place cells in Hippocampus [32].

For simulation, we fix $c = 2$, $\sigma = 1.5$, and $\lambda_0 = 2$. We assume that the target position x changes in time according to $x(t) = \sum_{j=1}^{N} \xi_j \chi(t \in I_j)$, where ξ_j are independent random variables uniformly distributed in $[0, L]$, and $I_j = \{t : (j-1)T_w < t < jT_w\}$, i.e., we consider a target hopping every T_w to a new random position between 0 and L. An outcome of our experiments is shown in Fig. 18.6.

The results above also provide a possible answer for another long-standing issue in neuroscience. It was pointed out in [31] that the time interval between sensory inputs and motor outputs is around 200 msec. It is then argued that only a few interspike intervals can be used to estimate the input in each layer and therefore a stochastic dynamics is implausible in the nervous system. Our results clearly show that within a very short time window (20 msec), the spikes generated from an array of neurons contain enough information for the central nervous system to decode the input information. Hence, even without the overlap of the processing time of each layer in the nervous system, within 200 msec it could reliably read out the information for around ten layers.

18.3.3 Decode Input Information in Networks with Interactions

So far we have assumed that neurons in the same column are regarded as independent. Certainly neurons in a microcolumn interact with each other, which might considerably change all conclusions in the previous sections. Can we or the central nervous system read out the input information from spike waves? In this section, let us further investigate such an issue. The purpose is to read out the information of an external stimulus even in the presence of interactions between neurons in a microcolumn.

The strategy adopted here (or possibly by the nervous system) is based on incorporating the lateral inhibition and the time delay of the synaptic inputs. As a result, all neurons in a network behave independently before the interactions kick in. The depolarization caused by the external inputs evokes the hyperpolarizing effect of inhibitory interactions between neurons, which subsequently shuts down the firing of all neurons (first epoch) and enables the neurons in the microcolumn act independently again.

The model we consider here consists of $P_E = 100$ excitatory and $P_I = 100$ inhibitory neurons with all-to-all connections. Biologically, a neuron has two compartments: soma and dendrite. For excitatory interactions, neurons receive inputs at its dendritic compartment; for inhibitory connection, neurons interact with each other via inhibitory neurons (not specifically modeled) and receive inputs at its somatic compartment. Here, for simplicity, we unify the somatic and dendritic compartments as the following single model described as

$$\frac{dV_j}{dt} = -\frac{V_j}{\gamma} + I_j^E(t) + I_j^S(t), \quad V_j \le V_{th}, \quad j = 1, 2, \ldots, N_n = P_E + P_I, \tag{18.23}$$

where

$$\begin{cases} I_j^E(t) = a\lambda_j(t) + a\sqrt{\lambda_j(t)}\xi_j(t), \\ I_j^S(t) = \sum_{k=1}^{P_E} \sum_{t_k^m + \tau < t} w_{jk}^E \delta(t - t_k^m - \tau) - r_{EI} \sum_{l=1}^{P_I} \sum_{t_l^m + \tau < t} w_{jl}^I \delta(t - t_l^m - \tau). \end{cases} \tag{18.24}$$

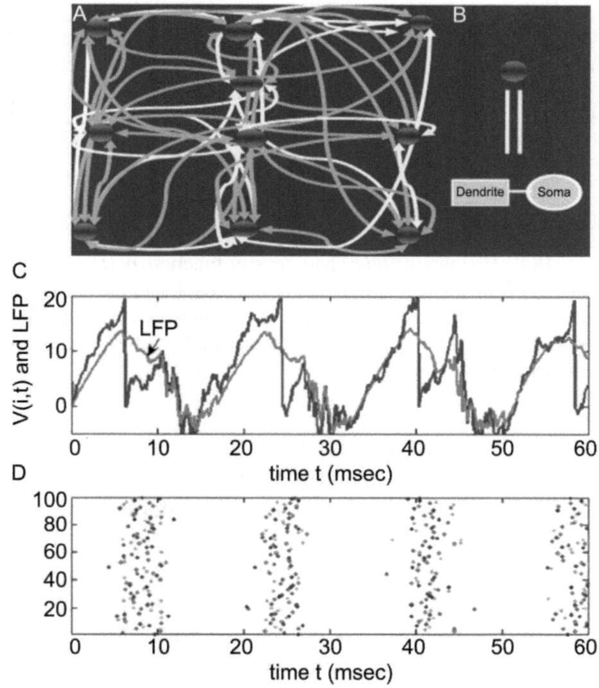

Fig. 18.7 (**A**) A schematic plot of the model setup with 10 neurons. *White arrows* are excitatory interactions, and *green arrows* are inhibitory interactions via inhibitory neurons (not specifically modelled). (**B**) A single neuron model of two compartments. (**C**) One trajectory of one single cell is depicted (*blue trace*), and the *red line* is the local field potentials for averaging all neurons in a microcolumn. The oscillatory in Gamma frequency is obvious. (**D**) The optimal time window for decoding the input (the window between two *thick vertical lines*) (for the color version, see Color Plates on p. 401)

Here $I_j^E(t)$ is the external input from the stimulus with an input rate $\lambda_j(t)$ and magnitude a, $\xi_j(t)$ are independent white noises, $I_j^S(t)$ is the spiking input from other neurons to the jth neuron, i.e., the intralayer interactions, w_{jk}^E and w_{jl}^I are EPSP and IPSP sizes that the jth neuron receives from the kth excitatory neuron and the lth inhibitory neuron, t_k^m is the mth spike generated from the kth neuron, and τ is the time delay due to the interaction. We suppose that w_{jk}^E and w_{jl}^I are uniformly distributed in $[0, D]$, and $\lambda_i(t) = \lambda_j(t) = \lambda$. Finally, r_{EI} is the ratio between intralayer inhibitory and excitatory interactions. In our numerical simulations, we fix $a = 1$, $D = 2$, $r_{EI} = 1.8$, and $\tau = 5$ msec. An explanation of our model setup is plotted in Figs. 18.7A and B.

Figures 18.7C and D show one simulation with the setup as above. As expected, with the application of the external inputs, all neurons start respond to it. Some neurons fire a few spikes. However, the inhibitory inputs kicks in when t is around 10 msec and inhibit all neuronal activities. Once all neurons are silent, the interactions between neurons disappear, and all neurons in the network act as independent units again. Hence the external inputs evoke the second epoch of the spikes. The procedure above repeats itself. We might ask ourselves why neurons use pulsed interactions to communicate between them rather than other forms of interactions such as gap junctions to interact. The difference between gap junctions and pulsed interaction is that the former exchanges information continuously and the later only react to each other when the membrane potential exceeds a threshold which results in the independent phase in Fig. 18.7D.

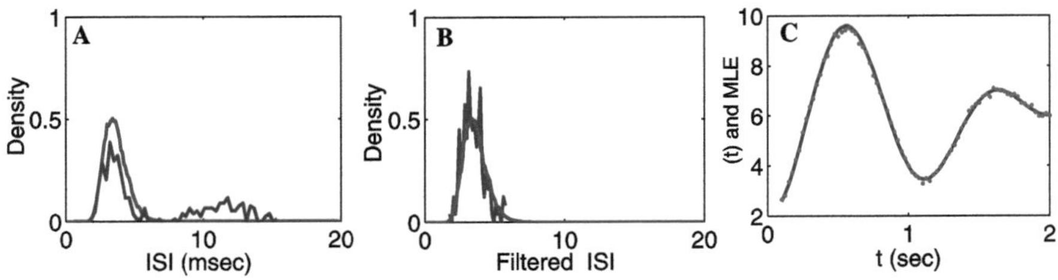

Fig. 18.8 Histogram (*blue*) and density function $p_\lambda(t)$ (*red*): (**A**) The histogram (*blue trace*) is obtained from a direct simulation of the model, and the *red line* is the theoretical density; (**B**) The histogram is from the filtered interspike intervals (the first modes in **A**). (**C**) Reading the input information (the *solid line*) from an ensemble of neurons, every *dot* is estimated within a fixed time window $T_w = 20$ msec

Based on such an inhibition-induced shutting down mechanism, let us now investigate how the afferent input rate $\lambda(t)$ can be read out accurately by applying the MLE strategy. The problem of introducing interactions in a fixed time window is that the decoding may have a significant bias. To resolve this issue, let us first have a look of the histogram of the ISIs of (18.23)–(18.24) in the case the input rate $\lambda(t)$ is time-independent, say $\lambda(t) = \lambda = 4$. It is indicated in Fig. 18.8A that the histogram now has two modes: the one with short ISIs corresponds to the actual ISIs driven by the external input, and the other is due to the interactions. After filtering out the second mode, the obtained histogram fits well with the theoretical density (the red trace in Fig. 18.8B. To decode the input information, we filter out the spikes corresponding to the second mode and use exclusively the ISIs of the first mode. It is shown in Fig. 18.8C that in a short time window, the input rate can be reliably decoded.

In Figs. 18.9A and B, we test our algorithm in a dynamic input with waveform $\lambda(t) = 2 + 4(\sin^2(2\pi t) + \sin^2(\frac{3}{2}\pi t))$, for different values of coupling strength. Here t is measured in the unit of second, which ensures the ISI distribution being adiabatically stationary in each time window of length around 25 msec. To show the network behavior under different coupling intensities, the ensemble voltage traces and the corresponding raster plots of 100 excitatory cells for $\lambda(t) = \lambda = 3$ are depicted in the above and the middle panels of Figs. 18.9A and B, respectively. Interestingly, it is shown that the network displays rhythmic activity when the coupling strength is strong and the decoding (shown in the bottom trace of Fig. 18.9B) is quite accurate; however, for weak values of the coupling strength, no rhythmic activities are observed, and as expected, the estimated values (see the bottom trace of Fig. 18.9A) are much less accurate than in the rhythmic case.

Figures 18.9A and B suggest that there should exist a critical value of the coupling strength D_c, after which the network can perform decoding accurately. To see this, we plot the relative decoding error $\frac{\hat{\lambda}-\lambda}{\lambda}$ vs. the coupling intensity D in Fig. 18.9C, for constant afferent inputs (here we choose $\lambda = 4$, $\lambda = 5$, and $\lambda = 8$). One can see that for $D = 0$, the decoding is very accurate, which is just the case discussed in Sect. 18.3.2. With D slightly increasing, the absolute decoding error increases. This is because the interactions between neurons cannot be shut down for

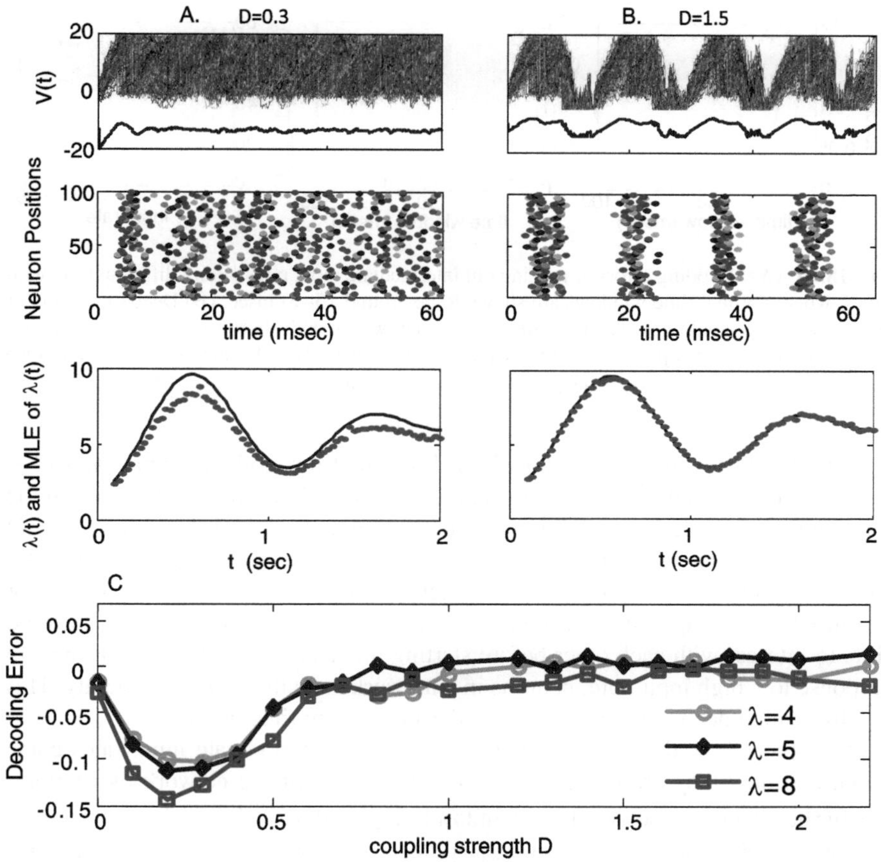

Fig. 18.9 MLE in a network with interactions. (**A**) and (**B**) The *above* traces: The trajectories of 100 excitatory neurons and the corresponding mean voltages for $\lambda = 3$ during 0–60 ms. The *middle* traces: Raster plots corresponding to the above trace. The *bottom* traces: Reading out the dynamic inputs (the *solid line*) from 100 excitatory neurons, every *dot* is estimated via the MLE approach within a fixed time window $T_w = 25$ msec. *Left column* is for $D = 0.3$, while *right column* is for $D = 1.5$. (**C**) The optimal coupling intensities for reading out the input information within a fixed time window $T_w = 25$ msec

weak coupling strengths. After about $D = 0.25$, the absolute decoding error gradually decreases, and after about $D = 0.8$, the input rate can be reliably read out, meanwhile, the network exhibits significant rhythmic activities.

If only interspike intervals are concerned, then for a too short time window, it is difficult to perform a decoding task as there are no enough spikes. On the other hand, if the time window is too long, then the informative interspike intervals generated via the external inputs are interfered with the ones due to interactions. We conclude that there should be an optimal decoding time window T_w^{optim} in which the input information is optimally decoded. To test this, we plot the relative decoding errors $\frac{|\hat{\lambda}-\lambda|}{\lambda}$ vs. different time windows in Fig. 18.10A for intermediate input rates and

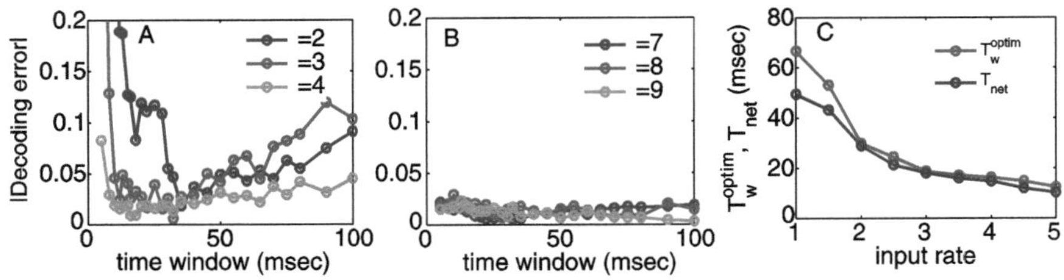

Fig. 18.10 (**A**) Decoding errors of reading out intermediate input rates within different time windows, where optimal time windows are shown to be around 15–40 msec. (**B**) Decoding errors of reading out high input rates are shown to be very small within both short and large time windows. (**C**) Relationship of the optimal decoding windows and the periods of the network oscillations. One can see that for moderate input rates, $T_w^{\text{optim}} \approx T_{\text{net}}$. Here we fix $D = 1.5$

in Fig. 18.10B for high input rates. It is shown that for moderate input rates ($2 \leq \lambda \leq 5$), such optimal windows exist and are around 15–40 msec, which are in the gamma range, while for a high input rate λ, the decoding errors are small (<5%) for both short and long time windows. The accuracy of decoding for high input rates is because in this case, spikes are generated within a very short time (~ or <5 msec, the synaptic delay time), which implies that neurons in a pool have no time to interact with each other before starting the next epoch of firings. Hence, in response to a high input rate, neurons in a network actually act independently. This is why a small decoding error is achieved even for long time windows.

To explore how the optimal decoding windows for moderate input rates come about and how they are related with the periods T_{net} of the network oscillations, we further depict these two time windows (T_{net} and T_w^{optim}) vs. the input rate λ in Fig. 18.10C. The definition of T_{net} is clear; for example, see Fig. 18.9B. One can see that for a reasonable range of input rates λ (about $2 \leq \lambda \leq 5$), the optimal decoding windows are approximately equal to the network periods. The consistence of the optimal decoding windows and the periods of the network oscillations for moderate input rates manifests that the spikes generated from a pool of neurons within the period of the network rhythm are sufficient to read out input rates. Too small or too large time windows will either include no enough input information or interfered redundant interspike intervals, both of which will introduce the bias in decoding. The results shown in Figs. 18.9 and 18.10 may serve as a good example to manifest the functional role of the Gamma rhythm (30 to 80 Hz) in information processing, which has been extensively discussed by many authors (see [7] and references therein).

18.4 Discussion

We have presented a study on how to perform MLE on spike waves [34]. First, a rigorous algorithm of the MLE based upon the ISI distribution is developed. The algorithm enabled us to address a few key issues in Neuroscience. We have shown that the MLE can be successfully applied to read out either steady or dynamic inputs in single or pools of neurons. Even within short time windows, the decoded information is of high accuracy. We have also addressed that even for spiking neural networks with interactions, we can still employ the strategy of the MLE to decode dynamic inputs, by properly incorporating the local inhibitory interactions in a microcolumn and the delay of synaptic currents.

We have only included a very brief account of our applications here and will publish more results such as tracking moving stimuli, decoding multistimuli, constructing IF-type visual cortex, and exploring the gamma rhythm in signal decoding elsewhere. Furthermore our approach may help us to answer another long-standing problem: what is the ratio between inhibitory and excitatory inputs in a biological neuron? Although it has been found that the number of inhibitory neurons is smaller than the number of excitatory neurons in the cortex [21, 29], it is generally agreed that inhibitory neurons send stronger signals than excitatory neurons [9]. Therefore the exact ratio between inhibitory and excitatory inputs remains elusive. With the maximum likelihood estimate developed in the current paper, we may be able to reliably estimate the ratio between inhibitory and excitatory inputs to a LIF neuron.

As a simplified phenomenological neuronal model, the LIF equation preserves spiking properties of a neuron, and the input information can be reliably read out, as manifested in the current paper. However, the LIF model fails to capture many biophysical details [5, 11]. In the literature, there are some models that are more biophysically accurate but still mathematically simple, such as quadratic IF neurons [6, 14], exponential IF neurons [16], and more recently adaptive exponential IF neurons [3]. It is shown that adaptive exponential IF neurons give an effective description of neuron activities, and can reliably predict the voltage trace of a naturalistic pyramidal neuron from a dynamic I–V curve [2]. We realized that generalizing the MLE strategy developed for the LIF neurons to these nonlinear IF neurons still needs more endeavor, as it is not so easy to derive exact expressions of the distributions of ISIs for these nonlinear IF neurons. Furthermore, we have not taken into account learning, here and it is certainly a more challenging issue since the estimated parameters are also dynamical variables [23].

Acknowledgements This work was supported by the National Natural Science Foundation of China (under Grant Nos. 10971196, 10771155) and a Foundation for the Author of National Excellent Doctoral Dissertation of P.R. China (FANEDD).

References

1. L. Alili, P. Patie, and J.L. Pedersen. Representations of the first hitting time density of an Ornstein–Uhlenbeck Process. *Stoch Models*, **21**:967, 2005.
2. L. Badel, S. Lefort, R. Brette, C.C.H. Petersen, W. Gerstner, and M.J.E. Richardson. Dynamic I–V curves are reliable predictors of naturalistic pyramidal-neuron voltage traces. *J Neurophysiol*, **99**(2):656–666, 2008.
3. R. Brette and W. Gerstner. Adaptive exponential integrate-and-fire model as an effective description of neuronal activity. *J Neurophysiol*, **94**:3637–3642, 2005.

4. K.H. Britten, M.N. Shadlen, W.T. Newsome, and J.A. Movshon. Responses of single neurons in macaque MT/V5 as a function of motion coherence in stochastic dot stimuli. http://www.neuralsignal.org/index_data.html.
5. D. Brown, J.F. Feng, and S. Feerick. Variability of firing of Hodgkin–Huxley and FitzHugh–Nagumo neurons with stochastic synaptic input. *Phys Rev Lett*, **82**:4731–4734, 1999.
6. N. Brunel and P.E. Latham. Firing rate of the noisy quadratic integrate-and-fire neuron. *Neural Comput*, **15**:2281–2306, 2003.
7. G. Buzsáki. *Rhythms of the Brain*. Oxford University Press, London, 2006.
8. S. Deneve, P.E. Latham, and A. Pouget. Reading population codes: a neural implementation of ideal observers. *Nat Neurosci*, **2**:740, 1999.
9. A. Destexhe and D. Contreras. Neuronal computations with stochastic network states. *Science*, **314**:85, 2006.
10. S. Ditlevsen and P. Lansky. Estimation of the input parameters in the Ornstein–Uhlenbeck neuronal model. *Phys Rev E*, **71**:011907, 2005.
11. J.F. Feng. Behaviours of spike output jitter in the integrate-and-fire model. *Phys Rev Lett*, **79**(21):4505–4508, 1997.
12. J.F. Feng. *Computational Neuroscience, A Comprehensive Approach*. Chapman & Hall/CRC, London/Boca Raton, 2003.
13. J.F. Feng and D. Brown. Coefficient of variation greater than .5. How and When? *Biol Cybern*, **80**:291–297, 1999.
14. J.F. Feng and D. Brown. Integrate-and-fire models with nonlinear leakage. *Bull Math Biol*, **62**(3):467, 2000.
15. J.F. Feng and M. Ding. Decoding spikes in a spiking neuronal network. *J Phys A, Math Gen*, **37**:5713, 2004.
16. N. Fourcaud-Trocme, D. Hansel, C. van Vreeswijk, and N. Brunel. How spike generation mechanisms determine the neuronal response to fluctuating inputs. *J Neurosci*, **23**:11628, 2003.
17. C.W. Gardiner. *Handbook of Stochastic Methods: for Physics, Chemistry and the Natural Sciences*. Springer Series in Synergetics. Springer, Berlin, 1985.
18. W. Gerstner and W. Kistler. *Spiking Neuron Models Single Neurons, Populations, Plasticity*. Cambridge University Press, Cambridge, 2002.
19. A. Göing-Jaeschke and M. Yor. A clarification note about hitting times densities for Ornstein–Uhlenbeck processes. *Finance Stoch*, **7**:413, 2003.
20. C.P. Hung, G. Kreiman, T. Poggio, and J.J. DiCarlo. Fast readout of object identity from macaque inferior temporal. *Science*, **310**:863, 2005.
21. G. Leng, C.H. Brown, P.M. Bull, D. Brown, S. Scullion, J. Currie, R.E. Blackburn-Munro, J.F. Feng, T. Onaka, J.G. Verbalis, J.A. Russell, and M. Ludwig. Responses of magnocellular neurons to osmotic stimulation involves coactivation of excitatory and inhibitory input: An experimental and theoretical analysis. *J Neurosci*, **21**:6967–6977, 2001.
22. R.M. Memmesheimer and M. Timme. Designing the dynamics of spiking neural networks. *Phys Rev Lett*, **97**:188101, 2006.
23. E.S. Nikitin, D.V. Vavoulis, I. Kemenes, V. Marra, Z. Pirger, M. Michel, J.F. Feng, M. O'Shea, P.R. Benjamin, and G. Kemenes. Persistent sodium current is a nonsynaptic substrate for long-term associative memory. *Curr Biol*, **18**(16):1221–1226, 2008.
24. L. Paninski, J.W. Pillow, and E.P. Simoncelli. Maximum likelihood estimation of a stochastic integrate-and-fire neural encoding model. *Neural Comput*, **16**:2533, 2004.
25. F. Rieke, D. Warland, and R. Steveninck. *Spikes: Exploring the Neural Code*. MIT Press, Cambridge, 1997.
26. H. Risken and T. Frank. *The Fokker–Planck Equation: Methods of Solutions and Applications*. Springer, Berlin, 1984.
27. E. Rossoni and J.F. Feng. Decoding spike ensembles: tracking a moving stimulus. *Biol Cyber*, **96**:99, 2007.
28. T.D. Sanger. Neural population codes. *Curr Opin Neurobiol*, **13**:238, 2003.
29. M.N. Shadlen and W.T. Newsome. The variable discharge of cortical neurons: Implications for connectivity, computation, and information coding. *J Neurosci*, **18**:3870, 1998.

30. R.L. Stratonovich. *Topics in the Theory of Random Noise*. Mathematics and Its Applications. Gordon & Breach, New York, 1967.
31. S. Thorpe. Speed of processing in the human visual system. *Nature*, **381**:520, 1996.
32. W. Truccolo and U. Eden. A point process framework for relating neural spiking activity to spiking history, neural ensemble and covariate effects. *J Neurophys*, **93**:1074, 2005.
33. H.C. Tuckwell. *Theoretical Neurobiology*. Cambridge University Press, Cambridge, 1998.
34. X.J. Zhang, G.Q. You, T.P. Chen, and J.F. Feng. Readout of spike waves in a microcolumn. *Neural Comput*, **21**:3079, 2009.

Chapter 19
False Positive Control for Genome-Wide ChIP-Chip Tiling Arrays

Yu Zhang

19.1 Introduction

Chromatin immunoprecipitation of DNA segments followed by microarray hybridization (ChIP-Chip) is a powerful tool for studying protein–DNA interactions in vivo. Applying this technology with high-density oligonucleotide tiling arrays allows precise localization of transcription factor binding in the whole-genome scale [1, 2]. A typical tiling array chip contains 4×10^5 DNA probes (oligonucleotide of 25–100 bp long) tiled at every 30–100 bp on the genome. ChIP-Chip data is thus an array of hybridization intensities measured from the ordered probes. Large intensity values are called peaks, and the probe locations of these peaks indicate the most likely positions of protein–DNA binding. The task is therefore to find statistically significantly large peaks and their corresponding locations on the genome, which is referred to as peak-calling.

Probe intensities of multiple adjacent probes are positively correlated due to the large size of sheared DNA fragments. As a result, regions containing many large values of hybridization intensities are more likely to cover a true protein–DNA binding site than regions containing just a single peak. The positive correlation structure of adjacent probes raises a statistically challenging problem for estimating the significance of peaks under the context of multiple testings. The well-known Bonferroni method is too conservative for estimating the family-wise type I error rate in peak-calling, because the method assumes independence between tests. Alternatively, one may want to control the false discovery rate (FDR) introduced by Benjamini and Hochberg [3]. However, when tests are positively correlated, the estimated proportion of false rejections can have large variability. More importantly, significant peaks tend to be clustered, making the usual FDR control on individual tests inappropriate.

Y. Zhang (✉)
Department of Statistics, The Pennsylvania State University, 325 Thomas Bldg, State College, PA 16803, USA
e-mail: yuzhang@stat.psu.edu

J. Feng et al. (eds.), *Frontiers in Computational and Systems Biology*, Computational Biology 15,
DOI 10.1007/978-1-84996-196-7_19,

Many computational methods have been developed for ChIP-chip peak-calling in recent years [4–7]. None of them, however, provided a rigorous control on the false discoveries adjusting for millions of simultaneous comparisons. A common practice is to report all significant peaks passing an arbitrarily chosen cutoff; the validity of called peaks will then be verified using rtPCR. A nested Bonferroni method [8] has been proposed that provides a better upper bound of type I error rate than the Bonferroni method. Alternatively, permutation tests may be used to estimate the statistical significance of peaks, provided that permutation preserves the covariance structure of the data [9]. However, it is common that a ChIP-chip experiment has none or just a few biological and technical replicates, which makes permutation not applicable.

We propose a new method that can accurately approximate the statistical significance of peaks adjusting for multiple testings. Our approach is to convert the test statistics of probe intensities into a declumped statistic such that the local correlation among the probes is compensated. The declumping approach has been previously used in approximating the significance of sequence matching [10]. For tiling arrays, we first use a Monte Carlo method called importance sampling [11] to estimate the significance of individual clumps. We then use a Poisson distribution to approximate the overall significance of a peak adjusting for multiple testings. We demonstrate that our approach can accurately approximate the true statistical significance of peaks in a very efficient way. For example, it takes less than 1 second to compute the family-wise statistical significance of an observed peak value. We further generalize the method to address two additional issues in tiling array analysis. First, we demonstrate how to combine peak-calling results from various window sizes to maximize the power for detecting subtle peaks. Using simulations, we show that peaks identified using a fixed window size can lose a substantial amount of power than combining results from a range of window sizes. Second, an FDR control without considering the positive correlation among probes can seriously underestimate the true FDR level of protein–DNA binding intervals. We therefore propose a modified FDR method to solve this problem.

19.2 Methods

Let $X = (X_1, \ldots, X_L)$ denote an array of standardized hybridization intensities of L probes ordered by their physical locations on the chromosome. For each probe, the null hypothesis is that the probe does not overlap with a binding site, and we assume that the probe intensities follow a normal distribution. A simple peak-calling approach is then to test the null hypothesis using a sliding window of k consecutive probes, where the window starts at each probe under testing and covers k consecutive probes to the right. There are a total of $L - k + 1$ such windows. Under the null, we test whether $\overline{X_i} = (X_i + \cdots + X_{i+k-1})/k$ is much larger than that expected from a normal distribution with mean 0. The task is therefore to determine a threshold for $\overline{X_i}$ such that the family-wise type I error rate is controlled at a desired level.

19.2.1 Poisson Approximation

Given n independent tests with test statistic $T_i, i = 1, \ldots, n$, and a stringent threshold t, if $T_i \geq t$ is a rare event under the null hypothesis, we have that the number of rejections, $W = \sum_{i=1}^{n} I\{T_i \geq t\}$, follows approximately a Poisson distribution [12]. That is, $P(W = w) = \mathrm{e}^{-\lambda}\lambda^{w}/w!$, where $\lambda = E(W)$ is the expected number of tests passing threshold t. As a result, the family-wise p-value is given by $P(\max_{i=1}^{n} T_i \geq t) = 1 - P(W = 0) = 1 - \mathrm{e}^{-\lambda}$. The Poisson approximation, however, will not work for positively correlated tests. We therefore compensate for the positive correlation by introducing the binary indicator $R_i(t) = I\{\overline{X}_i \geq t\}I\{\overline{X}_{i-1} < t, \forall l = 1, \ldots, k-1\}$ at each probe i, where $\overline{X}_{i-1} = 0$ for $i - l \leq 0$. We abbreviate $R_i(t)$ as R_i. It is easily checked that $R_i, R_{i+1}, \ldots, R_{i+k-1}$ are negatively correlated, i.e., at most one of $R_i, R_{i+1}, \ldots, R_{i+k-1}$ can take value 1. We can view the distribution of R_i as a mixture of a Bernoulli distribution and a unit mass on zero, with weights π and $1 - \pi$, respectively, where π is the probability that none of the $(k-1)$ windows proceeding the ith window passes the threshold t, i.e., $\pi = \Pr(\overline{X}_{i-1} < t, \forall l = 1, \ldots, k-1)$, and the Bernoulli parameter is $p = P(\overline{X}_i \geq t \mid \overline{X}_{i-1}, \forall l = 1, \ldots, k-1)$. If there are no other dependence structures, the Poisson approximation will work for the distribution of $W_R = \sum_{i=1}^{L-k+1} R_i$ with mean $\lambda = E(W_R) = (L - k + 1)\pi p$.

Instead of testing the window averages, we check whether $R_i = 1$ or not, where $R_i = 1$ indicates a significant peak at position i after declumping. We call this new test an R-test in contrast to the test of window averages, which we call X-test. We further define the clump size as the number of windows involved in an R-test. The clump size can be chosen by the user, where a larger clump size can compensate more to the positive correlation among probes, especially for small window sizes such as $k = 1$. A large clump size, however, will increase computation time. We generalize the definition of R_i to $R_i(t) = I\{\overline{X}_i \geq t\}I\{\overline{X}_{i-l} < t, \forall l = 1, \ldots, c-1\}$, with $c = \max(c_{\min}, k)$ denoting the clump size and $c_{\min}$ denoting a user-specified minimum clump size. The value of $c_{\min}$ can be determined from the covariance matrix of window averages, e.g., let $c_{\min}$ be the smallest distance between two windows whose covariance is smaller than a threshold (such as 0.05). Finally, an error bound of our Poisson approximation can be obtained by the Chen–Stein method [13]. According to Theorem 1 in Arratia et al. [14], we have $|P(W = w) - \mathrm{e}^{-\lambda}\lambda^{w}/w!| \leq 2(b_1 + b_2 + b_3)$. If probes are only locally dependent, $b_1 = \lambda^2/(L - k + 1)$ and $b_2 = b_3 = 0$, and thus the error bound for the probability of observing no significant binding among all tests is $|P(W_R = 0) - \mathrm{e}^{-\lambda}| \leq 2\lambda^2/(L - k + 1)$. Note that this error bound is very small for a large number of tests.

Our approach using the Poisson approximation reduces the family-wise p-value estimation to the Poisson mean estimation, i.e., estimating $\lambda = (L - k + 1)\pi p$. Importantly, only πp needs to be estimated, which is unrelated with the total number of tests. This enables us to compute the family-wise p-values at a relatively constant computational cost, and thus the method can be very efficient for whole-genome studies. Specifically, we estimate πp using a Monte Carlo method called importance sampling [11], which provides accurate and efficient calculations for any large values of peak thresholds. Importance sampling can also easily take into account the irregular probe tiling structures, such as gaps between probes masked by repeats.

Given a k-probe window and its $(c-1)$ preceding windows, let $X_1, \ldots, X_{c+k-1}$ denote the values of the $(c+k-1)$ involved probes. Under the null hypothesis of no binding, $X_1, \ldots, X_{c+k-1}$ follow a multivariate normal distribution with mean μ and covariance matrix Σ. The covariance matrix Σ captures the dependence structure among local probes. We estimate both μ and Σ from the data using the middle 90% of the data to avoid biased estimation from the true binding signals and outliers.

The importance sampling procedure works as follows:

1. Generate $\mathbf{X} = (X_1, \ldots, X_{c+k-1})$ according to a trial density function $g(\mathbf{X})$.
2. Calculate $\overline{X}_i = (X_i + \cdots + X_{i+k-1})/k$ for $i = 1, \ldots, c$; if $\overline{X}_c \geq t$ and $\overline{X}_i < t$, $\forall i = 1, \ldots, c-1$, calculate a weight $w = f(\mathbf{x})/g(\mathbf{x})$, otherwise $w = 0$; here, $f(\mathbf{x})$ denotes the multivariate normal density function with mean μ and covariance Σ; repeat the process and obtain weights $w_1, \ldots, w_n$.
3. Estimate πp by $(w_1 + \cdots + w_n)/n$, and thus the adjusted p-value of threshold t is approximately $1 - \mathrm{e}^{-(L-k+1)(w_1+\cdots+w_n)/n}$.

The trial density function $g(\mathbf{x})$ should be chosen properly to ensure the efficiency of the importance sampling algorithm. We choose $g(\mathbf{x})$ to be a multivariate normal with elevated means but the same covariance matrix Σ.

19.2.2 Varying Window Sizes

We can further generalize our method to combine peak-calling results from various window sizes. Suppose that we allow the window size to vary between $[k_{\min}, k_{\max}]$; the windows will then have a nested structure, i.e., smaller windows within larger windows. The problem is again to adjust for multiple testings but for both overlapping windows and nested windows. To do this, we define a new test, which we call an S-test, for simultaneous peak-calling of multiple window sizes.

Let $B_m(i, k)$ denote the starting positions of all m-probe windows that are at most $c_{\min}$ probes away from the k-probe window at the ith probe, for $m < k$. Let $\mathbf{t} = (t_{k_{\min}}, \ldots, t_{k_{\max}})$ denote the thresholds for each window sizes $k_{\min} \leq k \leq k_{\max}$, respectively. Let $\overline{X}_i(k)$ and $R_i(t, k)$ denote the window averages and the R values of a k-probe window. To compensate for the positive correlations among overlapping and nested windows, we declare significance of a k-probe window if and only if all overlapping and nested windows of smaller sizes are insignificant with respect to their own thresholds. In particular, we define the new indicator

$$S_i(\mathbf{t}) = \sum_{k=k_{\min}}^{k_{\max}} I\{R_i(t_k, k) = 1\}$$

$$\times \prod_{j=1}^{k-1} I\{\overline{X}_l(j) < t_j, \forall l \in B_j(i,k)\}, \quad \text{for } i = 1, \ldots, L-k+1.$$

We abbreviate $S_i(\mathbf{t})$ as S_i and call the corresponding test an S-test.

It is straightforward to show that S_i is binary and locally negatively correlated, and thus Poisson approximation can be used with an error bound given by the Chen–Stein method. The thresholds $\mathbf{t} = (t_{k_{\min}}, \ldots, t_{k_{\max}})$ can be calculated progressively for each window size in an increasing order. Let λ_k denote the expected number of false rejections for window size k, conditional on that no tests of smaller windows within the clump have been rejected, we can find each threshold t_k that gives λ_k rejections. The total number of expected false positives is therefore $\lambda = \lambda_{k_{\min}} + \cdots + \lambda_{k_{\max}}$.

19.3 Results

19.3.1 Simulation Study

We simulated two scenarios to evaluate our method: independent probes and dependent probes. For independent probes, we simulated 10,000 datasets with 20,000 evenly tiled probes with intensities following a standard normal distribution. For dependent probes, we simulated the same number of datasets, but the probe intensities were taken as the means of five adjacent probe intensities, which corresponded to a lag 5 correlation model.

We first checked the accuracy of our importance sampling algorithm (Fig. 19.1). Under both scenarios, the average number of significant R-tests for each window size agreed well to our estimation. We further checked the standard errors for approximating a significance of 10^{-8}, for window sizes from 1 to 10. The standard errors were consistently within 10^{-10} scale using 200,000 iterations. The computation time of our importance sampling is linearly related with the window sizes. For example, for $k = 10$, it takes $\sim$1 second to finish 200,000 importance samplings on a regular PC.

We next checked the assumed Poisson distributions for the number of significant R-tests (Fig. 19.2). We observed consistent agreement between the histograms of significant R-tests and the corresponding Poisson probability curves with the same means. This supported our Poisson heuristic for R-tests at large thresholds. In comparison, the number of significant X-tests, which are positively correlated, has a greatly inflated variance in comparison with that expected from a Poisson distribution. This is true even at stringent thresholds. Figure 19.2 further shows that both R-tests and X-tests have the same frequency at bin 0. This indicates that, ignoring some boundary effects, the two tests have the same family-wise type I error rate. This is why we can approximate the significance of peaks using the declumping method.

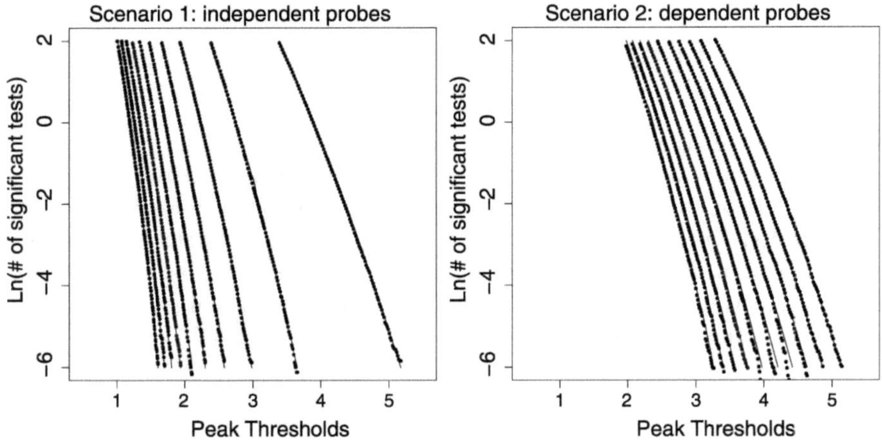

Fig. 19.1 Observed average number of significant R-tests (*dots*) versus the importance sampling estimated numbers (*lines*), under two simulated scenarios. *Lines* from *left* to *right* correspond to window sizes from 10 to 1

We finally checked the accuracy of the approximated p-values adjusting for multiple testings (Fig. 19.3). Our Poisson approximated p-values were plotted against the empirical p-values in the logarithm scale. We also plotted Bonferroni adjusted p-values as a comparison. Even for independent probes, we observed that the Bonferroni method is conservative, and it is more so when probes are dependent. This result is expected because the correlation among tests is stronger for dependent probes. For a numerical example, when the empirical p-value was 0.054 for $k = 10$, the Bonferroni adjusted p-values were 0.083 for independent probes and 0.139 for dependent probes, respectively, $\sim$60% and $\sim$170% more conservative than the actual significances. In comparison, our approximations agreed well with the empirical p-values.

19.3.2 Power of Various Window Sizes

We next performed a power study to check whether combining windows can obtain better power than using a fixed window size. We simulated probe intensities using multivariate normal distributions under two scenarios. In the first scenario, both the background and binding regions have the same covariance, but their mean intensities differs by 4. In the second scenario, the means between background and binding regions differ by 3, and their covariance matrices are also different (identity covariance for background and lag 5 correlation for binding). Under each scenario, 100 datasets were simulated containing 300,000 probes each, and 100 binding regions were also randomly simulated covering 4 to 7 probes. We calculated the powers of using fixed window sizes $k = 1, \ldots, 10$ and combining window sizes at the same significance level.

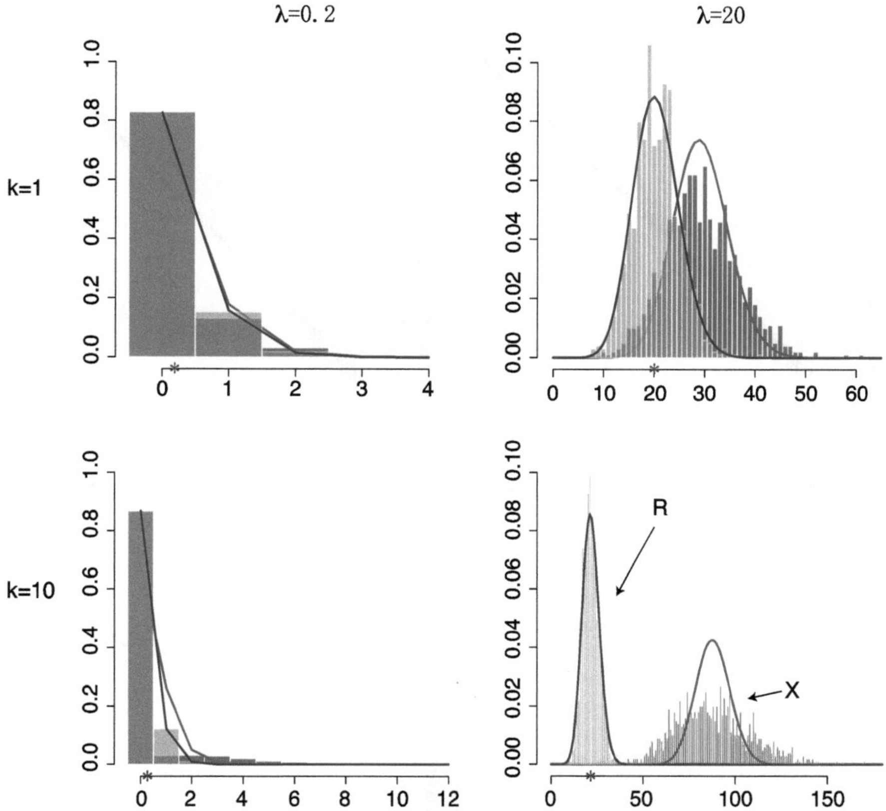

Fig. 19.2 Histogram of the number of significant tests of window size $k = 1$ and 10 compared to Poisson probability curves with the same means. λ denotes the expected number of false positives. *Light gray*: R-tests. *Dark gray*: X-tests. The same thresholds were used for R-tests and X-tests

As shown in Fig. 19.4, the power by using a fixed window size is strongly influenced by the choice of k. Under both scenarios, we observed significant losses of power when the window sizes deviated from the true widths of binding intervals. The power curves behaved also differently between the two scenarios. Smaller windows performed well for the first scenario, but larger windows performed well for the second scenario. In comparison, our method of combining all window sizes obtained the best powers under both scenarios. We therefore suggest combining window sizes in practice so to capture the protein–DNA binding signals with a maximized power.

19.3.3 FDR Control Accounting for Positive Correlations

Our method cannot only control family-wise error rate but also control FDR. Benjamini and Hochberg's FDR [3] is estimated as the ratio between the expected num-

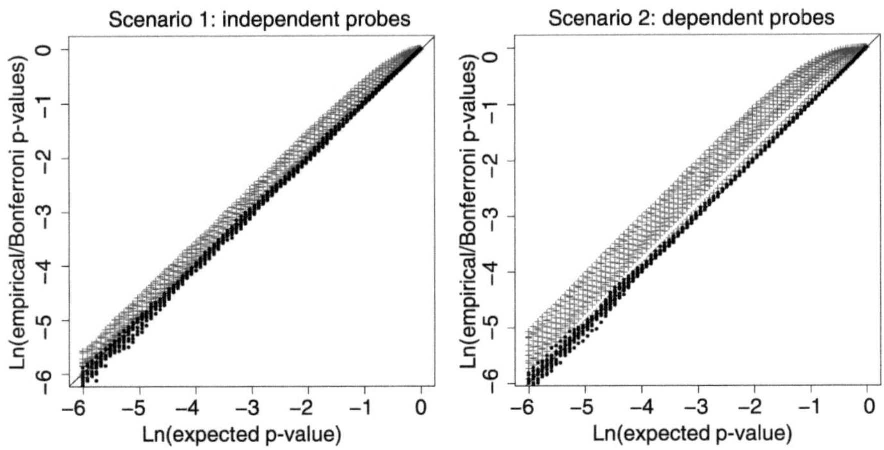

Fig. 19.3 Empirical p-values (*dots*) and Bonferroni adjusted p-values (*crosses*) compared to our approximated p-values (x-axis) in logarithm scale

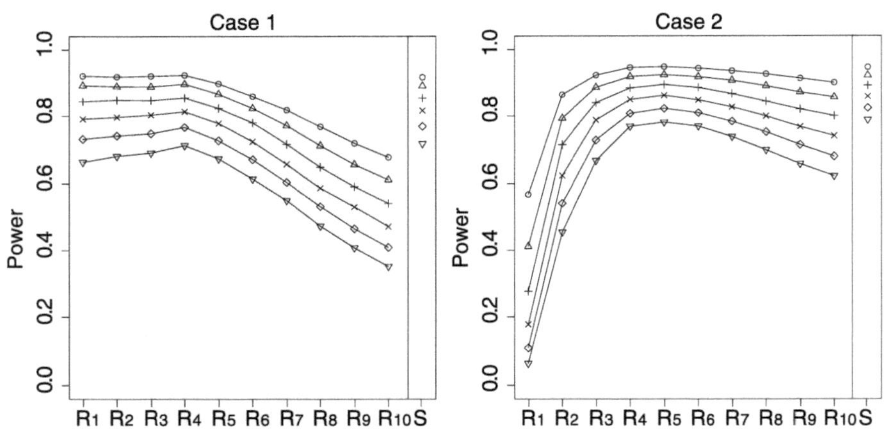

Fig. 19.4 Powers of using fixed window sizes (R-test) from 1 to 10 compared to the power of using combined window sizes (S-test). Case 1: same covariance between binding and background regions. Case 2: larger covariance in binding regions than background. *Lines* from *top* to *bottom* correspond to 50, 12.56, 3.15, 0.79, 0.20, and 0.05 expected false positives, respectively

ber of false rejections and the total number of rejections. If the number of true positives is relatively small compared to the total number of tests, the numerator can be approximated by the Bonferroni method. Most theoretical results for FDR are based on the independence assumption among tests. When applied to positively correlated tests, however, the estimated FDR can be much more variable. A more serious problem is that it is commonly practiced to merge overlapping peaks into a joint binding interval, where FDR method without taking this into account can be very misleading. Our method compensates the positive correlation among tests via declumping. As a result, we define a modified FDR as the ratio between the number of significant

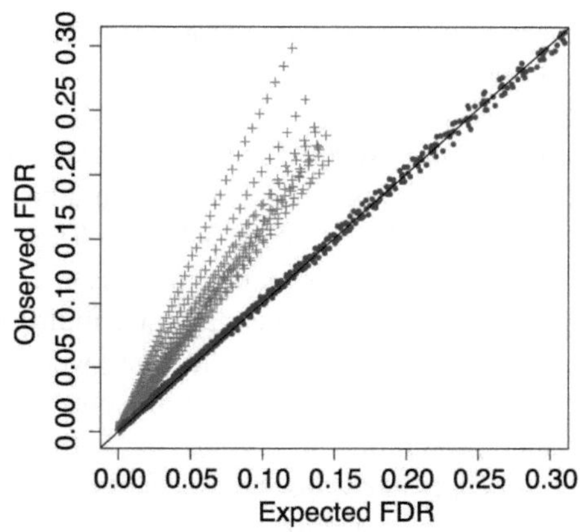

Fig. 19.5 Observed average FDR (y-axis) versus the expected FDR (x-axis). *Dots*: FDR controlled by our method of declumping (R-tests of $k = 1, \ldots, 10$ and S-test). *Crosses*: FDR controlled by Benjamini and Hochberg's method on window averages

clumps and the total number of rejected clumps. Since our significant clumps do not overlap, the FDR can be controlled at a proper level, which measures the proportion of nonoverlapping false positive binding intervals among all detected binding intervals.

We used the datasets in our power study under scenario 1 to illustrate our FDR control. We calculated the average FDR observed from R-tests and S-test using various window sizes and compared them with the expected FDR levels. As shown in Fig. 19.5, the empirical and theoretical FDR agreed well for both R-tests and S-test. We further used Benjamini and Hochberg's method to control FDR on the window averages (X-tests) assuming independence. We observed that the average FDR is significantly larger than that expected by the Benjamini and Hochberg's method. The reason is that the former measures the proportion of nonoverlapping false binding intervals, while the latter measures the proportion of false windows without accounting for the clumping effect. That is, the actual FDR of protein–DNA binding intervals can be seriously underestimated if we ignore the positive correlation among windows.

19.4 Discussion

We introduce a novel approach to control false positives in genome-wide ChIP-chip tiling arrays. There are three key components of our method. First, we assume that tests are locally correlated and the correlation can be compensated by clumps. Second, we use importance sampling to efficiently calculate the tail distribution of a single clump. Third, a Poisson distribution is used to approximate the statistical significance of one or a few clumps adjusting for multiple testings. Our approach can be easily applied to genome-scale multiple comparison problems without sacrificing accuracy and computation time. The method is a combination of theory and numerical calculation such that it is flexible to take into account complexities in real data analysis.

The assumption underlying our Poisson approximation is that the number of significant clumps appearing by chance is rare. This assumption is almost always satisfied in practice, because type I errors are usually tolerated at low levels in genome-wide studies. If the expected number of false positives is l, a rule of thumb is that the total number of tests $(L-k+1)$ should be much larger than $l(c+k-1)$. For example, $l(c+k-1) < 0.05(L-k+1)$. This criterion is necessary to reduce the edge effects and local interference in Poisson approximation. In addition, when estimating the Poisson mean, one may use a multiplier different from the total number of tests, where the latter is only accurate when the true signals are relatively rare (e.g., <2% of data are true signals). In some studies, there may be a large proportion of data representing the true signals, so that a smaller multiplier should be used. However, for cases like transcriptome arrays, one may want to use chromosome segmentations instead of window-based tests.

Although the detected binding intervals are statistically significant, they may not correspond to biological interesting sites. Possible factors confusing the peak detection include DNA repetitive sequences, genomic duplications, sequence-specific hybridizations, redundant probe designs, and pitfalls in raw data normalization. For this reason, sophisticated models may be used to take into account of those confounding factors. A particular advantage of our approach is that it can work with any models in peak-calling, as long as a test statistic can be summarized from the model using a local window of probes. In this paper, we only used window averages to demonstrate the method, but other test statistics can be easily applied. Particularly interesting models will be those incorporating biological knowledge [15, 16]. In summary, our method is a general approach that has a great potential to be applied to many current and future large-scale studies in computational biology.

References

1. B. Ren, et al. Genome-wide location and function of DNA binding proteins. *Science*, **290**:2306–2309, 2000.
2. V. Iyer, et al. Genomic binding sites of the yeast cell-cycle transcription factors SBF and MBF. *Nature*, **409**:533–538, 2001.
3. Y. Benjamini and Y. Hochberg. Controlling the false discovery rate: a practical and powerful approach to multiple testing. *J R Stat Soc, Ser B*, **57**:289–300, 1995.
4. S. Cawley, et al. Unbiased mapping of transcription factor binding sites along human chromosomes 21 and 22 points to widespread regulation of noncoding RNAs. *Cell*, **116**:499–509, 2004.
5. W. Li, C.A. Meyer, and X.S. Liu. A hidden Markov model for analyzing ChIP-chip experiments on genome tiling arrays and its application to p53 binding sequences. *Bioinfomatics*, **21**(1):i274–i282, 2005.
6. Y. Qi, et al. High-resolution computational models for genome binding events. *Nat Biotechnol*, **24**:963–970, 2006.
7. M. Zheng, L.O. Barrera, B. Ren, and Y.N. Wu. ChIP-chip: Data, model, and analysis. *Biometrics*, **63**:787–796, 2007.
8. S. Keles, M.J. van der Laan, S. Dudoit, and S.E. Cawley. Multiple testing methods for ChIP-chip high density oligonucleotide array data. *J Comput Biol*, **13**(3):579–613, 2006.

9. Y. Huang, H. Xu, V. Calian, and J.C. Hsu. To permute or not to permute. *Bioinformatics*, **22**:2244–2248, 2006.
10. M.S. Waterman and M. Vingron. Sequence comparison significance and Poisson approximation. *Stat Sci*, **9**:367–381, 1994.
11. J.S. Liu. *Monte Carlo Strategies in Scientific Computing*. Springer, New York, 2001.
12. D. Aldous. *Probability Approximations via the Poisson Clumping Heuristic*, volume 77 of Appl Math Sci. Springer, New York, 1989.
13. L.H.Y. Chen. Poisson approximation for dependent trials. *Ann Probab*, **3**:534–545, 1975.
14. R. Arratia, L. Goldstein, and L. Gordon. Poisson approximation and the Chen–Stein method. *Stat Sci*, **5**:403–424, 1990.
15. J. Du, et al. A supervised hidden Markov model framework for efficiently segmenting tiling array data in transcriptional and ChIP-chip experiments: systematically incorporating validated biological knowledge. *Bioinformatics*, **22**:3016–3024, 2006.
16. J.C. Marioni. BioHMM: a heterogeneous hidden Markov model for segmenting array CGH data. *Bioinformatics*, **22**:1144–1146, 2006.

Index

J. Feng et al. (eds.), *Frontiers in Computational and Systems Biology,*
Computational Biology 15,
DOI 10.1007/978-1-84996-196-7,

T

Color Plates

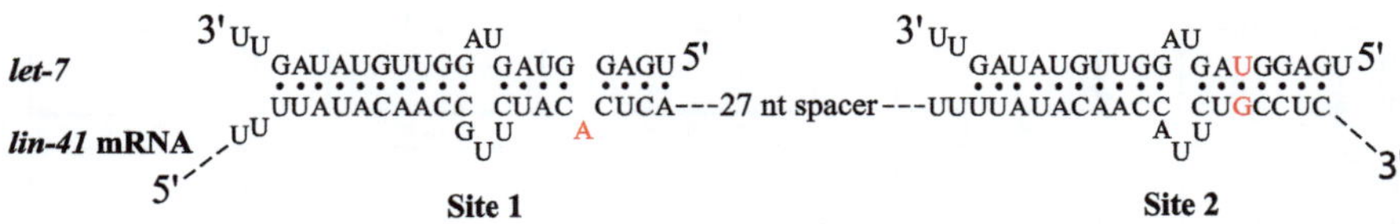

Fig. 2.4 *let-7* regulates *lin-41* by complementary base-pairing at two sites in the 3′ UTR of the *lin-41* mRNA [104, 105]. Neither the bulged A in the seed region for site 1 (in *red*, at position 5 from the 5′ end of the 27 nt spacer) nor the wobble G–U pair in the seed region for site 2 (in *red*, with U at position 6 of the 5′ end of *let-7*) meets the requirements of the seed model [56, 57] that bases 2 to 7 or 8 of the miRNA 5′ end must form Watson–Crick pairs with its target

J. Feng et al. (eds.), *Frontiers in Computational and Systems Biology*,
Computational Biology 15,
DOI 10.1007/978-1-84996-196-7, © Springer-Verlag London Limited 2010

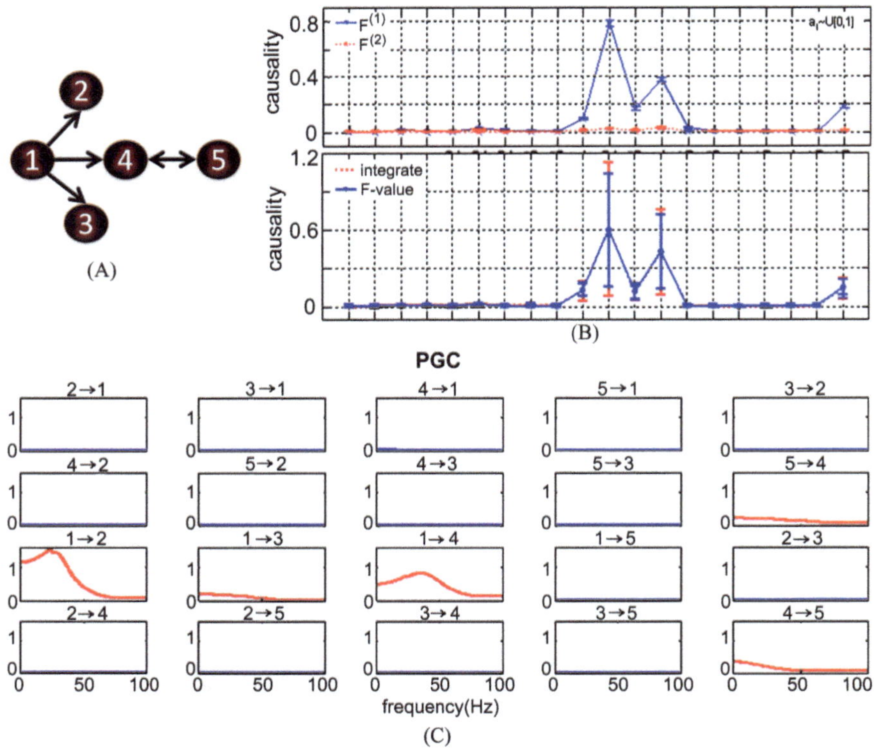

Fig. 5.2 Granger Causality applied to the system in Example 1. (**A**) The true network structure. (**B**) (*upper panel*) Comparison of the partial Granger causality F_1 and the conditional Granger causality F_2. F_2 fails to pick up any true connections, while the inferred links from F_1 are consistent with the correct structure (**A**). (*Bottom panel*) Comparison of the partial Granger causality in the time domain (*blue line*) and frequency domain (*red line*, the integral of the frequency domain formulation in the interval $[-\pi, \pi]$. (**C**) Results of the frequency domain decomposition of all 20 pairs of signals

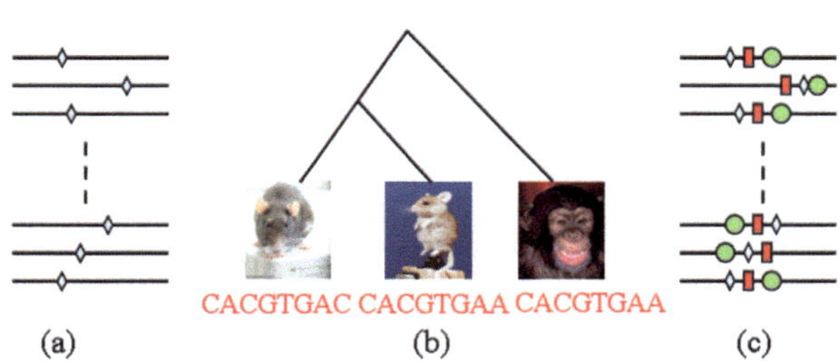

Fig. 6.2 Three types of information that are used for TFBS identification. (**a**) Over-representation. The *horizontal lines* are the sequences from coregulated genes in one species. The *small boxes* on the *line* are the TFBSs. (**b**) Conservation. The TFBSs in one group of orthologous chimpanzee, mouse and rat genes are similar. That is, the TFBSs in this gene are conserved across three species. (**c**) Clustering. Three different TFBSs often occur together in short regions in the input sequences. Such short regions are often called cis-regulatory modules (CRMs)

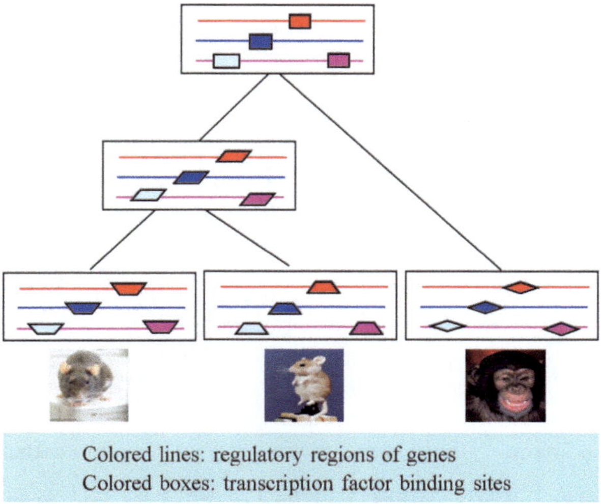

Fig. 6.6 A cartoon illustration of TGS. Each *rectangular box* represents one species. The *colored lines* are the regulatory regions of coregulated genes. The *small boxes* are the TFBSs in each sequence. TGS assumes that there is at most one TFBS in each sequence. The motifs may be different in different species although they evolve from the same ancestral motifs

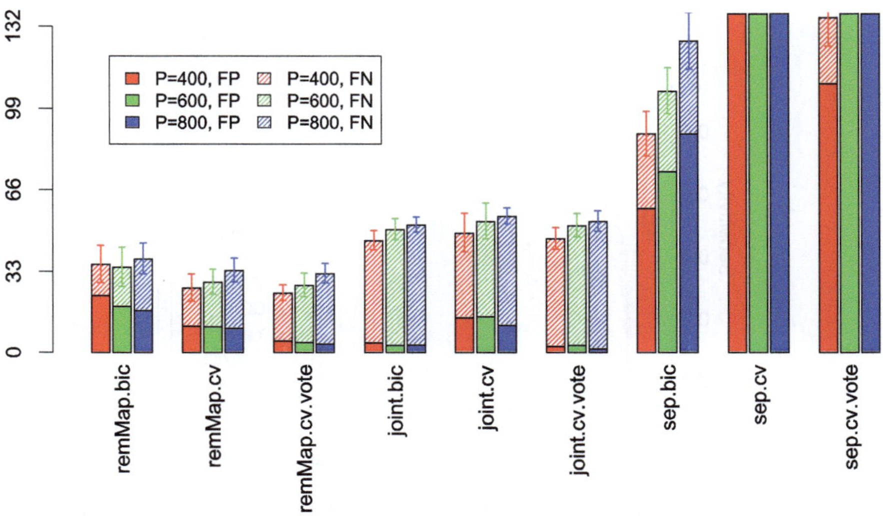

Fig. 7.4 Impact of dimensionality. Heights of *solid bars* represent numbers of false positive detections of `trans-edges` (FP); heights of *shaded bars* represent numbers of false negative detections of `trans-edges` (FN). All bars are truncated at height = 132 [28]

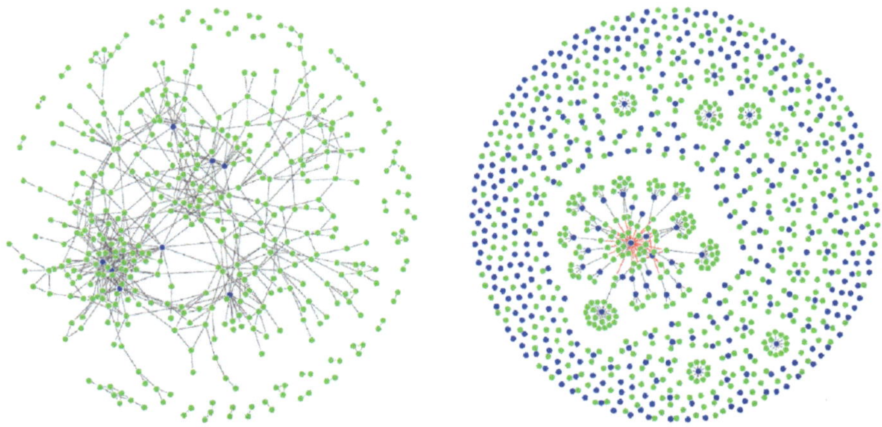

(a) space (b) remMap

Fig. 7.6 (**a**) Inferred network for the 654 breast cancer related genes (based on their expression levels) by space. Nodes with degrees greater than ten are drawn in *blue*. (**b**) Network of the estimated regulatory relationships between the copy numbers of the 384 CNAIs and the expressions of the 654 breast cancer related genes. Each *blue node* stands for one CNAI, and each *green node* stands for one gene. *Red edges* represent inferred transregulations (43 in total). *Gray edges* represent cis-regulations [28]

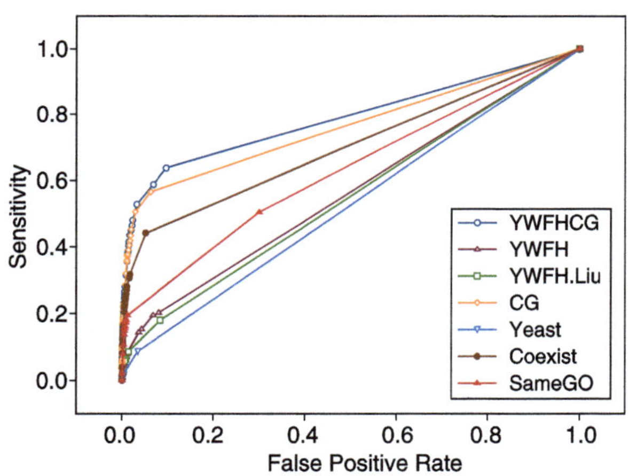

Fig. 8.2 The comparison of prediction accuracies by integrating multiple biological data sets using the naive Bayesian method. The letters Y, W, F, H, C, and G indicate domain interactions based on yeast, worm, fruitfly, humans, co-existence, and same GO function, respectively. YWFH. Liu shows the result of predicted domain interactions using the extended MLE method defined in Liu et al. [17] with protein interactions of yeast, worm, fruitfly, and humans. (This figure is excerpted from Lee et al. [16])

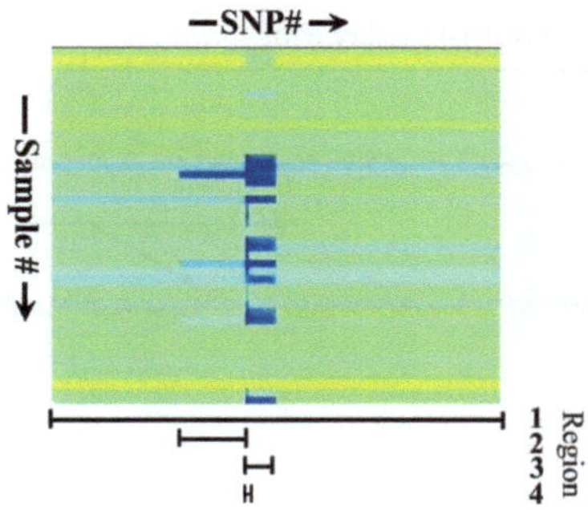

Region	Sample Number		
	1	...	N
1	X_{11}	...	X_{1N}
2	X_{21}	...	X_{2N}
3	X_{31}	...	X_{3N}
4	X_{41}	...	X_{4N}

Fig. 14.4 An example of a joint segmentation of a set of tumor samples. The segmentation outputs a set of common change-points to give the best sparse summary of the set of tumors

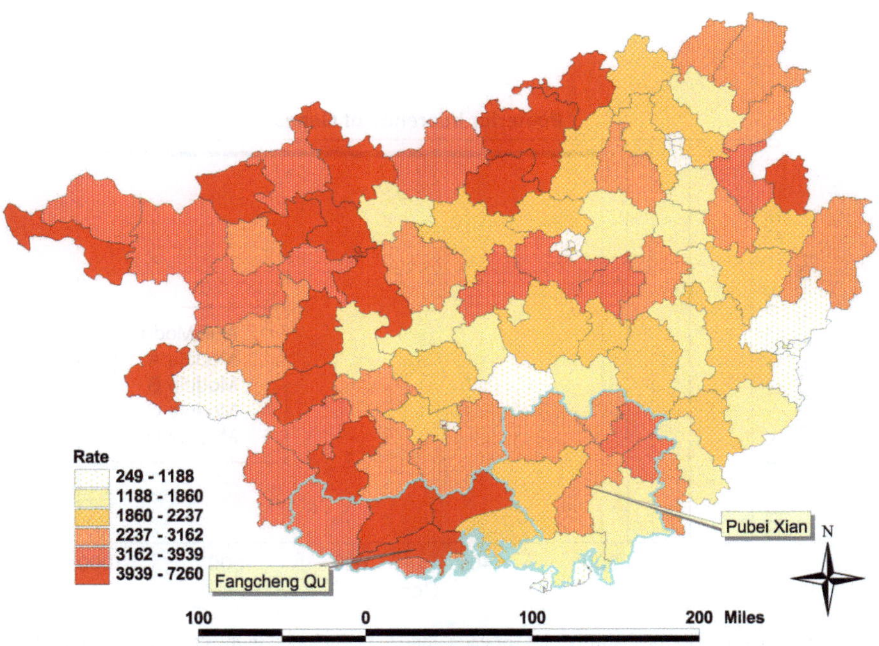

Fig. 15.1 County level infant mortality rate per 100,000 in Guangxi, China in 2000

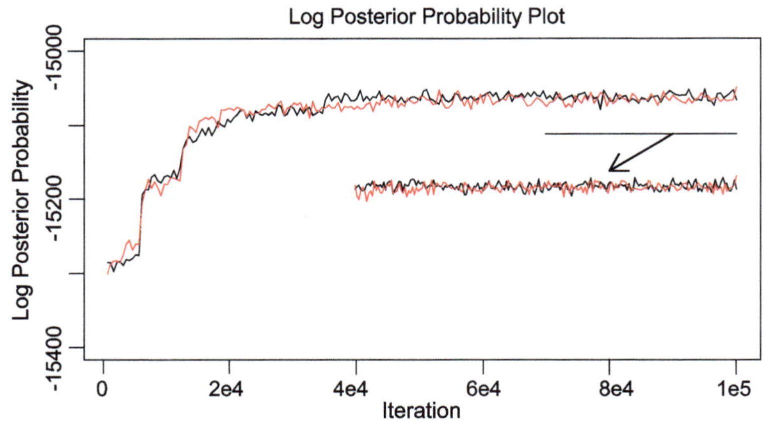

Fig. 16.4 Trace plots and autocorrelation plots of the log posterior probabilities for the simulated data set. The trace plot was generated from two independent runs, each having 100,000 iterations

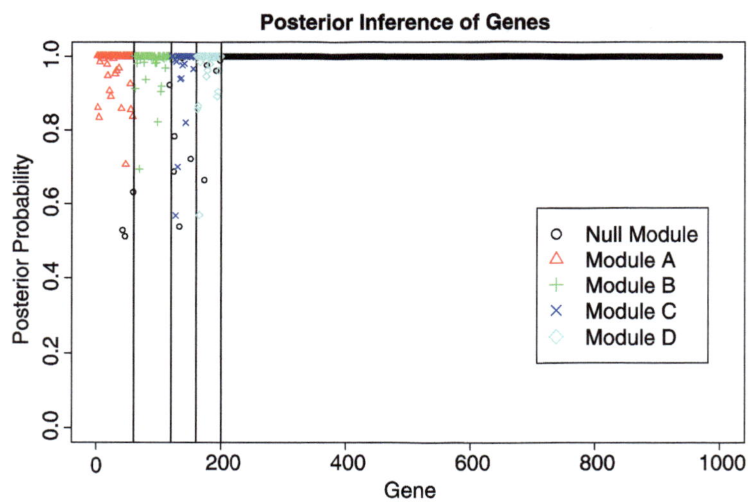

Fig. 16.5 The posterior probability plot for each gene to be included in the corresponding module. The first 200 genes are in one of the four modules, separated by the *red vertical line*. The module membership was determined by the majority vote based on the posterior samples from the last 25,000 iterations

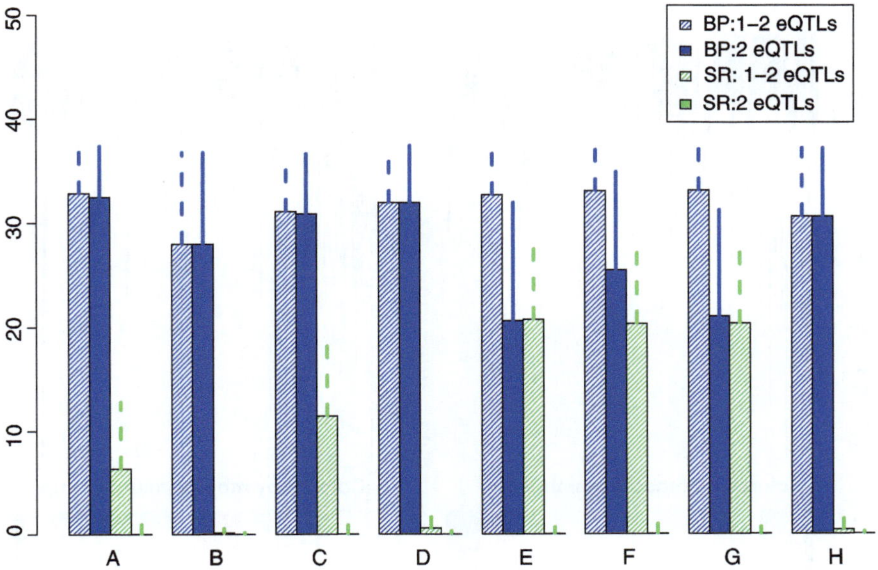

Fig. 16.7 Barplots of the number of true eQTLs detected in each module by the BP method (*blue*) and step-wise regression (SR) method (*green*). The *shaded bar* represents the number of genes detected as mapped to at least one of the true eQTLs, while the *solid bar* represents the number of genes detected as mapped to both eQTLs. The thresholds are 0.5 for both posterior probability (BP) and FDR (SR). From Fig. 16.6 we know that the total number of false positive gene-marker pairs is 11.41 and 38.04 for BP and SR, respectively. When the thresholds are relaxed to 0.1, more eQTLs were detected in each category, as indicated by the *vertical lines* above the *bars*. However, the total number of the false positive gene-marker pairs is still lower using BP (178.37) compared to that using SR (267.07)

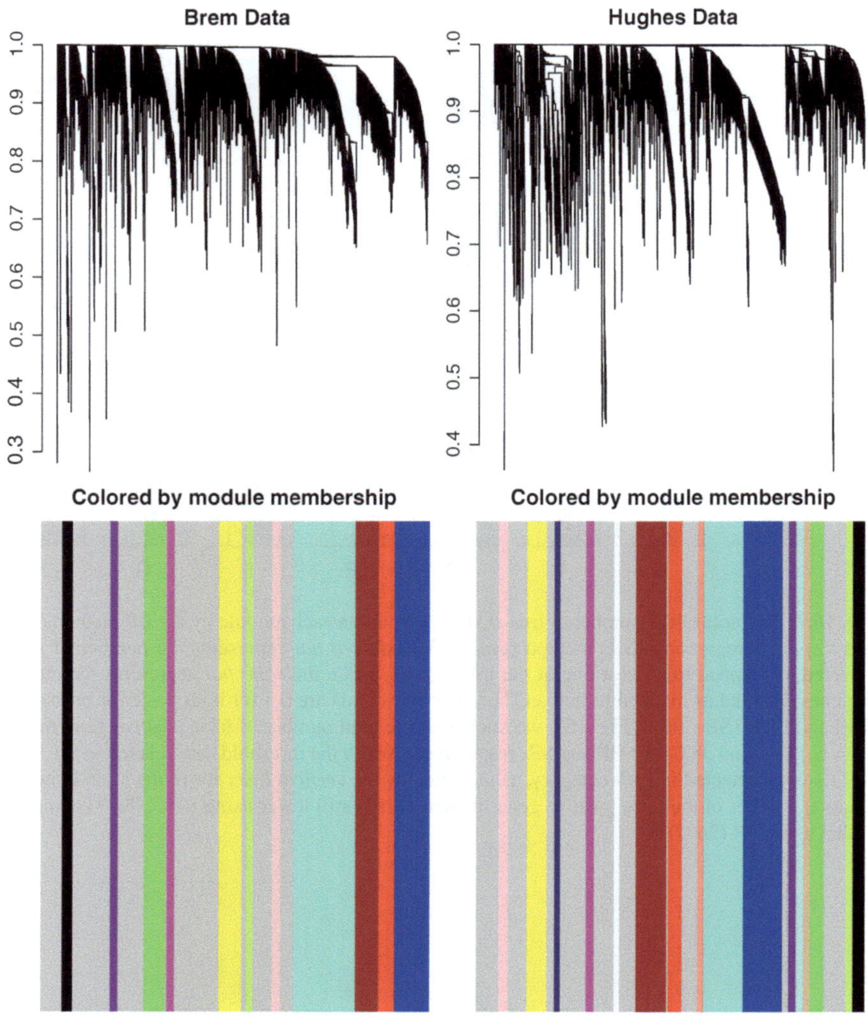

Fig. 17.1 Dendrogram view of hierarchical clustering results for the WGCNs constructed from the Brem data set and the Hughes data set

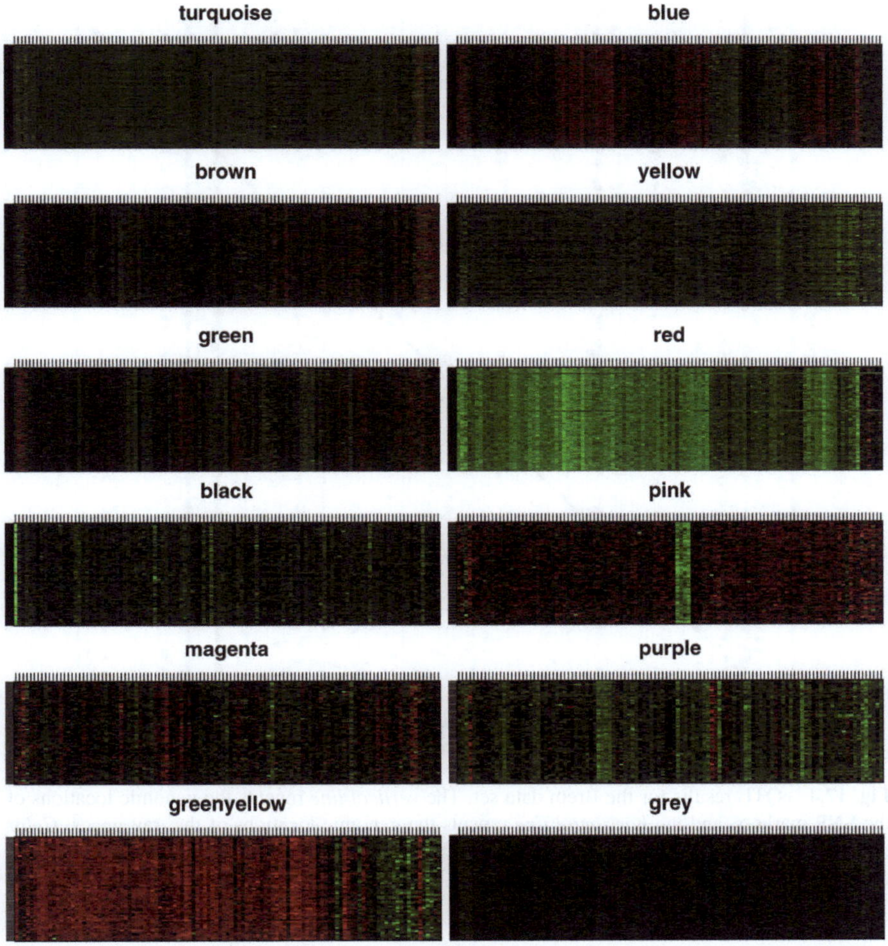

Fig. 17.2 The heatmap view of gene expression patterns of gene modules extracted from the WGCN for the Brem data set

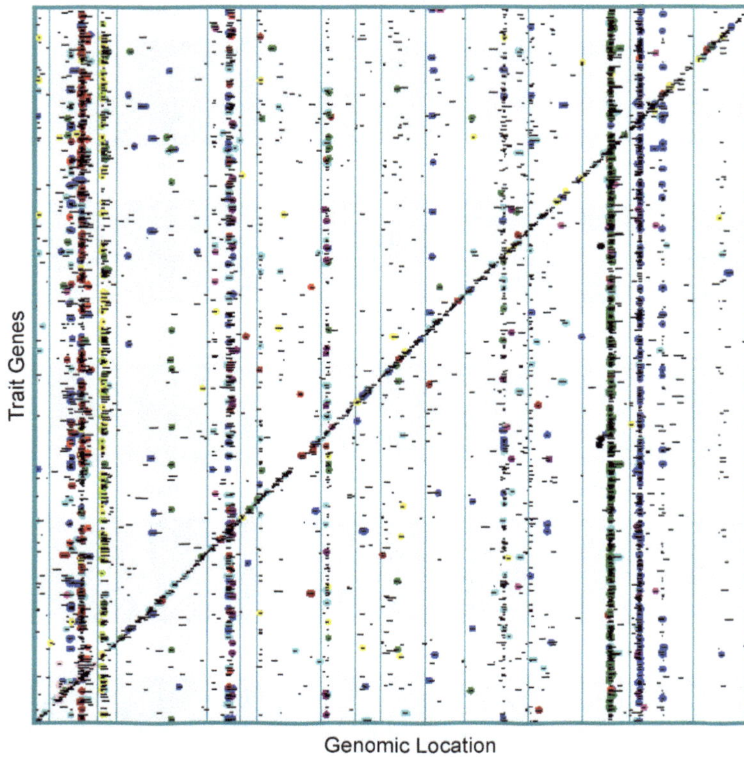

Fig. 17.4 eQTL results for the Brem data set. The *vertical line* repents the genomic locations of the SNP markers, and the *horizontal line* repents the genomic locations of the trait genes. *Color dots* with the same color correspond to the eQTLs of genes in the module with the corresponding color

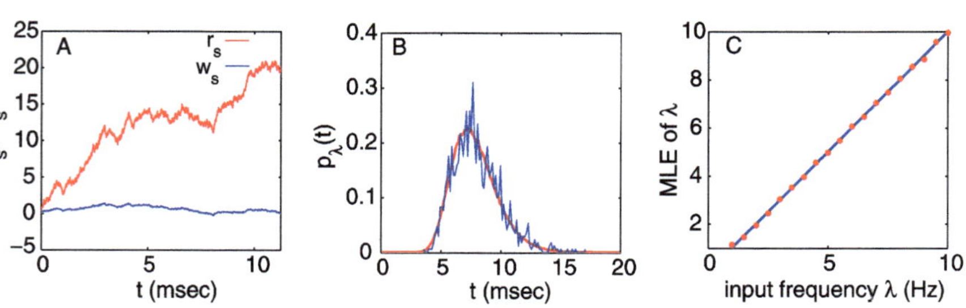

Fig. 18.1 MLE of a single neuron. (**A**) Examples of trajectories of r_s and w_s. (**B**) $p_\lambda(t)$ (*red*) and histogram (*blue*) from a direct simulation of the LIF model. (**C**) The estimated input rate $\hat{\lambda}$ (*dots*) obtained from 600 interspike intervals. Parameters are $V_{th} = 20$ mV and $\gamma = 20$ msec

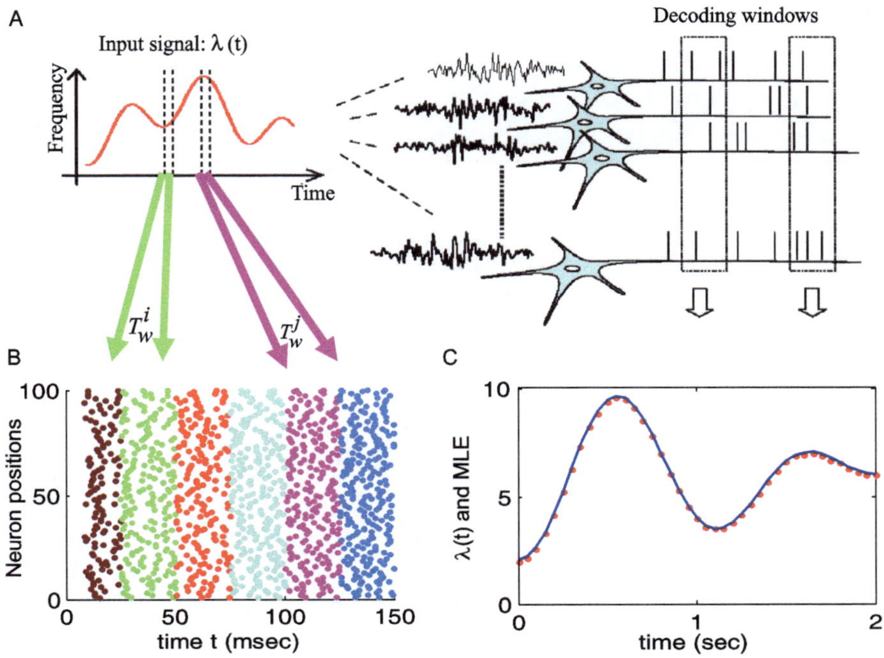

Fig. 18.5 MLE in a network without interactions. (**A**) Schematic plot of reading dynamic inputs from an ensemble of neurons. For a fixed time window (indicated by T_w^i, T_w^j), the spikes are collected, and the input is decoded by means of MLE. (**B**) Raster plots of spikes in five different decoding windows. The information is read out from the spikes in each window. (**C**) An example of reading the input information from ensemble of neurons is shown. The original signal $\lambda(t)$ is plotted in the *continuous line*, while *dots* are estimated values of $\lambda(t)$

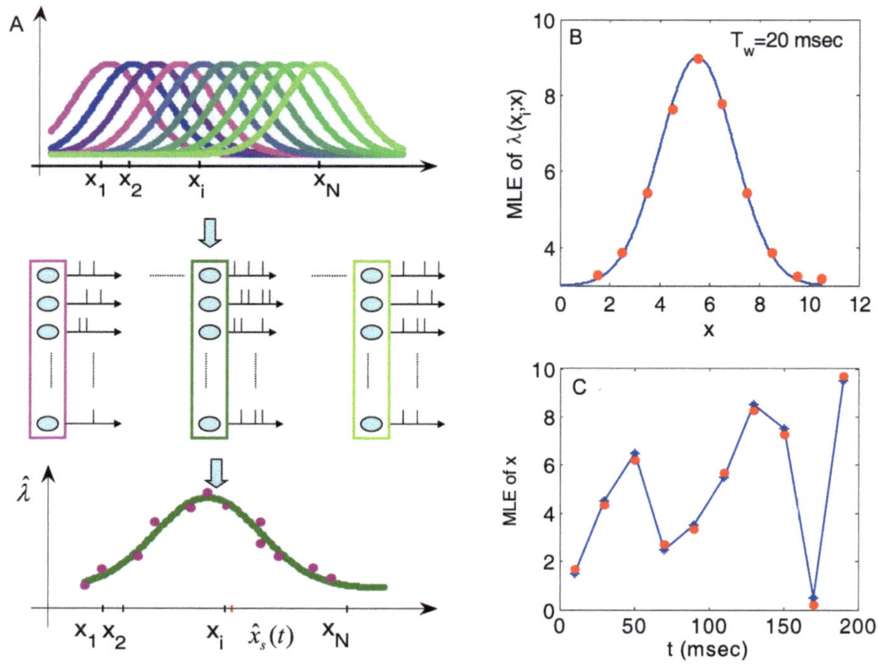

Fig. 18.6 MLE in position tracking. (**A**) The setup of our encoding and decoding. The place information $x(t)$ is feeded to 1000 neurons organized in 10 columns with different preferred positions x_i. For each column, spikes are collected in the window T_w, and the MLE is applied. The position is read out by first solving (18.22), obtaining the corresponding $\hat{x}$ and fitting a Gaussian curve to all 10 data points $\hat{x}$. The maximum point of the fitted Gaussian curve is the decoded position. (**B**) One example of $\hat{\lambda}$ vs. columns. (**C**) Examples of 10 decoded results

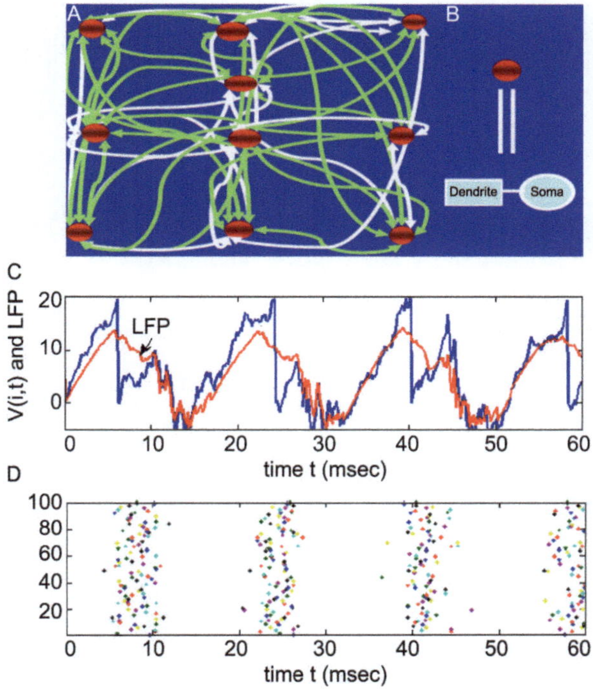

Fig. 18.7 (**A**) A schematic plot of the model setup with 10 neurons. *White arrows* are excitatory interactions, and *green arrows* are inhibitory interactions via inhibitory neurons (not specifically modelled). (**B**) A single neuron model of two compartments. (**C**) One trajectory of one single cell is depicted (*blue trace*), and the *red line* is the local field potentials for averaging all neurons in a microcolumn. The oscillatory in Gamma frequency is obvious. (**D**) The optimal time window for decoding the input (the window between two *thick vertical lines*)

MIX
Papier aus verantwortungsvollen Quellen
Paper from responsible sources
FSC® C105338

If you have any concerns about our products,
you can contact us on
ProductSafety@springernature.com

In case Publisher is established outside the EU,
the EU authorized representative is:
Springer Nature Customer Service Center GmbH
Europaplatz 3, 69115 Heidelberg, Germany

Printed by Libri Plureos GmbH
in Hamburg, Germany